DYNAMIC SYSTEMS

Modeling and Analysis

Hung V. Vu
California State University, Long Beach

Ramin S. Esfandiari
California State University, Long Beach

Boston   Burr Ridge, IL   Dubuque, IA   Madison, WI   New York   San Francisco   St. Louis
Bangkok   Bogotá   Caracas   Lisbon   London   Madrid
Mexico City   Milan   New Delhi   Seoul   Singapore   Sydney   Taipei   Toronto

Dynamic Systems
Modeling and Analysis

Copyright © 1997 by The McGraw-Hill Companies,Inc. All rights reserved. Printed in the United States of America. Except as permitted under the United States Copyright Act of 1976, no part of this publication may be reproduced or distributed in any form or by any means, or stored in a data base retrieval system, without prior written permission of the publisher

This book is a McGraw-Hill Custom Publishing textbook and contains select material from *Dynamic Systems: Modeling and Analysis* by Hung V. Vu and Ramin S. Esfandiari. Copyright © 1997 by The McGraw-Hill Companies, Inc. Reprinted with permission of the publisher. Many custom published texts are modified versions or adaptations of our best-selling textbooks. Some adaptations are printed in black and white to keep prices at a minimum, while others are in color.

5 6 7 8 9 0 QSR QSR 0 9 8 7 6

ISBN-13: 978-0-07-296661-9
ISBN-10: 0-07-296661-0

Editor: Mary Coman
Production Editor: Alice Link
Printer/Binder: Quebecor World

ABOUT THE AUTHORS

DR. HUNG VU is Professor of Mechanical Engineering at California State University, Long Beach. He received his Ph.D. in Applied Mechanics from the University of Michigan, Ann Arbor, having earned M.S. and B.S. degrees in Mechanical Engineering from Massachusetts Institute of Technology (M.I.T.) and the University of Washington, respectively. He has ten years of teaching and research experience in structural dynamics, vibrations, control systems, rigid-body dynamics, and biomechanical engineering. He has written three books and is the author of many publications in prestigious journals and proceedings. His numerous consulting experiences include such places as the National Aeronautics and Space Administration (NASA) and the Jet Propulsion Laboratory, California Institute of Technology; the United States Air Force Office of Scientific Research (AFOSR) at the Air Force Academy's Frank J. Seiler Research Laboratory and Edwards Air Force Base's Phillips Laboratory (Astronautics Laboratory); Newport Corporation; and Beckman Instruments, Inc. Dr. Vu is active as a member of Tau Beta Pi (National Engineering Honor Society) and as a member and faculty advisor of both ASME (American Society of Mechanical Engineers) and Pi Tau Sigma (National Mechanical Engineering Honor Society).

DR. RAMIN ESFANDIARI is Professor of Mechanical Engineering at California State University, Long Beach. He received his Ph.D. in Applied Mathematics with emphasis in optimal control theory from the University of California at Santa Barbara, having earned a B.S. degree in Mechanical Engineering and a M.S. in Applied Mathematics from the University of California at Santa Barbara. Dr. Esfandiari is the author of over 15 refereed research papers in optimal control of distributed structures and optimization in such journals as the *Journal of Optimization Theory and Applications* and *The Journal of Sound and Vibration*. He is the recipient of numerous teaching and research awards, such as the 1994-TRW Excellence in Teaching and Scholarship Award and the Distinguished Faculty Teaching Award.

To my family and my good friends.
H. V. V.

To my beloved sisters, Mandana and Roxana,
and my parents, to whom I owe it all.
R. S. E.

CONTENTS

Preface   xv

1 Complex Analysis, Differential Equations, and the Laplace Transform   1

   1.1 Complex Numbers in Rectangular Form   1
Addition and Multiplication of Complex Numbers: Rectangular Form / Complex Conjugation: Rectangular Form / Division of Complex Numbers: Rectangular Form

      Problem Set 1.1   3

   1.2 Representation of Complex Numbers in Polar Form   3
Multiplication and Division of Complex Numbers: Polar Form / Complex Conjugation: Polar Form / Powers of a Complex Number / Roots of a Complex Number

      Problem Set 1.2   8

   1.3 Complex Variables and Functions   8

      Problem Set 1.3   11

   1.4 Linear Differential Equations with Constant Coefficients   11
Standard Forms / Solution of Linear ODEs with Constant Coefficients / Homogeneous, Linear ODEs with Constant Coefficients / Free Response of First-Order Systems / Free Response of Second-Order Systems / Nonhomogeneous, Linear ODEs with Constant Coefficients

      Problem Set 1.4   21

   1.5 Laplace Transformation   22
Special Functions / Linearity of Laplace and Inverse Laplace Transforms

      Problem Set 1.5   29

   1.6 Laplace Transform of Derivatives and Integrals   30
Solving Initial-Value Problems

      Problem Set 1.6   34

   1.7 Inverse Laplace Transformation   34
Partial Fraction Method / Partial Fraction Method: An Alternative Approach / Convolution Method

	Problem Set 1.7	43
1.8	Shift on *t*-Axis	44
	Periodic Functions	
	Problem Set 1.8	49
1.9	Applications: System Response	50
	First-Order Systems / Second-Order Systems	
	Problem Set 1.9	61
1.10	Initial- and Final-Value Theorems	61
	Problem Set 1.10	64
	Summary	64
	Problems	69

2 Matrix Analysis — 76

2.1	Vectors and Matrices	76
	Matrix Operations / Matrix Transpose	
	Problem Set 2.1	81
2.2	Determinant, Inverse, and Rank of a Matrix	82
	Properties of Determinants / Determination of Inverse—Adjoint Matrix / Linear Systems of Algebraic Equations—Cramer's Rule	
	Problem Set 2.2	88
2.3	Eigenvalues and Eigenvectors	89
	Solving the Eigenvalue Problem / Special Matrices	
	Problem Set 2.3	95
2.4	Matrix Transformation and Diagonalization	95
	Exponential of a Matrix / Exponential of a Diagonal Matrix / Exponential of a Diagonalizable Matrix	
	Problem Set 2.4	101
	Summary	101
	Problems	105

3 System Model Representation — 108

3.1	Configuration Form	108
	Second-Order Matrix Form	
	Problem Set 3.1	112
3.2	State-Space Representation	113
	Methodology / General Formulation / Nonuniqueness of State Variables / Decoupling	
	Problem Set 3.2	124

	3.3	Input-Output (I/O) Equation	125
		Strategy	
		Problem Set 3.3	128
	3.4	Transfer Function	128
		Relation Between State-Space Form and Transfer Function / Single Input-Single Output (SISO) Systems / Multiple Input-Multiple Output (MIMO) Systems	
		Problem Set 3.4	137
	3.5	State-Space Representation from the Input-Output Equation	138
		Problem Set 3.5	143
	3.6	Linearization	144
		Graphical Interpretation / Taylor Series Expansion / Determination of the Operating Point / Numerical Solution of a Nonlinear Model / Fourth-Order Runge-Kutta Method	
		Problem Set 3.6	163
		Summary	164
		Problems	167
4	**Mechanical Systems**		**172**
	4.1	Introduction	172
	4.2	Mechanical Elements	173
		Mass Element / The Right-Hand Rule (RHR) / Spring Element / Damper Element	
	4.3	Equivalence	179
	4.4	Degrees of Freedom	181
		Rolling Constraints / Holonomic Constraints and Nonholonomic Constraints	
		Problem set 4.1–4.4	186
	4.5	Translational systems	190
		Newton's Second Law / Free-Body Diagram and Sign Convention / Modeling of a System with Viscous Damping / Modeling of a System with Coulomb Damping / D'Alembert's Principle / Gravity and Differential Equation of Spring-Mass Systems / Systems with Displacement Input / Transfer Functions and State-Space Representation for SDOF Systems / Two-Degree-of-Freedom (TDOF) Systems / Systems with Massless Junctions / Skeleton Approach	
		Problem Set 4.5	210

4.6	Rotational systems	213
	The Moment Equation / Angular Momentum and Moments of Inertia / Modeling of Rigid Bodies in Planar Motion	
4.7	Mixed Systems: Translational and Rotational	226
	Problem Set 4.6–4.7	236
4.8	Gear-Train Systems	238
	Problem set 4.8	243
4.9	Lagrange's Equations	244
	Kinetic Energy / Potential Energy / Nonconservative Forces	
	Problem Set 4.9	258
	Summary	261
	Problems	263

5 Electrical, Electronic, and Electromechanical Systems — 265

5.1	Introduction	265
5.2	Electrical Systems	265
	Passive Electrical Elements / Kirchhoff's Laws	
5.3	Electronic Systems: Operational Amplifiers	275
	Problem Set 5.1–5.3	277
5.4	Electromechanical Systems	279
	Elemental Relations of Electromechanical Systems / Armature-Controlled DC Motors / Field-Controlled DC Motors	
	Problem Set 5.4	288
5.5	Impedance Methods	290
	Electrical Systems / Equivalence / Voltage Divider and Transfer Function / Mechanical Systems	
	Problem Set 5.5	295
	Summary	295
	Problems	296

6 Fluid and Thermal Systems — 304

6.1	Introduction	304
6.2	Thermodynamics	305
	Chemistry / Ideal Gases / Gaseous Mixtures / Intensive and Extensive Properties / Conservation of Energy / Specific Heats	

	6.3	Fluid Mechanics	311
		Laminar and Turbulent Flows / Fluid Resistance	
	6.4	Pneumatic Systems	315
		Fundamentals / Mathematical Modeling	
	6.5	Liquid-Level Systems	320
		Fundamentals / Mathematical Modeling	
	6.6	Thermal Systems: Heat Transfer	323
		Fundamentals / Mathematical Modeling	
	6.7	Analogous Systems	327
		Summary	328
		Problems	331
7	**System Response**		**334**
	7.1	Types of Response	334
	7.2	Transient Response of First-Order Systems	335
		Free Response / Forced Response / Step Response / Ramp Response	
	7.3	Transient Response of Second-Order Systems	339
		Free Response / Forced Response / Response to a Unit Impulse / Unit-Impulse Response / Response to a Unit Step / Unit-Step Response	
		Problem Set 7.1–7.3	355
	7.4	Transient Response of Higher-Order Systems	356
		Problem Set 7.4	361
	7.5	Steady-State Response to Sinusoidal Input	362
		Bode Plot / First-Order System / Second-Order System	
	7.6	Response to an Arbitrary Input—Convolution Integral	370
		Problem Set 7.5–7.6	374
	7.7	Solving the State Equation	375
		Homogeneous State Equation / Laplace Transform Approach / State-Transition Matrix / Nonhomogeneous State Equation / Laplace Transform Approach	
		Problem Set 7.7	381
	7.8	Modal Decomposition	381
	7.9	Fourier Analysis	385
		Fourier Series / Fourier Cosine and Sine Series / Convergence of Fourier Series / Interval Extension / System Response via Fourier Series	

		Problem Set 7.8–7.9	394
		Summary	394
		Problems	396

8 Introduction to Vibrations — 401

8.1 Introduction — 401
8.2 Natural Frequencies and Mode Shapes — 403
Unsymmetrical Systems / Symmetrical Systems / Automobile Suspension
Problem Set 8.1–8.2 — 426
8.3 Modal Matrix — 427
Orthogonality / Orthonormality / Damping
Problem Set 8.3 — 436
8.4 Logarithmic Decrement — 437
Problem Set 8.4 — 440
8.5 Beat Phenomenon — 441
Problem Set 8.5 — 447
8.6 Frequency Response of Vibration Systems — 448
Simple Mechanical System / Vibration Isolation / System with Support Motion / Rotating Unbalanced Mass
Problem Set 8.6 — 459
8.7 Damped Vibration Absorber — 461
Frequency Response / Transient Response
Problem Set 8.7 — 469
8.8 Modal Analysis: System Response — 471
Eigenvalue Problem / Eigensolutions: Natural Frequencies and Mode Shapes / Orthonormality / System Response via Modal Analysis
8.9 Vibration Testing — 479
Summary — 479
Problems — 482

9 Block Diagram Representation — 486

9.1 Block Diagrams — 486
Block Diagram Operations / Closed-Loop Systems
9.2 Block Diagram from Governing Equations — 493
Block Diagram from State-Space Model / Block Diagram from Input-Output Equation

9.3	Block Diagram Reduction		499
	Moving a Branch Point / Moving a Summing Junction / Negative and Positive Feedback / Mason's Rule—Special Case / Mason's Rule—General Form		
	Summary		510
	Problems		510
10	**Introduction to Control Systems**		**516**
10.1	Introduction		516
10.2	Definitions		516
10.3	Essential Components of Control Systems		518
10.4	Historical Perspective		518
10.5	General Block Diagram for Control Systems		519
10.6	Further Examples of Control Systems		520
	Heating Control Systems: An Engineering Application / Heating Control Systems: Homeostasis, a Biological Condition		
10.7	Poles and Zeros		522
	First-Order Systems: Time Constant / Second-Order Systems: Natural Frequency and Damping Ratio / Higher-Order Systems / Damping Ratio and ζ-Line		
10.8	Transient Response Specifications		531
	First-Order Systems / Second-Order Systems		
10.9	Dominant Pole Concept		538
	First-Order and Second-Order Systems / Higher-Order Systems		
10.10	Routh-Hurwitz Stability Criterion		540
	General Systems / Fourth-Order Systems / Third-Order Systems / Second-Order Systems		
	Problem Set 10.1–10.10		543
10.11	Controller Types and Actions		545
	Proportional Control (P Control) / Derivative Control (D Control) / Integral Control (I Control) / Proportional-plus-Derivative Control (PD Control) / Proportional-plus-Integral Control (PI Control) / Proportional-plus-Integral-plus-Derivative Control (PID Control)		
10.12	Steady-State Error		548
	Problem Set 10.11–10.12		553
10.13	Methods of Ziegler-Nichols Tuning		554
	Optimizing PID Parameters in Practice		

xii Contents

10.14	Speed Control System	558
	Problem Set 10.13–10.14	561
	Summary	562
	Problems	564

Appendix A	Tables	567
Appendix B	Computer Simulation	571
	Useful Notes for MATRIX$_x$	571

Appendix C	Useful Formulas	582
	Euler's Identities	582
	Taylor's Series Expansion	582
	Trigonometry	582

Pythagorean Identities / Expansion Formulas / Product Formulas / Double-Angle/Half-Angle Formulas / Law of Cosines / Miscellaneous

| | Hyperbolic Functions and Relations | 584 |
| | Logarithm | 584 |

Common Logarithm / Natural Logarithm / Decibel (dB) / Change of Base / Logarithm of a Complex Number

| | Binomial Series | 585 |

| Appendix D | Answers to Odd-Numbered Problems | 586 |
| | Chapter 1 | 586 |

Problem Set 1.1 / Problem Set 1.2 / Problem Set 1.3 / Problem Set 1.4 / Problem Set 1.5 / Problem Set 1.6 / Problem Set 1.7 / Problem Set 1.8 / Problem Set 1.9 / Problem Set 1.10

| | Chapter 2 | 589 |

Problem Set 2.1 / Problem Set 2.2 / Problem Set 2.3 / Problem Set 2.4

| | Chapter 3 | 591 |

Problem Set 3.1 / Problem Set 3.2 / Problem Set 3.3 / Problem Set 3.4 / Problem Set 3.5 / Problem Set 3.6 / Problems

| | Chapter 4 | 594 |

Problem Set 4.1–4.4 / Problem Set 4.5 / Problem Set 4.6–4.7 / Problem Set 4.8 / Problem Set 4.9 / Problems

Chapter 5	598
Problem Set 5.1–5.3 / Problem Set 5.4 / Problem Set 5.5 / Problems	
Chapter 6	601
Chapter 7	602
Problem Set 7.1–7.3 / Problem Set 7.4 / Problem Set 7.5–7.6 / Problem Set 7.7 / Problem Set 7.8–7.9	
Chapter 8	604
Problem Set 8.1–8.2 / Problem Set 8.3 / Problem Set 8.4 / Problem Set 8.6 / Problem Set 8.7 / Problems	
Chapter 9	608
Chapter 10	610
Problem Set 10.1–10.10 / Problem Set 10.11–10.12 / Problem Set 10.13–10.14 / Problems	

Bibliography	612
Index	615

PREFACE

In the last decade, we have seen many advances and changes in engineering theory and applications. The modern trend is to combine modeling, theoretical analysis, and computer simulation. This book is intended to serve as a standard text for a junior-level engineering course in *system dynamics*. The text serves as a valuable source of information on mathematical modeling and analysis of dynamic systems, accompanied by selected computer simulation. The primary purpose of this text is to assist the student in acquiring a solid background in the principles and techniques of modeling and analysis, which are essential in subsequent courses of control systems and vibrations.

SCOPE AND SEQUENCE

The quintessential mathematical background, matrix analysis, and system model representation are thoroughly presented in Chapters 1–3. Chapter 1 contains complex variables, differential equations, and the Laplace transform. Chapter 2 provides a complete review of linear algebra and matrix analysis pertaining to system dynamics. System model representation, as well as linearization, are discussed in Chapter 3. Chapters 4 through 6 are concerned with modeling techniques of fundamental dynamic systems. Chapter 4 focuses on translational, rotational, and hybrid mechanical systems, whereas Chapter 5 covers electrical and electromechanical systems. Chapter 6 considers dynamic systems in thermal sciences: thermodynamics, pneumatic, liquid-level, and heat transfer systems. Chapter 7 contains time and frequency response of dynamic systems, as well as Fourier analysis. Chapter 8 provides a comprehensive introduction to mechanical vibrations, covering natural frequencies, mode shapes, the damped vibration absorber, and modal analysis. Block diagrams and their usage in model representation is introduced in Chapter 9. Chapter 10, an introduction to control systems, discusses fundamental concepts of stability, types of controllers, and control system design using the methods of Ziegler-Nichols tuning rules. Also included are the concepts of dominant poles and dominant time constants.

STUDY AIDS AND PEDAGOGY

To assist the students in learning, this book provides certain study aids and pedagogical techniques that emphasize and reinforce the fundamental concepts in modeling and analysis.

Essential mathematical background

Included in Chapters 1 and 2 is a review of the mathematical background that is essential for modeling and analysis. The topics include Laplace transforms, differential equations, complex algebra, and matrix analysis.

Free-body diagram

Extensive free-body diagrams are used in Newtonian mechanics to enhance the derivation of mathematical model. For translational systems, each primary mechanical component is accompanied by a free-body diagram to clearly show the relationship among the externally applied forces, mass, and acceleration. Similarly, for rotational motion, the free-body diagram shows the relation among the applied torques, mass moment inertia, and angular acceleration.

Velocity diagram

Lagrange's equations can also be used to obtain mathematical models of dynamics systems. When this energy approach is used, a velocity diagram is included for each complex mechanical system.

Applications

With a variety of applications, the book can be used in different engineering disciplines.

Examples

The examples are used to illustrate how a particular type of problem can be approached and solved.

Problems

The in-chapter problems provide immediate reinforcement of the material learned, whereas the end-of-chapter problems serve as the additional problems for further review and practice.

Summary

At the end of each chapter, a summary provides a concise review of the main points covered in the chapter.

Answers

To master the subject, one sure way is practicing in solving many problems. However, solving problems without knowing the correct answers, at the end, can be frustrating. Thus, the answers to nearly all odd-numbered problems are included at the end of the book so that the students can check their own work.

SUPPLEMENT

A solutions manual, with complete details of all problem solutions, is available to adopters of the book. For the convenience of the instructor, the manual provides a separate compiled list of the answers to all odd- and even-numbered problems. To ensure that the text materials, the problem statements, and the solutions are coherent and well integrated, the manual was prepared by the same authors who wrote the book. The approach as well as the solution techniques presented in the manual are consistent with those shown in the book. The major portion of the material has been class tested for about a year by the authors to improve on the accuracy and content.

ACKNOWLEDGMENTS

We would like to express our deepest gratitude to many of McGraw-Hill's personnel and to our colleagues and staff for their help at various stages in the development of this book. In this regard, we greatly appreciate Debra Riegert, senior editor in Mechanical and Aerospace Engineering; Michelle Sala, sales representative; Robert Christie, president of the College Division; and Ernest Mijares, Professor of Mechanical Engineering. We are indebted to Kris Engberg and the staff of Publication Services for their competence and patience in making this product technically accurate and appealing. Also, we would like to thank all of the reviewers who thoroughly reviewed our manuscript at various stages. Their invaluable suggestions and comments helped greatly in enhancing the clarity of the material presentation. Special thanks go to Professor Richard Scott, The University of Michigan at Ann Arbor; Dale A. Anderson, Louisiana Tech University; and Tsu-Chin Tsao, University of Illinois at Urbana-Champaign. Lastly, and most of all, we are grateful to our families and friends, whose constant support and encouragement have helped us in many ways throughout the entire process.

Hung V. Vu, Ph.D.
Ramin S. Esfandiari, Ph.D.

CHAPTER 1

Complex Analysis, Differential Equations, and the Laplace Transform

This chapter provides some preliminary mathematical concepts and methods that play essential roles in the analysis of dynamic systems. An ample portion of these background materials is supposed to have been covered in one or more courses such as applied mathematics, a prerequisite for a junior-level course on modeling and analysis of dynamic systems. Chapters 1 and 2 cover the background materials essential to an easier understanding of the topics in modeling and analysis. Although the first two chapters serve as a general review, an effort has been made to present the material as thoroughly as possible. The three main areas covered in the present chapter are *complex analysis; linear, time-invariant ordinary differential equations (ODEs);* and *Laplace transformation.*

1.1 COMPLEX NUMBERS IN RECTANGULAR FORM

DEFINITION 1.1. A number in the form $z = x + jy$, in which x and y are (constant) real numbers and $j = \sqrt{-1}$, is referred to as a **complex number.** Here, x and y are called the **real** and **imaginary parts** of z, and are denoted by Re$\{z\}$ and Im$\{z\}$, respectively. This representation of z is referred to as the **rectangular,** or **cartesian, form.**

Two complex numbers are said to be **equal** if and only if their respective real and imaginary parts are equal; i.e., if $z_1 = x_1 + jy_1$ and $z_2 = x_2 + jy_2$, then

$$z_1 = z_2 \quad \overset{\text{if and only if}}{\Longleftrightarrow} \quad x_1 = x_2 \quad \text{and} \quad y_1 = y_2$$

1

Addition and Multiplication of Complex Numbers: Rectangular Form

Complex numbers can be thought of as binomials. *Addition* is performed component-wise; i.e.,

$$z_1 = x_1 + jy_1 \quad \text{and} \quad z_2 = x_2 + jy_2 \implies z_1 + z_2 = (x_1 + x_2) + j(y_1 + y_2)$$

Multiplication is the same as that of two binomials with the provision that powers of the imaginary number, j, need to be fully reduced following the property of j, such as $j^2 = -1$, $j^3 = -j$, $j^4 = 1$, etc.

EXAMPLE 1.1. Perform the following multiplication and express the result in rectangular form.

$$(-2 + j5)(3 - j2)$$

Solution. Treating the two complex numbers as binomials, the product is obtained as

$$(-2 + j5)(3 - j2) = -6 + j4 + j15 - j^2 10 = -6 + j19 + 10 = 4 + j19$$

Complex Conjugation: Rectangular Form

DEFINITION 1.2. Given a complex number $z = x + jy$, its **conjugate**, denoted by $\bar{z}$, is defined as $\bar{z} = x - jy$ (Fig. 1.1). An immediate implication is that the product of a complex number and its conjugate is a positive, real number; i.e.,

$$z\bar{z} = (x + jy)(x - jy) = x^2 + y^2 \quad (1.1)$$

EXAMPLE 1.2. Given $z = -1 + 2j$, evaluate $z\bar{z}$.

Solution. Following the general formulation of Eq. (1.1), with $x = -1$ and $y = 2$, we have

$$z\bar{z} = (-1 + 2j)(-1 - 2j) = (-1)^2 + (2)^2 = 5$$

which is a positive, real number, as expected.

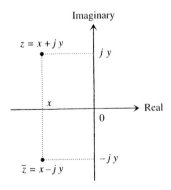

FIGURE 1.1 Locations of a complex number and its conjugate in the complex plane.

Division of Complex Numbers: Rectangular Form

Division of two complex numbers z_1 and z_2 ($z_2 \neq 0$) results in a complex number and is performed as follows: multiply the numerator and denominator of the fraction by the complex conjugate of the denominator. As a result of the property above, the newly constructed denominator will be a real number. In general, given two complex numbers $x_1 + jy_1$ and $x_2 + jy_2$, then

$$\frac{x_1 + jy_1}{x_2 + jy_2} = \frac{x_1 + jy_1}{x_2 + jy_2} \cdot \frac{x_2 - jy_2}{x_2 - jy_2}$$

Perform the multiplication and simplify the result to obtain

$$\frac{x_1 + jy_1}{x_2 + jy_2} = \frac{(x_1 x_2 + y_1 y_2) + j(y_1 x_2 - y_2 x_1)}{x_2^2 + y_2^2}$$
$$= \frac{x_1 x_2 + y_1 y_2}{x_2^2 + y_2^2} + j\left(\frac{y_1 x_2 - y_2 x_1}{x_2^2 + y_2^2}\right) \quad (1.2)$$

which is expressed in the *standard* rectangular form of a complex number.

EXAMPLE 1.3. Perform the following division of complex numbers and express the result in rectangular form.

$$\frac{-1 + j3}{2 + j5}$$

Solution. Using the strategy leading to Eq. (1.2), one obtains

$$\frac{-1 + j3}{2 + j5} \cdot \frac{2 - j5}{2 - j5} = \frac{-2 + j11 + 15}{4 + 25} = \frac{13 + j11}{29} = \frac{13}{29} + j\frac{11}{29}$$

PROBLEM SET 1.1

Perform the indicated operation and express the result in the standard rectangular form.

1.1. $(1 + j)(-2 + j3)$ **1.2.** $j(2 - j)$

1.3. $\dfrac{-1 + j}{2 - j}$ • **1.4.** $\dfrac{2 + j}{3j}$

1.2 REPRESENTATION OF COMPLEX NUMBERS IN POLAR FORM

A complex number, $z = x + jy \neq 0$, can be represented geometrically in the complex plane as shown in Fig. 1.2. Before the polar form representation of a complex number is introduced, a relation between rectangular and polar forms is presented here.

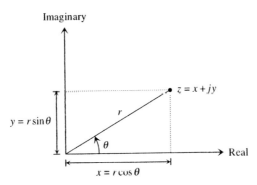

FIGURE 1.2 Rectangular and polar representation of a complex number.

Euler's formula
$$e^{j\theta} = \cos\theta + j\sin\theta \tag{1.3}$$

Using the relations between the rectangular and polar coordinates, as specified in Fig. 1.2, the complex number z can then be represented in **polar form** as

$$z = x + jy = \underbrace{r\cos\theta}_{x} + j\underbrace{(r\sin\theta)}_{y} = r\underbrace{(\cos\theta + j\sin\theta)}_{e^{j\theta}} = re^{j\theta} \tag{1.4}$$

in which $r = |z| = \sqrt{(\text{Re}\{z\})^2 + (\text{Im}\{z\})^2} = \sqrt{x^2 + y^2}$ denotes the **magnitude** of z and $\theta = \arg z = \tan^{-1}(y/x)$ represents the **argument**, or **phase**, of z. Angle θ is measured from the positive real axis and is regarded as positive in the counterclockwise (ccw) direction.

Notation

$$z = re^{j\theta} = |z|\angle\theta \tag{1.5}$$

Remark: Following the definition of the magnitude of z, i.e., $|z|$, the property of the complex conjugate, given by Eq. (1.1), can be interpreted as

$$z\bar{z} = x^2 + y^2 = |z|^2 \tag{1.1*}$$

EXAMPLE 1.4. Express the complex number $z = 3 + j\sqrt{3}$ in polar form.

Solution. This number can be represented using polar coordinates, as in Eqs. (1.4) and (1.5), as follows.

$$\left.\begin{array}{l} |z| = \sqrt{12} = 2\sqrt{3} \\ \theta = \tan^{-1}\left(\dfrac{\sqrt{3}}{3}\right) = \dfrac{\pi}{6} \end{array}\right\} \Rightarrow z = 2\sqrt{3}e^{j(\pi/6)} = 2\sqrt{3}\angle 30°$$

Note. In order to calculate the phase associated with a complex number, it is best to first determine the quadrant in which the complex number is located!

In our present example, the given complex number is located in the first quadrant (Fig. 1.3), which led to the conclusion that $\theta = \pi/6$ rad $= 30°$.

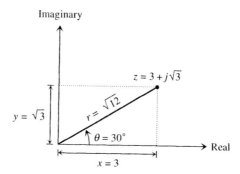

FIGURE 1.3 Geometrical interpretation of $z = 3 + j\sqrt{3}$.

EXAMPLE 1.5. Determine the location and the phase angle of the complex number $2/(-1 + j)$.

Solution. We first express this number in standard rectangular form by multiplying its numerator and denominator by the conjugate of the denominator; i.e.,

$$\frac{2}{-1+j}\frac{\overline{-1+j}}{\overline{-1+j}} = \frac{2}{-1+j}\left(\frac{-1-j}{-1-j}\right) = \frac{-2-2j}{1+1} = -1-j$$

The resulting complex number is clearly in the *third quadrant*, and direct calculation yields a phase angle of $+225°$ or $-135°$.

Multiplication and Division of Complex Numbers: Polar Form

Algebra of complex numbers, involving multiplication and division, can be greatly reduced via polar form representation. To this end, consider two complex numbers expressed in their respective polar forms, as

$$z_1 = r_1 e^{j\theta_1} = r_1 \angle \theta_1 \quad \text{and} \quad z_2 = r_2 e^{j\theta_2} = r_2 \angle \theta_2$$

Then the following relations hold.

Multiplication $\quad z_1 z_2 = r_1 r_2 e^{j(\theta_1 + \theta_2)} = r_1 r_2 \angle (\theta_1 + \theta_2)$ (1.6)

Division $\quad \dfrac{z_1}{z_2} = \dfrac{r_1}{r_2} e^{j(\theta_1 - \theta_2)} = \dfrac{r_1}{r_2} \angle (\theta_1 - \theta_2), \quad z_2 \neq 0$ (1.7)

Complex Conjugation: Polar Form

Given a complex number in polar form, $z = re^{j\theta}$ (Fig. 1.4), its complex conjugate is

$$\bar{z} = x - jy = r\cos\theta - j(r\sin\theta)$$
$$= r(\cos\theta - j\sin\theta) = r\underbrace{[\cos(-\theta) + j\sin(-\theta)]}_{e^{-j\theta}} = re^{-j\theta} \quad (1.8)$$

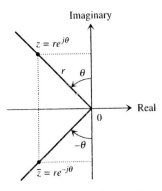

FIGURE 1.4 Polar form of a complex number and its conjugate.

where use has been made of Euler's formula, Eq. (1.3), and the fact that the cosine and sine functions are even and odd, respectively.

Remark: The property given by Eq. (1.1*) can be further verified using the polar form of a complex number, z, as follows.

$$z\bar{z} = \left(re^{j\theta}\right)\left(re^{-j\theta}\right) = r^2 = |z|^2 \tag{1.1**}$$

From the above discussion, it is therefore clear that the conjugate of $e^{j\theta}$ is simply $e^{-j\theta}$. Rewriting these two quantities via Euler's formula, one readily verifies that

$$\begin{cases} e^{j\theta} = \cos\theta + j\sin\theta \\ e^{-j\theta} = \cos\theta - j\sin\theta \end{cases} \xrightarrow{\text{combine}} \begin{cases} \cos\theta = (1/2)(e^{j\theta} + e^{-j\theta}) \\ \sin\theta = (1/2j)(e^{j\theta} - e^{-j\theta}) \end{cases} \tag{1.9}$$

Powers of a Complex Number

Given $z = re^{j\theta}$, then

$$z^n = \underbrace{r^n e^{jn\theta} = r^n \angle n\theta}_{\text{polar form}} = \underbrace{r^n(\cos n\theta + j\sin n\theta)}_{\text{cartesian form}} \tag{1.10}$$

Roots of a Complex Number

To determine the n different values $\sqrt[n]{z}$ ($z \neq 0$ and $n = 1, 2, \ldots$), first let $w = \sqrt[n]{z}$, which is equivalent to $z = w^n$. Then there exists one value of z corresponding to each value of w. Assuming that the polar form of w is $w = \rho e^{j\psi}$, then we have

$$w^n = z \implies \rho^n e^{jn\psi} = re^{j\theta} \implies \rho^n = r, \tag{1.11}$$
$$n\psi = \theta + 2k\pi, \quad k = \text{integer}$$

Note: To obtain n distinct values for w, only $k = 0, 1, \ldots, n-1$ are used! Higher values of k will correspond to points that have already been located using the first n values.

CHAPTER 1: Complex Analysis, Differential Equations, and the Laplace Transform   7

In order to completely describe w, its magnitude and phase (ρ and ψ) need to be determined. Solving Eq. (1.11) for ρ and ψ, the following quantities are obtained.

$$\rho = \sqrt[n]{r}, \qquad \psi = \frac{\theta}{n} + \frac{2k\pi}{n}, \qquad k = 0, 1, \ldots, n-1 \qquad (1.12)$$

Ultimately, using the information provided by Eq. (1.12) in the expression of w, the nth root of z can be presented as follows.

$$w = \sqrt[n]{z} = \underbrace{\rho e^{j\psi} = \sqrt[n]{r} e^{j(\theta/n + 2k\pi/n)} = r^{1/n} \angle \frac{1}{n}(\theta + 2k\pi)}_{\text{polar form}}$$

$$= \underbrace{\sqrt[n]{r}\left(\cos\frac{\theta + 2k\pi}{n} + j \sin\frac{\theta + 2k\pi}{n}\right)}_{\text{cartesian form}} \qquad (1.13)$$

with $k = 0, 1, \ldots, n-1$.

EXAMPLE 1.6. Determine the fourth roots of unity.

Solution. The equation $\sqrt[4]{z} = w$ must be solved in which $z = 1$. This implies that the magnitude and phase of z are 1 and 0, respectively. Inserting this information, $r = 1$, $\theta = 0$, and $n = 4$, into Eq. (1.13), one obtains

$$\rho = 1, \qquad \psi = \frac{2k\pi}{4} = \frac{k\pi}{2}, \qquad k = 0, 1, 2, 3 \qquad (1.14)$$

It is then readily seen that all four roots generated by Eq. (1.14) have a magnitude of 1 and correspond to arguments $\psi = 0, \pi/2, \pi$, and $3\pi/2$. These roots, in a more familiar cartesian form, are ± 1 and $\pm j$.

EXAMPLE 1.7. Evaluate $(2 - j3)^{1/3}$ and express the result in rectangular form.

Solution. The equation to be solved here is $\sqrt[3]{z} = w$ where $z = 2 - j3$. Direct calculation of the magnitude and phase of z yields

$$\left.\begin{array}{l} r = |2 - j3| = \sqrt{2^2 + 3^2} = 3.61 \\ \theta = \angle(2 - j3) = \tan^{-1}\left(\frac{-3}{2}\right) = -56.31° \end{array}\right\} \Rightarrow z = 2 - j3 = 3.61 e^{-j56.31}$$

Subsequently, following Eq. (1.13), with $n = 3$, we have

$$w = \sqrt[3]{z} = \sqrt[3]{r}\left(\cos\frac{\theta + 2k\pi}{3} + j \sin\frac{\theta + 2k\pi}{3}\right), \qquad k = 0, 1, 2$$

Substitution of r and θ results in

$$w = (2 - j3)^{1/3}$$

$$= \begin{cases} k = 0 \Rightarrow & 1.53(\cos 18.77 - j\sin 18.77) = 1.45 - j0.49 = w^{(1)} \\ k = 1 \Rightarrow & 1.53(\cos 101.23 + j\sin 101.23) = -0.30 + j1.50 = w^{(2)} \\ k = 2 \Rightarrow & 1.53(\cos 221.23 + j\sin 221.23) = -1.15 - j1.01 = w^{(3)} \end{cases}$$

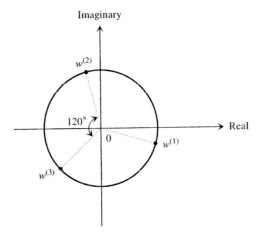

FIGURE 1.5 The three roots of $2 - j3$.

Geometrically, these three roots, $w^{(1)}$, $w^{(2)}$, and $w^{(3)}$, are located on the boundary of the circle of radius 1.53 and centered at the origin of the cartesian coordinate system. As shown in Fig. 1.5, these points are exactly $120°$ apart.

PROBLEM SET 1.2

Express each complex number in polar form, identifying the magnitude and phase.

1.5 $z = 1 + j$ **1.6** $z = 2 - j2\sqrt{3}$ • **1.7** $z = \dfrac{-1 + j2}{1 - j}$

Perform the operation, using polar coordinates, and express the result in the standard rectangular form.

1.8 $(3 - j\sqrt{3})^3$ **1.9** $(2 + j)^2$ • **1.10** $(3 + j)^{1/3}$

1.3 COMPLEX VARIABLES AND FUNCTIONS

Complex variables and functions play a particularly significant role in frequency response analysis of dynamic systems (see Chapter 7), and in what follows, the basic definitions and ideas are presented. Throughout most of this book, a *complex variable* is represented by s.

DEFINITION 1.3. A variable $s = \sigma + j\omega$ is referred to as a **complex variable** if σ, ω, or both are variables. Comparison with Definition 1.1 reveals that the main distinction between a complex number and a complex variable is that the real and imaginary parts of a complex number are constants.

CHAPTER 1: Complex Analysis, Differential Equations, and the Laplace Transform

DEFINITION 1.4. A function, $G(s)$, of a complex variable, s, is called a **complex function**. $G(s)$ consists of real and imaginary parts, with rectangular and polar forms.

$$G(s) = u + jv = M_G \angle \theta_G \quad (1.15)$$

where $u = \text{Re}\{G(s)\}$, $v = \text{Im}\{G(s)\}$, and M_G and θ_G denote the magnitude and phase of $G(s)$, respectively, and can be defined in a manner similar to those of a complex number (Section 1.2).

As mentioned earlier, these functions are of great importance while studying frequency response of a dynamic system. However, in this specific area, complex functions generally assume the form of a rational function as

$$G(s) = K \frac{s^m + a_1 s^{m-1} + \cdots + a_{m-1} s + a_m}{s^n + b_1 s^{n-1} + \cdots + b_{n-1} s + b_n} = \frac{N(s)}{D(s)}, \quad m < n \quad (1.16)$$

in which $K, a_1, \ldots, a_m, b_1, \ldots, b_n$ are (real) constants and m and n are nonnegative integers. Functions of this type will be referred to as *transfer functions* and will be studied in detail in Chapter 3.

DEFINITION 1.5. Those values of s for which $N(s) = 0$, i.e., *roots* of $N(s) = 0$, are referred to as **zeros** of $G(s)$. Similarly, those values of s for which $D(s) = 0$ are called **poles** of $G(s)$. Zeros and poles with multiplicity of 1 are called **simple**. *Multiplicity* refers to the number of times a certain root is repeated.

The complex function in Eq. (1.16) can be expressed in **pole-zero form**; i.e.,

$$G(s) = K \frac{(s - z_1)(s - z_2) \cdots (s - z_m)}{(s - p_1)(s - p_2) \cdots (s - p_n)} \quad (1.17)$$

from which it is seen that $s = z_i$ ($i = 1, 2, \ldots, m$) and $s = p_k$ ($k = 1, 2, \ldots, n$) are the zeros and poles of $G(s)$, respectively.

EXAMPLE 1.8. Express the given complex function in pole-zero form. Identify the zeros and the poles, as well as the multiplicity of each.

$$G(s) = \frac{2s + 1}{s(s + 2)^2 (10s + 3)}$$

Solution. $G(s)$ can be written in pole-zero form as

$$G(s) = \frac{2(s + 0.5)}{10s(s + 2)^2 (s + 0.3)} = \left(\frac{1}{5}\right) \frac{s + 0.5}{s(s + 2)^2 (s + 0.3)}$$

It is then observed that this function has a simple zero at $s = -0.5$, two simple poles at $s = 0$ and $s = -0.3$, and a *double* pole (multiplicity of 2) at $s = -2$. Moreover, in relation to Eq. (1.17), $K = \frac{1}{5}$.

Each linear term in Eq. (1.17) can be expressed in polar form of Eq. (1.4). To this end, let ψ_i and ϕ_k denote phases associated with the linear terms in the numerator and the denominator, respectively; i.e.,

$$s - z_i = |s - z_i| e^{j\psi_i}, \quad i = 1, 2, \ldots, m$$

$$s - p_k = |s - p_k| e^{j\phi_k}, \quad k = 1, 2, \ldots, n$$

As a result, Eq. (1.17) can be rewritten in polar form as follows.

$$G(s) = K \frac{|s - z_1| \cdots |s - z_m| e^{j(\psi_1 + \cdots + \psi_m)}}{|s - p_1| \cdots |s - p_n| e^{j(\phi_1 + \cdots + \phi_n)}}$$

$$= K \underbrace{\frac{|s - z_1| \cdots |s - z_m|}{|s - p_1| \cdots |s - p_n|}}_{\text{magnitude}} e^{j\overbrace{[(\psi_1 + \cdots + \psi_m) - (\phi_1 + \cdots + \phi_n)]}^{\text{phase}}}$$

and in relation to Eq. (1.15), one obtains the *magnitude* and *phase* of $G(s)$ as

$$M_G = K \frac{|s - z_1| \cdots |s - z_m|}{|s - p_1| \cdots |s - p_n|}, \qquad \theta_G = (\psi_1 + \cdots + \psi_m) - (\phi_1 + \cdots + \phi_n) \quad (1.18)$$

In frequency response analysis of dynamic systems, as will be studied in Chapter 7, the complex variable s assumes the form of a pure imaginary number, $s = j\omega$, in which ω is known as the **forcing frequency**. Evaluation of the magnitude and phase of the resulting complex function, $G(j\omega)$, can be performed either following the strategy leading to Eq. (1.18) or via direct substitution of $s = j\omega$ into the expression of $G(s)$. In Chapter 7, $G(j\omega)$ will be referred to as the *frequency response function* (FRF) and studied in detail.

EXAMPLE 1.9. Determine the magnitude and phase of the complex function $G(s) = (s + 2)/(s^2 + s + 1)$ when $s = j2$.

Solution. Substitution yields

$$G(j2) = G(s)\Big|_{s=j2} = \frac{j2 + 2}{(j2)^2 + j2 + 1} = \frac{2 + j2}{-3 + j2}$$

Multiply and divide by the complex conjugate of the denominator to obtain the standard rectangular form.

$$G(j2) = \frac{2 + j2}{-3 + j2} \cdot \frac{-3 - j2}{-3 - j2} = \frac{-2 - j10}{13} = -\frac{2}{13} - j\frac{10}{13}$$

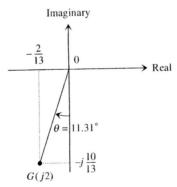

FIGURE 1.6 Location of $G(j2)$ in the complex plane.

indicating that $G(j2)$ is located in the third quadrant of the complex plane (Fig. 1.6). Then,

$$|G(j2)| = \frac{\sqrt{(-2)^2 + (-10)^2}}{13} = \frac{2\sqrt{26}}{13},$$

$$\angle G(j2) = \tan^{-1}\left(\frac{-10/13}{-2/13}\right) = \tan^{-1} 5 = 258.69° \text{ or } -101.31°$$

PROBLEM SET 1.3

Express each complex function in pole-zero form. Determine its poles and zeros and the multiplicity of each.

1.11. $F(s) = \dfrac{3s + 1}{s^2 + 2s + 1}$ **1.12.** $G(s) = \dfrac{s + 2}{s^2 + 2s + 2}$ **1.13.** $H(s) = \dfrac{1}{5s^2 + s}$

Evaluate each complex function at the indicated value of s and determine its magnitude and phase.

1.14. $F(s) = \dfrac{s + 1}{s(s^2 + s + 1)}$, $s = j$ **1.15.** $G(s) = \dfrac{1}{s^2 + 3s + 2}$, $s = j2$

1.4 LINEAR DIFFERENTIAL EQUATIONS WITH CONSTANT COEFFICIENTS

Mathematical models of many dynamic systems are represented by one or more ordinary differential equations. This section concerns itself with the *classical,* or *standard, forms* of the first- and second-order, ordinary differential equations that frequently arise in systems dynamics. Furthermore, we will focus our attention on a class of such equations that are linear and possess constant coefficients. We first make the following definition.

DEFINITION 1.6. An equation involving a dependent function of one independent variable and its derivative(s) is referred to as an **ordinary differential equation (ODE)**. The highest derivative of the dependent function is called the **order** of the ODE.

A **first-order,** ordinary differential equation is, in general, in the form

$$a_1 \dot{x} + a_0 x = f(t), \qquad \dot{x} = \frac{dx}{dt} \qquad (1.19)$$

where $f(t)$ is some specified time-varying function. If the coefficients, a_1 and a_0, are either constants or functions of t, the ODE is said to be **linear.** Otherwise, it is said to be **nonlinear.** In the event that these coefficients are all constants, Eq. (1.19) is called a **first-order, linear ODE with constant coefficients.**

Similarly, an ordinary differential equation expressed as

$$a_2\ddot{x} + a_1\dot{x} + a_0 x = f(t), \qquad \ddot{x} = \frac{d^2 x}{dt^2} \qquad (1.20)$$

is called a **second-order, linear ODE with constant coefficients** if all coefficients are constants.

EXAMPLE 1.10. The following is a list of examples of the types of ordinary differential equations just defined.

First-order, linear with constant coefficients: $\quad 2\dot{x} + x = 0$
Second-order, linear with constant coefficients: $\quad \ddot{x} + 3\dot{x} + 9x = 2\sin t$
Second-order, linear: $\quad \ddot{x} + (2t - 1)\dot{x} + 2x = 0$
First-order, linear: $\quad \dot{x} + (\sin t)x = \sin 3t$
First-order, nonlinear: $\quad 2\dot{x} + x^2 = t$
Second-order, nonlinear: $\quad \ddot{x} + (x + 1)\dot{x} + 9x = 0$

Standard Forms

DEFINITION 1.7. A first-order, linear ODE with constant coefficients is in **standard form** if it is expressed as

$$\dot{x} + \frac{1}{\tau} x = f(t), \qquad \tau > 0 \qquad (1.21)$$

Here τ is referred to as the **time constant** and provides a measure of how fast the system responds when subjected to a certain input (disturbance). The standard form of Eq. (1.21) represents the mathematical model of a wide range of dynamic systems that may be physically quite different.

EXAMPLE 1.11. Consider the single-tank, liquid-level system shown in Fig. 1.7. The mathematical model of this system will be shown in Chapter 6 to be given by the following first-order, linear ODE with constant coefficients.

$$\frac{RA}{g}\dot{h} + h = \frac{R}{g} q_i(t)$$

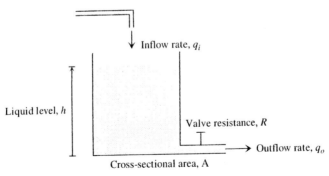

FIGURE 1.7 Single-tank, liquid-level system.

This can easily be expressed in the standard form of Eq. (1.21) as

$$\dot{h} + \frac{g}{RA}h = \frac{1}{A}q_i(t)$$

As a result, the system's *time constant* is identified as $\tau = RA/g$.

DEFINITION 1.8. A second-order, linear ODE with constant coefficients is in **standard form** if it is expressed as

$$\ddot{x} + 2\zeta\omega_n\dot{x} + \omega_n^2 x = f(t) \tag{1.22}$$

where ζ and ω_n are the **damping ratio** and **(undamped) natural frequency**, respectively. Mathematical models of many dynamic systems are described in the form of Eq. (1.22). These systems, although quite different in nature, exhibit similar types of behavior reflected through their governing equations.

EXAMPLE 1.12. Consider the RLC (resistance, inductance, and capacitance) circuit in Fig. 1.8. The mathematical model of this electrical system will be derived in Chapter 5 and shown to be described by the following second-order, linear ODE with constant coefficients

$$L\ddot{q} + R\dot{q} + \frac{1}{C}q = e(t) \tag{1.23}$$

where q denotes the electric charge and is defined by $i = dq/dt$. Rewrite Eq. (1.23) to resemble the standard form, Eq. (1.22), to obtain

$$\ddot{q} + \frac{R}{L}\dot{q} + \frac{1}{LC}q = \frac{1}{L}e(t) \tag{1.24}$$

Direct comparison of Eq. (1.24) with the standard form reveals

$$\omega_n^2 = \frac{1}{LC}, \quad 2\zeta\omega_n = \frac{R}{L}$$

the solution of which identifies the (undamped) natural frequency and the damping ratio, as

$$\omega_n = \sqrt{\frac{1}{LC}}, \quad \zeta = \frac{R}{2}\sqrt{\frac{C}{L}}$$

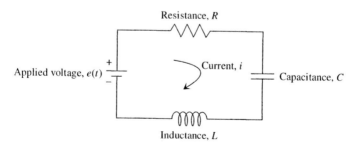

FIGURE 1.8 RLC circuit.

Solution of Linear ODEs with Constant Coefficients

Linear ordinary differential equations with constant coefficients have many attractive mathematical properties. One such property is that an analytical, and thus exact, solution to these equations can easily be obtained in closed form. Linear ODEs appear in one of two general forms: homogeneous or nonhomogeneous. This section concerns itself with the solutions to these equations. To this end, we first define what is meant by independent functions.

DEFINITION 1.9. Two functions, x_1 and x_2, are said to be **linearly independent** if

$$c_1 x_1 + c_2 x_2 = 0 \implies c_1 = 0 = c_2$$

which implies that neither one of the two functions may be expressible as a constant multiple of the other.

Homogeneous, Linear ODEs with Constant Coefficients

An nth-order, linear, homogeneous ODE with constant coefficients can, in general, be expressed in the form

$$x^{(n)} + a_{n-1} x^{(n-1)} + \cdots + a_2 \ddot{x} + a_1 \dot{x} + a_0 x = 0 \tag{1.25}$$

where $a_{n-1}, \ldots, a_0$ are constants and $x^{(n)} = d^n x/dt^n$. Assume that the solution to this equation is in the form $x(t) = e^{\lambda t}$ and substitute into Eq. (1.25) to obtain

$$(\lambda^n + a_{n-1}\lambda^{n-1} + \cdots + a_2\lambda^2 + a_1\lambda + a_0)e^{\lambda t} = 0 \xRightarrow{e^{\lambda t} \neq 0}$$

$$\underbrace{\lambda^n + a_{n-1}\lambda^{n-1} + \cdots + a_2\lambda^2 + a_1\lambda + a_0 = 0}_{\text{characteristic equation}}$$

which is known as the **characteristic equation**. The solutions to the characteristic equation are referred to as **characteristic values**, which completely determine the nature of the general solution to the original ODE. To each characteristic value, λ, corresponds an independent solution (function) in the form of $e^{\lambda t}$. Subsequently, a linear combination of these independent solutions constitutes the **general solution** to the ODE. The *strategy* to determine these independent solutions is as follows.

If a characteristic value, λ_i, is real and simple, the corresponding solution is $x_i = e^{\lambda_i t}$. If it is real and occurs m times (i.e., multiplicity $m \leq n$), then the m linearly independent solutions associated with it are $x_i, tx_i, \ldots, t^{m-1} x_i$. If λ_i is simple and complex, i.e., $\lambda_i = \sigma + j\omega$, it corresponds to $x_i = e^{\sigma t}(c_1 \cos \omega t + c_2 \sin \omega t)$. When it is complex with multiplicity m, the m linearly independent solutions are $x_i, tx_i, \ldots, t^{m-1} x_i$.

EXAMPLE 1.13. Solve the following homogeneous, second-order ODE: $\ddot{x} + 3\dot{x} + 2x = 0$.

CHAPTER 1: Complex Analysis, Differential Equations, and the Laplace Transform

Solution. The characteristic equation and values are obtained as

$$\lambda^2 + 3\lambda + 2 = 0 \implies (\lambda + 1)(\lambda + 2) = 0 \implies \lambda_{1,2} = -1, -2$$

Thus, the two real and distinct characteristic values correspond to e^{-t} and e^{-2t}, which are clearly linearly independent. The general solution to the ODE is then expressed as the linear combination of these functions; i.e.,

$$x(t) = c_1 e^{-t} + c_2 e^{-2t}$$

EXAMPLE 1.14. Find the general solution to the third-order ODE given by $\dddot{x} + 4\ddot{x} + 5\dot{x} + 2x = 0$.

Solution. The characteristic equation and values are obtained as

$$\lambda^3 + 4\lambda^2 + 5\lambda + 2 = 0 \implies (\lambda + 2)(\lambda + 1)^2 = 0$$
$$\implies \lambda_1 = -2, \quad \lambda_{2,3} = -1$$

The first characteristic value is real and simple; hence it corresponds to e^{-2t}. The second one, however, has a multiplicity of 2 (i.e., $m = 2$). Therefore, it corresponds to e^{-t} and te^{-t}. The general solution to the ODE is

$$x(t) = c_1 e^{-2t} + \underbrace{c_2 e^{-t} + c_3 t e^{-t}}_{\substack{\text{corresponding to} \\ \lambda = -1 \text{ of multiplicity } 2}}$$

Free Response of First-Order Systems

Recall that the standard form of a first-order system is given by Eq. (1.21). When $f(t) = 0$, the equation reduces to a homogeneous, first-order differential equation,

$$\dot{x} + \frac{1}{\tau}x = 0, \quad \tau > 0$$

and its solution is referred to as the **free response** of the system. The resulting characteristic equation is then

$$\lambda + \frac{1}{\tau} = 0 \implies \lambda = -\frac{1}{\tau}$$

with a corresponding solution

$$x(t) = c e^{-t/\tau}$$

Free Response of Second-Order Systems

The standard form of a second-order system is described by Eq. (1.22). When $f(t) = 0$, it then reduces to a homogeneous, second-order differential equation,

$$\ddot{x} + 2\zeta\omega_n \dot{x} + \omega_n^2 x = 0$$

the solution of which is known as its *free response*. The characteristic equation and values are found as

$$\lambda^2 + 2\zeta\omega_n\lambda + \omega_n^2 = 0 \quad \Longrightarrow \quad \lambda_{1,2} = -\zeta\omega_n \pm \sqrt{(\zeta\omega_n)^2 - \omega_n^2}$$
$$= -\zeta\omega_n \pm \omega_n\sqrt{\zeta^2 - 1}$$

It is then observed that the nature of a characteristic value depends on whether ζ is less than, equal to, or greater than 1. Whereas a thorough investigation of all possible cases is included in Chapter 7, in the present study, we simply assume $\zeta < 1$. Based on this assumption, the characteristic values are given by

$$\lambda_{1,2} = -\zeta\omega_n \pm j\omega_n\sqrt{1 - \zeta^2}, \qquad \zeta < 1 \qquad (1.26)$$

It is common to let $\sigma = \zeta\omega_n$ and $\omega_d = \omega_n\sqrt{1 - \zeta^2}$, where ω_d is called the **(damped) natural frequency**. Then

$$\lambda_{1,2} = -\sigma \pm j\omega_d$$

which generate the system's free response in the form

$$x(t) = Ce^{(-\sigma + j\omega_d)t} + \overline{C}e^{(-\sigma - j\omega_d)t}$$

where C is a complex number, in general, and $\overline{C}$ denotes its conjugate. Expansion of the exponential terms, and using Euler's formula, yields

$$x(t) = e^{-\sigma t}\left[Ce^{j\omega_d t} + \overline{C}e^{-j\omega_d t}\right]$$
$$= e^{-\sigma t}\left[C(\cos\omega_d t + j\sin\omega_d t) + \overline{C}(\cos\omega_d t - j\sin\omega_d t)\right]$$
$$= e^{-\sigma t}\left[(C + \overline{C})\cos\omega_d t + j(C - \overline{C})\sin\omega_d t\right]$$

Since C represents a complex number, let $C = a + jb$. Consequently, $C + \overline{C} = 2a$ and $j(C - \overline{C}) = -2b$, both of which are real numbers. Letting $A = 2a$ and $B = -2b$, we have

$$x(t) = e^{-\sigma t}(A\cos\omega_d t + B\sin\omega_d t), \qquad \zeta < 1 \qquad (1.27)$$

as the free response of a second-order system.

EXAMPLE 1.15. Solve the following homogeneous, second-order ODE: $\ddot{x} + 2\dot{x} + 2x = 0$

Solution. The characteristic equation and values are obtained as

$$\lambda^2 + 2\lambda + 2 = 0 \quad \Longrightarrow \quad \lambda_{1,2} = -1 \pm j$$

from which we obtain $\sigma = 1$ and $\omega_d = 1$. Following Eq. (1.27), the solution to the homogeneous ODE is $x(t) = e^{-t}(A\cos t + B\sin t)$.

Expressing $A\cos\alpha t + B\sin\alpha t$ as $D\sin(\alpha t + \phi)$ or $D\cos(\alpha t + \phi)$

It is often advantageous to express $A\cos\alpha t + B\sin\alpha t$ in terms of a single trigonometric function, in particular, as $D\sin(\alpha t + \phi)$. To this end, rewrite the latter via trigonometric expansion, as follows.

$$D\sin(\alpha t + \phi) = D[\sin \alpha t \cos \phi + \cos \alpha t \sin \phi]$$

Direct comparison with $A \cos \alpha t + B \sin \alpha t$ reveals that the respective coefficients of $\cos \alpha t$ and $\sin \alpha t$ in the two expressions must be proportional; i.e.,

$$\frac{\sin \phi}{\cos \phi} = \frac{A}{B} \implies \tan \phi = \frac{A}{B}$$

Next, construct a right triangle in which angle ϕ satisfies the above property (Fig. 1.9). Then, it is clear that

$$\cos \phi = \frac{B}{\sqrt{A^2 + B^2}} \quad \text{and} \quad \sin \phi = \frac{A}{\sqrt{A^2 + B^2}}$$

As a result,

$$D\sin(\alpha t + \phi) = D[\sin \alpha t \cos \phi + \cos \alpha t \sin \phi]$$

$$= D\left\{\frac{B}{\sqrt{A^2 + B^2}} \sin \alpha t + \frac{A}{\sqrt{A^2 + B^2}} \cos \alpha t\right\}$$

$$= \frac{D}{\sqrt{A^2 + B^2}} [A \cos \alpha t + B \sin \alpha t]$$

In order for this expression to be identical to $A \cos \alpha t + B \sin \alpha t$, we must have $D = \sqrt{A^2 + B^2}$. In conclusion,

$$A \cos \alpha t + B \sin \alpha t = D \sin(\alpha t + \phi) \quad \text{where } D = \sqrt{A^2 + B^2} \quad (1.28)$$

and

$$\phi = \tan^{-1}\left(\frac{A}{B}\right)$$

Similarly, it can be shown that

$$A \cos \alpha t + B \sin \alpha t = D \cos(\alpha t + \phi) \quad (1.29)$$

where

$$D = \sqrt{A^2 + B^2} \quad \text{and} \quad \phi = -\tan^{-1}\left(\frac{B}{A}\right)$$

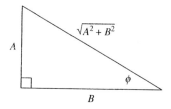

FIGURE 1.9 Construction of the shift angle.

Nonhomogeneous, Linear ODEs with Constant Coefficients

An nth-order, linear, nonhomogeneous ODE with constant coefficients is generally expressed in the form

$$x^{(n)} + a_{n-1}x^{(n-1)} + \cdots + a_2\ddot{x} + a_1\dot{x} + a_0x = f(t) \qquad (1.30)$$

in which the coefficients are constants and $f(t)$ is a prescribed function. The *general solution*, $x(t)$, to Eq. (1.30) is formed through the superposition of two types of solutions, homogeneous and particular.

Homogeneous solution, $x_h(t)$

The homogeneous solution satisfies the homogeneous version of Eq. (1.30),

$$x^{(n)} + a_{n-1}x^{(n-1)} + \cdots + a_2\ddot{x} + a_1\dot{x} + a_0x = 0$$

and is determined via the procedure outlined earlier.

Particular solution, $x_p(t)$

The particular solution is obtained via what is known as the **method of undetermined coefficients.** The nature of this solution is directly dependent on (1) the nature of the function, $f(t)$, which appears on the right-hand side of the ODE, and (2) the nature of the homogeneous solution in relation to $f(t)$. The applicability of the method of undetermined coefficients is limited in the sense that it can treat situations where $f(t)$ is a polynomial, an exponential function, sine or cosine functions, or possibly sums or products of the cited functions. The particular solution, $x_p(t)$, is determined as follows.

Step 1. Corresponding to each term in $f(t)$, pick the form suggested by Table 1.1. The particular solution is then selected as the linear combination of these.

Case 1. If none of the terms in $x_p(t)$, selected based on the right column of Table 1.1, appears to coincide with a homogeneous solution, the particular solution stands as it was originally selected.

Case 2. If any of the terms, say $(x_p)_i$, in the selected $x_p(t)$ coincides with a homogeneous solution, consider the following.

TABLE 1.1 Relations between the right-hand side function and the particular solution

$f(t)$	$x_p(t)$
kt^m, $\quad m = 0, 1, 2, \ldots$	$b_0t + b_1t + b_2t^2 + \cdots + b_mt^m$
ke^{at}	be^{at}
$k\sin\omega t$	$b_1\cos\omega t + b_2\sin\omega t$
$k\cos\omega t$	$b_1\cos\omega t + b_2\sin\omega t$
$ke^{\sigma t}\sin\omega t$	$e^{\sigma t}(b_1\cos\omega t + b_2\sin\omega t)$
$ke^{\sigma t}\cos\omega t$	$e^{\sigma t}(b_1\cos\omega t + b_2\sin\omega t)$

Case 2.1. If this homogeneous solution was obtained based on a *simple* characteristic value, modify that term to $t(x_p)_i$.

Case 2.2. If this homogeneous solution was obtained based on a characteristic value of multiplicity m, modify that term to $t^m(x_p)_i$.

Step 2. The particular solution in *Step 1* is a linear combination of the appropriately selected functions and hence involves constants to be determined. Substitute $x_p(t)$ into Eq. (1.30) for x and equate the coefficients of like terms on both sides. Subsequently, determine the unknown constants.

The general solution to the nonhomogeneous ODE is then $x(t) = x_h(t) + x_p(t)$. Although $x_p(t)$ is completely determined, $x_h(t)$ may involve constants that are yet to be evaluated. These are obtained via the application of additional conditions, such as initial and/or boundary conditions.

EXAMPLE 1.16. Find the solution to the following initial-value problem: $\dot{\omega} + 2\omega = t$, $\omega(0^-) = 1$.

Solution

Homogeneous solution, $\omega_h(t)$. The characteristic equation and the corresponding characteristic value are

$$\lambda + 2 = 0 \implies \lambda = -2$$

which results in $\omega_h(t) = ce^{-2t}$.

Particular solution, $\omega_p(t)$. The function on the right-hand side, known as the *forcing function*, is a first-degree polynomial and agrees with the first row of Table 1.1. Thus, the particular solution assumes the form $\omega_p(t) = At + B$. Furthermore, we notice that the forcing function does not coincide with the homogeneous solution. Insert $\omega_p(t)$ into the original differential equation to obtain

$$2At + (A + 2B) = t \underset{\substack{\text{equating coefficients}\\\text{of like terms}}}{\implies} \begin{cases} 2A = 1 \\ A + 2B = 0 \end{cases} \implies A = \frac{1}{2}, \quad B = -\frac{1}{4}$$

Consequently, $\omega_p(t) = \frac{1}{2}t - \frac{1}{4}$.

General solution. $\omega(t) = \omega_h(t) + \omega_p(t) = ce^{-2t} + \frac{1}{2}t - \frac{1}{4}$.

Initial conditions. Apply the given initial condition (I.C.) to the general solution to obtain

$$\underbrace{\omega(0^-) = 1}_{\text{given I.C.}} = c - \frac{1}{4} \implies c = \frac{5}{4}$$

As a result, the solution to this initial-value problem is $\omega(t) = \frac{1}{4}(5e^{-2t} + 2t - 1)$.

EXAMPLE 1.17. Solve the following initial-value problem: $\ddot{x} + 3\dot{x} + 2x = 2e^{-4t}$, $x(0^-) = 1$, $\dot{x}(0^-) = -1$.

Solution

Homogeneous solution, $x_h(t)$. The characteristic equation and values associated with the homogeneous equation are

$$\lambda^2 + 3\lambda + 2 = 0 \implies \lambda_{1,2} = -1, -2$$

which correspond to $x_h(t) = c_1 e^{-t} + c_2 e^{-2t}$.

Particular solution, $x_p(t)$. The function on the right-hand side agrees with the second row of Table 1.1; and hence $x_p(t) = ke^{-4t}$. Since this exponential function does not coincide with either of the homogeneous independent solutions, further modification is not needed. Substitute $x_p(t)$ into the original ODE to obtain

$$16ke^{-4t} - 12ke^{-4t} + 2ke^{-4t} = 2e^{-4t} \implies 6ke^{-4t} = 2e^{-4t}$$
$$\implies k = \tfrac{1}{3} \implies x_p(t) = \tfrac{1}{3} e^{-4t}$$

General solution, $x(t)$. $x(t) = x_h(t) + x_p(t) = c_1 e^{-t} + c_2 e^{-2t} + \tfrac{1}{3} e^{-4t}$.

Initial conditions. Apply the prescribed initial conditions to the general solution, $x(t)$, to evaluate c_1 and c_2.

$$\underbrace{x(0^-) = 1}_{\text{given I.C.}} = c_1 + c_2 + \tfrac{1}{3}, \qquad \underbrace{\dot{x}(0^-) = -1}_{\text{given I.C.}} = -c_1 - 2c_2 - \tfrac{4}{3}$$

Solving the two algebraic equations simultaneously, one obtains $c_1 = \tfrac{5}{3}$ and $c_2 = -1$. As a result, the specific form of the general solution is

$$x(t) = \tfrac{5}{3} e^{-t} - e^{-2t} + \tfrac{1}{3} e^{-4t}$$

EXAMPLE 1.18. Solve the following initial-value problem: $\ddot{x} + \omega^2 x = 2 \sin \omega t$, $x(0^-) = 0$, $\dot{x}(0^-) = 1$. It is often encountered in the frequency response study of undamped, second-order dynamic systems.

Solution

Homogeneous solution, $x_h(t)$. The characteristic equation and values associated with the homogeneous equation are

$$\lambda^2 + \omega^2 = 0 \implies \lambda_{1,2} = \pm j\omega$$

which correspond to $x_h(t) = c_1 \cos \omega t + c_2 \sin \omega t$.

Particular solution, $x_p(t)$. The forcing function agrees with the third row of Table 1.1; and hence $x_p(t) = A \cos \omega t + B \sin \omega t$. However, since it coincides with the homogeneous independent solutions, the particular solution is properly adjusted to $x_p(t) = t(A \cos \omega t + B \sin \omega t)$. Substitute $x_p(t)$ into the original ODE to obtain

$$-2A\omega \sin \omega t + 2B \cos \omega t = 2 \sin \omega t \implies A = -1, B = 0$$
$$\implies x_p(t) = -t \cos \omega t$$

General solution, $x(t)$. $x(t) = x_h(t) + x_p(t) = c_1 \cos \omega t + c_2 \sin \omega t - t \cos \omega t$.

Initial conditions. Apply the prescribed initial conditions to the general solution, $x(t)$, to evaluate c_1 and c_2.

$$\underbrace{x(0^-) = 0}_{\text{given I.C.}} = c_1, \qquad \underbrace{\dot{x}(0^-) = 1}_{\text{given I.C.}} = \omega c_2 - 1$$

Solving the two equations yields $c_1 = 0$ and $c_2 = 2/\omega$. As a result, the specific form of the general solution is

$$x(t) = \frac{2}{\omega} \sin \omega t - t \cos \omega t$$

EXAMPLE 1.19. Find the general solution to $\ddot{x} + 4\dot{x} + 5x = e^{-t} \sin t$, subjected to $x(0^-) = 0.2$ and $\dot{x}(0^-) = 0$.

Solution

Homogeneous solution, $x_h(t)$. Solving the characteristic equation yields

$$\lambda^2 + 4\lambda + 5 = 0 \implies \lambda_{1,2} = -2 \pm j \implies x_h(t) = e^{-2t}(c_1 \cos t + c_2 \sin t)$$

Particular solution, $x_p(t)$. The forcing function is in the form suggested by the fifth row of Table 1.1, so $x_p(t) = e^{-t}(A \cos t + B \sin t)$. Also, the two linearly independent homogeneous solutions are $e^{-2t} \cos t$ and $e^{-2t} \sin t$. It is then observed that neither of the two terms, $e^{-t} \cos t$ and $e^{-t} \sin t$, in x_p coincides with the homogeneous solutions. This implies that the suggested form of x_p is not subject to any modification. Insert x_p into the original ODE, and collect like terms, to obtain

$$(A + 2B)e^{-t} \cos t + (B - 2A)e^{-t} \sin t$$

$$= e^{-t} \sin t \implies \begin{cases} A + 2B = 0 \\ B - 2A = 1 \end{cases} \implies A = -\frac{2}{5}, \quad B = -\frac{1}{5}$$

General solution, $x(t)$. $x(t) = e^{-2t}(c_1 \cos t + c_2 \sin t) + \frac{1}{5}e^{-t}(-2 \cos t + \sin t)$. Application of the given initial conditions yields $c_1 = 0.6$ and $c_2 = 1$; and hence

$$x(t) = e^{-2t}(0.6 \cos t + \sin t) + \frac{1}{5}e^{-t}(-2 \cos t + \sin t)$$

Referring to Eq. (1.28), these two linear combinations are rewritten as follows.

$0.6 \cos t + \sin t = 1.1661 \sin(t + 0.5405), \quad -2 \cos t + \sin t = \sqrt{5} \sin(t - 1.1071)$

Subsequently,

$$x(t) = 1.1661 e^{-2t} \sin(t + 0.5404) + 0.4472 \sin(t - 1.1071)$$

PROBLEM SET 1.4

Decide whether the following ordinary differential equations are linear or nonlinear.

● **1.16.** $\dot{x} + x = e^{-t}$　　● **1.17.** $\dot{x} + x = e^x$　　● **1.18.** $\ddot{x} + \dot{x} + x^3 = \sin t$

22 Dynamic Systems: Modeling and Analysis

1.19. The rotational motion of a mechanical system, subjected to a sinusoidal applied torque, is described by $0.25\dot{\omega} + \omega = \sin 2t$. Express the first-order ODE in standard form and determine the time constant.

1.20. The mathematical model of a dynamic system is described by $2\ddot{x} + 6\dot{x} + 8x = f(t)$, where $f(t)$ is some specified time-varying function. Express the second-order ODE in standard form and determine the system's (undamped) natural frequency and damping ratio.

Solve the following homogeneous, linear ODEs with constant coefficients. In the event that initial conditions are provided, determine the constants associated with the solution.

1.21. $\ddot{x} + 2\dot{x} + 5x = 0$, $x(0^-) = 0$, $\dot{x}(0^-) = 2$

1.22. $\ddot{x} + 3\dot{x} = 0$, $x(0^-) = 0$, $\dot{x}(0^-) = -3$

● **1.23.** $\ddot{x} + 4\dot{x} = 0$

Determine the general solution to each ODE via the method of undetermined coefficients.

1.24 $\dot{x} + x = 2e^{-t}$, $x(0^-) = 1$ **1.25** $\ddot{x} + 2\dot{x} = e^{-2t}$

● **1.26** $\ddot{x} + 2\dot{x} + x = \sin 2t$

1.5 LAPLACE TRANSFORMATION

In systems analysis, one often encounters *time-invariant, linear* differential equations of second or higher orders. And it is generally difficult to obtain solutions to these equations in closed form via the solution methods in ordinary differential equations. One way to circumvent this difficulty, however, is to use *Laplace transformation,* the most significant feature of which is to transform a differential equation into an algebraic equation in terms of the transform function of the unknown quantity sought. The transform function is a function of a complex variable, denoted by s. This systematic procedure has a wide range of applications in many areas of dynamic systems. The technique is based on the transformation defined by

$$F(s) = L\{f(t)\} = \int_{0^-}^{\infty} e^{-st} f(t)\, dt \qquad (1.31)$$

in which $F(s)$ represents the **Laplace transform** of the function, $f(t)$, and where

$$f(t) = 0 \quad \text{for } t < 0$$

In Eq. (1.31), s is a *complex variable,* called the **Laplace variable,** and L is the Laplace transform **operator.** In order for a successful implementation of the method, it is imperative that the transform integral in Eq. (1.31) exists. The *necessary conditions for existence* of the integral are that

(C₁) $f(t)$ is piecewise continuous on every finite (time) interval of $[0, \infty)$.
(C₂) the magnitude of $f(t)$ is bounded by an exponential function:

$$|f(t)| \le Me^{at} \tag{1.32}$$

for some real constants a and M and all $t \in [0, \infty)$.

THEOREM 1.1. EXISTENCE OF LAPLACE TRANSFORM. Given that conditions (C₁) and (C₂) hold for $f(t)$, then its Laplace transform, $F(s)$, exists for all Re$\{s\} > a$.

Proof. Condition (C₁) guarantees that the integrand function, $e^{-st}f(t)$, is integrable on every finite interval. Using Condition (C₂), an upper bound for its magnitude is obtained as

$$\left|e^{-st}f(t)\right| = e^{-st}|f(t)| \le e^{-st}Me^{at} = Me^{(a-s)t}$$

$$\Longrightarrow \quad |F(s)| = \left|\int_{0^-}^{\infty} f(t)e^{-st}\,dt\right| \le \int_{0^-}^{\infty} Me^{(a-s)t}\,dt = \frac{M}{a-s}e^{(a-s)t}\Big|_{t=0^-}^{\infty} = \frac{M}{s-a}$$

where the assumption, $s - a > 0$, has been used for integral evaluation. Thus, the Laplace transform, $F(s)$, is bounded above in magnitude, and hence exists.

Note: Theorem 1.1 does *not* provide necessary and sufficient conditions for existence. For instance, there exist functions that do not satisfy (C₂) and still have a Laplace transform.

To *solve* a differential equation via the Laplace transformation method, one always finds it necessary to obtain the time history, $f(t)$, from the Laplace transform function, $F(s)$. This process is referred to as **inverse Laplace transformation,** and is represented by

$$f(t) = L^{-1}\{F(s)\} \tag{1.33}$$

In summary, the operations consisting of the Laplace transform and its inverse are interpreted, pictorially, as shown in Fig. 1.10.

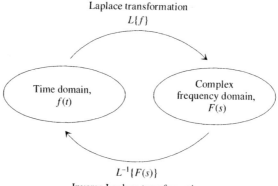

FIGURE 1.10 Operations involved in the Laplace transformation method.

24 Dynamic Systems: Modeling and Analysis

The inverse Laplace transform of $F(s)$ can be obtained through techniques such as the *partial fractions* method and the *convolution* method, which will be discussed in detail in a later section. Next, several special functions that are frequently encountered in systems analysis will be considered and their Laplace transforms will be derived. These include unit-step, unit-ramp, unit-impulse, sinusoidal, and exponential functions. For the Laplace transforms of several other functions, the reader is referred to Table 1.2 (p. 66–67).

Special Functions

Unit-step function, $u_s(t)$ or $u(t)$

The unit-step function can be physically interpreted as a constant force of magnitude 1 applied to a system at time $t \geq 0^+$ (Fig. 1.11). Using the definition of Laplace transform given by Eq. (1.31), one obtains

$$L\{u_s(t)\} = U_s(s) = \int_{0^-}^{\infty} u_s(t)e^{-st}\,dt$$

$$= \underbrace{\int_{0^-}^{0^+} u_s(t)e^{-st}\,dt}_{\substack{\downarrow \\ 0 \\ \text{the integrand is finite}}} + \int_{0^+}^{\infty} u_s(t)e^{-st}\,dt = \int_{0^+}^{\infty} e^{-st} = \frac{1}{s}$$

In general, a *step function* behaves in the same manner as the *unit-step* with the exception that the amplitude is, in general, some constant $A \neq 1$. In this event, the Laplace transform of the step function is readily seen to be A/s. In the work that follows, we generally study functions of unit amplitude—with the understanding that similar treatment is still valid for amplitudes other than 1.

Unit-ramp function, $u_r(t)$

$$L\{u_r(t)\} = U_r(s) = \int_{0^-}^{\infty} te^{-st}\,dt = \left\{t\left(\frac{e^{-st}}{-s}\right)\right\}_{t=0^-}^{\infty} - \int_{0^-}^{\infty}\left(\frac{e^{-st}}{-s}\right)dt = \left.\frac{e^{-st}}{-s^2}\right|_{t=0^-}^{\infty} = \frac{1}{s^2}$$

Figure 1.12 graphs the unit-ramp function.

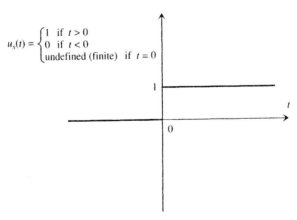

$$u_s(t) = \begin{cases} 1 & \text{if } t > 0 \\ 0 & \text{if } t < 0 \\ \text{undefined (finite)} & \text{if } t = 0 \end{cases}$$

FIGURE 1.11 Unit-step function.

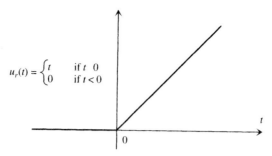

FIGURE 1.12 Unit-ramp function.

Unit-pulse function, $u_p(t)$

$$L\{u_p(t)\} = U_p(s) = \int_{0^-}^{t_1} e^{-st}\, dt = \left(\frac{e^{-st}}{-s}\right)\bigg|_{t=0^-}^{t_1} = \frac{1}{s}(1 - e^{-st_1}) \quad (1.34)$$

Note: From Fig. 1.13, it is observed that as $t_1 \to \infty$, the unit-pulse function behaves similarly to the unit-step function. So, it is expected that, in the limiting case, its Laplace transform agrees with that of the unit-step function. Taking the limit of the transform of the unit-pulse, given by Eq. (1.34), yields

$$\lim_{t_1 \to \infty} \left\{ \frac{1}{s}(1 - e^{-st_1}) \right\} = \frac{1}{s}$$

which is indeed the transform of the unit-step function.

A more general form of pulse function, other than the unit-pulse, is described by

$$u_p(t) = \begin{cases} 1/t_1 & \text{if } 0 < t < t_1 \\ 0 & \text{if } t < 0 \text{ and } t > t_1 \end{cases}$$

Laplace transformation, via Eq. (1.31), of the above yields

$$L\{u_p(t)\} = \frac{1}{st_1}(1 - e^{-st_1}) \quad (1.35)$$

Based on the structure of this function, the unit-impulse function may now be defined.

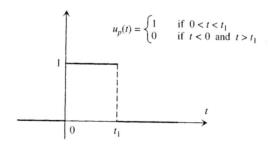

FIGURE 1.13 Unit-pulse function.

Unit-impulse (Dirac delta) function

A unit-impulse function, denoted by $\delta(t)$ also known as a **Dirac delta** function, can be physically realized as a force, very large in magnitude, and applied for a very short period of time (see Fig. 1.14). It is referred to as a unit-impulse because its area is unity. The Laplace transform of a unit-impulse is calculated as follows.

$$L\{\delta(t)\} = \Delta(s) = \lim_{t_1 \to 0}\left\{\frac{1}{st_1}(1 - e^{-st_1})\right\} = \lim_{t_1 \to 0}\left\{\frac{se^{-st_1}}{s}\right\} = 1$$

where use has been made of *l'Hôpital's rule*. An impulse function is generally expressed as $A\delta(t)$, where A denotes its area. In general, $\delta(t-\tau)$ denotes a unit-impulse applied at time $t = \tau$. This function makes a desirable system input owing to its attractive mathematical properties, in particular, a **filtering property**, defined as

$$\int_{-\infty}^{\infty} f(\tau)\delta(t - \tau)\,d\tau = f(t) \tag{1.36}$$

provided that $f(t)$ is continuous at $t = \tau$. In the event that the unit-impulse occurs at $t = 0$, Eq. (1.36) yields

$$\int_{-\infty}^{\infty} f(\tau)\delta(\tau)\,d\tau = f(0)$$

Following the definition of the Laplace transform, it is readily seen that

$$L\{\delta(t - \tau)\} = e^{-\tau s} \tag{1.37}$$

Unit-sinusoidal function

$$f(t) = \begin{cases} \sin\omega t & \text{if } t \geq 0 \\ 0 & \text{if } t < 0 \end{cases} \implies F(s) = L\{\sin\omega t\} \tag{1.38}$$

$$= \int_{0^-}^{\infty} e^{-st}\sin\omega t\,dt \underset{\text{integration by parts}}{=} \frac{\omega}{s^2 + \omega^2}$$

$$f(t) = \begin{cases} \cos\omega t & \text{if } t \geq 0 \\ 0 & \text{if } t < 0 \end{cases} \implies F(s) = L\{\cos\omega t\} \tag{1.39}$$

$$= \int_{0^-}^{\infty} e^{-st}\cos\omega t\,dt \underset{\text{integration by parts}}{=} \frac{s}{s^2 + \omega^2}$$

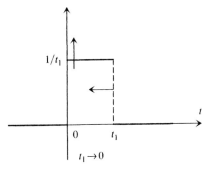

FIGURE 1.14 Unit-impulse (Dirac delta) function.

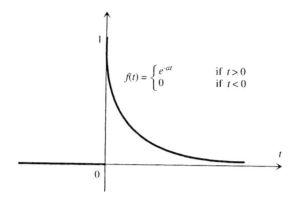

FIGURE 1.15 Exponential function.

Exponential function
See Fig.1.15.

$$L\{f(t)\} = \int_{0^-}^{\infty} e^{-at}e^{-st}\,dt = \underbrace{\int_{0^-}^{0^+} e^{-at}e^{-st}\,dt}_{\substack{0 \\ \text{integrand is finite} \\ \text{in the interval}}} + \int_{0^+}^{\infty} e^{-at}e^{-st}\,dt \quad (1.40)$$

$$= \frac{1}{-(s+a)} e^{-(s+a)t}\Big|_{t=0^+}^{\infty} = \frac{1}{s+a}$$

EXAMPLE 1.20. Find the Laplace transform of the signal, $g(t)$, geometrically defined in Fig. 1.16.

Solution. The analytical description of this signal is defined as

$$g(t) = \begin{cases} 2t & \text{if } 0 < t < 0.5 \\ 0 & \text{otherwise} \end{cases}$$

Following the definition of the Laplace transform, given by Eq. (1.31), and using integration by parts, one obtains

$$G(s) = \int_{0^-}^{0.5} 2te^{-st}\,dt \stackrel{\text{integration by parts}}{=} \left(2t\frac{e^{-st}}{-s}\right)_{t=0^-}^{0.5} - \int_{0^-}^{0.5} 2\left(\frac{e^{-st}}{-s}\right)dt$$

$$= -\frac{e^{-0.5s}}{s^2}(2+s) + \frac{2}{s^2}$$

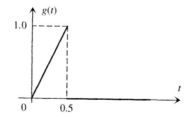

FIGURE 1.16 Geometrical description of a signal.

Linearity of Laplace and Inverse Laplace Transforms

The operation of the Laplace transformation is **linear**, i.e., given that the Laplace transforms of two functions, $f_1(t)$ and $f_2(t)$ exist, and denoted by $F_1(s)$ and $F_2(s)$, and that α and β are constant scalars, then

$$L\{\alpha f_1(t) + \beta f_2(t)\} = \alpha L\{f_1(t)\} + \beta L\{f_2(t)\}$$
$$= \alpha F_1(s) + \beta F_2(s) \qquad (1.41)$$

This follows directly from the definition of the Laplace transform, given by Eq. (1.31). Moreover, the linearity of the L^{-1} operator can also be established as follows: take the inverse Laplace transform of both sides of Eq. (1.41) to obtain

$$\alpha f_1(t) + \beta f_2(t) = L^{-1}\{\alpha F_1(s) + \beta F_2(s)\}$$

However, recall that $f_1(t) = L^{-1}\{F_1(s)\}$ and $f_2(t) = L^{-1}\{F_2(s)\}$. Using these in the above equation, we have

$$\alpha L^{-1}\{F_1(s)\} + \beta L^{-1}\{F_2(s)\} = L^{-1}\{\alpha F_1(s) + \beta F_2(s)\}$$

which clearly indicates that the inverse Laplace transformation is a linear operation.

EXAMPLE 1.21. Determine the Laplace transform of the function $f(t) = 3e^{-t} + \sin 5t$.

Solution. By linearity of the Laplace transform, defined by Eq. (1.41), it is readily seen that

$$L\{3e^{-t} + \sin 5t\} = 3L\{e^{-t}\} + L\{\sin 5t\} = 3\left(\frac{1}{s+1}\right) + \frac{5}{s^2 + 25}$$

THEOREM 1.2. SHIFT ON s-AXIS. Suppose that the Laplace transform of $f(t)$ exists and is denoted by $F(s)$ where $s > a$, and that α is a positive constant (Fig. 1.17). Then,

$$L\{e^{-\alpha t} f(t)\} = F(s + \alpha) \qquad (1.42)$$

where $s + \alpha > a$.

EXAMPLE 1.22. Determine the Laplace transform of $e^{-\alpha t} u_r(t)$ where $u_r(t)$ denotes the unit-ramp function and α is a constant.

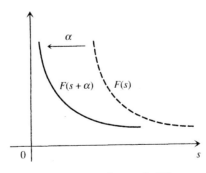

FIGURE 1.17 Shift on the s-axis (Theorem 1.2).

Solution. From earlier results of this section, the Laplace transform of the unit-ramp function is $U_r(s) = 1/s^2$. Subsequently, applying Theorem 1.2 with $f(t) = u_r(t)$, one obtains

$$L\left\{e^{-\alpha t}\underbrace{f(t)}_{u_r(t)}\right\} = \underbrace{F(s+\alpha)}_{F(s)=1/s^2} = \frac{1}{(s+\alpha)^2}$$

which requires considerably less algebraic manipulation as compared to a direct evaluation via Eq. (1.31).

THEOREM 1.3. Given that $L\{f(t)\} = F(s)$ exists, then at any point, except at the poles of the transform function $F(s)$, we have

$$L\{tf(t)\} = -\frac{d}{ds}F(s) \tag{1.43}$$

And in general, for $n = 1, 2, \ldots,$

$$L\{t^n f(t)\} = (-1)^n \frac{d^n}{ds^n} F(s) \tag{1.44}$$

EXAMPLE 1.23. Determine $L\{t \sin 3t\}$.

Solution. It appears that Eq. (1.43) is applicable. First, by comparison, identify the function $f(t)$ in Eq. (1.43), i.e., $f(t) = \sin 3t$, and obtain its Laplace transform, as

$$f(t) = \sin 3t \xRightarrow{\text{Eq. (1.38)}} F(s) = \frac{3}{s^2+9}$$

Consequently, based on the result of Theorem 1.3, we have

$$L\left\{t\underbrace{\sin 3t}_{f(t)}\right\} = -\frac{d}{ds}\underbrace{\left(\frac{3}{s^2+9}\right)}_{F(s)} = \frac{6s}{(s^2+9)^2}$$

PROBLEM SET 1.5

Evaluate the Laplace transform of each of the following functions.

1.27. $e^{j\omega t}$, $\omega = $ constant. Use Euler's formula.

1.28. $\sin(\omega t + \phi)$, ω and ϕ are constants. First expand the function.

1.29. $e^{-\alpha t + \beta}$, α and β are constants.

● **1.30.** $\sin^2 at$. Use the double-angle formula, $\sin^2 \alpha = \frac{1}{2}(1 - \cos 2\alpha)$.

1.31. t^2. Use Theorem 1.3.

1.32. $f(t)$, geometrically defined in Fig. P1.32

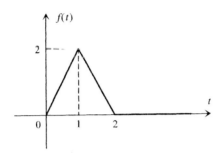

FIGURE P1.32 Geometric description of $f(t)$, Problem 1.32.

1.33. $e^{-t}\sin 2t$

Using the linearity of L^{-1}, determine the inverse Laplace of each of the following transform functions.

1.34. $\dfrac{2}{s} + \dfrac{3}{s+1}$ **1.35.** $\dfrac{6}{s^2+9}$ **1.36.** $\dfrac{s+2}{s^2+4}$

1.6 LAPLACE TRANSFORM OF DERIVATIVES AND INTEGRALS

Mathematical models of dynamic systems are generally described in terms of differential equations of various orders. As discussed briefly in Section 1.5, these differential equations are then transformed into algebraic ones via application of the Laplace transform operator. Naturally, in this process, Laplace transforms of different orders of differentiation of functions need to be identified properly. In what follows, the details of such transitions are studied. Prior to the presentation of the main results, however, we make the following definition.

DEFINITION 1.10. A function, $f(t)$, is said to be **piecewise continuous** on an interval, $[t_1, t_2]$, if

(PC$_1$) $f(t)$ is defined on $[t_1, t_2]$.
(PC$_2$) $[t_1, t_2]$ may be divided into a finite number of sub-intervals in each of which $f(t)$ is continuous.
(PC$_3$) if t approaches either endpoint of a sub-interval from the inside, $f(t)$ attains a finite limit.

THEOREM 1.4. Given that $L\{f(t)\} = F(s)$ exists and $\dot{f}(t)$ is piecewise continuous on every finite interval of $[0, \infty)$, then $L\{\dot{f}(t)\}$ exists and is given by

$$L\{\dot{f}(t)\} = sF(s) - f(0^-) \qquad (1.45)$$

where dot represents differentiation with respect to time, t, and $f(0^-)$ denotes the **initial condition** of $f(t)$.

Note: Throughout this book, the notations of $f(0^-)$ and $f(0)$ are used interchangeably to represent the initial condition of a time-varying function, $f(t)$. Moreover, the value of $f(t)$ immediately after the initial time is known as the **initial value** of the function and is denoted by $f(0^+)$.

Since mathematical models of physical systems are normally governed by differential equations, the result of Theorem 1.4 is of considerable importance in their analysis. Of course, this result may be extended to higher-order derivatives of $f(t)$, specifically

$$L\{\ddot{f}(t)\} = s^2 F(s) - s f(0^-) - \dot{f}(0^-) \tag{1.46}$$

where the knowledge of a set of two initial conditions, $f(0^-)$ and $\dot{f}(0^-)$, is required. The general form of Theorem 1.4 and Eqs. (1.45) and (1.46) is presented below.

THEOREM 1.5. Given that $f(t)$ and its first $n-1$ derivatives, $\dot{f}, \ddot{f}, \ldots, f^{(n-1)}$, are continuous on the interval $[0, \infty)$, and satisfy condition (C_2), and that $f^{(n)}(t)$ is piecewise continuous on every finite interval of $[0, \infty)$, then

$$L\{f^{(n)}(t)\} = s^n F(s) - s^{n-1} f(0^-) - s^{n-2} \dot{f}(0^-) - \cdots - f^{(n-1)}(0^-) \tag{1.47}$$

where $f^{(n)}(t)$ denotes the nth-derivative of f with respect to t.

EXAMPLE 1.24. Find $L\{\dot{f}(t)\}$ where $f(t) = \cos 2t$.

Solution. Prior to the application of Eq. (1.45), we note that

$$F(s) \stackrel{\text{Eq. (1.39)}}{=} \frac{s}{s^2 + 4}, \qquad f(0^-) = \cos 2t \big|_{t=0^-} = 1$$

Subsequently, following the result of Theorem 1.4, one obtains

$$L\{\dot{f}(t)\} = sF(s) - f(0^-) = s\left(\frac{s}{s^2 + 4}\right) - 1 = \frac{-4}{s^2 + 4}$$

This result can easily be verified directly, i.e.,

$$\dot{f}(t) = -2 \sin 2t \implies L\{\dot{f}(t)\} = L\{-2 \sin 2t\} = -2L\{\sin 2t\}$$

$$= -2\left(\frac{2}{s^2 + 4}\right) = \frac{-4}{s^2 + 4}$$

Theorems 1.4 and 1.5, and Eq. (1.47) can sometimes be applied indirectly for determining the Laplace transform of a given function. This approach proves to be an efficient alternative to the direct implementation of Eq. (1.31), as illustrated in the following example.

EXAMPLE 1.25. Determine the Laplace transform of $g(t) = \cos^2 t$.

Solution. Clearly, the direct approach using Eq. (1.31) could involve tedious and time-consuming algebraic manipulations of the integral. However, an alternative technique to find $G(s)$ is as follows:

$$\dot{g}(t) = -2 \cos t \sin t \underset{\text{double-angle formula}}{=} -\sin 2t$$

Since $\dot{g}(t)$ is of a simpler nature than $g(t)$, let us take its Laplace transform via Eq. (1.45), as

$$\underbrace{L\{\dot{g}(t)\} = sG(s) - g(0^-)}_{\text{by Theorem 1.4}} = sG(s) - 1$$

Furthermore, the Laplace transform of $\dot{g}(t)$ may be obtained directly, as

$$L\{\dot{g}(t)\} = L\{-\sin 2t\} = -\frac{2}{s^2 + 4}$$

Equating the right-hand sides of these two equations yields $G(s) = L\{g(t)\}$,

$$sG(s) - 1 = -\frac{2}{s^2 + 4} \underset{\text{solve for } G(s)}{\Longrightarrow} G(s) = \frac{s^2 + 2}{s(s^2 + 4)}$$

THEOREM 1.6. Given that $f(t)$ is piecewise continuous and satisfies existence condition (C$_2$), then

$$L\left\{\int_0^t f(\tau)\,d\tau\right\} = \frac{1}{s}F(s) \quad (1.48)$$

where $L\{f(t)\} = F(s)$. Alternatively, this result may be expressed as

$$L^{-1}\left\{\frac{1}{s}F(s)\right\} = \int_0^t f(\tau)\,d\tau \quad (1.49)$$

EXAMPLE 1.26. Given $H(s) = 1/[s(s+1)]$, determine $h(t) = L^{-1}\{H(s)\}$.

Solution. First, we rewrite $H(s)$ as

$$h(t) = L^{-1}\{H(s)\} = L^{-1}\left\{\frac{1}{s(s+1)}\right\} = L^{-1}\left\{\frac{1}{s} \cdot \frac{1}{s+1}\right\}$$

so that the last expression resembles the left-hand side of Eq. (1.49). Direct comparison then reveals that $F(s) = 1/(s+1)$, the inverse Laplace of which was previously shown to be $f(t) = e^{-t}$. Applying Eq. (1.49), one obtains

$$h(t) = L^{-1}\left\{\frac{1}{s} \cdot \frac{1}{s+1}\right\} = \int_0^t e^{-\tau}\,d\tau = 1 - e^{-t}$$

Solving Initial-Value Problems

The role of the Laplace transform of derivatives and integrals of time-varying functions is most significant when solving ordinary differential equations subjected to prescribed initial conditions. A more detailed version of the process involved, suggested in Fig. 1.10, may now be presented as shown in Fig. 1.18.

EXAMPLE 1.27. Solve the following initial-value problem via Laplace transformation.

$$\ddot{x} + 4x = 2u_s(t), \qquad x(0^-) = 0, \qquad \dot{x}(0^-) = 0$$

CHAPTER 1: Complex Analysis, Differential Equations, and the Laplace Transform 33

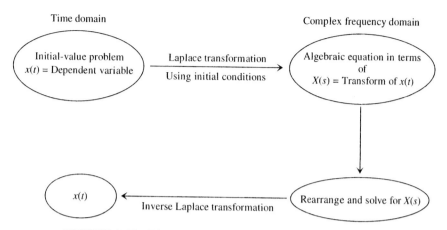

FIGURE 1.18 The solution method for initial-value problems.

Solution. As a first step, take the Laplace transform of both sides of the differential equation. In doing so, employ the general result of Theorem 1.5 for each term. More specifically, use Eqs. (1.45) and (1.46) for the first and second derivatives of a function.

$$[s^2 X(s) - sx(0^-) - \dot{x}(0^-)] + 4X(s) = \frac{2}{s}$$

Next, take into account the given initial conditions—in this case, zero. Then, rearrange terms and solve for $X(s)$

$$X(s) = \frac{2}{s(s^2 + 4)}$$

The final step involves seeking $x(t)$ through the inverse Laplace transformation of $X(s)$, i.e.,

$$x(t) = L^{-1}\{X(s)\} = L^{-1}\left\{\frac{2}{s(s^2+4)}\right\}$$

Although our current knowledge of the inverse transform is rather limited, we can handle this particular inverse using Theorem 1.6. As suggested by Example 1.26, rewrite the above so that it resembles the left-hand side of Eq. (1.49), as follows.

$$x(t) = L^{-1}\left\{\frac{1}{s} \cdot \frac{2}{s^2+4}\right\}$$

Comparison reveals that $F(s) = 2/(s^2 + 4)$ and hence, $f(t) = \sin 2t$. Finally, application of Eq. (1.49) yields

$$x(t) = \int_0^t f(\tau)\, d\tau = \int_0^t \sin 2\tau\, d\tau = \frac{1}{2}(1 - \cos 2t)$$

PROBLEM SET 1.6

Determine the following Laplace transforms with the aid of Theorem 1.5.

1.37. $L\{\sin^2 t\}$ **1.38.** $L\{t \sin \omega t\}$ **1.39.** $L\{te^{-t}\}$

Identify the following inverse Laplace transforms via Theorem 1.6.

1.40. $L^{-1}\left\{\dfrac{1}{s(s+\alpha)}\right\}$ **1.41.** $L^{-1}\left\{\dfrac{\omega}{s^3+\omega^2 s}\right\}$ **1.42.** $L^{-1}\left\{\dfrac{1}{s^2(s+1)}\right\}$

Solve the following initial-value problems.

1.43. $\ddot{x} + \omega^2 x = 0,$ $x(0^-) = 0,$ $\dot{x}(0^-) = 1$

1.44. $\ddot{x} + \omega^2 x = 0,$ $x(0^-) = 1,$ $\dot{x}(0^-) = 0$

1.45. $\dot{x} + x = u_r(t),$ $x(0^-) = 0$

1.46. $\ddot{x} + 2\dot{x} + x = 0,$ $x(0^-) = 0,$ $\dot{x}(0^-) = 1$

1.7 INVERSE LAPLACE TRANSFORMATION

As briefly discussed in Section 1.5 and illustrated by Fig. 1.10, the idea behind Laplace transformation is to transfer information from the time-domain to the s-domain (frequency domain) through application of the Laplace operator, L, and eventually back to the time-domain with the aid of the inverse Laplace operator, L^{-1}. It is the latter that this section is mainly concerned with, namely, determining $f(t)$ from $F(s)$. Two methods, the *partial fraction method* and *convolution*, will be presented here, accompanied by examples. Of course, in certain cases, the results of Section 1.6 may be implemented whenever possible, as was demonstrated in Example 1.18.

Partial Fraction Method

Suppose a Laplace transform function, $F(s)$, can be expressed as $F(s) = F_1(s) + F_2(s) + \cdots + F_k(s)$, where each $F_i(s)$ denotes the transform of $f_i(t)$, $1 \le i \le k$. Then, using the linear property of the inverse Laplace transform, we have

$$f(t) = L^{-1}\{F_1(s)\} + \cdots + L^{-1}\{F_k(s)\} = f_1(t) + \cdots + f_k(t) \quad (1.50)$$

A specific form of $F(s)$ that one frequently encounters in systems analysis is

$$F(s) = \frac{N(s)}{D(s)}, \quad m = \deg N(s) < \deg D(s) = n \quad (1.51)$$

CHAPTER 1: Complex Analysis, Differential Equations, and the Laplace Transform   35

Note: In the event that the degree of the numerator is greater than (or equal to) that of the denominator, then polynomial division must be performed so that the remainder polynomial is of a lower degree than $D(s)$. For instance,

$$\frac{s^3 + 2}{s^2 + 1} = s + \frac{-s + 2}{s^2 + 1}, \qquad \deg(-s + 2) < \deg(s^2 + 1)$$

The denominator, $D(s)$, may always be factored into linear terms as

$$D(s) = (s - p_1) \cdots (s - p_n)$$

where, as mentioned in Section 1.3, each p_i is referred to as a *pole* of $F(s)$. Then, depending on whether these poles are real, complex, distinct, or repeated, the treatment is slightly different. These cases are discussed below.

Case 1. Distinct real poles ($p_i \neq p_j$ for $i \neq j$). In the event that all poles of $F(s)$ are distinct and real, $F(s)$ can be expressed as follows:

$$F(s) = \frac{A_1}{s - p_1} + \frac{A_2}{s - p_2} + \cdots + \frac{A_n}{s - p_n} \qquad (1.52)$$

where each A_i ($1 \leq i \leq n$) is called the **residue** of $F(s)$ at the corresponding pole p_i, and is given by

$$A_i = \operatorname*{Res}_{s = p_i} F(s) = \left[(s - p_i) F(s)\right]_{s = p_i} \qquad (1.53)$$

EXAMPLE 1.28. Find the inverse Laplace transform of $F(s) = (s + 1)/[(s + 3) \times (s + 4)]$.

Solution. We notice that the numerator has a lower degree than the denominator, and that the two poles, -3 and -4, are real and distinct. Thus we can proceed as follows:

$$F(s) = \frac{s + 1}{(s + 3)(s + 4)} = \frac{A_1}{s + 3} + \frac{A_2}{s + 4} \qquad (1.54)$$

where the constants A_1 and A_2 are obtained via Eq. (1.53), as

$$A_1 = \left[(s + 3) F(s)\right]_{s = -3} = \left(\frac{s + 1}{s + 4}\right)_{s = -3} = -2$$

$$A_2 = \left[(s + 4) F(s)\right]_{s = -4} = 3$$

Substitution into Eq. (1.54) yields

$$F(s) = \frac{-2}{s + 3} + \frac{3}{s + 4}$$

Taking the inverse Laplace transform of both sides, one obtains

$$f(t) = L^{-1}\left\{\frac{-2}{s + 3}\right\} + L^{-1}\left\{\frac{3}{s + 4}\right\} = -2 L^{-1}\left\{\frac{1}{s + 3}\right\} + 3 L^{-1}\left\{\frac{1}{s + 4}\right\}$$

$$= -2e^{-3t} + 3e^{-4t}, \qquad t \geq 0$$

where, once again, the linearity of L^{-1} has been used.

Case 2. Repeated (multiple) real poles. If a certain real pole, p_i, of a Laplace transform function has a multiplicity of $k > 1$, then $D(s)$ will contain $(s - p_i)^k$ as one of the terms in its factored form. The fractions that correspond to this term in the partial fraction method are then

$$\frac{A_k}{(s - p_i)^k} + \frac{A_{k-1}}{(s - p_i)^{k-1}} + \cdots + \frac{A_1}{(s - p_i)} \tag{1.55}$$

in which constants, $A_1, \ldots, A_k$, are determined via

$$A_l = \frac{1}{(k - l)!} \left(\frac{d^{k-l}}{ds^{k-l}} \left[(s - p_i)^k F(s) \right] \right)_{s = p_i}, \quad l = 1, 2, \ldots, k \tag{1.56}$$

EXAMPLE 1.29. Find the inverse Laplace transform of $F(s) = (2s + 3)/[(s + 4) \times (s + 1)^2]$.

Solution. This function possesses a simple pole at -4 and a pole of order (multiplicity) 2 at -1. The fractions corresponding to the multiple pole are then [according to Eq. (1.55)]

$$\frac{A_2}{(s + 1)^2} + \frac{A_1}{s + 1}$$

and the one corresponding to the simple pole is simply $A_3/(s + 4)$. Combining these yields

$$F(s) = \frac{2s + 3}{(s + 4)(s + 1)^2} = \frac{A_2}{(s + 1)^2} + \frac{A_1}{s + 1} + \frac{A_3}{s + 4} \tag{1.57}$$

where the first two constants are calculated via Eq. (1.56)

$$A_1 = \frac{1}{(2 - 1)!} \left\{ \frac{d}{ds} \left[(s + 1)^2 F(s) \right] \right\}_{s = -1} = \left\{ \frac{d}{ds} \left[\frac{2s + 3}{s + 4} \right] \right\}_{s = -1} = \frac{5}{9}$$

$$A_2 = \frac{1}{(2 - 2)!} \left[(s + 1)^2 F(s) \right]_{s = -1} = \left[\frac{2s + 3}{s + 4} \right]_{s = -1} = \frac{1}{3}$$

and the last constant can be found via Eq. (1.53), as

$$A_3 = \left[(s + 4) F(s) \right]_{s = -4} = \left[\frac{2s + 3}{(s + 1)^2} \right]_{s = -4} = -\frac{5}{9}$$

Substitute into Eq. (1.57) to obtain

$$F(s) = \frac{2s + 3}{(s + 4)(s + 1)^2} = \left(\frac{1}{3}\right) \frac{1}{(s + 1)^2} + \left(\frac{5}{9}\right) \frac{1}{s + 1} - \left(\frac{5}{9}\right) \frac{1}{s + 4}$$

The final step involves inverse Laplace transformation of each individual term, i.e.,

$$f(t) = \frac{1}{3} L^{-1} \left\{ \frac{1}{(s + 1)^2} \right\} + \frac{5}{9} L^{-1} \left\{ \frac{1}{s + 1} \right\} - \frac{5}{9} L^{-1} \left\{ \frac{1}{s + 4} \right\}$$

$$= \frac{1}{3} t e^{-t} + \frac{5}{9} e^{-t} - \frac{5}{9} e^{-4t}, \quad t \geq 0$$

in which Theorem 1.3 (or Theorem 1.2) was used to determine the first term.

Case 3. Complex conjugate poles. Often, the denominator of a Laplace transform function contains one or more **irreducible** second-order polynomials. A polynomial is referred to as *irreducible* if it has a pair of complex conjugate roots. The techniques, outlined in cases (1) and (2) above, also apply to such transform functions, with simple or repeated complex poles, respectively. Each irreducible polynomial, $s^2 + as + b$, with roots $\alpha \pm j\beta$ corresponds to two fractions

$$\frac{A}{s - \alpha - j\beta} + \frac{B}{s - \alpha + j\beta}$$

where A and B are calculated via Eq. (1.53) and are complex conjugates of each other, i.e., $B = \bar{A}$. In the event that complex poles have multiplicities of higher than 1, the partial fractions are constructed as in case (2) and the residues are computed via Eq. (1.56).

EXAMPLE 1.30. Find the inverse Laplace transform of $G(s) = (s-1)/(s^2+2s+2)$.

Solution. The denominator of $G(s)$ is an *irreducible* second-order polynomial and possesses a pair of simple complex conjugate roots at $s_{1,2} = -1 \pm j$. Following the procedure of case (1), $G(s)$ is rewritten as

$$G(s) = \frac{s - 1}{(s + 1 + j)(s + 1 - j)} = \frac{A}{s + 1 + j} + \frac{B}{s + 1 - j}$$

where A and B, the residues, are calculated via Eq. (1.53) as

$$A = \big[(s + 1 + j)G(s)\big]_{s=-1-j} = \frac{2 + j}{2j} = \frac{1}{2} - j$$

$$B = \big[(s + 1 - j)G(s)\big]_{s=-1+j} = \frac{-2 + j}{2j} = \frac{1}{2} + j$$

Note that A and B are indeed conjugates of each other. Substitution into the expression of $G(s)$ yields

$$G(s) = \left(\frac{1}{2} - j\right)\frac{1}{s + 1 + j} + \left(\frac{1}{2} + j\right)\frac{1}{s + 1 - j}$$

Term-by-term inverse Laplace transformation results in

$$g(t) = \left(\tfrac{1}{2} - j\right)e^{(-1-j)t} + \left(\tfrac{1}{2} + j\right)e^{(-1+j)t}$$

$$= \left(\tfrac{1}{2} - j\right)e^{-t}e^{-jt} + \left(\tfrac{1}{2} + j\right)e^{-t}e^{jt}$$

$$= e^{-t}\left\{\left(\tfrac{1}{2} - j\right)(\cos t - j\sin t) + \left(\tfrac{1}{2} + j\right)(\cos t + j\sin t)\right\}$$

$$= e^{-t}(\cos t - 2\sin t), \qquad t \geq 0$$

where use has been made of Euler's formula, Eq. (1.3).

Partial Fraction Method: An Alternative Approach

The alternative approach has two distinguishing features in relation to those outlined above; (1) the irreducible polynomials are treated differently, and (2) the residues are not calculated directly.

1. Treatment of irreducible polynomials

Each irreducible polynomial is treated as one entity and is not expressed in factored form. The partial fraction corresponding to each irreducible polynomial, $s^2 + as + b$, takes the form

$$\frac{As + B}{s^2 + as + b}$$

with A and B to be determined. Once these constants have been identified, completion of the square in the denominator is performed and the inverse Laplace transform of this term can be found via the following relations

$$L^{-1}\left\{\frac{\omega}{(s+\sigma)^2 + \omega^2}\right\} = e^{-\sigma t}\sin\omega t, \qquad L^{-1}\left\{\frac{s+\sigma}{(s+\sigma)^2 + \omega^2}\right\} = e^{-\sigma t}\cos\omega t \tag{1.58}$$

In the event that the denominator of the transform function contains a term $(s^2 + as + b)^k$, where $s^2 + as + b$ is *irreducible*, its corresponding fractions are then

$$\frac{A_1 s + B_1}{s^2 + as + b} + \frac{A_2 s + B_2}{(s^2 + as + b)^2} + \cdots + \frac{A_k s + B_k}{(s^2 + as + b)^k} \tag{1.59}$$

where constants $A_1, \ldots, A_k, B_1, \ldots, B_k$ are to be determined.

2. Determination of constants

Once appropriate fractions are formed corresponding to different terms in the denominator, they are combined into one single fraction. Naturally, the resulting fraction has a denominator that agrees with that of the original transform function. Subsequently, the numerator of the newly constructed fraction is set *identical* to that of the transform function. This identity holds (for all values of s) if and only if the coefficients of like powers of s are equal. As a result, a system of algebraic equations in terms of the unknown constants (such as $A_1, B_1, \ldots$) is obtained and solved.

First, to demonstrate the effectiveness of the treatment of an irreducible polynomial in the denominator, outlined in (1) above, consider the transform function of Example 1.30.

EXAMPLE 1.31. Find the inverse Laplace transform of

$$G(s) = \frac{s - 1}{s^2 + 2s + 2}$$

Solution. The denominator of $G(s)$ is an *irreducible* second-order polynomial that has a pair of complex conjugate roots at $s_{1,2} = -1 \pm j$. Therefore, an effort is made to complete the square in the denominator and rewrite $G(s)$ so that one or both of the identities in Eq. (1.58) may apply.

$$G(s) = \frac{s-1}{s^2+2s+2} = \frac{s-1}{(s+1)^2+1} = \frac{s+1}{(s+1)^2+1} - \frac{2}{(s+1)^2+1}$$

$$\Longrightarrow \quad g(t) = L^{-1}\left\{\frac{s+1}{(s+1)^2+1}\right\} - 2L^{-1}\left\{\frac{1}{(s+1)^2+1}\right\}$$

$$= e^{-t}\cos t - 2e^{-t}\sin t = e^{-t}(\cos t - 2\sin t), \qquad t \geq 0$$

where $\sigma = 1$ and $\omega = 1$ are used in the identities of Eq. (1.58). As expected, the result agrees with that of Example 1.30.

EXAMPLE 1.32. Determine the inverse Laplace transform of

$$F(s) = \frac{s+1}{(s+3)(s+4)}$$

Solution. This is the function considered earlier in Example 1.28. Since $F(s)$ does not involve an irreducible polynomial in the denominator, treatment (1) is not needed. Applying the procedure of (2), we have

$$F(s) = \frac{s+1}{(s+3)(s+4)} = \frac{A}{s+3} + \frac{B}{s+4} = \frac{A(s+4)+B(s+3)}{(s+3)(s+4)}$$

Equating the numerators of the original function and the last fraction, collecting like terms, one obtains

$$s+1 \equiv s(A+B) + 4A + 3B$$

The identity holds if the coefficients of like powers of s on both sides are equal; i.e.,

$$\begin{cases} 1 = A+B \\ 1 = 4A+3B \end{cases} \Longrightarrow \quad A = -2, B = 3$$

Substitution into the partial fraction expansion yields

$$F(s) = \frac{s+1}{(s+3)(s+4)} = \frac{-2}{s+3} + \frac{3}{s+4}$$

Term-by-term inversion gives the time-dependent function, $f(t)$,

$$f(t) = -2e^{-3t} + 3e^{-4t}, \qquad t \geq 0$$

which is in agreement with Example 1.28.

EXAMPLE 1.33. Determine the inverse Laplace transform of

$$F(s) = \frac{2}{(s+3)(s^2+2s+5)}$$

Solution. $F(s)$ has a (simple) real pole at -3 and a pair of complex conjugate poles at $-1 \pm j2$. Rewrite $F(s)$ as

$$F(s) = \frac{2}{(s+3)(s^2+2s+5)} = \frac{A}{s+3} + \frac{Bs+C}{s^2+2s+5}$$

where constants A, B, and C are to be determined. Combine the two fractions on the right-hand side, and equate with the expression of $F(s)$; i.e.,

$$\frac{2}{(s+3)(s^2+2s+5)} = \frac{A}{s+3} + \frac{Bs+C}{s^2+2s+5}$$

$$= \frac{A(s^2+2s+5) + (Bs+C)(s+3)}{(s+3)(s^2+2s+5)}$$

Setting the numerators equal, while collecting like terms, we have

$$2 = (A+B)s^2 + (2A+3B+C)s + 5A + 3C$$

In order for this identity to hold, coefficients of like powers of s on both sides must be equal; i.e.,

$$\begin{cases} A+B=0 \\ 2A+3B+C=0 \\ 5A+3C=2 \end{cases} \implies A = \tfrac{1}{4} = C, B = -\tfrac{1}{4}$$

Back substitution results in

$$F(s) = \frac{1}{4}\left\{\frac{1}{s+3} - \frac{s-1}{s^2+2s+5}\right\} = \frac{1}{4}\left\{\frac{1}{s+3} - \frac{(s+1)-2}{(s+1)^2+2^2}\right\}$$

$$= \frac{1}{4}\left\{\frac{1}{s+3} - \frac{(s+1)}{(s+1)^2+2^2} + \frac{2}{(s+1)^2+2^2}\right\}$$

Term-by-term inverse Laplace transformation, using identities in Eq. (1.58), yields

$$f(t) = \tfrac{1}{4}\left\{e^{-3t} - e^{-t}\cos 2t + e^{-t}\sin 2t\right\}$$

EXAMPLE 1.34. Determine the inverse Laplace transform of

$$F(s) = \frac{-10s + 34}{(s+3)(s^2+2s+5)^2}$$

Solution. $F(s)$ has a real (simple) pole at -3 and a pair of complex conjugate (repeated) poles at $-1 \pm j2$. Since the complex poles are repeated, their corresponding partial fractions are formed via Eq. (1.59) with $k = 2$, i.e.,

$$\frac{A_1 s + B_1}{s^2+2s+5} + \frac{A_2 s + B_2}{(s^2+2s+5)^2}$$

Therefore,

$$\frac{-10s+34}{(s+3)(s^2+2s+5)^2} = \frac{A}{s+3} + \frac{A_1 s + B_1}{s^2+2s+5} + \frac{A_2 s + B_2}{(s^2+2s+5)^2}$$

$$= \frac{A(s^2+2s+5)^2 + (A_1 s + B_1)(s+3)(s^2+2s+5) + (A_2 s + B_2)(s+3)}{(s+3)(s^2+2s+5)^2}$$

(1.60)

CHAPTER 1: Complex Analysis, Differential Equations, and the Laplace Transform   41

As expected, the denominators are in agreement, and hence the numerators are set identical; i.e.,

$$-10s + 34 = A(s^2 + 2s + 5)^2 + (A_1s + B_1)(s + 3)(s^2 + 2s + 5)$$
$$+ (A_2s + B_2)(s + 3)$$

Collecting terms on the right-hand side, and equating the coefficients of like powers of s on both sides, one obtains

$$\begin{cases} A + A_1 = 0 \\ 4A + 5A_1 + B_1 = 0 \\ 14A + 11A_1 + 5B_1 + A_2 = 0 \\ 20A + 15A_1 + 11B_1 + 3A_2 + B_2 = -10 \\ 25A + 15B_1 + 3B_2 = 34 \end{cases}$$

Solution of this 5×5 system of equations is

$$A = 1, \quad A_1 = -1, \quad B_1 = 1, \quad A_2 = -8, \quad B_2 = -2$$

Substitution into the expression of $F(s)$ yields

$$F(s) = \frac{1}{s + 3} + \frac{-s + 1}{s^2 + 2s + 5} + \frac{-8s - 2}{(s^2 + 2s + 5)^2} \tag{1.61}$$

Before a term-by-term inverse Laplace transformation is performed, the second and third fractions must be rewritten in a convenient form so that previous results or Table 1.2 can be utilized. To this end, noting that $s^2 + 2s + 5 = (s + 1)^2 + 2^2$, we rewrite Eq. (1.61) as

$$F(s) = \frac{1}{s + 3} - \frac{(s + 1) - 2}{(s + 1)^2 + 2^2} - \frac{8(s + 1) - 6}{\left[(s + 1)^2 + 2^2\right]^2}$$

The inverse Laplace of the *first term* is simply e^{-3t}. Algebraic manipulation of the *second term* yields

$$-\frac{s + 1}{(s + 1)^2 + 2^2} + \frac{2}{(s + 1)^2 + 2^2}$$

the inverse of which (rows 18 and 17 of Table 1.2) is $-e^{-t} \cos 2t + e^{-t} \sin 2t$. The *third term* is expressed as

$$-8 \frac{s + 1}{\left[(s + 1)^2 + 2^2\right]^2} + 6 \frac{1}{\left[(s + 1)^2 + 2^2\right]^2} \tag{1.62}$$

If we let

$$G(s) = \frac{s}{(s^2 + 2^2)^2} \quad \text{and} \quad H(s) = \frac{1}{(s^2 + 2^2)^2}$$

then the first fraction in Eq. (1.62) is $G(s + 1)$, and the second fraction is $H(s + 1)$. Using rows 25 and 26 of Table 1.2, it is observed that

$$g(t) = L^{-1}\{G(s)\} = \tfrac{1}{4}t \sin 2t \quad \text{and} \quad h(t) = L^{-1}\{H(s)\} = \tfrac{1}{16}(\sin 2t - 2t \cos 2t)$$

Recall from Theorem 1.2 (shift on s-axis) that $L^{-1}\{Y(s+\alpha)\} = e^{-\alpha t}y(t)$ for some function, $y(t)$. Applying this result, with $\alpha = 1$, to Eq. (1.62), its inverse Laplace transform is given by

$$-8L^{-1}\{G(s+1)\} + 6L^{-1}\{H(s+1)\} = -8e^{-t}g(t) + 6e^{-t}h(t)$$

$$= -8e^{-t}\left[\tfrac{1}{4}t\sin 2t\right] + 6e^{-t}\left[\tfrac{1}{16}(\sin 2t - 2t\cos 2t)\right]$$

Combine the inverse Laplace transforms of all three terms of Eq. (1.61) to obtain

$$f(t) = e^{-3t} - e^{-t}\cos 2t + e^{-t}\sin 2t - 2te^{-t}\sin 2t + \tfrac{3}{8}e^{-t}(\sin 2t - 2t\cos 2t)$$

$$= e^{-3t} - e^{-t}\left[\left(1 + \tfrac{3}{4}t\right)\cos 2t - \left(\tfrac{11}{8} - 2t\right)\sin 2t\right]$$

Convolution Method

A second method of significant practical importance is the **convolution method**. Often, in systems analysis, a product of two transform functions, $G(s)$ and $H(s)$, is encountered, where their respective inverses, $g(t)$ and $h(t)$, are known. The convolution method is used to determine the inverse of the product, $F(s) = G(s)H(s)$, directly from the knowledge of $g(t)$ and $h(t)$.

Notation: $f(t) = L^{-1}\{F(s)\} = L^{-1}\{G(s)H(s)\} = (g*h)(t) = (h*g)(t)$ read "convolution of g and h."

THEOREM 1.7. CONVOLUTION. Suppose functions $g(t)$ and $h(t)$ satisfy existence conditions (C_1) and (C_2) of Theorem 1.1. Let $G(s) = L\{g(t)\}$, $H(s) = L\{h(t)\}$, $F(s) = G(s)H(s)$. Then

$$L^{-1}\{F(s)\} = f(t) = (g*h)(t) = \int_0^t g(\tau)h(t-\tau)\,d\tau$$

$$= \int_0^t h(\tau)g(t-\tau)\,d\tau = (h*g)(t)$$

EXAMPLE 1.35. Consider the transform function $F(s) = 1/[s(s+1)]$ studied in Example 1.26. Determine the inverse function via the convolution method.

Solution. Rewrite $F(s)$ so that Theorem 1.7 can be applied; i.e., rewrite it as the product of two recognizable transform functions,

$$F(s) = \frac{1}{s} \cdot \frac{1}{s+1}$$

so that

$$G(s) = \frac{1}{s} \quad \text{and} \quad H(s) = \frac{1}{s+1}$$

Based on previous work, the inverse Laplace transforms of $G(s)$ and $H(s)$ are given as

$$g(t) = 1 \quad \text{and} \quad h(t) = e^{-t}, \quad t \geq 0$$

Subsequently, the application of Theorem 1.7 yields

$$f(t) = (g * h)(t) = \int_0^t e^{-(t-\tau)} d\tau = e^{-(t-\tau)}\Big|_{\tau=0}^t = 1 - e^{-t}$$

confirming the result obtained earlier via Theorem 1.6. Furthermore, the *symmetry* of the convolution operation can easily be verified as follows:

$$(h * g)(t) = \int_0^t e^{-\tau} d\tau = 1 - e^{-t}$$

EXAMPLE 1.36. Determine the inverse Laplace transform of

$$F(s) = \frac{1}{s^2(s+2)}$$

Solution. Rewrite $F(s)$ in the form of a product of two transform functions with known inverses, as

$$F(s) = \frac{1}{s^2} \cdot \frac{1}{s+2}$$

so that

$$G(s) = \frac{1}{s^2} \quad \text{and} \quad H(s) = \frac{1}{s+2}$$

The inverse Laplace transforms are

$$g(t) = t \quad \text{and} \quad h(t) = e^{-2t}$$

Then, by the convolution theorem, one obtains

$$f(t) = (g * h)(t) = \int_0^t \tau \cdot e^{-2(t-\tau)} d\tau \underset{\substack{\text{integration}\\\text{by parts}}}{=} \left\{\tau \cdot \frac{e^{-2(t-\tau)}}{2}\right\}_{\tau=0}^t - \int_0^t \frac{e^{-2(t-\tau)}}{2} d\tau$$

$$= \tfrac{1}{2}t - \tfrac{1}{4}(1 - e^{-2t}) = \tfrac{1}{4}(e^{-2t} + 2t - 1)$$

The practical significance of convolution, specifically in the study of the response of linear dynamic systems, will be thoroughly discussed in Chapter 7. There, it will be shown that the response of a time-invariant, linear system to an arbitrary input can be expressed in terms of what is known as the *convolution integral*.

PROBLEM SET 1.7

Using the partial fraction method, determine the inverse Laplace transform of each function given below.

1.47. $\dfrac{2s+3}{s^2+4s+3}$ **1.48.** $\dfrac{6}{s(s+3)^3}$ **1.49.** $\dfrac{5}{(s+1)(s^2+4s+8)}$

Solve the following initial-value problems.

1.50. $\ddot{x} + 4\dot{x} + 5x = 2,\ x(0^-) = 0,\ \dot{x}(0^-) = 0$

1.51. $\dot{x} + 2x = t,\ x(0^-) = 1$

1.52. $\ddot{x} + 4\dot{x} + 3x = 0,\ x(0^-) = 1,\ \dot{x}(0^-) = 1$

1.53. Prove that

$$L\left\{\int_0^t g(\tau)h(t-\tau)\,d\tau\right\} = G(s)H(s)$$

where $G(s) = L\{g(t)\}$ and $H(s) = L\{h(t)\}$.

1.8 SHIFT ON t-AXIS

In many cases, the input signal to a dynamic system may exhibit various types of behavior during different time intervals of the process. Since dynamic systems are typically modeled by initial-value problems, the methodology illustrated in Fig. 1.18 can be employed. In this process, it is then required that the Laplace transform of the forcing function be sought. To this end, (1) the input signal is expressed analytically in terms of unit-step functions, and (2) its Laplace transform is obtained subsequently. This section concerns itself with both of the tasks mentioned. Before the main result is presented, however, we introduce a useful notation.

Notation: $u(t-a)$ [or $u_a(t)$] denotes a *unit-step* function that assumes a value of zero for all $t < a$ and a value of 1 for $t > a$, as in Fig. 1.19.

Next, consider any function that may have a graphical representation as shown in Fig. 1.20, and also its positive shift on the t-axis. Combining the unit-step, $u_a(t)$, of Fig. 1.19 with $f(t-a)$ of Fig. 1.20, one obtains the product function in Fig. 1.21.

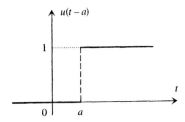

FIGURE 1.19 Unit-step function.

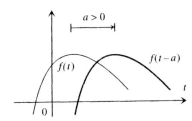

FIGURE 1.20 Positive shift of $f(t)$ on t-axis.

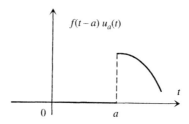

FIGURE 1.21 Function multiplied by the unit step.

THEOREM 1.8. SHIFT ON t-AXIS. Given that $L\{f(t)\} = F(s)$, then

$$L\{f(t-a)u(t-a)\} = e^{-as}F(s) \tag{1.63}$$

Conversely,

$$L^{-1}\{e^{-as}F(s)\} = f(t-a)u(t-a) \tag{1.63*}$$

As an immediate application of this theorem, the Laplace transform of the unit-step function, $u(t-a)$, can be determined by choosing $f(t) = 1$ in Eq. (1.63), as

$$L\{u(t-a)\} = \frac{e^{-as}}{s} \tag{1.64}$$

The result provided by Theorem 1.8 is particularly useful when the forcing function is expressed in terms of unit-step functions. Each term can be modified appropriately to resemble $f(t-a)u(t-a)$ for suitable choices of $f(t)$ and a. Subsequently, the Laplace transform of the function can be determined through transformation of individual terms via Theorem 1.8. A forcing function can be physically realized as an applied force to a mechanical system, or an applied voltage to an electrical system. The example that follows demonstrates an application of the procedure just mentioned.

EXAMPLE 1.37. Suppose a certain system is subjected to a forcing function, $f(t)$, represented graphically as in Fig. 1.22. Express $f(t)$ in terms of unit-step functions, and obtain its Laplace transform, $F(s)$, via Theorem 1.8.

Solution. The graph of $f(t)$ can be constructed following a step-by-step procedure, as outlined below.

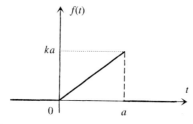

FIGURE 1.22 Forcing function in Example 1.37.

Step 1. First, construct the graph shown in Fig.1.23.

Step 2. Multiplication by the unit-step function, $u(t)$, ensures that the resulting function is zero for $t < 0$. This is depicted in Fig. 1.24.

Step 3. On the line with slope k, the *ray* starting at the point (a, ka) is eliminated as follows. Note that this *ray* is described as in Fig. 1.25.

If we now subtract this graph from that in step 2, Fig. 1.24, we obtain the graph in Fig. 1.26. Therefore, it is observed that $f(t) = ktu(t) - ktu(t-a)$. The next task is then to find the Laplace transform of this function. To this end, we look at each of the two terms individually and combine the results eventually.

$$L\{f(t)\} = L\{ktu(t) - ktu(t-a)\} = L\{ktu(t)\} - L\{ktu(t-a)\}$$
$$= kL\{tu(t)\} - kL\{tu(t-a)\} \qquad (1.65)$$

The first Laplace transform on the right-hand side of Eq. (1.65) can easily be obtained via Theorem 1.8 with $f(t) = t$ and $a = 0$; i.e.,

$$L\{tu(t)\} = \frac{1}{s^2} \qquad (1.66)$$

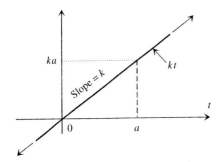

FIGURE 1.23

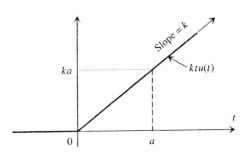

FIGURE 1.24

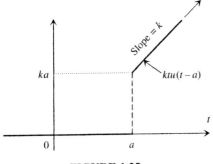

FIGURE 1.25

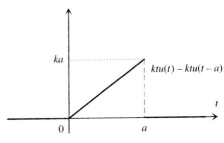

FIGURE 1.26

To determine the second Laplace transform in Eq. (1.65), we need to rewrite the expression of the time function so that it agrees with that in Theorem 1.8, as

$$tu(t - a) = (t - a)u(t - a) + au(t - a)$$

Subsequently,

$$L\{tu(t - a)\} = \underbrace{L\{(t - a)u(t - a)\}}_{\substack{\text{Theorem 1.8 with } f(t)=t, a=a \\ \downarrow \\ e^{-as} \cdot \frac{1}{s^2}}} + \underbrace{aL\{u(t - a)\}}_{\substack{f(t)=1, a=a \\ \downarrow \\ e^{-as} \cdot \frac{1}{s}}} = e^{-as}\frac{1}{s^2} + ae^{-as}\frac{1}{s} \quad (1.67)$$

Substitution of Eqs. (1.66) and (1.67) into Eq. (1.65) yields the Laplace transform of $f(t)$:

$$L\{f(t)\} = kL\{tu(t)\} - kL\{tu(t - a)\}$$

$$= k\frac{1}{s^2} - k\left[e^{-as}\frac{1}{s^2} + ae^{-as}\frac{1}{s}\right] = k\left[\frac{1 - e^{-as}}{s^2} - \frac{ae^{-as}}{s}\right]$$

EXAMPLE 1.38. The governing equation of a certain dynamic system is given as $\dot{x} + x = f(t)$ where $f(t)$ is the function considered in Example 1.37 with $k = 1$ and $a = 1$. Assume zero initial conditions. Determine $x(t)$.

Solution. Taking the Laplace transform of both sides of the differential equation, and using zero initial conditions, yields

$$(s + 1)X(s) = \frac{1 - e^{-s}}{s^2} - \frac{e^{-s}}{s}$$

where the result of the previous example, with $k = 1$ and $a = 1$, has been used. Solving for $X(s)$, one obtains

$$X(s) = \frac{1}{s^2(s + 1)}(1 - e^{-s}) - \frac{1}{s(s + 1)}e^{-s}$$

In this expression, let

$$H(s) = \frac{1}{s^2(s + 1)} \quad \text{and} \quad G(s) = \frac{1}{s(s + 1)}$$

so that $h(t) = t - 1 + e^{-t}$ and $g(t) = 1 - e^{-t}$. Rewrite

$$X(s) = (1 - e^{-s})H(s) - e^{-s}G(s) = H(s) - e^{-s}H(s) - e^{-s}G(s)$$

Term-by-term inverse Laplace transformation yields

$$x(t) = \underbrace{L^{-1}\{H(s)\}}_{h(t)} - \underbrace{L^{-1}\{e^{-s}H(s)\}}_{\substack{\text{Theorem 1.8} \\ \text{Eq. (1.63*)}, a=1}} - \underbrace{L^{-1}\{e^{-s}G(s)\}}_{\substack{\text{Theorem 1.8} \\ \text{Eq. (1.63*)}, a=1}}$$

$$= h(t) - h(t - 1)u(t - 1) - g(t - 1)u(t - 1)$$

$$= h(t) - [h(t - 1) + g(t - 1)]u(t - 1)$$

However, because $g(t) + h(t) = t$, then $g(t-1) + h(t-1) = t - 1$. Consequently,

$$x(t) = h(t) - (t-1)u(t-1) = t - 1 + e^{-t} - (t-1)u(t-1)$$
$$= \begin{cases} t - 1 + e^{-t} & \text{if } 0 \leq t < 1 \\ e^{-t} & \text{if } t > 1 \end{cases}$$

Periodic Functions

DEFINITION 1.11. A function $f(t)$ is called **periodic** with *period* $P > 0$ if it is defined for all t and

$$f(t + P) = f(t) \qquad \text{for all } t$$

The signal, representing the forcing function $f(t)$, may be periodic; thus, it is important to know its Laplace transform when analyzing a dynamic system. Assuming $f(t)$ is a periodic function with period P, then following the definition of the Laplace transform, Eq. (1.31), we have

$$F(s) = \int_{0^-}^{\infty} f(t)e^{-st}\,dt = \int_{0^-}^{P} f(t)e^{-st}\,dt + \int_{P}^{2P} f(t)e^{-st}\,dt + \cdots$$

$$= \sum_{k=0}^{\infty}\left\{\int_{kP}^{(k+1)P} f(t)e^{-st}\,dt\right\}$$

In order to construct a definite integral, whose lower and upper limits are independent of the period, P, we introduce a new dummy variable, τ, such that

$$t = \tau + kP \quad \Longrightarrow \quad \tau = t - kP$$

As a result, the limits of integration corresponding to the new variable, τ, are 0 and P. Substitution for t and the new limits yields

$$F(s) = \sum_{k=0}^{\infty}\left\{\int_{0}^{P} f(\tau)e^{-s(\tau+kP)}\,d\tau\right\} = \sum_{k=0}^{\infty} e^{-skP}\left\{\int_{0}^{P} f(\tau)e^{-s\tau}\,d\tau\right\}$$

in which we note that the integral term is independent of the summation index, k. Hence, the summation only affects the exponential term,

$$\sum_{k=0}^{\infty} e^{-skP} = \sum_{k=0}^{\infty}\left(e^{-sP}\right)^k \underset{\text{geometric series}}{=} \frac{1}{1 - e^{-sP}}$$

Using this information, one obtains

$$F(s) = \left\{\int_{0}^{P} f(\tau)e^{-s\tau}\,d\tau\right\}\sum_{k=0}^{\infty} e^{-skP} = \left\{\int_{0}^{P} f(\tau)e^{-s\tau}\,d\tau\right\}\frac{1}{1 - e^{-sP}}$$

$$\Longrightarrow F(s) = \frac{\int_{0}^{P} f(t)e^{-st}\,dt}{1 - e^{-Ps}} \qquad (1.68)$$

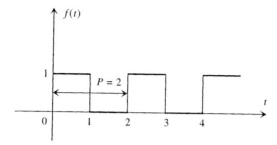

FIGURE 1.27 Periodic function of Example 1.39.

EXAMPLE 1.39. Find the Laplace transform of the periodic function in Fig. 1.27.

Solution. It is clear that the period is $P = 2$, and hence by Eq. (1.68), we have

$$F(s) = \frac{\int_0^P f(t)e^{-st}\,dt}{1 - e^{-Ps}} = \frac{\int_0^2 f(t)e^{-st}\,dt}{1 - e^{-2s}}$$

$$= \frac{1}{1 - e^{-2s}}\left\{\int_0^1 e^{-st}\,dt\right\} = \frac{1 - e^{-s}}{s(1 - e^{-2s})}$$

Noting that $1 - e^{-2s} = 1 - (e^{-s})^2 = (1 - e^{-s})(1 + e^{-s})$, the above expression reduces to

$$F(s) = \frac{1}{s(1 + e^{-s})}$$

PROBLEM SET 1.8

Express each of the following functions in terms of the unit-step function, and then find the corresponding Laplace transform.

1.54. The function $g(t)$ defined in Fig. P1.54

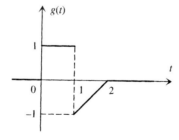

FIGURE P1.54

1.55. The function $f(t)$ defined in Fig. P1.55

50 Dynamic Systems: Modeling and Analysis

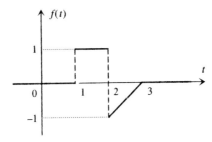

FIGURE P1.55

1.56. The function $h(t)$ defined in Fig. P1.56

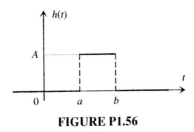

FIGURE P1.56

1.57. The function $f(t)$ defined in Fig. P1.57

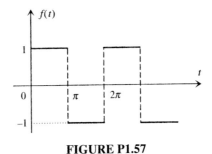

FIGURE P1.57

1.58. $f(t) = \sin t$

1.9 APPLICATIONS: SYSTEM RESPONSE

In Section 1.7, two methods were introduced for determination of the inverse Laplace of a given transform function. In particular, the methods were implemented to solve initial-value problems involving first- and second-order differential equations. In the present section, we focus our attention on seeking certain types of responses of dynamic systems that are governed by first- and second-order differential equations and are subjected to prescribed initial conditions. More specifically, we seek the

CHAPTER 1: Complex Analysis, Differential Equations, and the Laplace Transform 51

responses of such systems when they are subjected to inputs that are modeled as impulse- and step-functions, as defined in Section 1.5. A system's response to an impulse input is known as **impulse response,** and to a step input as **step response.** Additional information on these types of responses, and an introduction to others, will be presented in Chapter 7.

First-Order Systems

In order to investigate the behavior of a linear, first-order system when it is subjected to a certain input signal, we recall from Section 1.4 that its standard form is described as

$$\dot{x} + \frac{1}{\tau}x = f(t), \qquad \tau > 0 \qquad (1.21)$$

where τ is known as the time constant. We are concerned with the determination of the response, $x(t)$, when the input, $f(t)$, is either an impulse- or a step-function. For reasons of simplicity, however, we assume that the input has an amplitude of unity.

Unit-impulse response

Case 1. $f(t) = \delta(t)$, $x(0^-) = 0$. Prior to the determination of the unit-impulse response in the general case, we first consider the simplest possible scenario. That is, assume that the input is a unit-impulse function, $\delta(t)$, applied at $t = 0$, and zero initial condition. Then, Eq. (1.21) becomes

$$\dot{x} + \frac{1}{\tau}x = \delta(t), \qquad \underset{\text{initial condition}}{x(0^-) = 0}$$

Taking the Laplace transform of the equation, solving for $X(s)$, and performing an inverse Laplace transformation, we have

$$\left(s + \frac{1}{\tau}\right)X(s) = 1 \quad \Longrightarrow \quad X(s) = \frac{1}{s + 1/\tau} \quad \Longrightarrow \quad x(t) = e^{-t/\tau}$$

Time variations of response, $x(t)$, are presented in Fig. 1.28 and correspond to $\tau = 0.5$.

Case 2. $f(t) = \delta(t - a)$, $x(0^-) \neq 0$. We now consider the extension of case 1. To this end, assume that the input is modeled as a unit-impulse function, applied at $t = a$, and with a nonzero initial condition. Then, Eq. (1.21) takes the more specific form

$$\dot{x} + \frac{1}{\tau}x = \delta(t - a), \qquad \underset{\text{initial condition}}{x(0^-) = x_0}$$

A Laplace transformation, taking into account the prescribed initial condition, yields

$$\left(s + \frac{1}{\tau}\right)X(s) = e^{-as} + x_0 \quad \Longrightarrow \quad X(s) = \left(\frac{1}{s + 1/\tau}\right)e^{-as} + \frac{x_0}{s + 1/\tau} \qquad (1.69)$$

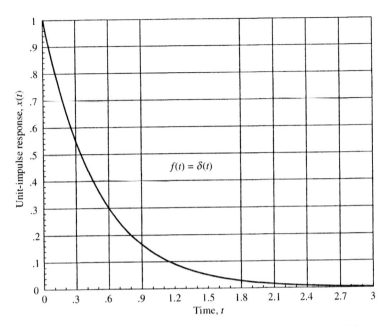

FIGURE 1.28 Unit-impulse response of a first-order system with $f(t) = \delta(t)$ and zero initial conditions.

Letting $G(s) = 1/[s + (1/\tau)]$, we have $g(t) = e^{-t/\tau}$, and an inverse Laplace transformation of Eq. (1.69) results in

$$X(s) = e^{-as}G(s) + x_0 G(s) \quad \Longrightarrow \quad x(t) = u(t-a)g(t-a) + x_0 g(t)$$
$$\text{Theorem 1.8} \qquad\qquad\qquad\qquad \text{Theorem 1.8}$$

$$\Longrightarrow x(t) = u(t-a)e^{-(t-a)/\tau} + x_0 e^{-t/\tau} = \begin{cases} x_0 e^{-t/\tau} & \text{if } 0 \le t < a \\ e^{-(t-a)/\tau} + x_0 e^{-t/\tau} & \text{if } t > a \end{cases}$$

It is then clear that the response curve starts at x_0 initially, and approaches zero as $t \to \infty$. Figure 1.29 shows the graphical representation of the response for $a = 1$, $\tau = 0.5$, and $x_0 = 1$.

It is readily seen that if $a = 0$ and $x_0 = 0$, then case 2 agrees with case 1 and the response curve in Fig. 1.29 reduces to that in Fig. 1.28.

Unit-step response

Case 1. $f(t) = u(t)$, $x(0^-) = 0$. Suppose the input signal is a unit-step function, occurring at $t = 0$, and that the initial condition is zero. In this event, Eq. (1.21) becomes

$$\dot{x} + \frac{1}{\tau}x = u(t), \qquad \underset{\text{initial condition}}{x(0^-) = 0}$$

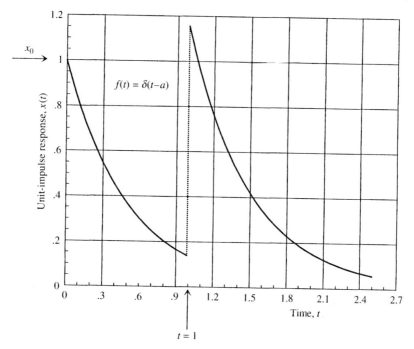

FIGURE 1.29 Response of a first-order system with $f(t) = \delta(t-1)$ and $x_0 = 1$.

Taking the Laplace transform of the equation, solving for $X(s)$, and performing an inverse Laplace transformation, one obtains

$$\left(s + \frac{1}{\tau}\right)X(s) = \frac{1}{s} \implies X(s) = \frac{1}{s\left(s + \frac{1}{\tau}\right)} \implies x(t) = \tau\left[1 - e^{-\frac{t}{\tau}}\right]$$

The response, $x(t)$, is presented graphically in Fig. 1.30 and corresponds to $\tau = 1$.

Case 2. $f(t) = u(t - a)$, $x(0^-) \neq 0$. We now consider the extension of case 1. That is, assume that the input is modeled as a unit-step function, applied at $t = a$, and that it has a nonzero initial condition. Then, Eq. (1.21) takes the following specific form

$$\dot{x} + \frac{1}{\tau}x = u(t - a), \qquad \underset{\text{initial condition}}{x(0^-) = x_0}$$

Laplace transformation, taking into account the prescribed initial condition, yields

$$\left(s + \frac{1}{\tau}\right)X(s) = \frac{e^{-as}}{s} + x_0 \implies X(s) = \frac{1}{s(s + 1/\tau)}e^{-as} + \frac{x_0}{s + 1/\tau} \quad (1.70)$$

Letting $G(s) = 1/\{s[s + (1/\tau)]\}$, then $g(t)$ is determined as $g(t) = \tau\left(1 - e^{-t/\tau}\right)$. Subsequently, an inverse Laplace transformation of Eq. (1.70) yields

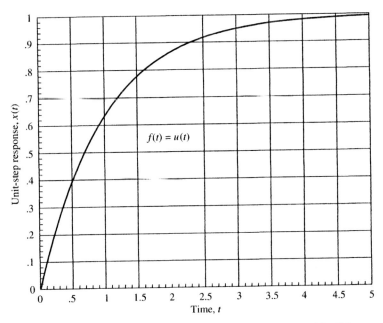

FIGURE 1.30 Unit-step response of a first-order system with $f(t) = u(t)$ and zero initial condition.

$$x(t) = \underbrace{L^{-1}\{e^{-as}G(s)\}}_{\text{Theorem 1.8}} + L^{-1}\left\{x_0\left(\frac{1}{s+1/\tau}\right)\right\}$$

$$\Longrightarrow \quad x(t) = \underbrace{u(t-a)g(t-a)}_{\text{Theorem 1.8}} + x_0 e^{-t/\tau}$$

$$\Longrightarrow \quad x(t) = \begin{cases} x_0 e^{-t/\tau} & \text{if } 0 \le t < a \\ \tau[1 - e^{-(t-a)/\tau}] + x_0 e^{-t/\tau} & \text{if } t > a \end{cases}$$

Figure 1.31 shows the time variations of the response, $x(t)$. It is observed that the response curve starts at x_0 initially, and approaches 1 as $t \to \infty$. The numerical results correspond to $a = 1$, $\tau = 1$, and $x_0 = 0.5$.

As observed in the study of unit-impulse response, in the event that $a = 0$ and $x_0 = 0$, cases 1 and 2 are in agreement, and the response curve of Fig. 1.31 reduces to that of Fig. 1.30.

Second-Order Systems

Recall from Section 1.4 that the standard form of a linear, second-order system is given by

$$\ddot{x} + 2\zeta\omega_n\dot{x} + \omega_n^2 x = f(t) \tag{1.22}$$

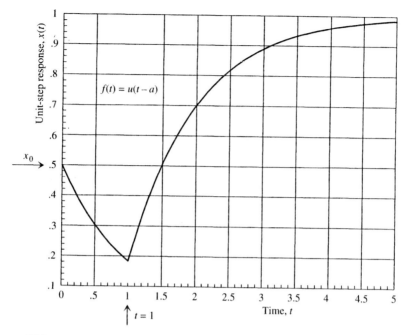

FIGURE 1.31 Response of a first-order system with $f(t) = u(t-1)$ and $x_0 = 0.5$.

where ζ and ω_n denote the damping ratio and the (undamped) natural frequency, respectively. As in the case of a first-order system, we are concerned with the response of the system when subjected to either a unit-impulse or unit-step input.

Unit-impulse response

Case 1. $f(t) = \delta(t)$, $x(0^-) = 0, \dot{x}(0^-) = 0$. Assuming zero initial conditions and that the input signal is a unit impulse, occurring at $t = 0$, Eq. (1.22) becomes

$$\ddot{x} + 2\zeta\omega_n\dot{x} + \omega_n^2 x = \delta(t), \qquad \underbrace{x(0^-) = 0, \dot{x}(0^-) = 0}_{\text{initial conditions}}$$

Taking the Laplace transform of the equation, using zero initial conditions, we have

$$(s^2 + 2\zeta\omega_n s + \omega_n^2)X(s) = 1 \quad \Longrightarrow \quad X(s) = \frac{1}{s^2 + 2\zeta\omega_n s + \omega_n^2} \qquad (1.71)$$

The inverse Laplace transform of $X(s)$ is dependent on the nature of its poles, which in turn depend on the damping ratio, ζ, through

$$s_{1,2} = -\zeta\omega_n \pm \sqrt{(\zeta\omega_n)^2 - \omega_n^2} = -\zeta\omega_n \pm \omega_n\sqrt{\zeta^2 - 1}$$

For the purpose of illustration of the results, as in Section 1.4, assume that $\zeta < 1$. Based on this assumption, the poles of $X(s)$ are $s_{1,2} = -\sigma \pm j\omega_d$ where ω_d denotes the (damped) natural frequency. As a result, $s^2 + 2\zeta\omega_n s + \omega_n^2$ is what we referred to as irreducible, and following the remarks of Section 1.7, we complete the square as

$$s^2 + 2\zeta\omega_n s + \omega_n^2 = (s + \zeta\omega_n)^2 - (\zeta\omega_n)^2 + \omega_n^2$$
$$= (s + \zeta\omega_n)^2 + \omega_n^2(1 - \zeta^2) = (s + \sigma)^2 + \omega_d^2$$

Substituting into the expression of $X(s)$ in Eq. (1.71), one obtains

$$X(s) = \frac{1}{(s + \sigma)^2 + \omega_d^2}$$

Manipulating this expression and referring to Table 1.2 results in the unit-impulse response

$$x(t) = L^{-1}\left\{ \frac{1}{\omega_d} \cdot \frac{\omega_d}{(s + \sigma)^2 + \omega_d^2} \right\}$$
$$= \frac{1}{\omega_d} e^{-\sigma t} \sin \omega_d t = \frac{1}{\omega_d} e^{-\zeta\omega_n t} \sin \omega_d t, \qquad \zeta < 1 \qquad (1.72)$$

The unit-impulse response, defined in Eq. (1.72), is presented graphically in Fig. 1.32. The numerical results are based on $\zeta = 0.5 < 1$ and $\omega_n = 1$.

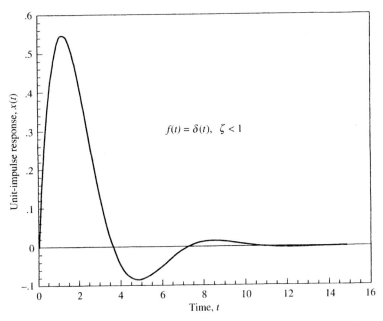

FIGURE 1.32 Unit-impulse response of a second-order system with $f(t) = \delta(t)$ and zero initial conditions.

CHAPTER I: Complex Analysis, Differential Equations, and the Laplace Transform 57

Case 2. $f(t) = \delta(t - a)$, $x(0^-) \neq 0$, $\dot{x}(0^-) \neq 0$. If initial conditions are not necessarily zero, and the input is a unit impulse, occurring at $t = a$, Eq. (1.22) takes a more general form

$$\ddot{x} + 2\zeta\omega_n\dot{x} + \omega_n^2 x = \delta(t - a), \qquad \underbrace{x(0^-) = x_0, \; \dot{x}(0^-) = \dot{x}_0}_{\text{initial conditions}}$$

Performing a Laplace transformation on the equation, and solving for $X(s)$, results in

$$[s^2 X(s) - sx_0 - \dot{x}_0] + 2\zeta\omega_n[sX(s) - x_0] + \omega_n^2 X(s) = e^{-as}$$

$$\implies X(s) = \frac{\dot{x}_0 + (s + 2\zeta\omega_n)x_0 + e^{-as}}{s^2 + 2\zeta\omega_n s + \omega_n^2}$$

As in the previous case, we simply assume that $\zeta < 1$ and proceed to complete the square in the denominator of the above expression. In the meantime, rearrange the numerator in such a way that Table 1.2 can be used, as

$$X(s) = \frac{\dot{x}_0 + \sigma x_0 + (s + \sigma)x_0 + e^{-as}}{(s + \sigma)^2 + \omega_d^2}$$

$$= \frac{\dot{x}_0 + \sigma x_0}{(s + \sigma)^2 + \omega_d^2} + \frac{(s + \sigma)x_0}{(s + \sigma)^2 + \omega_d^2} + \frac{e^{-as}}{(s + \sigma)^2 + \omega_d^2}$$

$$= \frac{\dot{x}_0 + \sigma x_0}{\omega_d} \frac{\omega_d}{(s + \sigma)^2 + \omega_d^2} + x_0 \frac{s + \sigma}{(s + \sigma)^2 + \omega_d^2} + \frac{1}{\omega_d} e^{-as} \frac{\omega_d}{(s + \sigma)^2 + \omega_d^2}$$

A term-by-term inverse Laplace transformation, using Theorem 1.8 for the last term, yields

$$x(t) = \frac{\dot{x}_0 + \sigma x_0}{\omega_d} e^{-\sigma t} \sin \omega_d t + x_0 e^{-\sigma t} \cos \omega_d t$$

$$+ \frac{1}{\omega_d}\left[e^{-\sigma(t-a)} \sin \omega_d(t - a)\right] u(t - a)$$

Thus, switching from σ back to $\zeta\omega_n$, the description of the unit-impulse response of a second-order system in the general case is given as

$$x(t) = \begin{cases} e^{-\zeta\omega_n t}\left[\dfrac{\dot{x}_0 + \zeta\omega_n x_0}{\omega_d} \sin \omega_d t + x_0 \cos \omega_d t\right] & \text{if } 0 \leq t < a \\[2mm] e^{-\zeta\omega_n t}\left[\dfrac{\dot{x}_0 + \zeta\omega_n x_0}{\omega_d} \sin \omega_d t + x_0 \cos \omega_d t\right] & \\[2mm] \quad + \dfrac{1}{\omega_d} e^{-\zeta\omega_n(t-a)} \sin \omega_d(t - a) & \text{if } t > a \end{cases} \qquad (1.73)$$

for $\zeta < 1$. The response curve is presented graphically in Fig. 1.33. The numerical results are based on the following parameter values: $\zeta = 0.5 < 1$, $\omega_n = 1$, $x_0 = 1$, $\dot{x}_0 = -1$, and $a = 1$.

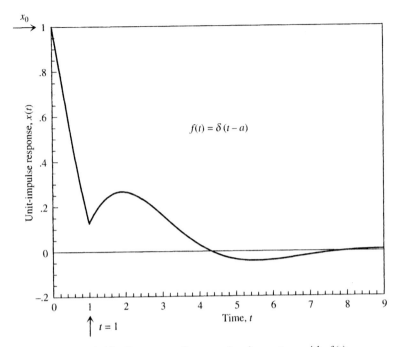

FIGURE 1.33 Response of a second-order system with $f(t) = \delta(t - 1)$ and nonzero initial conditions.

Unit-step response

Case 1. $f(t) = u(t)$, $x(0^-) = 0$, $\dot{x}(0^-) = 0$. The governing equation, subjected to zero initial conditions, is then given by

$$\ddot{x} + 2\zeta\omega_n\dot{x} + \omega_n^2 x = u(t), \qquad \underbrace{x(0^-) = 0, \; \dot{x}(0^-) = 0}_{\text{initial conditions}}$$

Laplace transformation yields

$$\left(s^2 + 2\zeta\omega_n s + \omega_n^2\right)X(s) = \frac{1}{s} \quad \Longrightarrow \quad X(s) = \frac{1}{s\left(s^2 + 2\zeta\omega_n s + \omega_n^2\right)}$$

Assuming, once again, that $\zeta < 1$, and implementing the partial-fraction method, one obtains

$$X(s) = \frac{1}{\omega_n^2}\left[\frac{1}{s} - \frac{s + 2\zeta\omega_n}{s^2 + 2\zeta\omega_n s + \omega_n^2}\right]$$

$$= \frac{1}{\omega_n^2}\left[\frac{1}{s} - \frac{s + \sigma}{(s + \sigma)^2 + \omega_d^2} - \frac{\sigma}{(s + \sigma)^2 + \omega_d^2}\right]$$

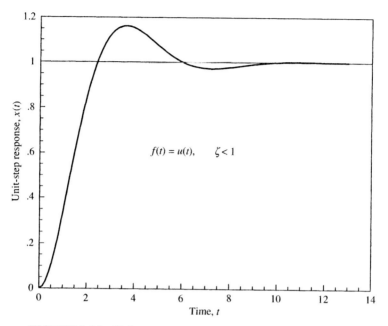

FIGURE 1.34 Unit-step response of a second-order system with $f(t) = u(t)$ and zero initial conditions.

Term-by-term inverse Laplace transformation yields

$$x(t) = \frac{1}{\omega_n^2}\left[1 - e^{-\sigma t}\cos\omega_d t - \frac{\sigma}{\omega_d}e^{-\sigma t}\sin\omega_d t\right]$$

$$= \frac{1}{\omega_n^2}\left\{1 - e^{-\zeta\omega_n t}\left[\cos\omega_d t - \frac{\zeta}{\sqrt{1-\zeta^2}}\sin\omega_d t\right]\right\}$$

Rewriting the quantity within square brackets via Eq. (1.28), we have

$$x(t) = \frac{1}{\omega_n^2}\left[1 - \frac{1}{\sqrt{1-\zeta^2}}e^{-\zeta\omega_n t}\sin(\omega_d t + \phi)\right] \quad \text{where } \phi = \tan^{-1}\left(\frac{\sqrt{1-\zeta^2}}{\zeta}\right)$$

(1.74)

The response curve, corresponding to $\zeta = 0.5 < 1$ and $\omega_n = 1$, is illustrated in Fig. 1.34.

Case 2. $f(t) = u(t - a)$, $x(0^-) \neq 0$, $\dot{x}(0^-) \neq 0$. The governing equation, subjected to nonzero initial conditions, is then given by

$$\ddot{x} + 2\zeta\omega_n\dot{x} + \omega_n^2 x = u(t - a), \quad \underbrace{x(0^-) = x_0, \ \dot{x}(0^-) = \dot{x}_0}_{\text{initial conditions}}$$

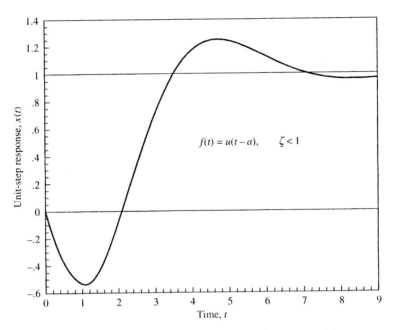

FIGURE 1.35 Response of a second-order system with $f(t) = u(t-1)$ and nonzero initial conditions.

Proceeding as before, the response for the general case is obtained as

$$x(t) = e^{-\zeta\omega_n t}\left[\frac{\dot{x}_0 + \zeta\omega_n x_0}{\omega_d}\sin\omega_d t + x_0\cos\omega_d t\right]$$
$$+ \frac{1}{\omega_n^2}\left\{1 - \frac{1}{\sqrt{1-\zeta^2}}e^{-\zeta\omega_n(t-a)}\sin[\omega_d(t-a) + \phi]\right\}u(t-a),\ \zeta < 1 \quad (1.75)$$

The response is presented in Fig. 1.35 for specific parameter values: $\zeta = 0.5$, $\omega_n = 1$, $x_0 = 0$, $\dot{x}_0 = -1$, $a = 1$.

EXAMPLE 1.40. Consider the mechanical system shown in Fig. 1.36, consisting of a block of mass m, a linear spring of stiffness k, and where the coefficient

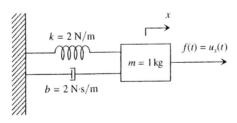

FIGURE 1.36 Mass-spring-damper system.

CHAPTER 1: Complex Analysis, Differential Equations, and the Laplace Transform   61

of viscous damping is b. The system is subjected to an applied force $f(t) = u_s(t)$ and zero initial conditions. Determine the displacement, $x(t)$, at any time $t > 0$.

Solution. The *equation of motion* for this system, derived in Chapter 4, is $m\ddot{x} + b\dot{x} + kx = f(t)$. Substituting for the numerical values of the physical parameters, the initial-value problem is

$$\ddot{x} + 2\dot{x} + 2x = u_s(t), \qquad x(0^-) = 0 = \dot{x}(0^-)$$

Comparing with the standard form of second-order systems, defined by Eq. (1.22), we have

$$\left. \begin{array}{l} 2\zeta\omega_n = 2 \\ \omega_n^2 = 2 \end{array} \right\} \implies \left\{ \begin{array}{l} \omega_n = \sqrt{2} \\ \zeta = \dfrac{\sqrt{2}}{2} < 1 \end{array} \right.$$

Following the results of case (1) above, the unit-step response is given by Eq. (1.74), as

$$x(t) = \frac{1}{\omega_n^2}\left[1 - \frac{1}{\sqrt{1-\zeta^2}}e^{-\zeta\omega_n t}\sin(\omega_d t + \phi)\right] \qquad \text{where } \phi = \tan^{-1}\left(\frac{\sqrt{1-\zeta^2}}{\zeta}\right)$$

Substitution of the damping ratio, as well as the damped and undamped natural frequencies, yields

$$x(t) = \frac{1}{2}\left[1 - \sqrt{2}e^{-t}\sin\left(t + \frac{\pi}{4}\right)\right]$$

PROBLEM SET 1.9

1.59. The governing equation for a first-order dynamic system is given as $\dot{x} + 2x = \delta(t-1)$. Assuming that the system is subjected to zero initial conditions, determine the response, $x(t)$. Roughly sketch the graph of the response curve.

1.60. Repeat Problem 1.59 for $\dot{x} + 2x = \delta(t)$ subjected to $x(0^-) = 0.5$.

1.61. Repeat Problem 1.59 for $\dot{x} + x = u(t-1)$ subjected to zero initial conditions.

1.62. The governing equation for a second-order dynamic system is given as $\ddot{x} + 2\dot{x} + 3x = \delta(t-1)$. Assuming zero initial conditions, determine the response, $x(t)$. Roughly sketch the graph of the response curve.

✦ 1.63. Repeat Problem 1.62 for $\ddot{x} + 2\dot{x} + 3x = u(t)$ subject to $x(0^-) = 0$ and $\dot{x}(0^-) = -1$.

1.10 INITIAL- AND FINAL-VALUE THEOREMS

Consider a function $f(t)$ such that $\lim_{t\to\infty} f(t)$ exists and has a finite value; i.e., it settles down after a sufficiently long time. This finite value, denoted by f_{ss}, is referred to as the **steady-state value** or **final value** of $f(t)$. One possible way to obtain this value, of course, is to inspect the behavior of $f(t)$ as $t \to \infty$. However, if complete information on the time history, $f(t)$, is not available, its steady-state value may alternatively be obtained using the knowledge of $F(s)$ instead. This relation is provided by the **final-value theorem** (Theorem 1.9). One particular area of application of the final-value theorem (FVT) is in steady-state accuracy analysis of feedback control systems where steady-state value of the error signal is sought. There, although sufficient knowledge of the error signal is not available in the time domain, its Laplace transform function can be used to determine the steady-state value.

THEOREM 1.9. FINAL-VALUE THEOREM (FVT). Suppose that none of the poles of $F(s)$ lies in the right-half plane (RHP) and on the imaginary axis, except possibly a *simple pole* at the origin. Then, $f(t)$ has a definite steady-state value and it is given by

$$\lim_{t \to \infty} f(t) = f_{ss} = \lim_{s \to 0} \{sF(s)\} \qquad (1.76)$$

EXAMPLE 1.41. Given

$$F(s) = \frac{s+1}{s(s+3)}$$

find the final value of the function $f(t)$.

Solution. As mentioned earlier, one way to find f_{ss} is to determine $f(t)$ through inverse Laplace transformation, and then let $t \to \infty$. Using the partial fraction method, $F(s)$ is rewritten as

$$F(s) = \frac{A}{s} + \frac{B}{s+3}$$

where

$$A = \{sF(s)\}_{s=0} = \left\{\frac{s+1}{s+3}\right\}_{s=0} = \frac{1}{3}, \qquad B = \{(s+3)F(s)\}_{s=-3} = \frac{2}{3}$$

As a result,

$$F(s) = \frac{1}{3}\left(\frac{1}{s}\right) + \frac{2}{3}\left(\frac{1}{s+3}\right)$$

and inverse Laplace transformation yields

$$f(t) = \frac{1}{3}\left(1 + 2e^{-3t}\right) \stackrel{as\ t \to \infty}{\longrightarrow} \frac{1}{3} = f_{ss}$$

On the other hand, since the conditions of Theorem 1.9 are met, using FVT

$$f_{ss} = \lim_{t \to \infty} f(t) = \underbrace{\lim_{s \to 0}\{sF(s)\}}_{\text{FVT}} = \lim_{s \to 0}\left\{\frac{s+1}{s+3}\right\} = \frac{1}{3}$$

which agrees with the earlier result.

It should be mentioned that the final-value theorem *does **not** apply to functions that exhibit oscillatory behavior* due to the fact that these functions do not possess a limit as $t \to \infty$.

A similar result, known as the **initial-value theorem** (Theorem 1.10), provides the value of $f(t)$ at time $t = 0^+$ directly from the Laplace transform function, $F(s)$. The initial-value theorem (IVT) holds under less restrictive conditions than the final-value theorem in the sense that the poles of $F(s)$ are not limited to specific regions in the complex plane. This means that the initial-value theorem is applicable to functions of oscillatory nature, even though they do not have a limit.

THEOREM 1.10. INITIAL-VALUE THEOREM (IVT). Given that $\lim_{s \to \infty} \{sF(s)\}$ exists, then $\lim_{s \to \infty} \{sF(s)\} = f(0^+)$.

EXAMPLE 1.42. Determine the initial value of the function $f(t)$ where

$$F(s) = \frac{s+1}{s(s+3)}$$

Solution. Notice that $F(s)$ is the transform function used in Example 1.41. Thus, inverse Laplace transformation yields

$$f(t) = \frac{1}{3}\left(1 + 2e^{-3t}\right) \stackrel{\text{as } t \to 0^+}{\to} 1 = f(0^+)$$

On the other hand, application of the initial-value theorem results in

$$f(0^+) = \lim_{s \to \infty} \{sF(s)\} = \lim_{s \to \infty}\left\{\frac{s+1}{s+3}\right\} = 1$$

which is in agreement with the earlier result.

EXAMPLE 1.43. Consider the mechanical (mass-spring-damper) system in Fig. 1.37 subjected to a unit-impulse function and with initial conditions $x(0^-) = 0 = \dot{x}(0^-)$. Determine the initial values of displacement $x(t)$ and velocity $\dot{x}(t)$.

Solution. The equation of motion of this system (Chapter 4) is expressed as

$$m\ddot{x} + b\dot{x} + kx = \delta(t)$$

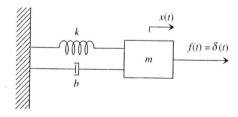

FIGURE 1.37 Mass-spring-damper system.

Taking the Laplace transform and taking into account the *(zero)* initial conditions above yields

$$X(s) = \frac{1}{ms^2 + bs + k}$$

In order to examine the *initial values*, $x(0^+)$ and $\dot{x}(0^+)$, we employ the initial-value theorem as follows:

$$x(0^+) = \lim_{s \to \infty}\{sX(s)\} = \lim_{s \to \infty}\left\{\frac{s}{ms^2 + bs + k}\right\} = 0$$

which agrees with the initial condition $x(0^-) = 0$. To determine the initial value $\dot{x}(0^+)$, the IVT *cannot* be applied in its present form. To this end, from Theorem 1.4 (with zero initial conditions), recall that $\dot{x}$ corresponds to $sX(s)$. Thus, the initial-value theorem can be used to provide the value of $\dot{x}(0^+)$ if $sX(s)$ is used in place of $X(s)$; i.e.,

$$\dot{x}(0^+) = \lim_{s \to \infty}\{s[sX(s)]\} = \lim_{s \to \infty}\left\{\frac{s^2}{ms^2 + bs + k}\right\} = \frac{1}{m} \neq 0$$

which clearly shows that $\dot{x}(0^+) \neq \dot{x}(0^-)$.

PROBLEM SET 1.10

In each case, decide whether the final-value theorem is applicable, and explain. If it is applicable, evaluate f_{ss}. Also, apply the initial-value theorem to evaluate $f(0^+)$. Verify your results by determining $f(t) = L^{-1}\{F(s)\}$ and inspecting its behavior as $t \to \infty$ and at $t = 0^+$.

1.64. $F(s) = \dfrac{1}{s^2(s + 2)}$ **1.65.** $F(s) = \dfrac{1}{s(s^2 + 4)}$ **1.66.** $F(s) = \dfrac{s + 1}{s(2s + 3)}$

SUMMARY

The rectangular and polar forms of a complex number z are related through

$$z = x + jy = re^{j\theta}, \quad r = |z| = \sqrt{x^2 + y^2}, \quad \theta = \tan^{-1}\frac{y}{x}$$

where r and θ are the *magnitude* and *phase* of z, respectively. *Euler's formula* provides a relation between rectangular and polar forms, as

$$e^{j\theta} = \cos\theta + j\sin\theta$$

Given that a complex function is in the form $G(s) = N(s)/D(s)$, its *zeros* and *poles* are the roots of $N(s)$ and $D(s)$, respectively. The *frequency responce function* (FRF) is obtained via replacement of s by $j\omega$, that is, $G(j\omega)$.

The *standard form* of a first-order, linear ODE is defined as

$$\dot{x} + \frac{1}{\tau}x = f(t), \qquad \tau > 0$$

where τ is known as the *time constant* and provides a measure of how fast a system responds to an input. For second-order, linear ODEs, the standard form is given by

$$\ddot{x} + 2\zeta\omega_n\dot{x} + \omega_n^2 x = f(t)$$

where ζ and ω_n are the *damping ratio* and *undamped natural frequency*, respectively.

Assuming that $x(t) = e^{\lambda t}$ is the solution of the nth-order, linear, homogeneous ODE

$$x^{(n)} + a_{n-1}x^{(n-1)} + \cdots + a_0 x = 0$$

then the *characteristic equation* is

$$\lambda^n + a_{n-1}\lambda^{n-1} + \cdots + a_0 = 0$$

whose solutions are known as the *characteristic values*. If a characteristic value, λ_k, is simple (multiplicity 1), then the corresponding independent solution is $x_k(t) = e^{\lambda_k t}$. If λ_k has a multiplicity of m, the corresponding independent solutions are $x_k, tx_k, t^2 x_k, \ldots, t^{m-1}x_k$. When $\lambda_k = \sigma_k + j\omega_k$ is simple, the independent solution is $x_k(t) = e^{\sigma_k t}(c_1 \cos\omega_k t + c_2 \sin\omega_k t)$. For the case of multiplicity m, the rule cited for the real values applies. A linear combination of all independent solutions associated with all the characteristic values is called the *general solution* to the homogeneous equation.

The general solution to the nth-order, nonhomogeneous, linear ODE

$$x^{(n)} + a_{n-1}x^{(n-1)} + \cdots + a_0 x = f(t)$$

consists of the homogeneous and particular solutions. The *particular solution* is chosen based on the nature of the *forcing function* $f(t)$ (Table 1.1) as well as the relation of $f(t)$ to the independent homogeneous solutions, as in cases (1) and (2).

To transform a linear, time-invariant differential equation into an algebraic equation, we use a *linear* operation known as the Laplace transformation. Given that $f(t) = 0$ for $t < 0$, its *Laplace transform* is defined as

$$F(s) = L\{f(t)\} = \int_{0^-}^{\infty} e^{-st} f(t)\, dt$$

The Laplace transform of the first and second *derivatives* of a function $f(t)$ are derived as

$$L\{\dot{f}(t)\} = sF(s) - f(0^-)$$
$$L\{\ddot{f}(t)\} = s^2 F(s) - sf(0^-) - \dot{f}(0^-)$$

The Laplace transformation obeys two *shifting* theorems, one on the ω axis and the other on the t axis, as

$$L\{e^{-\alpha t} f(t)\} = F(s + \alpha)$$
$$L\{f(t-a)u(t-a)\} = e^{-as} F(s)$$

TABLE 1.2 Table of Laplace transforms

No.	Function $f(t)$	Laplace $F(s)$
1	Unit-impulse, $\delta(t)$	1
2	$\delta(t - \alpha)$	$e^{-\alpha s}$
3	Unit step, $u(t)$	$\dfrac{1}{s}$
4	$u(t - \alpha)$	$\dfrac{e^{-\alpha s}}{s}$ (Theorem 1.8)
5	t	$\dfrac{1}{s^2}$
6	$t^{n-1}, \quad n = 1, 2, \ldots$	$\dfrac{(n-1)!}{s^n}$
7	$t^{\alpha-1}, \quad \alpha > 0, \alpha \text{ real}$	$\dfrac{\Gamma(\alpha)}{s^\alpha}$, $\Gamma(\cdot)$ is the gamma function
8	$e^{-\alpha t}$	$\dfrac{1}{s + \alpha}$
9	$te^{-\alpha t}$	$\dfrac{1}{(s + \alpha)^2}$ (Theorem 1.3)
10	$t^n e^{-\alpha t}, \quad n = 1, 2, \ldots$	$\dfrac{n!}{(s + \alpha)^{n+1}}$ (Theorem 1.3)
11	$\ln \alpha t$	$-\dfrac{1}{s}\ln\left(\dfrac{s}{\alpha}\right) - \dfrac{\gamma}{s}$ $\gamma \approx 0.5772157$ is Euler's constant
12	$\dfrac{1}{\alpha}(1 - e^{-\alpha t})$	$\dfrac{1}{s(s + \alpha)}$
13	$\dfrac{1}{\beta - \alpha}(e^{-\alpha t} - e^{-\beta t})$	$\dfrac{1}{(s + \alpha)(s + \beta)}, \quad \alpha \neq \beta$
14	$\dfrac{1}{\alpha - \beta}(\alpha e^{-\alpha t} - \beta e^{-\beta t})$	$\dfrac{s}{(s + \alpha)(s + \beta)}, \quad \alpha \neq \beta$
15	$\sin \omega t$	$\dfrac{\omega}{s^2 + \omega^2}$
16	$\cos \omega t$	$\dfrac{s}{s^2 + \omega^2}$
17	$e^{-\sigma t} \sin \omega t$	$\dfrac{\omega}{(s + \sigma)^2 + \omega^2}$ (Theorem 1.2)
18	$e^{-\sigma t} \cos \omega t$	$\dfrac{s + \sigma}{(s + \sigma)^2 + \omega^2}$ (Theorem 1.2)
19	$\dfrac{1}{\alpha\beta}\left[1 + \dfrac{1}{\alpha - \beta}(\beta e^{-\alpha t} - \alpha e^{-\beta t})\right]$	$\dfrac{1}{s(s + \alpha)(s + \beta)}$
20	$\dfrac{1}{\alpha^2}(-1 + \alpha t + e^{-\alpha t})$	$\dfrac{1}{s^2(s + \alpha)}$

TABLE 1.2 *(Continued)*

No.	Function $f(t)$	Laplace $F(s)$	
21	$\dfrac{1}{\alpha^2}(1 - e^{-\alpha t} - \alpha t e^{-\alpha t})$	$\dfrac{1}{s(s+\alpha)^2}$	
22	$1 - \cos\omega t$	$\dfrac{\omega^2}{s(s^2+\omega^2)}$	
23	$\omega t - \sin\omega t$	$\dfrac{\omega^3}{s^2(s^2+\omega^2)}$	
24	$t\cos\omega t$	$\dfrac{s^2 - \omega^2}{(s^2+\omega^2)^2}$	(Theorem 1.3)
25	$\dfrac{1}{2\omega}(t\sin\omega t)$	$\dfrac{s}{(s^2+\omega^2)^2}$	(Theorem 1.3)
26	$\dfrac{1}{2\omega^3}(\sin\omega t - \omega t\cos\omega t)$	$\dfrac{1}{(s^2+\omega^2)^2}$	
27	$1 - \dfrac{1}{\sqrt{1-\zeta^2}}e^{-\zeta\omega_n t}\sin(\omega_d t + \phi)$ $\omega_d = \omega_n\sqrt{1-\zeta^2},\ \phi = \tan^{-1}\dfrac{\sqrt{1-\zeta^2}}{\zeta}$	$\dfrac{\omega_n^2}{s^2 + 2\zeta\omega_n s + \omega_n^2} \cdot \dfrac{1}{s},$	$\zeta < 1$
28	$-\dfrac{1}{\sqrt{1-\zeta^2}}e^{-\zeta\omega_n t}\sin(\omega_d t - \phi)$ $\omega_d = \omega_n\sqrt{1-\zeta^2},\ \phi = \tan^{-1}\dfrac{\sqrt{1-\zeta^2}}{\zeta}$	$\dfrac{s}{s^2 + 2\zeta\omega_n s + \omega_n^2},$	$\zeta < 1$
29	$\dfrac{\omega_n}{\sqrt{1-\zeta^2}}e^{-\zeta\omega_n t}\sin\omega_d t$ $\omega_d = \omega_n\sqrt{1-\zeta^2}$	$\dfrac{\omega_n^2}{s^2 + 2\zeta\omega_n s + \omega_n^2},$	$\zeta < 1$
30	$1 - \dfrac{1}{\sqrt{1-\zeta^2}}e^{-\zeta\omega_n t}\{2\zeta\omega_n\sin(\omega_d t - \phi) + \sin(\omega_d t + \phi)\}$ $\omega_d = \omega_n\sqrt{1-\zeta^2},\ \phi = \tan^{-1}\dfrac{\sqrt{1-\zeta^2}}{\zeta}$	$\dfrac{2\zeta\omega_n s + \omega_n^2}{s^2 + 2\zeta\omega_n s + \omega_n^2},$	$\zeta < 1$
31	$e^{-\alpha t}f(t)$	$F(s+\alpha)$	(Theorem 1.2)
32	$tf(t)$	$-\dfrac{d}{ds}F(s)$	(Theorem 1.3)
33	$t^n f(t),\quad n = 1, 2, \ldots$	$(-1)^n \dfrac{d^n}{ds^n}F(s)$	(Theorem 1.3)
34	$\dot{f}(t)$	$sF(s) - f(0^-)$	(Theorem 1.4)
35	$\ddot{f}(t)$	$s^2 F(s) - sf(0^-) - \dot{f}(0^-)$	
36	$\displaystyle\int_0^t f(\tau)\,d\tau$	$\dfrac{1}{s}F(s)$	(Theorem 1.6)
37	$f(t-a)u(t-a)$	$e^{-as}F(s)$	(Theorem 1.8)

Inverse Laplace transformation is used to transform the information from the complex frequency domain back to the time domain. This can be achieved through (1) the *partial fraction* method, or (2) the *convolution* method.

1. Partial fraction method

If the transform function $F(s)$ has distinct (real or complex) poles, $p_1, \ldots, p_n$, then

$$F(s) = \frac{A_1}{s - p_1} + \cdots + \frac{A_n}{s - p_n}$$

where the constants are called the *residues* and are obtained as

$$A_i = \operatorname*{Res}_{s = p_i} F(s) = [(s - p_i)F(s)]_{s = p_i}, \qquad i = 1, 2, \ldots, n$$

In the event that a certain pole p_i has a multiplicity of $k > 1$, then the term $(s - p_i)^k$ will appear in the denominator of $F(s)$. The fractions corresponding to this term are formed as

$$\frac{A_k}{(s - p_i)^k} + \cdots + \frac{A_1}{s - p_i}$$

where

$$A_m = \frac{1}{(k - m)!} \left(\frac{d^{k-m}}{ds^{k-m}} [(s - p_i)^k F(s)]_{s = p_i} \right), \qquad m = 1, 2, \ldots, k$$

The *alternative approach* is to form the partial fraction as before and combine them to generate a single fraction whose denominator agrees with that of $F(s)$. Subsequently, set the numerators identical and find the constants.

2. Convolution method

If $G(s)$ and $H(s)$ are the Laplace transforms of function $g(t)$ and $h(t)$, respectively, then the inverse Laplace transform of the product $G(s)H(s)$ is given by the convolution of g and h, as

$$L^{-1}\{G(s)H(s)\} = (g * h)(t) = \int_0^t g(\tau)h(t - \tau)\, d\tau$$

$$= (h * g)(t) = \int_0^t h(\tau)g(t - \tau)\, d\tau$$

A function $f(t)$ is said to be *periodic* with period $P > 0$ if it is defined for all t and $f(t + P) = f(t)$ for all t. The Laplace transform of this type of a function is given by

$$F(s) = \frac{1}{1 - e^{-Ps}} \int_0^P e^{-st} f(t)\, dt$$

When a dynamic system is driven by its initial condition(s) only, the response is called *free* response. If the system is subject to an applied forcing function, the response is known as *forced* response. In particular, when the system is subject to an impulse or step input, we speak of impulse response and step response. If, in addition, the forcing function is of unit magnitude and the initial conditions are zero, these responses are referred to as *unit-impulse* and *unit-step* responses.

If a function $f(t)$ settles down after a sufficient amount of time, then we say that its limit, as $t \to \infty$, exists and is finite. This limit is called the final value (or steady-state value) of $f(t)$, denoted by $f(\infty)$ or f_{ss}. In that case,

$$f_{ss} = \lim_{t \to \infty} f(t)$$

In the event that a complete knowledge of $f(t)$ in the time domain is not available, its Laplace transform may by used to determine f_{ss}. The *final-value theorem* (FVT) states that if none of the poles of $F(s)$ lies in the right-half plane (RHP) and on the imaginary axis, except possibly a simple pole at the origin, then the steady-state value of f is given by

$$f_{ss} = \lim_{s \to 0} \{sF(s)\}$$

Thus, the FVT does not apply to functions that exhibit oscillatory behavior. The *initial-value theorem* does apply to oscillatory functions and states that

$$f(0^+) = \lim_{s \to \infty} \{sF(s)\}$$

PROBLEMS

Express each of the following complex numbers in rectangular form. Also, express them in polar form, identifying their magnitude and phase.

1.67. $\dfrac{-3 + j2}{1 + j3}$ **1.68.** $\dfrac{(1 + j)(2 + j)}{3 + j}$ **1.69.** $\dfrac{s + 1}{s^2 + s + 1}$, $s = j2$

1.70. Evaluate $(3 + j2)^{1/2}$ and express in rectangular form.

1.71. Find the Laplace transform of the function given in Fig. P1.71
 (a) Using the definition of Laplace transform, given by Eq. (1.31).
 (b) By expressing $f(t)$ as a linear combination of step functions and using Theorem 1.8.

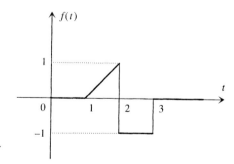

FIGURE P1.71

1.72. Find the Laplace transform of the periodic function shown in Fig. P1.72.

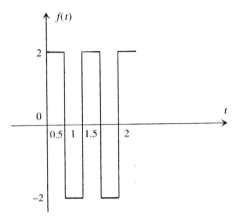

FIGURE P1.72

1.73. Determine the Laplace transform of $f(t) = e^{-t} \sin 2t$.

1.74. Determine the Laplace transform of $f(t) = t^2$ indirectly using Theorem 1.5.

1.75. Find $L^{-1}\{1/[s^2(s+1)]\}$ using
  (a) The result of Theorem 1.6. *Hint:* Apply Theorem 1.6 twice!
  (b) The partial fraction method.
  (c) The convolution method.

1.76. Given

$$G(s) = \frac{2s+3}{s^2+2s+2}$$

find its inverse Laplace transform, $g(t)$.

1.77. Find the inverse Laplace transform of

$$F(s) = \frac{s-2}{s(s+2)^2(s+1)}$$

using the partial-fraction method.

1.78. The equation of motion of a mechanical system is given by $\ddot{x} + 4\dot{x} + 3x = \delta(t)$ subject to initial conditions, $x(0^-) = 1$ and $\dot{x}(0^-) = 0$. Through Laplace transformation, find $x(t)$.

1.79. Decide whether the final-value theorem applies to

$$G(s) = \frac{2s+3}{s^2+2s+2}$$

of Problem 1.76. Explain!

1.80. Find the steady-state value of $f(t)$ where

$$F(s) = \frac{2s^2 + 3s}{s^2(s+2)}$$

(a) Using the final-value theorem.
(b) By determining $f(t)$ as $t \to \infty$.

1.81. Consider a function with the graphical representation given in Fig. P1.81.
(a) Express $h(t)$ as a linear combination of step functions, and determine its Laplace transform via Theorem 1.8.
(b) Determine $H(s)$ using Eq. (1.31) directly.

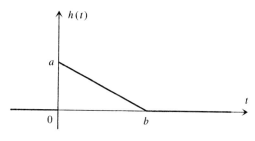

FIGURE P1.81

1.82. Repeat Problem 1.81 for the function $g(t)$, graphically described in Fig. P1.82.

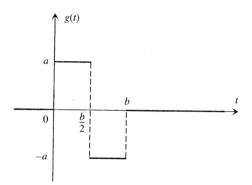

FIGURE P1.82

1.83. The governing equation for the mechanical system shown in Fig. P1.83 is $m\ddot{x} + b\dot{x} + kx = f(t)$, where $x(t)$ and $f(t)$ represent the displacement of the block and the applied force, respectively. The system is subject to initial conditions $x(0^-) = 0$ and $\dot{x}(0^-) = 2$.
(a) Assume the following numerical values: $m = 1$ kg, $b = 2$ N·s/m, $k = 1$ N/m. Given that the applied force (input) is a unit-step function, find the response (displacement) via the method of partial fractions.
(b) Having found the expression of $X(s)$, decide whether the FVT applies. If so, find the final value of $x(t)$.
(c) Assuming zero initial conditions, obtain the response via the convolution method.

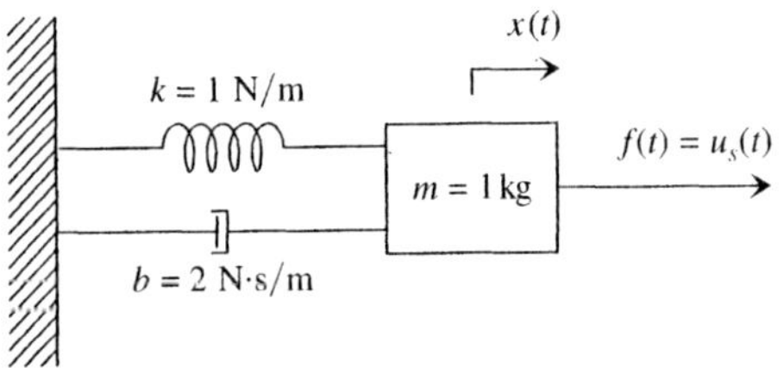

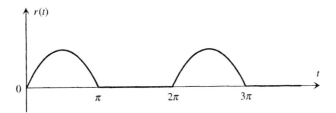

FIGURE P1.83 Mechanical system.

1.84. Repeat Problem 1.83 for a unit-ramp input.

1.85. Consider the half-wave rectification of the sinusoidal function sin t, with a period of 2π, as shown in Fig. P1.85. Determine its Laplace transform.

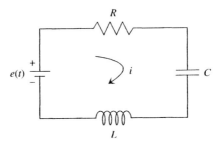

FIGURE P1.85 Half-wave rectifier.

1.86. The equation of motion of a rotational mechanical system is given by

$$J\dot{\omega} + B\omega = T_a(t), \qquad \omega(0^-) = 1 \text{ rad/s}$$

where $\omega(t)$ and $T_a(t)$ represent the angular velocity and applied torque, respectively. Assume that the applied torque is a unit-step, and that $J = 1 \text{ kg} \cdot \text{m}^2$ and $B = 2 \text{ N·m·s}$. Determine the angular velocity at any time.

1.87. Consider the rotational system of Problem 1.86 subject to zero initial conditions. Find $\omega(t)$ via the convolution method.

1.88. In the electrical circuit shown in Fig. P1.88, R, L, C, $i(t)$, and $e(t)$ represent resistance, inductance, capacitance, current, and the applied voltage, respectively. Assume $R =$

FIGURE P1.88 RLC circuit.

$4\,\Omega$, $L = 3$ H, $C = 1$ F. The governing equation is determined (see Chapter 5) to be

$$L\frac{di}{dt} + Ri + \frac{1}{C}\int_0^t i(\tau)\,d\tau = e(t)$$

Assuming that the applied voltage is a unit-step function, and zero initial conditions, find the current at any time, i.e., $i(t)$.

1.89. Repeat Problem 1.88 with $e(t) = u_r(t)$. Decide whether the final-value theorem (FVT) can be applied to determine the steady-state value of the current. If so, find i_{ss}.

1.90. Find the inverse Laplace transform of the function $2/(s + 3)(s^2 + 2s + 5)$ via the convolution method. Compare your result with that obtained in Example 1.33.

1.91. Determine the magnitude and phase of the transform function of Problem 1.90 when $s = j3$.

1.92. Prove the identity

$$L\{\sin \omega t\} = \frac{\omega}{s^2 + \omega^2}$$

1.93. Consider the liquid-level system shown in Fig. P1.93. The system is a storage tank of cross-sectional area A, whose liquid level (or height) is h. The liquid enters the tank from the top and leaves the tank at the bottom through a valve of fluid resistance R. The fluid density ρ is assumed to be constant. q_i and q_o represent the volume flow rate in and out, respectively. The system is governed by a first-order differential equation (see Chapter 6)

$$\frac{RA}{g}\dot{h} + h = \frac{R}{g}q_i(t)$$

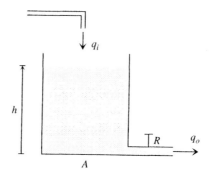

FIGURE P1.93 Liquid-level system.

(a) Assuming the inflow rate q_i is a unit step, and zero initial conditions, determine the time history of system output, $h(t)$, through Laplace transformation.
(b) Decide whether the final-value theorem applies to the transform function $H(s)$ in part (a). If so, determine the final (steady-state) value of h.

1.94. Consider the rotational system of Problem 1.86. Suppose the applied torque is a unit-pulse function; i.e., $\tau_a(t) = u_p(t)$, as shown in Fig. P1.94. Assume zero initial conditions. Determine the angular velocity, $\omega(t)$.

$$u_p(t) = \begin{cases} 1 & \text{if } 0 < t < t_1 \\ 0 & \text{if } t < 0 \text{ and } t > t_1 \end{cases}$$

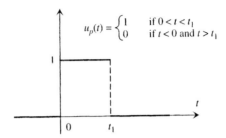

FIGURE P1.94

1.95. The RL circuit in Fig. P1.95 is subjected to a sinusoidal applied voltage, defined by $e(t) = \sin 2t$. The governing equation of the circuit is $L(di/dt) + Ri = e(t)$. Assuming $R = 2\,\Omega$ and $L = 1$ H, and zero initial current, determine the current at any time, i.e., $i(t)$.

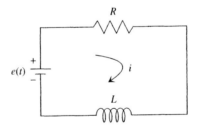

FIGURE P1.95

1.96. Prove the following identity, listed in Table 1.2.

$$\mathcal{L}^{-1}\left\{\frac{1}{s} \cdot \frac{\omega_n^2}{s^2 + 2\zeta\omega_n s + \omega_n^2}\right\} = 1 - e^{-\zeta\omega_n t}\left[\cos\omega_d t + \frac{\zeta}{\sqrt{1-\zeta^2}}\sin\omega_d t\right], \quad \zeta < 1$$

where

$$\omega_d = \omega_n\sqrt{1-\zeta^2}$$

1.97. Consider the mechanical system shown in Fig. P1.97, subjected to zero initial conditions. Through Laplace transformation and the result of Problem 1.96 determine the displacement $x(t)$.

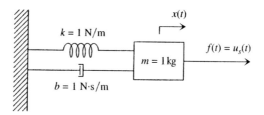

FIGURE P1.97

1.98. Determine

$$L^{-1}\left\{\frac{2s+3}{(s+2)(s+1)(s^2+2s+3)}\right\}$$

CHAPTER 2

Matrix Analysis

This chapter is primarily concerned with fundamental concepts in linear algebra and matrix analysis. The methods of linear algebra are particularly useful while treating linear systems of algebraic and differential equations. As mentioned earlier, mathematical models of dynamic systems are described by differential equations. The number of differential equations involved in a model of a dynamic system depends on the number of independent coordinates required to describe the position of all its elements. Whereas elementary ideas of matrix analysis are needed for system model representations (see Chapter 3), a slightly more advanced knowledge of it is crucial in the determination of system response. To this end, concepts will be introduced in this chapter that enable one to seek response (see Chapter 7) while working entirely in the time domain. Occasionally, this is better than using the Laplace transform discussed in Chapter 1.

2.1 VECTORS AND MATRICES

DEFINITION 2.1. An ***n*th-order vector**, **v**, is an ordered set of n scalars, $v_1, v_2, \ldots, v_n$, denoted by

$$\mathbf{v} = \begin{Bmatrix} v_1 \\ v_2 \\ \vdots \\ v_n \end{Bmatrix}$$

where $v_1, v_2, \ldots,$ and v_n are known as the **components** of **v**. The **norm** (or **length**) of **v** is defined as

$$|\mathbf{v}| = (v_1^2 + \cdots + v_n^2)^{1/2}$$

A vector of length 1 is referred to as a **unit vector**.

DEFINITION 2.2. The **dot product** of two nth-order vectors **u** and **v** is defined as $\mathbf{u} \cdot \mathbf{v} = u_1v_1 + \cdots + u_nv_n$, where u_i and v_i are the components of **u** and **v**, respectively. Two vectors, **u** and **v**, are said to be **orthogonal** if their dot product is zero; i.e., $\mathbf{u} \cdot \mathbf{v} = 0$. They are called **orthonormal** if they are orthogonal and each has a length of 1.

DEFINITION 2.3. A set of m vectors (n-dimensional) is said to be **linearly independent** if none of them may be expressed as a linear combination of the others; i.e.,

$$c_1\mathbf{w}_1 + c_2\mathbf{w}_2 + \cdots + c_m\mathbf{w}_m = \mathbf{0} \iff c_1 = c_2 = \cdots = c_m = 0$$

DEFINITION 2.4. A collection of real or complex numbers—as well as functions—arranged in a rectangular array and enclosed by square brackets is referred to as a **matrix.**

Simple examples of matrices include the following:

$$\begin{bmatrix} -1 & 0 \\ 2 & 5.3 \end{bmatrix}, \quad \begin{bmatrix} 1 & -4 \end{bmatrix}, \quad \begin{bmatrix} 2 \\ 7 \end{bmatrix}, \quad \begin{bmatrix} 1 & 0 & -5 \\ 2.1 & 3 & 5 \end{bmatrix}$$

The horizontal and vertical lines are referred to as **rows** and **columns** of the matrix, respectively. Each of the elements in a matrix is called an **entry** (or **element**) of the matrix. The **size** of a matrix is determined by the number of its rows and columns. If a matrix **A** has m rows and n columns, then it is of size $m \times n$ and is denoted by

$$\mathbf{A} = \begin{bmatrix} a_{11} & a_{12} & \cdots & a_{1n} \\ a_{21} & a_{22} & \cdots & a_{2n} \\ \vdots & & & \vdots \\ a_{m1} & a_{m2} & \cdots & a_{mn} \end{bmatrix}_{m \times n}$$

The $m \times n$ **zero matrix**, denoted by $\mathbf{0}_{m \times n}$, is defined as an $m \times n$ matrix all of whose entries are zero. **A** is said to be a **square matrix** if it has the same number of rows as columns. The $n \times n$ **identity matrix**, denoted by $\mathbf{I}_n$ or simply **I**, consists of 1's along the main diagonal and 0's everywhere else. Two matrices are said to be **equal** if and only if they are of the same size and have equal entries in respective locations.

Notation:

$$\mathbf{A}_{m \times n} = [a_{ij}] \quad 1 \leq i \leq m, \quad 1 \leq j \leq n$$

a_{ij} is the (i, j) entry of **A**, located at the intersection of the ith row and jth column of **A**.

A **submatrix** of matrix **A** is obtained as a result of omitting some rows and columns of **A**. If a matrix consists of only one row, it is called a **row matrix** (or *row vector*), and if it consists of only one column, it is called a **column matrix** (or *column vector*). In vector notation

$$[a_1, \cdots, a_n], \quad \begin{bmatrix} a_1 \\ \vdots \\ a_n \end{bmatrix}$$

(row vector)   (column vector)

DEFINITION 2.5. The **trace** of an $n \times n$ matrix $\mathbf{A}$, tr($\mathbf{A}$), is defined as the sum of the diagonal entries of $\mathbf{A}$; i.e., tr($\mathbf{A}$) = $\sum_{i=1}^{n} a_{ii}$.

EXAMPLE 2.1. The 2×2 submatrix $\mathbf{S}$ is obtained by deleting the third row and second column of the 3×3 matrix $\mathbf{A}$:

$$\mathbf{A} = \begin{bmatrix} 1 & 2 & -2 \\ 3 & 0 & 4 \\ -3 & 1 & -1 \end{bmatrix}, \quad \mathbf{S} = \begin{bmatrix} 1 & -2 \\ 3 & 4 \end{bmatrix}$$

Matrix Operations

The **sum** of two matrices of the same size, $\mathbf{A} = [a_{ij}]_{m \times n}$ and $\mathbf{B} = [b_{ij}]_{m \times n}$, is

$\mathbf{A} + \mathbf{B} = \mathbf{C} = [c_{ij}]$ where $c_{ij} = a_{ij} + b_{ij}$ for $1 \leq i \leq m, 1 \leq j \leq n$

with the following properties:

Commutative law $\mathbf{A} + \mathbf{B} = \mathbf{B} + \mathbf{A}$

Associative law $\mathbf{A} + (\mathbf{B} + \mathbf{D}) = (\mathbf{A} + \mathbf{B}) + \mathbf{D}$

$\mathbf{A}_{m \times n} + \mathbf{0}_{m \times n} = \mathbf{A}$

$\mathbf{A}_{m \times n} + (-\mathbf{A}) = \mathbf{0}_{m \times n}$

The product of a scalar k and a matrix $\mathbf{A} = [a_{ij}]_{m \times n}$ is an $m \times n$ matrix $k\mathbf{A} = [ka_{ij}]_{m \times n}$. **Multiplication** of matrices is only possible if their sizes are **compatible**; i.e., given $\mathbf{A} = [a_{ij}]_{m \times n}$ and $\mathbf{B} = [b_{ij}]_{n \times p}$, their **product** is an $m \times p$ matrix $\mathbf{C} = [c_{ij}]_{m \times p}$ whose entries are defined by

$$c_{ij} = \sum_{k=1}^{n} a_{ik} b_{kj}, \quad 1 \leq i \leq m, \quad 1 \leq j \leq p \qquad (2.1)$$

Assuming all matrices involved are of compatible sizes, then

Distributive law $\mathbf{A}(\mathbf{B} + \mathbf{D}) = \mathbf{AB} + \mathbf{AD}, \quad (\mathbf{A} + \mathbf{B})\mathbf{D} = \mathbf{AD} + \mathbf{BD}$

Associative law $\mathbf{A}(\mathbf{BD}) = (\mathbf{AB})\mathbf{D}$

$\mathbf{IA} = \mathbf{AI} = \mathbf{A}$

Noncommutativity $\mathbf{AB} \neq \mathbf{BA}$

Furthermore, unlike the product of real numbers (scalars), the product of two nonzero matrices could result in a zero matrix:

Nonzero factor property $AB = 0$ does not imply $A = 0$ or $B = 0$

EXAMPLE 2.2. Regarding the noncommutative and the nonzero factor properties of matrix multiplication, consider the following:

$$A = \begin{bmatrix} 1 & -2 & 3 \\ 0 & 1 & 4 \end{bmatrix}_{2\times 3}, \quad B = \begin{bmatrix} -2 & -1 & 4 \\ 1 & 2 & 0 \\ 3 & 5 & 1 \end{bmatrix}_{3\times 3} \implies AB = \begin{bmatrix} 5 & 10 & 7 \\ 13 & 22 & 4 \end{bmatrix}_{2\times 3} \quad (i)$$

Note that the product BA is undefined, because multiplication of a 3×3 matrix by a 2×3 matrix is impossible. It is then clear that $AB \neq BA$.

$$A = \begin{bmatrix} 1 & 1 \\ 2 & 2 \end{bmatrix}, \quad B = \begin{bmatrix} -1 & 1 \\ 1 & -1 \end{bmatrix} \implies AB = 0_{2\times 2} = \begin{bmatrix} 0 & 0 \\ 0 & 0 \end{bmatrix} \quad (ii)$$

Matrix multiplication may also be performed through dot products of vectors. To that end, consider matrices $A_{m \times n}$ and $B_{n \times p}$, and express them as

$$A = \begin{bmatrix} a_1 \\ a_2 \\ \vdots \\ a_m \end{bmatrix}_{\text{row partition}}, \quad B = [\underbrace{b_1 \quad b_2 \quad \ldots \quad b_p}_{\text{column partition}}]$$

where a_i ($1 \leq i \leq m$) denotes the ith row of A and is $1 \times n$, and b_j ($1 \leq j \leq p$) denotes the jth column of B and is $n \times 1$. Suppose the product of A and B is denoted by $C = [c_{ij}]_{m \times p}$. Each entry of C is calculated as a dot product:

$$c_{ij} = a_i \cdot b_j, \quad 1 \leq i \leq m, \quad 1 \leq j \leq p \quad (2.2)$$

Alternatively, AB may be obtained using the column partition of B only:

$$AB = [Ab_1 \quad Ab_2 \quad \ldots \quad Ab_p] \quad (2.3)$$

EXAMPLE 2.3. Consider again the two matrices of Example 2.2(*i*). Row and column partitioning yields

$$a_1 = \begin{bmatrix} 1 & -2 & 3 \end{bmatrix}, \quad a_2 = \begin{bmatrix} 0 & 1 & 4 \end{bmatrix},$$

$$b_1 = \begin{bmatrix} -2 \\ 1 \\ 3 \end{bmatrix}, \quad b_2 = \begin{bmatrix} -1 \\ 2 \\ 5 \end{bmatrix}, \quad b_3 = \begin{bmatrix} 4 \\ 0 \\ 1 \end{bmatrix}$$

Following the procedure suggested by Eq. (2.2), entries of the matrix product are computed as follows:

$$c_{11} = a_1 \cdot b_1 = 5, \quad c_{12} = a_1 \cdot b_2 = 10, \quad c_{13} = a_1 \cdot b_3 = 7,$$
$$c_{21} = a_2 \cdot b_1 = 13, \quad c_{22} = a_2 \cdot b_2 = 22, \quad c_{23} = a_2 \cdot b_3 = 4$$

which agree with the earlier result. Furthermore, using the strategy of Eq. (2.3), the same result may be obtained:

$$AB = [Ab_1 \quad Ab_2 \quad Ab_3] = \begin{bmatrix} 5 & 10 & 7 \\ 13 & 22 & 4 \end{bmatrix}$$

Matrix Transpose

DEFINITION 2.6. Given an $m \times n$ matrix A, its **transpose** (denoted by A^T) is an $n \times m$ matrix generated by interchanging rows and columns of matrix A. The following rules hold for the transpose of sums and products of matrices:

$$(A + B)^T = A^T + B^T, \quad A \text{ and } B \text{ of equal size}$$
$$(AB)^T = B^T A^T, \quad A \text{ and } B \text{ of compatible sizes}$$

EXAMPLE 2.4.

$$A = \begin{bmatrix} -2 & 0 & 1 \\ 6 & -12 & 5 \end{bmatrix}, \quad B = \begin{bmatrix} 1 \\ -1 \\ 0 \end{bmatrix} \implies \left(\underbrace{\underset{2 \times 3}{A} \underset{3 \times 1}{B}}_{2 \times 1} \right)^T = [-2 \quad 18]_{1 \times 2} = \underset{1 \times 3}{B^T} \underset{3 \times 2}{A^T}$$

DEFINITION 2.7. A square matrix A (complex in general) is said to be **Hermitian** if $A = \overline{A}^T$, where the overbar denotes complex conjugation. In the event that matrix A is real, $A^T = A$, and it is known as **symmetric**. Furthermore, a square matrix A is **skew-Hermitian** if $\overline{A}^T = -A$. If A is real, then $A^T = -A$, and it is called **skew-symmetric**. Any square matrix A may be expressed as the sum of a Hermitian and a skew-Hermitian matrix; i.e.,

$$A = M + N \quad \text{where} \quad M = \tfrac{1}{2}\left(A + \overline{A}^T\right), \quad N = \tfrac{1}{2}\left(A - \overline{A}^T\right) \quad (2.4)$$
$$\phantom{A = M + N \quad \text{where} \quad} \underset{\text{Hermitian}}{} \quad\quad\quad \underset{\text{skew-Hermitian}}{}$$

EXAMPLE 2.5. Consider a 3×3 matrix:

$$A = \begin{bmatrix} j & 2+j & 1 \\ -2+j & -j3 & 1+j \\ -1 & -1+j & 0 \end{bmatrix}$$

It is readily seen that

$$\overline{A}^T = \begin{bmatrix} -j & -2-j & -1 \\ 2-j & j3 & -1-j \\ 1 & 1-j & 0 \end{bmatrix} = -A$$

Thus, A is skew-Hermitian.

EXAMPLE 2.6.

$$A = \begin{bmatrix} 1 & 3 \\ -4 & 2 \end{bmatrix} = M + N$$

where

$$M = \tfrac{1}{2}(A + A^T) = \begin{bmatrix} 1 & -\tfrac{1}{2} \\ -\tfrac{1}{2} & 2 \end{bmatrix}, \quad N = \tfrac{1}{2}(A - A^T) = \begin{bmatrix} 0 & \tfrac{7}{2} \\ -\tfrac{7}{2} & 0 \end{bmatrix}$$

It is readily seen that **M** and **N** are symmetric and skew-symmetric, respectively.

DEFINITION 2.8. A square matrix $A = [a_{ij}]_{n \times n}$ is said to be **upper-triangular** if $a_{ij} = 0$ for all $i > j$. It is said to be **lower-triangular** if $a_{ij} = 0$ for all $i < j$. It is said to be **diagonal** if $a_{ij} = 0$ for all $i \neq j$.

Examples are

$$\text{Upper-triangular} \quad U = \begin{bmatrix} -2 & 1 & 2 \\ 0 & 5 & 0 \\ 0 & 0 & 3 \end{bmatrix}$$

$$\text{Lower-triangular} \quad L = \begin{bmatrix} 1 & 0 & 0 \\ 2 & 5 & 0 \\ 4 & 7 & -1 \end{bmatrix}$$

$$\text{Diagonal} \quad D = \begin{bmatrix} 3 & 0 & 0 \\ 0 & -4 & 0 \\ 0 & 0 & 1 \end{bmatrix}$$

PROBLEM SET 2.1

Determine whether the following pairs of vectors are orthogonal, orthonormal, or neither.

2.1. $u = \begin{Bmatrix} 1 \\ 0 \\ -2 \end{Bmatrix}, \quad v = \begin{Bmatrix} 6 \\ 1 \\ 3 \end{Bmatrix}$

2.2. $u = \begin{Bmatrix} -1/\sqrt{3} \\ 1/\sqrt{3} \\ 1/\sqrt{3} \end{Bmatrix}, \quad v = \begin{Bmatrix} 1/\sqrt{2} \\ 0 \\ 1/\sqrt{2} \end{Bmatrix}$

2.3. $u = \begin{Bmatrix} 1 \\ -3 \end{Bmatrix}, \quad v = \begin{Bmatrix} 3 \\ -1 \end{Bmatrix}$

2.4. Given

$$A = \begin{bmatrix} 1 & 0 \\ -3 & -2 \end{bmatrix} \quad \text{and} \quad B = \begin{bmatrix} 0 & 1 & 3 \\ 1 & -2 & 2 \end{bmatrix}$$

determine their product, **AB**, via column partitioning of **B** as suggested by Eq. (2.3). Furthermore, letting $\mathbf{b}_1$, $\mathbf{b}_2$, and $\mathbf{b}_3$ denote the column vectors of **B**, verify that

$$(\mathbf{AB})^T = \begin{bmatrix} \mathbf{b}_1^T \mathbf{A}^T \\ \mathbf{b}_2^T \mathbf{A}^T \\ \mathbf{b}_3^T \mathbf{A}^T \end{bmatrix}$$

2.5. Suppose **A** is a square matrix with complex entries, in general. Prove that matrices **M** and **N** in Eq. (2.4) are Hermitian and skew-Hermitian, respectively.

2.6. Given that m is a positive integer, the mth **power of a square matrix A** is defined as

$$\mathbf{A}^m = \underbrace{\mathbf{AA}\ldots\mathbf{A}}_{m \text{ times}}$$

Consider the lower-triangular and diagonal matrices, labeled **A** and **D**, respectively. Compute $\mathbf{A}^2$, $\mathbf{A}^3$, $\mathbf{D}^2$, and $\mathbf{D}^3$. What general conclusions can one draw from these results?

$$\mathbf{A} = \begin{bmatrix} 1 & 0 & 0 \\ -1 & 2 & 0 \\ 0 & 1 & -2 \end{bmatrix}, \quad \mathbf{D} = \begin{bmatrix} -1 & 0 & 0 \\ 0 & 2 & 0 \\ 0 & 0 & 3 \end{bmatrix}$$

2.2 DETERMINANT, INVERSE, AND RANK OF A MATRIX

DEFINITION 2.9. Given a square matrix $\mathbf{A} = [a_{ij}]_{n \times n}$, the **determinant** of **A** is a (real) scalar denoted by $\det(\mathbf{A})$ or $|\mathbf{A}|$. For the special case when $n = 1$, **A** is simply 1×1; i.e., $\mathbf{A} = [a_{11}]$ and $|\mathbf{A}| = a_{11}$. For $n \geq 2$, however, the determinant is defined as

$$|\mathbf{A}| = \sum_{k=1}^{n} a_{ik}(-1)^{i+k} M_{ik}, \quad i = 1, 2, \ldots, n \quad (2.5)$$

using the ith row or

$$|\mathbf{A}| = \sum_{k=1}^{n} a_{kj}(-1)^{k+j} M_{kj}, \quad j = 1, 2, \ldots, n \quad (2.6)$$

using the jth column.

In Eq. (2.5), M_{ik} is referred to as the **minor** of a_{ik} and is defined as the determinant of the $(n-1) \times (n-1)$ submatrix obtained by deleting the ith row and kth column of **A**. The quantity $(-1)^{i+k} M_{ik}$ is known as the **cofactor** of a_{ik} and is denoted by C_{ik}. It is apparent by Eqs. (2.5) and (2.6) that the determinant of a (square) matrix may be calculated using any row or column of the matrix. The row (or column) containing the most zeros is usually selected for the computation of the determinant. A square matrix with a determinant of zero is referred to as a **singular matrix**. Otherwise, it is **nonsingular**.

EXAMPLE 2.7. Find the determinant of

$$A = \begin{bmatrix} 1 & 2 & -3 \\ 4 & -1 & 1 \\ 2 & 0 & 1 \end{bmatrix}$$

Solution. The determinant can be computed using the third row because it contains a zero, and hence fewer calculations will be involved. Following the operations in Eq. (2.5):

$$|A| = 2(-1)^{3+1} M_{31} + 0 \cdot M_{32} + 1(-1)^{3+3} M_{33}$$

$$= 2 \begin{vmatrix} 2 & -3 \\ -1 & 1 \end{vmatrix} + \begin{vmatrix} 1 & 2 \\ 4 & -1 \end{vmatrix} = 2(2-3) + (-1-8) = -11$$

indicating that **A** is nonsingular. The same result could have been obtained using the second column, which also contains a zero.

Properties of Determinants

(D_1) A square matrix and its transpose have the same determinant; i.e., $|A| = |A^T|$.

(D_2) The determinant of a diagonal, an upper-triangular, and a lower-triangular matrix is the product of the entries along the main diagonal.

(D_3) If an entire row (or column) of a matrix is zero, then $|A| = 0$.

(D_4) If a row (or column) of **A** is multiplied by a scalar k, the determinant of the resulting matrix is $k|A|$. If **A** is an $n \times n$ matrix and k is a scalar, then $|kA| = k^n|A|$.

(D_5) If any two rows (or columns) of **A** are interchanged, the determinant of the resulting matrix is $-|A|$.

(D_6) $|A||B| = |AB|$

(D_7) If any two rows (or columns) of **A** are linearly dependent, then $|A| = 0$.

(D_8) If the ith row of **A** is multiplied by a scalar k and then added to the jth row, and the jth row is replaced by the result, then the determinant of the new matrix is equal to the determinant of **A**.

EXAMPLE 2.8. Investigate some of the properties listed above for the following square matrix:

$$A = \begin{bmatrix} 1 & 2 & -3 \\ 4 & -1 & 1 \\ 2 & 0 & 1 \end{bmatrix}$$

Solution. The determinant of **A** was calculated in Example 2.7 and shown to be -11. Suppose the second row of **A** is multiplied by -3 to generate

$$\mathbf{B} = \begin{bmatrix} 1 & 2 & -3 \\ -12 & 3 & -3 \\ 2 & 0 & 1 \end{bmatrix}$$

Then, using the third row together with Eq. (2.5), the determinant of $\mathbf{B}$ is computed as

$$|\mathbf{B}| = 2\begin{vmatrix} 2 & -3 \\ 3 & -3 \end{vmatrix} + \begin{vmatrix} 1 & 2 \\ -12 & 3 \end{vmatrix} = 2(3) + 27 = 33 = -3|\mathbf{A}|$$

which is in agreement with property (D_4) above. Moreover, if the first and third rows of $\mathbf{A}$ are interchanged to create a new matrix:

$$\mathbf{C} = \begin{bmatrix} 2 & 0 & 1 \\ 4 & -1 & 1 \\ 1 & 2 & -3 \end{bmatrix}$$

then $|\mathbf{C}| = 11 = -|\mathbf{A}|$, as suggested by property (D_5).

EXAMPLE 2.9. Investigate linear dependence of the rows (or columns) of the following matrix:

$$\mathbf{A} = \begin{bmatrix} 1 & -1 & 2 \\ -2 & 4 & 1 \\ -5 & 13 & 10 \end{bmatrix}$$

Solution. Linear dependence of the rows (or columns) of a matrix may be investigated by directly examining the determinant of the matrix. In the event that the determinant is zero, the rows (or columns) must be linearly dependent, by property (D_7). For the matrix under consideration, using the first row, we have

$$|\mathbf{A}| = \begin{vmatrix} 4 & 1 \\ 13 & 10 \end{vmatrix} + \begin{vmatrix} -2 & 1 \\ -5 & 10 \end{vmatrix} + 2\begin{vmatrix} -2 & 4 \\ -5 & 13 \end{vmatrix} = 0$$

which implies that the rows (or columns) of $\mathbf{A}$ are dependent. A closer inspection of the rows of $\mathbf{A}$ reveals that

third row = 3[first row] + 4[second row]

indicating that the rows are indeed dependent (Definition 2.3).

DEFINITION 2.10. In the event that a square matrix $\mathbf{A}$ is nonsingular, it is referred to as **invertible**. Then a unique matrix exists, denoted by $\mathbf{A}^{-1}$ and known as the **inverse** of $\mathbf{A}$ with the property that $\mathbf{A}\mathbf{A}^{-1} = \mathbf{I} = \mathbf{A}^{-1}\mathbf{A}$.

Determination of Inverse—Adjoint Matrix

Given that a square matrix $\mathbf{A}$ is invertible, its inverse is determined by what is known as the *adjoint matrix* of $\mathbf{A}$, denoted by adj($\mathbf{A}$). Assuming that $\mathbf{A} = [a_{ij}]$ is $n \times n$, its

adjoint is defined as

$$\text{adj}(\mathbf{A}) = \begin{bmatrix} (-1)^{1+1}M_{11} & (-1)^{2+1}M_{21} & \cdots & (-1)^{n+1}M_{n1} \\ (-1)^{1+2}M_{12} & (-1)^{2+2}M_{22} & \cdots & (-1)^{n+2}M_{n2} \\ \vdots & & & \vdots \\ (-1)^{1+n}M_{1n} & (-1)^{2+n}M_{2n} & \cdots & (-1)^{n+n}M_{nn} \end{bmatrix}$$

$$= \begin{bmatrix} C_{11} & C_{21} & \cdots & C_{n1} \\ C_{12} & C_{22} & \cdots & C_{n2} \\ \vdots & & & \vdots \\ C_{1n} & C_{2n} & \cdots & C_{nn} \end{bmatrix}$$

(2.7)

in which M_{ij} and C_{ij} denote the minor and cofactor of a_{ij}, respectively, as previously defined. Note that each minor M_{ij} occupies the (j, i) slot in the adjoint matrix. Then the inverse matrix is simply defined by

$$\mathbf{A}^{-1} = \frac{1}{|\mathbf{A}|}\text{adj}(\mathbf{A}) \qquad (2.8)$$

It is then clear from Eq. (2.8) that if $\mathbf{A}$ is singular (i.e., $|\mathbf{A}| = 0$), its inverse does not exist. As a special case, when $\mathbf{A}$ is diagonal its inverse is a diagonal matrix whose diagonal entries are the reciprocals of the diagonal entries of $\mathbf{A}$.

EXAMPLE 2.10. Determine the inverse of the following matrix via the adjoint matrix:

$$\mathbf{A} = \begin{bmatrix} -1 & 1 & 2 \\ 3 & -1 & 1 \\ -1 & 3 & 4 \end{bmatrix}$$

Solution. First calculate the determinant using the first row via Eq. (2.5) as follows:

$$|\mathbf{A}| = -\begin{vmatrix} -1 & 1 \\ 3 & 4 \end{vmatrix} - \begin{vmatrix} 3 & 1 \\ -1 & 4 \end{vmatrix} + 2\begin{vmatrix} 3 & -1 \\ -1 & 3 \end{vmatrix} = 10$$

Next identify the minor corresponding to each entry to form the adjoint matrix, given by Eq. (2.7), as

$$\text{adj}(\mathbf{A}) = \begin{bmatrix} -7 & 2 & 3 \\ -13 & -2 & 7 \\ 8 & 2 & -2 \end{bmatrix}$$

Ultimately, making use of Eq. (2.8), one obtains

$$\mathbf{A}^{-1} = \frac{1}{10}\text{adj}(\mathbf{A}) = \begin{bmatrix} -0.7 & 0.2 & 0.3 \\ -1.3 & -0.2 & 0.7 \\ 0.8 & 0.2 & -0.2 \end{bmatrix}$$

DEFINITION 2.11. The **rank** of an $n \times n$ matrix $\mathbf{A}$, denoted by $rk(\mathbf{A})$, is the maximum number of linearly independent rows (or columns) of $\mathbf{A}$. Alternatively, the rank of $\mathbf{A}$ is defined as the size of the largest submatrix of $\mathbf{A}$ with a nonzero determinant. That is, $rk(\mathbf{A}) = k$ if there exists a $k \times k$ submatrix of $\mathbf{A}$ that has a nonzero determinant and if any other $m \times m$ submatrix, with $m > k$, has a zero determinant. A square matrix $\mathbf{A}_{n \times n}$ is said to have a **full rank** if $rk(\mathbf{A}) = n$.

EXAMPLE 2.11. Determine the rank of the following square matrix:

$$\mathbf{A} = \begin{bmatrix} 1 & -1 & 2 \\ -2 & 4 & 1 \\ -5 & 13 & 10 \end{bmatrix}$$

Solution. This is the matrix discussed in Example 2.9 and shown to have a zero determinant. Thus, its rank must be 2 or less. The 2×2 submatrix obtained by deleting the third row and third column has a determinant of $2 \neq 0$; hence, $rk(\mathbf{A}) = 2$.

Linear Systems of Algebraic Equations—Cramer's Rule

Consider a system of n linear algebraic equations in n unknowns in the general form

$$\begin{cases} a_{11}x_1 + a_{12}x_2 + \cdots + a_{1n}x_n = b_1 \\ a_{21}x_1 + a_{22}x_2 + \cdots + a_{2n}x_n = b_2 \\ \quad \vdots \\ a_{n1}x_1 + a_{n2}x_2 + \cdots + a_{nn}x_n = b_n \end{cases} \qquad (2.9)$$

where a_{ij} $(1 \leq i, j \leq n)$ are constant coefficients, x_i $(1 \leq i \leq n)$ are the unknown quantities, and b_i $(1 \leq i \leq n)$ represent the known quantities on the right-hand sides. The ith unknown, x_i, is given by

$$x_i = \frac{\Delta_i}{\Delta}, \qquad i = 1, 2, \ldots, n \qquad (2.10)$$

where Δ represents the determinant of the coefficient matrix in Eq. (2.9), and Δ_i is obtained by replacing the ith column in Δ by the vector of the right-hand side quantities; i.e.,

$$\Delta = \begin{vmatrix} a_{11} & a_{12} & \cdots & a_{1n} \\ a_{21} & a_{22} & \cdots & a_{2n} \\ \vdots & & & \vdots \\ a_{n1} & a_{n2} & \cdots & a_{nn} \end{vmatrix}, \qquad \Delta_i = \begin{vmatrix} a_{11} & a_{12} & \cdots & b_1 & \cdots & a_{1n} \\ a_{21} & a_{22} & \cdots & b_2 & \cdots & a_{2n} \\ \vdots & & & \vdots & & \vdots \\ a_{n1} & a_{n2} & \cdots & b_n & \cdots & a_{nn} \end{vmatrix}$$

(ith column of Δ)

EXAMPLE 2.12. Solve the following system of three equations in three unknowns:

$$\begin{cases} 2x_1 + 3x_2 - x_3 = 1 \\ -x_1 + 2x_2 + x_3 = 8 \\ x_1 - 3x_2 - 2x_3 = -13 \end{cases}$$

Solution. Following the general formulation in Eq. (2.10) and the information on the determinant of a matrix we proceed as

$$\Delta = \begin{vmatrix} 2 & 3 & -1 \\ -1 & 2 & 1 \\ 1 & -3 & -2 \end{vmatrix} = -6, \quad \Delta_1 = \begin{vmatrix} 1 & 3 & -1 \\ 8 & 2 & 1 \\ -13 & -3 & -2 \end{vmatrix} = 6,$$

$$\Delta_2 = \begin{vmatrix} 2 & 1 & -1 \\ -1 & 8 & 1 \\ 1 & -13 & -2 \end{vmatrix} = -12, \quad \Delta_3 = \begin{vmatrix} 2 & 3 & 1 \\ -1 & 2 & 8 \\ 1 & -3 & -13 \end{vmatrix} = -18$$

Subsequently, the three unknown quantities are determined as

$$x_1 = \frac{\Delta_1}{\Delta} = \frac{6}{-6} = -1, \quad x_2 = \frac{\Delta_2}{\Delta} = \frac{-12}{-6} = 2, \quad x_3 = \frac{\Delta_3}{\Delta} = \frac{-18}{-6} = 3$$

Special Case: In the event that the coefficient matrix is nonsingular, a unique set of solutions exists and are determined by applying Cramer's rule. However, if the coefficient matrix has a zero determinant, then Eq. (2.10) involves division by zero and hence is not valid. In this case, either the system has no solution or infinitely many solutions, and the decision is made based on the nature of the equations.

EXAMPLE 2.13. Solve the following system of three equations in three unknowns:

$$\begin{cases} x_1 + x_2 - x_3 = 2 \\ 2x_1 + 2x_2 - 3x_3 = -4 \\ -3x_1 - 3x_2 + 4x_3 = 2 \end{cases}$$

Solution. The determinant of the coefficient matrix is first calculated as

$$\Delta = \begin{vmatrix} 1 & 1 & -1 \\ 2 & 2 & -3 \\ -3 & -3 & 4 \end{vmatrix} = 0$$

Direct examination of the original equations reveals that there are infinitely many solutions, defined by

$$x_3 = 8, \quad x_1 + x_2 = 10$$

PROBLEM SET 2.2

Find the determinant of each given square matrix.

2.7. $A = \begin{bmatrix} 1 & -1 & 0 \\ 2 & 0 & 3 \\ -1 & -1 & -3 \end{bmatrix}$

2.8. $A = \begin{bmatrix} 1 & 0 & 0 \\ -3 & 8 & 0 \\ 5 & -4 & 10 \end{bmatrix}$

2.9. $A = \begin{bmatrix} 2 & 6 & 4 \\ 1 & 3 & 0 \\ -1 & 0 & 2 \end{bmatrix}$

2.10. Consider the matrix of Problem 2.9. Determine $|2A|$ without direct computation.

2.11. Prove that any square matrix and its transpose have the same determinant.

Find the inverse of each given square matrix.

2.12. $A = \begin{bmatrix} 2 & 1 \\ -3 & -1 \end{bmatrix}$

2.13. $A = \begin{bmatrix} 1 & 2 & 3 \\ 0 & -1 & -2 \\ 0 & 0 & -1 \end{bmatrix}$

2.14. $A = \begin{bmatrix} s & -1 & 0 \\ 0 & s & -1 \\ 1 & 2 & s+3 \end{bmatrix}$

2.15. Prove that an $n \times n$ matrix with a rank of $k < n$ is singular and hence has linearly dependent rows.

2.16. Prove that any square matrix and its transpose have the same rank.

Solve the following linear systems via Cramer's rule.

2.17. $\begin{cases} 2x_1 - x_2 + 3x_3 = 7 \\ x_1 + 2x_2 - x_3 = -4 \\ -x_1 + 2x_2 + x_3 = 0 \end{cases}$

2.18. $\begin{cases} x_1 - x_3 = 3 \\ 2x_1 + x_2 - x_3 = 7 \\ 3x_2 + x_3 = 7 \end{cases}$

2.3 EIGENVALUES AND EIGENVECTORS

Let $\mathbf{A}$ be an $n \times n$ matrix, $\mathbf{v}$ a nonzero $n \times 1$ vector, and λ a scalar (complex in general). Consider

$$\mathbf{A}\mathbf{v} = \lambda\mathbf{v} \tag{2.11}$$

DEFINITION 2.12. A number λ for which Eq. (2.11) has a nontrivial solution ($\mathbf{v} \neq \mathbf{0}_{n\times 1}$) is known as an **eigenvalue** (or **characteristic value**) of matrix $\mathbf{A}$. The corresponding solution ($\mathbf{v} \neq \mathbf{0}_{n\times 1}$) of Eq. (2.11) is referred to as the **eigenvector** of $\mathbf{A}$ corresponding to λ. The problem of determining eigenvalues and the corresponding eigenvectors of a matrix is called an **eigenvalue problem**.

Eigenvalue problems play a significant role in systems analysis, particularly in the determination of responses of dynamic systems.

Solving the Eigenvalue Problem

Consider the eigenvalue problem described by Eq. (2.11) and rewrite it as

$$\mathbf{A}\mathbf{v} = \lambda\mathbf{v} \implies (\mathbf{A} - \lambda\mathbf{I}_n)\mathbf{v} = \mathbf{0}_{n\times 1} \tag{2.12}$$

This equation has a nontrivial solution, $\mathbf{v} \neq \mathbf{0}$, if and only if the coefficient matrix, $\mathbf{A} - \lambda\mathbf{I}$, is singular; i.e.,

$$\text{Characteristic equation} \quad |\mathbf{A} - \lambda\mathbf{I}| = 0 \tag{2.13}$$

This equation is referred to as the **characteristic equation** of $\mathbf{A}$. The determinant $|\mathbf{A} - \lambda\mathbf{I}|$ is an nth-degree polynomial in λ and is known as the **characteristic polynomial** of $\mathbf{A}$ whose roots are precisely the eigenvalues of $\mathbf{A}$. Once the eigenvalues have been identified, each eigenvector corresponding to each of the eigenvalues can be obtained by solving the matrix equation

$$(\mathbf{A} - \lambda\mathbf{I}_n)\mathbf{v} = \mathbf{0}_{n\times 1} \tag{2.14}$$

THEOREM 2.1. Eigenvalues of diagonal, upper-triangular, and lower-triangular matrices are simply the elements along the main diagonal of the matrix.

THEOREM 2.2. Let $\mathbf{A}$ be an $n \times n$ matrix. Then $\mathbf{A}$ and $\mathbf{A}^T$ have the same eigenvalues.

The procedure to obtain the eigenvalues of a square matrix and their corresponding eigenvectors is illustrated by the examples that follow.

EXAMPLE 2.14. Consider the upper-triangular matrix

$$\mathbf{A} = \begin{bmatrix} 1 & 4 & -2 \\ 0 & -3 & -1 \\ 0 & 0 & 2 \end{bmatrix}$$

90 Dynamic Systems: Modeling and Analysis

Its eigenvalues are solutions of the corresponding characteristic equation, namely:

$$|\mathbf{A} - \lambda \mathbf{I}| = 0 \implies \begin{vmatrix} 1-\lambda & 4 & -2 \\ 0 & -3-\lambda & -1 \\ 0 & 0 & 2-\lambda \end{vmatrix} = (1-\lambda)(-3-\lambda)(2-\lambda) = 0$$

$$\implies \lambda_{1,2,3} = 1, -3, 2$$

which are indeed the entries along the main diagonal and are in agreement with the statement of Theorem 2.1. Furthermore, we note that $\mathbf{A}^T$ is lower-triangular and by inspection has the same eigenvalues as $\mathbf{A}$, which agrees with the result of Theorem 2.2.

EXAMPLE 2.15. Determine the eigenvalues and the corresponding eigenvectors of the 2×2 matrix given below.

$$\mathbf{A} = \begin{bmatrix} -1 & -3 \\ 0 & 2 \end{bmatrix}$$

Solution. To find the eigenvalues of $\mathbf{A}$, solve the characteristic equation given by Eq. (2.13),

$$|\mathbf{A} - \lambda \mathbf{I}| = 0 \implies \begin{vmatrix} -1-\lambda & -3 \\ 0 & 2-\lambda \end{vmatrix} = 0 \implies (\lambda+1)(\lambda-2) = 0$$

$$\implies \lambda_{1,2} = -1, 2$$

First, to obtain the eigenvector corresponding to $\lambda_1 = -1$, solve Eq. (2.14):

$$(\mathbf{A} - \lambda_1 \mathbf{I})\mathbf{v}_1 = \mathbf{0} \implies (\mathbf{A} + \mathbf{I})\mathbf{v}_1 = \mathbf{0} \tag{2.15}$$

where $\mathbf{v}_1$ is the 2×1 eigenvector corresponding to $\lambda_1 = -1$, expressed as

$$\mathbf{v}_1 = \begin{bmatrix} v_{11} \\ v_{21} \end{bmatrix}$$

Rewrite Eq. (2.15) as

$$\begin{bmatrix} 0 & -3 \\ 0 & 3 \end{bmatrix} \begin{bmatrix} v_{11} \\ v_{21} \end{bmatrix} = \begin{bmatrix} 0 \\ 0 \end{bmatrix} \tag{2.16}$$

Eq. (2.16) clearly suggests that the two rows of the matrix $\mathbf{A} + \mathbf{I}$ are linearly dependent and that there is a **free variable**. Interpret the first row of Eq. (12.16) as

$$-3v_{21} = 0 \implies v_{21} = 0$$

Let v_{11} be the free variable and hence arbitrary. As a result, eigenvector $\mathbf{v}_1$ is given by

$$\mathbf{v}_1 = \begin{bmatrix} v_{11} \\ 0 \end{bmatrix}$$

Because v_{11} is arbitrary, assign to it a value of 1. Thus

$$\mathbf{v}_1 = \begin{bmatrix} 1 \\ 0 \end{bmatrix}$$

Similarly, the eigenvector corresponding to $\lambda_1 = 2$ may be denoted by

$$\mathbf{v}_2 = \begin{bmatrix} v_{12} \\ v_{22} \end{bmatrix}$$

and obtained as

$$(\mathbf{A} - \lambda_2 \mathbf{I})\mathbf{v}_2 = \mathbf{0} \implies (\mathbf{A} - 2\mathbf{I})\mathbf{v}_2 = \mathbf{0} \implies \begin{bmatrix} -3 & -3 \\ 0 & 0 \end{bmatrix}\begin{bmatrix} v_{12} \\ v_{22} \end{bmatrix} = \begin{bmatrix} 0 \\ 0 \end{bmatrix}$$

$$\implies v_{12} = -v_{22}$$

Let v_{22} be the free variable, and assign a value of 1 to it to obtain the eigenvector corresponding to λ_2, as

$$\mathbf{v}_2 = \begin{bmatrix} -1 \\ 1 \end{bmatrix}$$

The set

$$\left\{ \begin{bmatrix} 1 \\ 0 \end{bmatrix}, \begin{bmatrix} -1 \\ 1 \end{bmatrix} \right\}$$

of these two vectors is referred to as the **basis** of all eigenvectors of matrix **A**.

Whereas 2×2 systems such as the ones studied in the last two examples are relatively easy to work with, the same is generally not true for larger systems. This becomes particularly evident when solving Eq. (2.14) to determine the eigenvectors. To circumvent this problem we introduce what are known as **elementary row operations.** These operations transform a generally full matrix into one that is much more convenient to work with, such as diagonal, upper-triangular, or lower-triangular. If a system undergoes a finite number of elementary row operations to generate a new system, the nature of the original system is preserved and the two systems are said to be **row-equivalent.** There are three types of elementary row operations:

(ERO$_1$)   Multiply a row by a nonzero scalar.

(ERO$_2$)   Interchange any two rows.

(ERO$_3$)   Multiply a row by a nonzero scalar and add to another row, and replace the latter by the result.

EXAMPLE 2.16. Find the eigenvalues and the corresponding eigenvectors of

$$\mathbf{A} = \begin{bmatrix} -2 & -6 & 84 \\ 1 & 3 & -32 \\ 0 & 0 & 4 \end{bmatrix}$$

Solution. The characteristic equation is solved to determine the eigenvalues, as follows:

$$|A - \lambda I| = 0 \implies \begin{vmatrix} -2-\lambda & -6 & 84 \\ 1 & 3-\lambda & -32 \\ 0 & 0 & 4-\lambda \end{vmatrix} = 0 \implies \lambda(\lambda^2 - 5\lambda + 4) = 0$$

$$\implies \lambda(\lambda - 1)(\lambda - 4) = 0 \implies \lambda_{1,2,3} = 0, 1, 4$$

Let $\lambda_1 = 0$ with corresponding eigenvector, v_1. Solve $(A - \lambda_1 I)v_1 = 0 \implies Av_1 = 0$ to obtain

$$\begin{bmatrix} -2 & -6 & 84 \\ 1 & 3 & -32 \\ 0 & 0 & 4 \end{bmatrix} \begin{bmatrix} v_{11} \\ v_{21} \\ v_{31} \end{bmatrix} = \begin{bmatrix} 0 \\ 0 \\ 0 \end{bmatrix} \xRightarrow{\text{divide 1st row by } -2} \begin{bmatrix} 1 & 3 & -42 \\ 1 & 3 & -32 \\ 0 & 0 & 1 \end{bmatrix} \begin{bmatrix} v_{11} \\ v_{21} \\ v_{31} \end{bmatrix}$$

$$= \begin{bmatrix} 0 \\ 0 \\ 0 \end{bmatrix} \xRightarrow{\text{EROs}} \begin{bmatrix} 1 & 3 & -42 \\ 0 & 0 & 1 \\ 0 & 0 & 0 \end{bmatrix} \begin{bmatrix} v_{11} \\ v_{21} \\ v_{31} \end{bmatrix} = \begin{bmatrix} 0 \\ 0 \\ 0 \end{bmatrix}$$

The second row suggests that $v_{31} = 0$, and the third row indicates a free variable, say v_{21}. The first row reads

$$v_{11} = -3v_{21} + 42\underset{\underset{0}{\downarrow}}{v_{31}} = -3v_{21}$$

Assigning a value of 1 to v_{21}, one obtains

$$v_1 = \begin{bmatrix} -3 \\ 1 \\ 0 \end{bmatrix}$$

Similarly, for $\lambda_2 = 1$ the eigenvector is determined by solving $(A - \lambda_2 I)v_2 = 0$ $\implies (A - I)v_2 = 0$ as

$$\begin{bmatrix} -3 & -6 & 84 \\ 1 & 2 & -32 \\ 0 & 0 & 3 \end{bmatrix} \begin{bmatrix} v_{21} \\ v_{22} \\ v_{32} \end{bmatrix} = \begin{bmatrix} 0 \\ 0 \\ 0 \end{bmatrix} = \xRightarrow[\text{subtract 2nd from 1st}]{\text{divide 1st row by } -3} \begin{bmatrix} 1 & 2 & -28 \\ 0 & 0 & -4 \\ 0 & 0 & 3 \end{bmatrix} \begin{bmatrix} v_{12} \\ v_{22} \\ v_{32} \end{bmatrix} = \begin{bmatrix} 0 \\ 0 \\ 0 \end{bmatrix}$$

$$\xRightarrow[\text{subtract 3rd from 2nd}]{\substack{\text{divide 2nd by } -4 \\ \text{and 3rd by 3}}} \begin{bmatrix} 1 & 2 & -28 \\ 0 & 0 & 1 \\ 0 & 0 & 0 \end{bmatrix} \begin{bmatrix} v_{12} \\ v_{22} \\ v_{32} \end{bmatrix} = \begin{bmatrix} 0 \\ 0 \\ 0 \end{bmatrix}$$

The second row implies that $v_{32} = 0$. Letting v_{22} be the free variable (with a value of 1), we have

$$v_2 = \begin{bmatrix} -2 \\ 1 \\ 0 \end{bmatrix}$$

Finally, for $\lambda_3 = 4$, solving $(\mathbf{A} - \lambda_3\mathbf{I})\mathbf{v}_3 = \mathbf{0}$ $\implies$ $(\mathbf{A} - 4\mathbf{I})\mathbf{v}_3 = \mathbf{0}$ yields

$$\begin{bmatrix} -6 & -6 & 84 \\ 1 & -1 & -32 \\ 0 & 0 & 0 \end{bmatrix} \begin{bmatrix} v_{13} \\ v_{23} \\ v_{33} \end{bmatrix} = \begin{bmatrix} 0 \\ 0 \\ 0 \end{bmatrix} \xRightarrow{\text{elementary row operations}} \begin{bmatrix} 1 & 1 & -14 \\ 0 & 1 & 9 \\ 0 & 0 & 0 \end{bmatrix} \begin{bmatrix} v_{13} \\ v_{23} \\ v_{33} \end{bmatrix}$$

$$= \begin{bmatrix} 0 \\ 0 \\ 0 \end{bmatrix} \implies \mathbf{v}_3 = \begin{bmatrix} 23 \\ -9 \\ 1 \end{bmatrix}$$

EXAMPLE 2.17. Determine eigenvalues and eigenvectors of the following matrix:

$$\mathbf{A} = \begin{bmatrix} 2 & 0 & -115 \\ 0 & 2 & 45 \\ 0 & 0 & -3 \end{bmatrix}$$

Solution. **A** is upper-triangular, so by Theorem 2.1 its eigenvalues are simply the entries along the diagonal; i.e., $\lambda_{1,2,3} = 2, 2, -3$. To determine the eigenvector corresponding to $\lambda_1 = 2$, solve $(\mathbf{A} - 2\mathbf{I})\mathbf{v}_1 = \mathbf{0}$, as

$$\begin{bmatrix} 0 & 0 & -115 \\ 0 & 0 & 45 \\ 0 & 0 & -5 \end{bmatrix} \mathbf{v}_1 = \mathbf{0} \xRightarrow{\text{elementary row operations}} \begin{bmatrix} 0 & 0 & 1 \\ 0 & 0 & 0 \\ 0 & 0 & 0 \end{bmatrix} \begin{bmatrix} v_{11} \\ v_{21} \\ v_{31} \end{bmatrix} = \begin{bmatrix} 0 \\ 0 \\ 0 \end{bmatrix}$$

From the structure of the first row, we conclude that $v_{31} = 0$. However, unlike previous examples there exist two zero rows and hence two free variables. Clearly, these are v_{11} and v_{21}. As a result, two independent eigenvectors corresponding to $\lambda_1 = 2$ can be obtained as follows:

Set $v_{11} = 1$ and $v_{21} = 0$:

$$\mathbf{v}_1 = \begin{bmatrix} 1 \\ 0 \\ 0 \end{bmatrix}$$

Set $v_{11} = 0$ and $v_{21} = 1$:

$$\mathbf{v}_2 = \begin{bmatrix} 0 \\ 1 \\ 0 \end{bmatrix}$$

The eigenvector corresponding to $\lambda_3 = -3$ is the solution of $(\mathbf{A} + 3\mathbf{I})\mathbf{v}_3 = \mathbf{0}$ and can be shown to be

$$\mathbf{v}_3 = \begin{bmatrix} 23 \\ -9 \\ 1 \end{bmatrix}$$

Remark: The square matrix in Example 2.17 possessed an eigenvalue with a multiplicity of 2 and corresponded to exactly 2 linearly independent eigenvectors. This, however, does not always occur. In the event that an eigenvalue of multiplicity $m > 1$ corresponds to k linearly independent eigenvectors where $k < m$, then there exist exactly $m - k$ corresponding **generalized eigenvectors.**

Special Matrices

There are special matrices that possess special eigenvalue properties associated with them. We first define such matrices; the above-mentioned properties will follow subsequently.

DEFINITION 2.13. A square matrix $\mathbf{A}$ (complex in general) is said to be **unitary** if the following holds:

$$\mathbf{A}^T = \overline{\mathbf{A}}^{-1} \quad \text{or} \quad \overline{\mathbf{A}}^T = \mathbf{A}^{-1} \qquad (2.17)$$

A *real* unitary matrix is called an **orthogonal** matrix, and from Eq. (2.17), it is defined by $\mathbf{A}^T = \mathbf{A}^{-1}$.

THEOREM 2.3

i. All eigenvalues of a *unitary* matrix have absolute values of 1.
ii. All eigenvalues of a *Hermitian* matrix are real.
iii. All eigenvalues of a *skew-Hermitian* matrix are zero or pure imaginary.

In the case of real matrices, the three results stated in Theorem 2.3 will be valid for orthogonal, symmetric, and skew-symmetric matrices, respectively.

EXAMPLE 2.18. Consider the skew-Hermitian matrix of Example 2.5. A direct calculation of its eigenvalues reveals that they are all pure imaginary. More specifically,

$$-0.5916j, \quad 2.8071j, \quad -4.2155j$$

This is in complete agreement with Theorem 2.3(*iii*).

EXAMPLE 2.19. The 2×2 matrix

$$\mathbf{U} = \frac{1}{\sqrt{2}} \begin{bmatrix} 1 & j \\ -j & -1 \end{bmatrix}$$

is unitary because

$$\overline{\mathbf{U}}^{-1} = \frac{1}{\sqrt{2}} \begin{bmatrix} 1 & -j \\ j & -1 \end{bmatrix} = \mathbf{U}^T$$

Eigenvalue calculation results in

$$|\mathbf{U} - \lambda \mathbf{I}| = \left(\tfrac{1}{2} - \lambda^2\right) - \tfrac{1}{2} = 0 \implies \lambda^2 - 1 = 0 \implies \lambda_{1,2} = \pm 1$$

which was expected from Theorem 2.3(*i*).

PROBLEM SET 2.3

Determine the eigenvalues and corresponding eigenvectors of each given matrix.

2.19. $A = \begin{bmatrix} 1 & 0 & 0 \\ 0 & 1 & 0 \\ -3 & 4 & 0 \end{bmatrix}$

2.20. $A = \begin{bmatrix} 0 & 1 & 0 \\ 1 & 0 & 0 \\ 0 & 0 & 1 \end{bmatrix}$

2.21. $A = \begin{bmatrix} 16 & 9 \\ -30 & -17 \end{bmatrix}$

Identify the special type of each given matrix and verify Theorem 2.3 on the nature of its eigenvalues.

2.22. $A = \begin{bmatrix} (1+j)/2 & (-1+j)/2 \\ (1+j)/2 & (1-j)/2 \end{bmatrix}$

2.23. $A = \begin{bmatrix} 1 & 2+j \\ 2-j & 3 \end{bmatrix}$

2.4 MATRIX TRANSFORMATION AND DIAGONALIZATION

Often, while treating linear systems, the coefficient matrix **A** is a *full* matrix in the sense that it does not occur in any special form such as diagonal or triangular. In this event, it is desirable to transform matrix **A** into one that is more convenient and easier to work with. Performing such a transformation depends solely on the nature of the eigenvalues and eigenvectors of matrix **A**. This section is concerned with a special type of matrix transformation that, under specific conditions, transforms a full matrix into one that is diagonal. The advantages, of course, are enormous, as will be demonstrated here and in Chapter 7. First, we introduce what is known as a **similarity transformation,** under which matrix eigenvalues are preserved.

DEFINITION 2.14. Two $n \times n$ matrices, **A** and **B**, are said to be **similar** if there exists a nonsingular $n \times n$ matrix **S** such that $B = S^{-1}AS$ (or $A = SBS^{-1}$). This transformation is referred to as a **similarity transformation.**

THEOREM 2.4. Similar matrices have identical eigenvalues.

Proof. Let **A** and **B** be similar, and suppose λ is an eigenvalue of **B**. It then suffices to show λ is also an eigenvalue of **A**. For λ to be an eigenvalue of **B**, it must be a solution to the characteristic equation associated with **B**. Then we have

$$|B - \lambda I| = 0 \quad \overset{B = S^{-1}AS}{\Longrightarrow} \quad |S^{-1}AS - \lambda I| = 0$$

Noting that $S^{-1}S = I$, the above equation may be rewritten as

$$|S^{-1}(A - \lambda I)S| = 0 \quad \underset{\text{of determinants}}{\overset{\text{property } (D_6)}{\Longrightarrow}} \quad |S^{-1}||A - \lambda I||S| = 0$$

$$\overset{|S||S^{-1}|=1}{\Longrightarrow} \quad |A - \lambda I| = 0$$

The last equation indicates that λ satisfies the characteristic equation associated with A; hence it is an eigenvalue of A.

EXAMPLE 2.20. Show that the eigenvalues of A and $B = S^{-1}AS$ are the same, where

$$A = \begin{bmatrix} -1 & -3 \\ 0 & 2 \end{bmatrix} \quad \text{and} \quad S = \begin{bmatrix} 2 & -1 \\ -1 & 1 \end{bmatrix}$$

Solution. It is readily seen that

$$S^{-1} = \begin{bmatrix} 1 & 1 \\ 1 & 2 \end{bmatrix} \implies B = S^{-1}AS = \begin{bmatrix} -1 & 0 \\ -3 & 2 \end{bmatrix}$$

Matrix B is clearly the transpose of A and thus has the same eigenvalues as A (Theorem 2.2).

THEOREM 2.5. DIAGONALIZATION. Let A be an $n \times n$ matrix with n linearly independent eigenvectors, $v_1, v_2, \ldots, v_n$. Then there exists a nonsingular $n \times n$ matrix P such that

$$P^{-1}AP = \Lambda \tag{2.18}$$

where $P = [v_1 \quad v_2 \quad \ldots \quad v_n]$ is known as the **modal matrix** and is composed of the n eigenvectors of A. Furthermore, Λ is a diagonal matrix whose diagonal entries are the eigenvalues of A corresponding to $v_1, v_2, \ldots, v_n$ in that order; i.e.,

$$\Lambda = \text{diag}(\lambda_1, \ldots, \lambda_n)$$

In this event, matrix A is said to be **diagonalizable.**

Equation (2.18) represents a similarity transformation. That is, A and Λ are similar and hence by Theorem 2.4 have the same eigenvalues. However, because Λ is diagonal, by Theorem 2.1 its eigenvalues are its main diagonal elements, $\lambda_1, \lambda_2, \ldots, \lambda_n$, which are indeed the eigenvalues of A. Furthermore, the column vectors of the modal matrix P, which are the eigenvectors of A, are linearly independent by assumption. Therefore, by property (D_7) of determinants, P is nonsingular. Finally, Theorem 2.5 does not necessarily require that all eigenvalues of matrix A be distinct. It is, however, concerned with linear independence of all its eigenvectors. For instance, in Example 2.17 the 3×3 matrix had an eigenvalue of multiplicity 2, which corresponded to two linearly independent eigenvectors. Thus, although the three eigenvalues were shown not to be distinct, they corresponded to three linearly independent eigenvectors. In the event that all eigenvalues are distinct, their corresponding eigenvectors are linearly independent.

EXAMPLE 2.21. Investigate the diagonalizability of the following:

$$\mathbf{A} = \begin{bmatrix} 1 & 2 \\ 0 & 3 \end{bmatrix}$$

Solution. The eigenvalues of **A** can be shown to be $\lambda_1 = 1$ and $\lambda_2 = 3$ (distinct) with corresponding eigenvectors

$$\mathbf{v}_1 = \begin{bmatrix} 1 \\ 0 \end{bmatrix}, \quad \mathbf{v}_2 = \begin{bmatrix} 1 \\ 1 \end{bmatrix}$$

which are linearly independent, as expected. The modal matrix is then formed as

$$\mathbf{P} = [\mathbf{v}_1 \quad \mathbf{v}_2] = \begin{bmatrix} 1 & 1 \\ 0 & 1 \end{bmatrix}$$

and

$$\mathbf{P}^{-1}\mathbf{A}\mathbf{P} = \mathbf{\Lambda} = \begin{bmatrix} 1 & 0 \\ 0 & 3 \end{bmatrix}$$

EXAMPLE 2.22. Consider the matrix studied in Example 2.17:

$$\mathbf{A} = \begin{bmatrix} 2 & 0 & -115 \\ 0 & 2 & 45 \\ 0 & 0 & -3 \end{bmatrix}$$

Although there are only two distinct eigenvalues, we learned that there exist two linearly independent eigenvectors $\mathbf{v}_1$ and $\mathbf{v}_2$ corresponding to the repeated eigenvalue $\lambda = 2$. Then

$$\mathbf{P} = [\mathbf{v}_1 \quad \mathbf{v}_2 \quad \mathbf{v}_3] = \begin{bmatrix} 1 & 0 & 23 \\ 0 & 1 & -9 \\ 0 & 0 & 1 \end{bmatrix} \implies \mathbf{P}^{-1}\mathbf{A}\mathbf{P} = \begin{bmatrix} 2 & 0 & 0 \\ 0 & 2 & 0 \\ 0 & 0 & -3 \end{bmatrix} = \mathbf{\Lambda}$$

Notice the order in which the diagonal entries appear. Because $\mathbf{v}_1$ and $\mathbf{v}_2$ were picked as the first two columns of **P**, the corresponding eigenvalues occupy the respective positions in $\mathbf{\Lambda}$.

Exponential of a Matrix

The idea of the exponential of a square matrix plays a significant role in the treatment of systems of linear first-order differential equations. Direct involvement of the exponential of a matrix in treatment of such systems will be studied in detail in Chapter 7. In this section, we present the basic definition and fundamental properties of the exponential of a matrix, as well as its connection with the diagonalization procedure. To that end, our focus is on a class of $n \times n$ matrices that possess n linearly independent eigenvectors. Consequently, previous results of this section may be used.

Suppose **M** is an $n \times n$ matrix. The **exponential** of **M** is also $n \times n$ and is defined by

$$e^{\mathbf{M}} = \sum_{k=0}^{\infty} \frac{\mathbf{M}^k}{k!} = \mathbf{I}_n + \mathbf{M} + \frac{1}{2!}\mathbf{M}^2 + \cdots \tag{2.19}$$

and for a fixed, finite scalar t, can be extended to

$$e^{\mathbf{M}t} = \sum_{k=0}^{\infty} \frac{(\mathbf{M}t)^k}{k!} = \mathbf{I}_n + t\mathbf{M} + \frac{t^2}{2!}\mathbf{M}^2 + \cdots \tag{2.20}$$

The infinite series in Eq. (2.19) can be shown to be *absolutely convergent*. The same is true for Eq. (2.20) for finite values of t. This implies that, at least numerically, the series may be calculated to within reasonable accuracy by taking into account a sufficient number of terms. Also, manual calculation of the matrix exponential may be an overwhelming task unless **M** possesses some special properties—more specifically, diagonalizability. First, some fundamental properties associated with the matrix exponential of **M** are presented. These are crucial in solving a system of linear differential equations.

(E_1) Suppose **M** is an $n \times n$ matrix and t and τ are scalars. Then

$$e^{\mathbf{M}(t+\tau)} = e^{\mathbf{M}t}e^{\mathbf{M}\tau} \tag{2.21}$$

This identity can be verified as follows: Starting with the left-hand side and using the formulation of Eq. (2.20), we have

$$e^{\mathbf{M}(t+\tau)} = \sum_{k=0}^{\infty} \frac{1}{k!}\mathbf{M}^k(t+\tau)^k = \sum_{k=0}^{\infty} \frac{1}{k!}\mathbf{M}^k \sum_{m=0}^{\infty} \binom{k}{m} t^m \tau^{k-m}$$

$$= \sum_{k=0}^{\infty} \frac{1}{k!}\mathbf{M}^k \sum_{m=0}^{\infty} \frac{k!}{m!(k-m)!} t^m \tau^{k-m}$$

$$= \sum_{m=0}^{\infty}\sum_{k=m}^{\infty} \mathbf{M}^k \frac{t^m \tau^{k-m}}{m!(k-m)!} \underset{\text{let } k-m=r}{=} \sum_{m=0}^{\infty}\sum_{r=0}^{\infty} \mathbf{M}^{r+m} \frac{t^m \tau^r}{m!r!}$$

$$= \left(\sum_{m=0}^{\infty} \frac{1}{m!}\mathbf{M}^m t^m\right)\left(\sum_{r=0}^{\infty} \frac{1}{r!}\mathbf{M}^r \tau^r\right) = e^{\mathbf{M}t}e^{\mathbf{M}\tau}$$

where

$$(t+\tau)^k = \sum_{m=0}^{\infty} \binom{k}{m} t^m \tau^{k-m}$$

represents the **binomial series** and

$$\binom{k}{m} = \frac{k!}{m!(k-m)!}$$

is the mth **binomial coefficient**.

(E$_2$) Given that **M** and **P** are $n \times n$ matrices and t is a scalar, then

$$e^{(M+P)t} = e^{Mt}e^{Pt} \quad \text{if } \mathbf{MP} = \mathbf{PM}$$

$$e^{(M+P)t} \neq e^{Mt}e^{Pt} \quad \text{if } \mathbf{MP} \neq \mathbf{PM}$$

(E$_3$) Because the infinite series in Eq. (2.20) is absolutely convergent for finite values of t, term-by-term differentiation of the series is possible to give

$$\frac{d}{dt}e^{Mt} = \mathbf{M}e^{Mt} \tag{2.22}$$

In order to verify the identity, start with the left-hand side and use the expression of Eq. (2.20), as

$$\frac{d}{dt}e^{Mt} = \frac{d}{dt}\left[\mathbf{I}_n + t\mathbf{M} + \frac{t^2}{2!}\mathbf{M}^2 + \cdots\right] = \mathbf{M} + t\mathbf{M}^2 + \frac{1}{2!}t^2\mathbf{M}^3 + \cdots$$

$$= \mathbf{M}\left[\mathbf{I} + t\mathbf{M} + \frac{1}{2!}t^2\mathbf{M}^2 + \cdots\right] = \mathbf{M}e^{Mt}$$

Scalar t commutes with **M**, so this last result can also be shown to be equal to $e^{Mt}\mathbf{M}$.

Exponential of a Diagonal Matrix

If $\mathbf{\Lambda}$ is an $n \times n$ diagonal matrix with main diagonal entries denoted by λ_i ($1 \leq i \leq n$), the exponential of $\mathbf{\Lambda}$, denoted by $e^{\mathbf{\Lambda}}$, is an $n \times n$ diagonal matrix whose main diagonal entries are exponentials of λ_i ($1 \leq i \leq n$); i.e.,

$$\mathbf{\Lambda} = \begin{bmatrix} \lambda_1 & & & \\ & \lambda_2 & & \\ & & \ddots & \\ & & & \lambda_n \end{bmatrix} \xRightarrow{\text{exponential}}$$

$$e^{\mathbf{\Lambda}} = \begin{bmatrix} e^{\lambda_1} & & & \\ & e^{\lambda_2} & & \\ & & \ddots & \\ & & & e^{\lambda_n} \end{bmatrix} \xRightarrow{t \text{ is finite}} e^{\mathbf{\Lambda}t} = \begin{bmatrix} e^{\lambda_1 t} & & & \\ & e^{\lambda_2 t} & & \\ & & \ddots & \\ & & & e^{\lambda_n t} \end{bmatrix}$$

Exponential of a Diagonalizable Matrix

By Theorem 2.5, if an $n \times n$ matrix **A** has n linearly independent eigenvectors, then it is similar to a diagonal matrix $\mathbf{\Lambda}$ and is called diagonalizable. Because the exponential of a diagonal matrix is calculated quite easily, it would be ideal if $e^{\mathbf{A}}$ could be conveniently expressed in terms of e^{Λ}. First, the following general result on the exponentials of similar matrices is presented.

THEOREM 2.6. Suppose two square matrices $\mathbf{A}$ and $\mathbf{B}$ are similar; i.e., $\mathbf{S}^{-1}\mathbf{AS} = \mathbf{B}$. Then

$$e^{\mathbf{A}} = \mathbf{S}e^{\mathbf{B}}\mathbf{S}^{-1} \qquad (2.23)$$

Proof. Premultiplication of $\mathbf{S}^{-1}\mathbf{AS} = \mathbf{B}$ by $\mathbf{S}$ and postmultiplication by $\mathbf{S}^{-1}$ yields

$$\mathbf{A} = \mathbf{SBS}^{-1}$$

Equating the matrix exponentials of both sides and using Eq. (2.19), we have

$$e^{\mathbf{A}} = e^{\mathbf{SBS}^{-1}} = \mathbf{I} + \mathbf{SBS}^{-1} + \frac{1}{2!}(\mathbf{SBS}^{-1})^2 + \cdots$$

Note that $(\mathbf{SBS}^{-1})^2 = (\mathbf{SBS}^{-1})(\mathbf{SBS}^{-1}) = \mathbf{SB}^2\mathbf{S}^{-1}$, $(\mathbf{SBS}^{-1})^3 = \mathbf{SB}^3\mathbf{S}^{-1}$, and so on. Then

$$e^{\mathbf{A}} = \mathbf{I} + \mathbf{SBS}^{-1} + \frac{1}{2!}\mathbf{SB}^2\mathbf{S}^{-1} + \cdots = \mathbf{S}\left(\mathbf{I} + \mathbf{B} + \frac{1}{2!}\mathbf{B}^2 + \cdots\right)\mathbf{S}^{-1} = \mathbf{S}e^{\mathbf{B}}\mathbf{S}^{-1}$$

COROLLARY 2.1. Suppose $\mathbf{A}$ is $n \times n$ with n linearly independent eigenvectors. Then $\mathbf{P}^{-1}\mathbf{AP} = \Lambda$ where all matrices are as defined in Theorem 2.5, and

$$e^{\mathbf{A}} = \mathbf{P}e^{\Lambda}\mathbf{P}^{-1} \qquad (2.24)$$

which, for a scalar t, may be extended to

$$e^{\mathbf{A}t} = \mathbf{P}e^{\Lambda t}\mathbf{P}^{-1} \qquad (2.25)$$

EXAMPLE 2.23. Calculate $e^{\mathbf{A}t}$ where

$$\mathbf{A} = \begin{bmatrix} 1 & 2 \\ 0 & 3 \end{bmatrix}$$

and t is scalar.

Solution. This is the matrix considered in Example 2.21. Its eigenvalues are $\lambda_1 = 1$ and $\lambda_2 = 3$, and

$$\mathbf{P}^{-1}\mathbf{AP} = \Lambda = \begin{bmatrix} 1 & 0 \\ 0 & 3 \end{bmatrix} \quad \text{where} \quad \mathbf{P} = \begin{bmatrix} 1 & 1 \\ 0 & 1 \end{bmatrix}$$
$$\text{modal matrix}$$

Following the result of Corollary 2.1 in Eq. (2.24) and taking t into account, we have

$$e^{\mathbf{A}t} = \mathbf{P}e^{\Lambda t}\mathbf{P}^{-1} = \begin{bmatrix} 1 & 1 \\ 0 & 1 \end{bmatrix}\begin{bmatrix} e^t & 0 \\ 0 & e^{3t} \end{bmatrix}\begin{bmatrix} 1 & 1 \\ 0 & 1 \end{bmatrix}^{-1}$$

$$= \begin{bmatrix} 1 & 1 \\ 0 & 1 \end{bmatrix}\begin{bmatrix} e^t & -e^t \\ 0 & e^{3t} \end{bmatrix} = \begin{bmatrix} e^t & e^{3t} - e^t \\ 0 & e^{3t} \end{bmatrix}$$

PROBLEM SET 2.4

Find the modal matrix associated with each given matrix and diagonalize.

2.24. $A = \begin{bmatrix} -2 & 2 & 1 \\ 2 & 1 & 2 \\ 1 & 2 & 6 \end{bmatrix}$

2.25. $A = \begin{bmatrix} 1 & 0 & 0 \\ 0 & 1 & 0 \\ -3 & 4 & 0 \end{bmatrix}$

2.26. $A = \begin{bmatrix} 0 & 1 \\ 2 & 1 \end{bmatrix}$

2.27. Show that if M is $n \times n$ and t is scalar, then $e^{Mt}e^{-Mt} = I_n$.

2.28. Show that if M is $n \times n$ and t is scalar, then

$$\frac{d}{dt}e^{Mt} = e^{Mt}M$$

For each given square matrix, calculate e^{At} where t is scalar.

2.29. Matrix A of Problem 2.25

2.30. Matrix A of Problem 2.26

SUMMARY

An *nth-order vector*, v, is an ordered set of n scalars, $v_1, v_2, \ldots, v_n$, each known as a component of v. The *norm* (or *length*) of v is defined as

$$|v| = (v_1^2 + \cdots + v_n^2)^{1/2}$$

The *dot product* of two nth-order vectors, u and v, is defined as $u \cdot v = u_1v_1 + \cdots + u_nv_n$ where u_i and v_i are the components of u and v, respectively. Two vectors are orthogonal if their dot product is zero. A set of m vectors (n-dimensional each) is said to be *linearly independent* if none of them is expressible as a linear combination of the others, i.e.,

$$c_1w_1 + c_2w_2 + \cdots + c_mw_m = 0 \iff c_1 = c_2 = \cdots = c_m = 0$$

A collection of real or complex numbers, arranged in a rectangular array, is referred to as a *matrix*. The *size* of a matrix is determined by its number of rows and columns. If a matrix A has m rows and n columns, then it is of size $m \times n$. The $m \times n$ zero matrix, denoted by $0_{m \times n}$, is defined as an $m \times n$ matrix all of whose entries are zero. A is said to be a *square matrix* if it has the same number of rows as columns. The $n \times n$ identity matrix, denoted by I_n or I, consists of ones along the main diagonal and zeros everywhere else. The *trace* of an $n \times n$ matrix **A** is defined as the sum of the diagonal entries of **A**.

Multiplication of matrices is defined if their sizes are compatible; i.e., given $\mathbf{A} = [a_{ij}]_{m \times n}$ and $\mathbf{B} = [b_{ij}]_{n \times p}$, their product is an $m \times p$ matrix, $\mathbf{C} = [c_{ij}]_{m \times p}$, whose entries are defined by

$$c_{ij} = \sum_{k=1}^{n} a_{ik} b_{kj}, \qquad 1 \le i \le m, \qquad 1 \le j \le p$$

Matrix multiplication may also be performed via dot products of vectors. Let $\mathbf{A}_{m \times n}$ and $\mathbf{B}_{n \times p}$ be expressed as

$$\mathbf{A} = \begin{bmatrix} \mathbf{a}_1 \\ \mathbf{a}_2 \\ \vdots \\ \mathbf{a}_m \end{bmatrix}_{\text{row partition}}, \qquad \mathbf{B} = \underbrace{[\mathbf{b}_1 \ \mathbf{b}_2 \ \cdots \ \mathbf{b}_p]}_{\text{column partition}}$$

where $\mathbf{a}_i$ ($1 \le i \le m$) denotes the ith row of **A** and is $1 \times n$, and $\mathbf{b}_j$ ($1 \le j \le p$) denotes the jth column of **B** and is $n \times 1$. Then, each entry of their product, **C**, is calculated as

$$c_{ij} = \underset{\text{dot product}}{\mathbf{a}_i \cdot \mathbf{b}_j}, \qquad 1 \le i \le m, \qquad 1 \le j \le p$$

Alternatively, **AB** may be obtained using a column partition of **B** only, as

$$\mathbf{AB} = [\mathbf{Ab}_1 \ \mathbf{Ab}_2 \ \cdots \ \mathbf{Ab}_p]$$

Given an $m \times n$ matrix **A**, its *transpose*, $\mathbf{A}^T$, is an $n \times m$ matrix generated by interchanging rows and columns of matrix **A**. The following rules hold for the transpose of sums and products of matrices:

$$\underbrace{(\mathbf{A} + \mathbf{B})^T = \mathbf{A}^T + \mathbf{B}^T}_{\text{A and B of equal size}}, \qquad \underbrace{(\mathbf{AB})^T = \mathbf{B}^T \mathbf{A}^T}_{\text{A and B of compatible sizes}}$$

A square matrix **A** is said to be Hermitian if $\mathbf{A} = \overline{\mathbf{A}}^T$. If **A** is real, then $\mathbf{A}^T = \mathbf{A}$, and it is known as *symmetric*. A square matrix **A** is skew-Hermitian if $\overline{\mathbf{A}}^T = -\mathbf{A}$. If **A** is real, then $\mathbf{A}^T = -\mathbf{A}$, and it is called *skew-symmetric*.

A square matrix **A** is said to be upper-triangular if all its entries below the main diagonal are zero. It is lower-triangular if all its elements above the main diagonal are zero. It is diagonal if all its off-diagonal entries are zero. The *determinant* of a square matrix **A** is a real scalar defined as

Using the ith row
$$|\mathbf{A}| = \sum_{k=1}^{n} a_{ik}(-1)^{i+k} M_{ik}, \qquad i = 1, 2, \ldots, n$$

or

Using the jth column
$$|\mathbf{A}| = \sum_{k=1}^{n} a_{kj}(-1)^{k+j} M_{kj}, \qquad j = 1, 2, \ldots, n$$

M_{ik} is the minor of a_{ik}, and is defined as the determinant of the $(n-1) \times (n-1)$ submatrix obtained by deleting the ith row and kth column of **A**. The quantity $(-1)^{i+k} M_{ik}$ is the cofactor of a_{ik} and is denoted by C_{ik}. A square matrix with a determinant of zero is referred to as singular.

Properties of determinants

(D$_1$) A square matrix and its transpose have the same determinant, i.e., $|\mathbf{A}| = |\mathbf{A}^T|$.

(D$_2$) The determinant of a diagonal, an upper-triangular, and a lower-triangular matrix is the product of the entries along the main diagonal.

(D$_3$) If an entire row (or column) of a matrix is zero, then $|\mathbf{A}| = 0$.

(D$_4$) If a row (or column) of **A** is multiplied by a scalar k, the determinant of the resulting matrix is $k|\mathbf{A}|$. If **A** is an $n \times n$ matrix and k is a scalar, then $|k\mathbf{A}| = k^n |\mathbf{A}|$.

(D$_5$) If any two rows (or columns) of **A** are interchanged, the determinant of the resulting matrix is $-|\mathbf{A}|$.

(D$_6$) $|\mathbf{A}||\mathbf{B}| = |\mathbf{AB}|$

(D$_7$) If any two rows (or columns) of **A** are linearly dependent, then $|\mathbf{A}| = 0$.

(D$_8$) If the ith row of **A** is multiplied by a scalar k and then added to the jth row, and the jth row is replaced by the result, then the determinant of the new matrix is equal to the determinant of **A**.

The *adjoint* of matrix **A** is given by

$$\text{adj}(\mathbf{A}) = \begin{bmatrix} (-1)^{1+1}M_{11} & (-1)^{2+1}M_{21} & \cdots & (-1)^{n+1}M_{n1} \\ (-1)^{1+2}M_{12} & (-1)^{2+2}M_{22} & \cdots & (-1)^{n+2}M_{n2} \\ \vdots & & & \vdots \\ (-1)^{1+n}M_{1n} & (-1)^{2+n}M_{2n} & \cdots & (-1)^{n+n}M_{nn} \end{bmatrix}$$

Then, the *inverse* of **A** is defined by

$$\mathbf{A}^{-1} = \frac{1}{|\mathbf{A}|} \text{adj}(\mathbf{A})$$

The *rank* of an $n \times n$ matrix **A**, denoted by $rk(\mathbf{A})$, is the maximum number of linearly independent rows (or columns) of **A**. Alternatively, the rank of **A** is defined as the size of the largest submatrix of **A** with a nonzero determinant.

A system of n linear algebraic equations in n unknowns is solved via *Cramer's rule*. Consider, in general,

$$\begin{cases} a_{11}x_1 + a_{12}x_2 + \cdots + a_{1n}x_n = b_1 \\ a_{21}x_1 + a_{22}x_2 + \cdots + a_{2n}x_n = b_2 \\ \quad \vdots \\ a_{n1}x_1 + a_{n2}x_2 + \cdots + a_{nn}x_n = b_n \end{cases}$$

where a_{ij} $(1 \leq i, j \leq n)$ are the constant coefficients, x_i $(1 \leq i \leq n)$ are the unknown quantities, and b_i $(1 \leq i \leq n)$ represent the known quantities on the right-hand sides. The ith unknown, x_i, is given by

$$x_i = \frac{\Delta_i}{\Delta}, \quad i = 1, 2, \ldots, n$$

where Δ is the determinant of the coefficient matrix and Δ_i is obtained by replacing the ith column in Δ by the vector of the right-hand-side quantities.

The *eigenvalue problem* associated with a matrix $\mathbf{A}$ is defined as

$$\mathbf{Av} = \lambda \mathbf{v}$$

A number λ for which this equation has a nontrivial solution is known as an *eigenvalue* of $\mathbf{A}$. The corresponding solution ($\mathbf{v} \neq \mathbf{0}_{n \times 1}$) is referred to as the *eigenvector* of $\mathbf{A}$ corresponding to λ. The eigenvalues of $\mathbf{A}$ are the solutions of the *characteristic equation*,

$$|\mathbf{A} - \lambda \mathbf{I}| = 0$$

Once the eigenvalues have been identified, each eigenvector corresponding to each of the eigenvalues can be obtained by solving the matrix equation

$$(\mathbf{A} - \lambda \mathbf{I}_n)\mathbf{v} = \mathbf{0}_{n \times 1}$$

This equation is normally solved via *elementary row operations*, defined as

(ERO$_1$)   Multiply a row by a nonzero scalar.
(ERO$_2$)   Interchange any two rows.
(ERO$_3$)   Multiply a row by a nonzero scalar and add to another row, and replace the latter by the result.

Eigenvalues of diagonal, upper-triangular, and lower-triangular matrices are the main diagonal elements. A matrix and its transpose have the same eigenvalues.

A square matrix $\mathbf{A}$ is called unitary if

$$\mathbf{A}^T = \overline{\mathbf{A}}^{-1} \quad \text{or} \quad \overline{\mathbf{A}}^T = \mathbf{A}^{-1}$$

A *real* unitary matrix is called an orthogonal matrix, and it is defined by $\mathbf{A}^T = \mathbf{A}^{-1}$.

1. All eigenvalues of a *unitary* matrix have absolute values of 1.
2. All eigenvalues of a *Hermitian* matrix are real.
3. All eigenvalues of a *skew-Hermitian* matrix are zero or pure imaginary.

Two $n \times n$ matrices, $\mathbf{A}$ and $\mathbf{B}$, are said to be *similar* if there exists a nonsingular $n \times n$ matrix $\mathbf{S}$ such that $\mathbf{B} = \mathbf{S}^{-1}\mathbf{AS}$ (or, $\mathbf{A} = \mathbf{SBS}^{-1}$). This transformation is referred to as a *similarity transformation*. Similar matrices have the same eigenvalues.

Let $\mathbf{A}$ be an $n \times n$ matrix with n linearly independent eigenvectors, $\mathbf{v}_1, \mathbf{v}_2, \ldots, \mathbf{v}_n$. The *modal matrix* of $\mathbf{A}$ is the matrix composed of the eigenvectors along its columns. The modal matrix diagonalizes $\mathbf{A}$, i.e.,

$$\mathbf{P}^{-1}\mathbf{AP} = \Lambda$$

where Λ is a diagonal matrix whose diagonal entries are the eigenvalues of $\mathbf{A}$ corresponding to $\mathbf{v}_1, \mathbf{v}_2, \ldots, \mathbf{v}_n$ in that order, i.e.,

$$\Lambda = \text{diag}(\lambda_1, \ldots, \lambda_n)$$

The *exponential* of an $n \times n$ matrix $\mathbf{M}$ is also $n \times n$ and is defined by

$$e^{\mathbf{M}} = \sum_{k=0}^{\infty} \frac{\mathbf{M}^k}{k!} = \mathbf{I}_n + \mathbf{M} + \frac{1}{2!}\mathbf{M}^2 + \cdots$$

Suppose $\mathbf{A}$ is $n \times n$ with n linearly independent eigenvectors. In that case,

$$e^{\mathbf{A}} = \mathbf{P} e^{\mathbf{\Lambda}} \mathbf{P}^{-1}$$

PROBLEMS

Determine whether each set of vectors given below forms a linearly independent set. Combine the vectors to form a square matrix and use an appropriate property of determinants.

2.31. $\mathbf{u} = \begin{Bmatrix} 1 \\ -1 \\ 3 \end{Bmatrix}$, $\mathbf{v} = \begin{Bmatrix} 1 \\ 1 \\ 2 \end{Bmatrix}$, $\mathbf{w} = \begin{Bmatrix} 0 \\ 2 \\ -1 \end{Bmatrix}$

2.32. $\mathbf{u} = \begin{Bmatrix} 1 \\ -1 \\ 2 \end{Bmatrix}$, $\mathbf{v} = \begin{Bmatrix} 2 \\ 0 \\ 1 \end{Bmatrix}$, $\mathbf{w} = \begin{Bmatrix} 1 \\ -1 \\ 1 \end{Bmatrix}$

2.33. $\mathbf{u} = \begin{Bmatrix} 1 \\ 2 \end{Bmatrix}$, $\mathbf{v} = \begin{Bmatrix} 2 \\ -1 \end{Bmatrix}$

2.34. Express the following matrix as the sum of a symmetric and a skew-symmetric matrix.

$$\mathbf{A} = \begin{bmatrix} 1 & 5 & 2 \\ -3 & 6 & 0 \\ 2 & -3 & 1 \end{bmatrix}$$

Find the inverse, if possible, of each given matrix.

2.35. $\mathbf{A} = \begin{bmatrix} 3 & -1 & 1 \\ 1 & -3 & -4 \\ -1 & 1 & 2 \end{bmatrix}$

2.36. $\mathbf{B} = \begin{bmatrix} 4 & -3 \\ 9 & -7 \end{bmatrix}$

2.37. $\mathbf{C} = \begin{bmatrix} -3 & 0 & 0 \\ 1 & 4 & 0 \\ 2 & -5 & 0 \end{bmatrix}$

2.38. $\mathbf{D} = \begin{bmatrix} 1 & 0 & 0 \\ 0 & -3 & 0 \\ 0 & 0 & 4 \end{bmatrix}$

2.39. Prove that the determinant of a lower-triangular matrix is the product of its entries along the main diagonal.

2.40. Show that if $\mathbf{D} = \text{diag}(d_{11}, d_{22}, \ldots, d_{nn})$, then

$$\mathbf{D}^{-1} = \text{diag}\left(\frac{1}{d_{11}}, \frac{1}{d_{22}}, \ldots, \frac{1}{d_{nn}}\right)$$

Determine the rank of each matrix below.

2.41. $\mathbf{A} = \begin{bmatrix} 1 & -2 & 1 & 3 \\ 2 & 5 & 3 & 4 \\ 6 & -1 & 2 & 8 \\ 4 & 3 & 0 & 2 \end{bmatrix}$

2.42. $\mathbf{B} = \begin{bmatrix} -1 & 2 & 0 \\ 0 & 0 & -3 \\ 0 & 0 & 4 \end{bmatrix}$

2.43. $\mathbf{C} = \begin{bmatrix} 0 & 1 \\ -3 & -2 \end{bmatrix}$

2.44. Prove that $rk(\mathbf{AB}) = rk(\mathbf{B}^T \mathbf{A}^T)$.

2.45. Show that if a square matrix has a full rank, it is nonsingular.

2.46. Given that $\mathbf{A}$ is $n \times n$ and nonsingular, prove that $(\mathbf{A}^T)^{-1} = (\mathbf{A}^{-1})^T$.

For each system of linear equations, decide whether Cramer's rule is applicable. If so, find the solution.

2.47. $\begin{bmatrix} -2 & 1 \\ 3 & -1 \end{bmatrix} \begin{Bmatrix} x_1 \\ x_2 \end{Bmatrix} = \begin{Bmatrix} -7 \\ 9 \end{Bmatrix}$

2.48. $\begin{bmatrix} 0 & 2 & 7 \\ 1 & -3 & -4 \\ -1 & 1 & 2 \end{bmatrix} \begin{Bmatrix} x_1 \\ x_2 \\ c_3 \end{Bmatrix} = \begin{Bmatrix} 12 \\ -4 \\ 2 \end{Bmatrix}$

Solve each linear system below in the form $\mathbf{Ax} = \mathbf{b}$ using the inverse of the coefficient matrix.

2.49. $\mathbf{A} = \begin{bmatrix} 1 & -1 & 2 \\ 2 & 0 & -1 \\ 1 & 1 & 1 \end{bmatrix}$, $\mathbf{x} = \begin{Bmatrix} x_1 \\ x_2 \\ x_3 \end{Bmatrix}$, $\mathbf{b} = \begin{bmatrix} 4 \\ 0 \\ 2 \end{bmatrix}$

2.50. The system in Problem 2.48

2.51. $\mathbf{A} = \begin{bmatrix} 0 & 1 \\ -1 & -2 \end{bmatrix}$, $\mathbf{x} = \begin{Bmatrix} x_1 \\ x_2 \end{Bmatrix}$, $\mathbf{b} = \begin{bmatrix} 0 \\ 2 \end{bmatrix}$

2.52. Given that

$$A = \begin{bmatrix} -2 & 0 & 0 \\ -1 & -4 & 0 \\ -1 & -1 & -5 \end{bmatrix} \quad \text{and} \quad B = \begin{bmatrix} 1 \\ 1 \\ 1 \end{bmatrix}$$

form the matrix $[B \quad AB \quad A^2B]$ and determine its rank.

2.53. Given that

$$A = \begin{bmatrix} -2 & 0 & 0 \\ -1 & -4 & 0 \\ -1 & -1 & -5 \end{bmatrix} \quad \text{and} \quad C = \begin{bmatrix} 0 & 0 & 1 \end{bmatrix}$$

form the matrix $[C^T \quad A^T C^T \quad (A^T)^2 C^T]$ and determine its rank.

Determine the eigenvalues and eigenvectors of each matrix.

2.54. $A = \begin{bmatrix} -2 & 2 & -1 \\ 2 & 1 & -2 \\ -3 & -6 & 0 \end{bmatrix}$

2.55. $A = \begin{bmatrix} 2 & 0 & 0 \\ 4 & 0 & 0 \\ 1 & 2 & -3 \end{bmatrix}$

In each case, find the modal matrix and diagonalize.

2.56. Matrix A of Problem 2.54

2.57. Matrix A of Problem 2.55

2.58. Calculate e^{At} where t is scalar and

$$A = \begin{bmatrix} 0 & 1 \\ -4 & -5 \end{bmatrix}$$

2.59. Prove that if a square, diagonalizable matrix is singular, then at least one of its eigenvalues must be zero.

2.60. For a nonsingular matrix A, show that $(A^{-1})^2 = (A^2)^{-1}$.

CHAPTER 3

System Model Representation

This chapter is about the most commonly used forms of representing the mathematical model of dynamic systems. While the actual derivation of a system model is not covered by the present chapter, this chapter emphasizes the treatment of the model once the system's governing equations have been obtained successfully. Based on the type of the dynamic system under consideration, governing equations are derived (see Chapters 4 through 6) appropriately by applications of fundamental laws such as Newton's laws, conservation of mass, and Kirchhoff's voltage and current laws. The mathematical model of a dynamic system should consist of the same number of unknown (independent) variables as equations, and these equations should be expressed in a form that is convenient to work with. To this end, four different such forms will be introduced and applied to systems that are relatively simple in nature:

1. Configuration form
2. State-space representation
3. Input-output equation
4. Transfer function

Throughout most of this chapter, it will be assumed that systems under consideration are linear. Nonlinear systems, as well as linearization procedures, will be discussed later in this chapter.

3.1 CONFIGURATION FORM

DEFINITION 3.1. A set of independent coordinates that completely describes the motion of a system is referred to as a set of **generalized coordinates.** For a given system, any set of coordinates that can be used to completely describe its motion may be chosen as generalized coordinates. Thus, this set of coordinates is *not unique.*

Notation: $q_i(i = 1, 2, \ldots, n)$ denote the generalized coordinates for a system with n degrees of freedom, where the **number of degrees of freedom** (DOF) is equal to the minimum number of independent coordinates required to describe the position of all elements of a system (see Chapter 4 for more details).

Geometrically speaking, generalized coordinates, $q_i(i = 1, 2, \ldots, n)$, define an n-dimensional cartesian space that is referred to as the **configuration space.**

Consider an n-degree-of-freedom system whose governing (second-order) differential equations are given as

$$\begin{cases} \ddot{q}_1 = f_1(q_1, q_2, \ldots, q_n, \dot{q}_1, \dot{q}_2, \ldots, \dot{q}_n, t) \\ \ddot{q}_2 = f_2(q_1, q_2, \ldots, q_n, \dot{q}_1, \dot{q}_2, \ldots, \dot{q}_n, t) \\ \quad \vdots \\ \ddot{q}_n = f_n(q_1, q_2, \ldots, q_n, \dot{q}_1, \dot{q}_2, \ldots, \dot{q}_n, t) \end{cases} \quad (3.1)$$

where q_i and $\dot{q}_i$ ($i = 1, 2, \ldots, n$) denote the generalized coordinates and **generalized velocities,** respectively. Functions $f_i(i = 1, 2, \ldots, n)$, generally nonlinear, denote the **generalized forces** and are algebraic functions of q_i, $\dot{q}_i$, and time t. The system of differential equations represented by Eq. (3.1) is subject to *initial conditions*

Initial generalized coordinates $\quad q_1(0), \ldots, q_n(0)$
Initial generalized velocities $\quad \dot{q}_1(0), \ldots, \dot{q}_n(0)$ $\quad (3.2)$

Equation (3.1) together with Eq. (3.2) describes the system's configuration form. Again, we are mainly concerned with systems in which the functions f_i ($i = 1, 2, \ldots, n$) are linear. Such functions are expressible as linear combinations of q_i and $\dot{q}_i$, as illustrated in the following examples.

EXAMPLE 3.1. Consider a simple mechanical system, shown in Fig. 3.1, subject to initial conditions

$$x(0) = x_0, \qquad \dot{x}(0) = \dot{x}_0$$

where x_0 and $\dot{x}_0$ denote the prescribed initial displacement and velocity, respectively. Express the system's equation of motion in configuration form, as defined by Eq. (3.1).

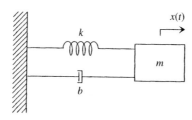

FIGURE 3.1 Single-degree-of-freedom mechanical system.

Solution. The governing differential equation (see Chapter 4), describing the motion of the system, is given by

$$m\ddot{x} + b\dot{x} + kx = 0, \qquad \underbrace{x(0) = x_0, \qquad \dot{x}(0) = \dot{x}_0}_{\text{initial conditions}} \qquad (3.3)$$

Dividing by m and rearranging terms in Eq. (3.3), one obtains

$$\underbrace{\ddot{x}}_{\ddot{q}_1} = \underbrace{-\frac{b}{m}\dot{x} - \frac{k}{m}x,}_{f_1(q_1,\dot{q}_1,t)} \qquad x(0) = x_0, \qquad \dot{x}(0) = \dot{x}_0 \qquad (3.4)$$

Observe that only one generalized coordinate exists, $q_1 = x$, and Eq. (3.4) is in agreement with the general form of Eq. (3.1). Moreover, the generalized function f_1 is expressed as a linear combination of x and $\dot{x}$, and hence is linear.

EXAMPLE 3.2. Consider the two-degree-of-freedom system shown in Fig. 3.2, subject to initial conditions

$$x_1(0) = x_{10}, \qquad \dot{x}_1(0) = \dot{x}_{10}, \qquad x_2(0) = x_{20}, \qquad \dot{x}_2(0) = \dot{x}_{20}$$

where double indices are used. The first index indicates the variable, displacement, or velocity; the second index, 0, refers to the initial condition of the variable. Express the governing equations in configuration form, as defined by Eq. (3.1).

Solution. The differential equations (see Chapter 4) describing the motion of this mechanical system are given as

$$\begin{cases} m_1\ddot{x}_1 + b_1\dot{x}_1 + k_1 x_1 - k_2(x_2 - x_1) - b_2(\dot{x}_2 - \dot{x}_1) = 0 & (3.5a) \\ m_2\ddot{x}_2 + b_2(\dot{x}_2 - \dot{x}_1) + k_2(x_2 - x_1) = 0 & (3.5b) \end{cases}$$

Dividing Eq. (3.5a and b) by m_1 and m_2 separately and rearranging yields

$$\begin{cases} \ddot{x}_1 = \dfrac{1}{m_1}\left[-b_1\dot{x}_1 - k_1 x_1 + k_2(x_2 - x_1) + b_2(\dot{x}_2 - \dot{x}_1)\right] = f_1(x_1, x_2, \dot{x}_1, \dot{x}_2) \\ \ddot{x}_2 = \dfrac{1}{m_2}\left[-b_2(\dot{x}_2 - \dot{x}_1) - k_2(x_2 - x_1)\right] = f_2(x_1, x_2, \dot{x}_1, \dot{x}_2) \end{cases}$$

$$(3.6)$$

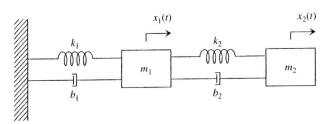

FIGURE 3.2 Two-degree-of-freedom mechanical system.

Observe from this formulation that there are two generalized coordinates, $q_1 = x_1$ and $q_2 = x_2$, and that Eq. (3.6) is in the general form of Eq. (3.1). Moreover, the generalized forces f_1 and f_2 are expressed as linear combinations of generalized coordinates x_1 and x_2 and generalized velocities $\dot{x}_1$ and $\dot{x}_2$, and hence are linear.

Second-Order Matrix Form

A convenient and commonly used form of representing an $n \times n$ system of second-order differential equations is standard **second-order matrix form**, defined as

$$\mathbf{m\ddot{x} + c\dot{x} + kx = f} \qquad (3.7)$$

where $\mathbf{m}$ = mass matrix ($n \times n$), $\mathbf{c}$ = damping matrix ($n \times n$), $\mathbf{k}$ = stiffness matrix ($n \times n$), $\mathbf{x}$ = configuration vector ($n \times 1$) = vector of generalized coordinates, and $\mathbf{f}$ = vector of external forces ($n \times 1$).

EXAMPLE 3.3. Consider the single-degree-of-freedom system of Example 3.1 when subjected to an applied force $f(t)$, as shown in Fig. 3.3. Represent the equation of motion in second-order matrix form.

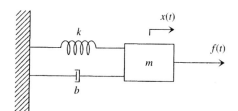

FIGURE 3.3 Mechanical system subjected to an applied force.

Solution. The equation of motion of this system is given as

$$m\ddot{x} + b\dot{x} + kx = f(t)$$

Because there is only one generalized coordinate, x, matrices and vectors in Eq. (3.7) are scalars. In particular, comparing the equation of motion with Eq. (3.7), yields

$$\mathbf{m} = m, \qquad \mathbf{c} = b, \qquad \mathbf{k} = k, \qquad \mathbf{f} = f, \qquad \mathbf{x} = x$$

EXAMPLE 3.4. Suppose the mechanical system of Example 3.2 is subjected to an applied force $f(t)$, as shown in Fig. 3.4. Express the system's equations of motion in second-order matrix form as given by Eq. (3.7).

Solution. The equations of motion, a slight modification of Eq. (3.5), are

$$\begin{cases} m_1 \ddot{x}_1 + b_1 \dot{x}_1 + k_1 x_1 - k_2(x_2 - x_1) - b_2(\dot{x}_2 - \dot{x}_1) = 0 \\ m_2 \ddot{x}_2 + b_2(\dot{x}_2 - \dot{x}_1) + k_2(x_2 - x_1) = f(t) \end{cases} \qquad (3.8)$$

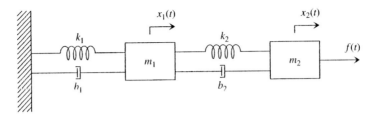

FIGURE 3.4 Mechanical system of Example 3.2 subjected to an applied force.

The system of equations in Eq. (3.8) can then be expressed using matrix notation into the standard second-order matrix form, Eq. (3.7), as

$$m\ddot{x} + c\dot{x} + kx = f$$

or, equivalently,

$$\begin{bmatrix} m_1 & 0 \\ 0 & m_2 \end{bmatrix} \begin{Bmatrix} \ddot{x}_1 \\ \ddot{x}_2 \end{Bmatrix} + \begin{bmatrix} b_1 + b_2 & -b_2 \\ -b_2 & b_2 \end{bmatrix} \begin{Bmatrix} \dot{x}_1 \\ \dot{x}_2 \end{Bmatrix} + \begin{bmatrix} k_1 + k_2 & -k_2 \\ -k_2 & k_2 \end{bmatrix} \begin{Bmatrix} x_1 \\ x_2 \end{Bmatrix} = \begin{Bmatrix} 0 \\ f \end{Bmatrix}$$

PROBLEM SET 3.1

3.1. Consider the two-degree-of-freedom mechanical system in Fig. P3.1. The governing equations are given as

$$\begin{cases} m_1 \ddot{x}_1 + b_1 \dot{x}_1 + k_1 x_1 - k_2 (x_2 - x_1) = 0 \\ m_2 \ddot{x}_2 + k_2 (x_2 - x_1) = 0 \end{cases}$$

Assume the following parameter values: $m_1 = 1 \text{ kg} = m_2$, $b_1 = 2 \text{ N} \cdot \text{s/m}$, $k_1 = 1 \text{ N/m} = k_2$. Express the equations of motion in configuration form and identify the generalized forces.

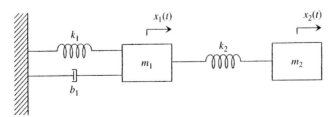

FIGURE P3.1 Mechanical system.

3.2. Suppose that in the system of Problem 3.1, block m_1 is subjected to an applied force u, as shown in Fig. P3.2. The equations of motion are then

$$\begin{cases} m_1 \ddot{x}_1 + b_1 \dot{x}_1 + k_1 x_1 - k_2(x_2 - x_1) = u(t) \\ m_2 \ddot{x}_2 + k_2(x_2 - x_1) = 0 \end{cases}$$

Using the parameter values in Problem 3.1, express the governing equations in the standard second-order matrix form.

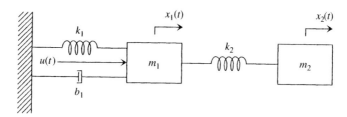

FIGURE P3.2

3.3. A mechanical system experiencing translational and rotational motion is governed by

$$\begin{cases} J\ddot{\theta} + B\dot{\theta} + K\theta + Rk(R\theta - x) = Ru(t) \\ m\ddot{x} + b\dot{x} - k(R\theta - x) = 0 \end{cases}$$

where $J, K, B, R, m, k,$ and b denote physical parameters and are regarded as constants. x and θ represent displacement and angular displacement, respectively, and $u(t)$ is an applied force. Express the equations in second-order matrix form.

3.2 STATE-SPACE REPRESENTATION

The concept of state-space relies on what are known as *state variables*.

DEFINITION 3.2. The smallest possible set of independent variables that completely describes the state of a system is referred to as the set of **state variables**. These variables at some fixed time ($t = t_0$) and system inputs for all $t \geq t_0$ will provide a complete description of system behavior at any time $t \geq t_0$. Because *independence* is essential, state variables cannot be expressible as algebraic functions of each other and the inputs. Moreover, the set of state variables for a certain system is *not unique*.

The objective, through introduction of suitable state variables, is to rewrite a system's equations of motion as a larger system of first-order differential equations. Each of these differential equations consists of the time derivative of one of the state variables on the left-hand side and an algebraic function of the state variables, as well as the system inputs, on the right-hand side. These differential equations are then referred to as **state-variable equations**.

Methodology

Given a set of differential equations describing a certain dynamic system, the two key questions are

(Q_1) How many state variables are there?
(Q_2) What are the state variables?

(Q_1) How many state variables are there?

The number of state variables is equal to the number of initial conditions required in order to completely solve the differential equations of motion. For instance, if a single second-order differential equation describes a dynamic system, two initial conditions are required. Hence, there are two state variables.

(Q_2) What are the state variables?

Those variables for which initial conditions are required [in (Q_1)] are chosen as state variables.

EXAMPLE 3.5. For the mechanical system discussed in Example 3.3, determine the state variables.

Solution. The equation of motion is given as $m\ddot{x} + b\dot{x} + kx = f(t)$. In order to completely solve this second-order differential equation, knowledge of two initial conditions is required, namely,

$$x(0) \quad \text{and} \quad \dot{x}(0)$$

Hence, there are two state variables (according to Q_1). Furthermore, because initial conditions correspond to x and $\dot{x}$, state variables should be chosen, according to Q_2, as x and $\dot{x}$; i.e.,

$$x_1 = x, \qquad x_2 = \dot{x}$$

EXAMPLE 3.6. Suppose that the governing equations for a certain dynamic system are found to be

$$\begin{cases} \ddot{x}_1 + \dot{x}_1 + x_1 - x_2 = 0 \\ \dot{x}_2 - 2x_1 + x_2 = 0 \end{cases}$$

where x_1 and x_2 represent displacements, and $\dot{x}_1$ and $\dot{x}_2$ denote velocities. Determine the state variables.

Solution. Because the highest order of differentiation is two for x_1 and one for x_2, the required initial conditions are $x_1(0)$, $x_2(0)$, and $\dot{x}_1(0)$. Therefore, there are a total of three state variables, and they are chosen to be x_1, x_2, and $\dot{x}_1$. Throughout this textbook, state variables are denoted by x_i with an appropriate index i. It is important to distinguish between state variables and physical variables such as displacement and velocity. State variables are generally mathematical quantities that are employed to represent a system's governing equations in a convenient form. Sometimes the conventional notation of x_i, which is reserved for state variables, may indeed coincide with some of the physical variables involved in a system model. For instance, in the current example there are three state variables which, by convention, will be denoted by x_1, x_2, and x_3. However, we note that x_1 and x_2 also represent displacements of the blocks. This should not cause any concern because we have determined that

$$\underbrace{x_1, x_2, \dot{x}_1}_{\substack{\text{physical quantities} \\ \text{chosen as state variables}}} \quad \Leftrightarrow \quad \underbrace{x_1, x_2, x_3}_{\substack{\text{mathematical quantities} \\ \text{denoting state variables}}}$$

All that remains is to make one-to-one assignments between the elements of the two sets. In this process, it is customary to use up the physical variables in increasing order of derivatives. In the current problem, this means x_1 and x_2, followed by $\dot{x}_1$. To that end, the assignments are made as follows:

$$\underset{\text{first state}}{x_1} = \underset{\text{first displacement}}{x_1}$$

$$\underset{\text{second state}}{x_2} = \underset{\text{second displacement}}{x_2}$$

$$\underset{\text{third state}}{x_3} = \underset{\text{first velocity}}{\dot{x}_1}$$

General Formulation

Once the state variables are appropriately selected, the next task is to construct the state-variable equations. As mentioned earlier, these are first-order differential equations, each of which contains the first derivative of one of the state variables on the left-hand side and an algebraic function of the state variables, inputs, and (possibly) time on the right-hand side. In general, let us consider a multi-input/multi-output (MIMO) system with n state variables, $x_1, x_2, \ldots, x_n$, m inputs, $u_1, u_2, \ldots, u_m$, and p outputs, $y_1, y_2, \ldots, y_p$. Then, the state-variable equations are in the following general form:

State-variable equations
$$\begin{cases} \dot{x}_1 = f_1(x_1, \ldots, x_n; u_1, \ldots, u_m; t) \\ \dot{x}_2 = f_2(x_1, \ldots, x_n; u_1, \ldots, u_m; t) \\ \quad \vdots \\ \dot{x}_n = f_n(x_1, \ldots, x_n; u_1, \ldots, u_m; t) \end{cases} \quad (3.9)$$

where $f_1, f_2, \ldots, f_n$ are nonlinear, in general. Similarly, system **outputs** may also be expressed as follows:

System outputs
$$\begin{cases} y_1 = h_1(x_1, \ldots, x_n; u_1, \ldots, u_m; t) \\ y_2 = h_2(x_1, \ldots, x_n; u_1, \ldots, u_m; t) \\ \quad \vdots \\ y_p = h_p(x_1, \ldots, x_n; u_1, \ldots, u_m; t) \end{cases} \quad (3.10)$$

where $h_1, h_2, \ldots, h_p$ are nonlinear, in general. In the event that nonlinear elements are present in the system, the algebraic functions $f_i (i = 1, 2, \ldots, n)$ and $h_k (k = 1, 2, \ldots, p)$ turn out to be nonlinear and quite complex in nature, thereby complicating the analysis. Equations (3.9) and (3.10) may be represented more conveniently through matrix notation. To that end, define

$$\mathbf{x} = \begin{bmatrix} x_1 \\ x_2 \\ \vdots \\ x_n \end{bmatrix}_{n \times 1}, \quad \mathbf{f} = \begin{bmatrix} f_1 \\ f_2 \\ \vdots \\ f_n \end{bmatrix}_{n \times 1}, \quad \mathbf{u} = \begin{bmatrix} u_1 \\ u_2 \\ \vdots \\ u_m \end{bmatrix}_{m \times 1}, \quad \mathbf{y} = \begin{bmatrix} y_1 \\ y_2 \\ \vdots \\ y_p \end{bmatrix}_{p \times 1}, \quad \mathbf{h} = \begin{bmatrix} h_1 \\ h_2 \\ \vdots \\ h_p \end{bmatrix}_{p \times 1}$$

so that the state-variable equations are expressed as

$$\dot{\mathbf{x}} = \mathbf{f}(\mathbf{x}, \mathbf{u}, t)$$

and the system outputs are

$$\mathbf{y} = \mathbf{h}(\mathbf{x}, \mathbf{u}, t)$$

The complexity associated with the general formulation reduces considerably for the case of linear systems. In the event that all elements in the model of a dynamic system are linear, the algebraic functions in Eqs. (3.9) and (3.10) will take the following special forms:

Linear state-variable equations
$$\begin{cases} \dot{x}_1 = a_{11} x_1 + \cdots + a_{1n} x_n + b_{11} u_1 + \cdots + b_{1m} u_m \\ \dot{x}_2 = a_{21} x_1 + \cdots + a_{2n} x_n + b_{21} u_1 + \cdots + b_{2m} u_m \\ \quad \vdots \\ \dot{x}_n = a_{n1} x_1 + \cdots + a_{nn} x_n + b_{n1} u_1 + \cdots + b_{nm} u_m \end{cases} \quad (3.11)$$

Linear system outputs
$$\begin{cases} y_1 = c_{11} x_1 + \cdots + c_{1n} x_n + d_{11} u_1 + \cdots + d_{1m} u_m \\ y_2 = c_{21} x_1 + \cdots + c_{2n} x_n + d_{21} u_1 + \cdots + d_{2m} u_m \\ \quad \vdots \\ y_p = c_{p1} x_1 + \cdots + c_{pn} x_n + d_{p1} u_1 + \cdots + d_{pm} u_m \end{cases} \quad (3.12)$$

In order to represent Eqs. (3.11) and (3.12) in matrix form, define the following quantities

$$\mathbf{x} = \begin{bmatrix} x_1 \\ x_2 \\ \vdots \\ x_n \end{bmatrix}_{n \times 1} = \text{state vector}, \quad \mathbf{u} = \begin{bmatrix} u_1 \\ u_2 \\ \vdots \\ u_m \end{bmatrix}_{m \times 1} = \text{input vector}$$

$$\mathbf{y} = \begin{bmatrix} y_1 \\ y_2 \\ \vdots \\ y_p \end{bmatrix}_{p \times 1} = \text{output vector}$$

$$\mathbf{A} = \begin{bmatrix} a_{11} & a_{12} & \cdots & a_{1n} \\ a_{21} & a_{22} & \cdots & a_{2n} \\ \vdots & \vdots & & \vdots \\ a_{n1} & a_{n2} & \cdots & a_{nn} \end{bmatrix}_{n \times n} = \text{state matrix}$$

$$\mathbf{B} = \begin{bmatrix} b_{11} & b_{12} & \cdots & b_{1m} \\ b_{21} & b_{22} & \cdots & b_{2m} \\ \vdots & \vdots & & \vdots \\ b_{n1} & b_{n2} & \cdots & b_{nm} \end{bmatrix}_{n \times m} = \text{input matrix}$$

$$\mathbf{C} = \begin{bmatrix} c_{11} & c_{12} & \cdots & c_{1n} \\ c_{21} & c_{22} & \cdots & c_{2n} \\ \vdots & \vdots & & \vdots \\ c_{p1} & c_{p2} & \cdots & c_{pn} \end{bmatrix}_{p \times n} = \text{output matrix}$$

$$\mathbf{D} = \begin{bmatrix} d_{11} & d_{12} & \cdots & d_{1m} \\ d_{21} & d_{22} & \cdots & d_{2m} \\ \vdots & \vdots & & \vdots \\ d_{p1} & d_{p2} & \cdots & d_{pm} \end{bmatrix}_{p \times m} = \text{direct transmission matrix}$$

As a result, Eqs. (3.11) and (3.12) are expressed in matrix form as

$$\begin{cases} \dot{\mathbf{x}} = \mathbf{A}\mathbf{x} + \mathbf{B}\mathbf{u} & \text{state equation} \\ \mathbf{y} = \mathbf{C}\mathbf{x} + \mathbf{D}\mathbf{u} & \text{output equation} \end{cases} \quad (3.13a)$$
$$(3.13b)$$

Equation (3.13) is known as the **state-space representation** or **state-space form** of the system model. This convenient form of representing a system model is particularly useful in the analysis and control of dynamic systems. Figure 3.5 illustrates the procedure to obtain the state-space representation from the governing equations of a dynamic system.

EXAMPLE 3.7. Once again, consider the simple mechanical system of Example 3.3. Assume that the displacement of the block, $x(t)$, is the output of the system. Obtain the state-space form of the system model.

Solution. The governing differential equation is given by

$$m\ddot{x} + b\dot{x} + kx = f(t) \qquad (3.14)$$

As discussed in Example 3.5, the state variables are $x_1 = x$ and $x_2 = \dot{x}$. Because there are two state variables, two first-order differential equations are expected. The first of these equations simply gives a relation between the two state variables:

$$\dot{x}_1 = x_2 \qquad (3.15)$$

which does <u>not</u> depend on the dynamics of the system. The second equation, however, is obtained directly from the equation of motion, Eq. (3.14), by sub-

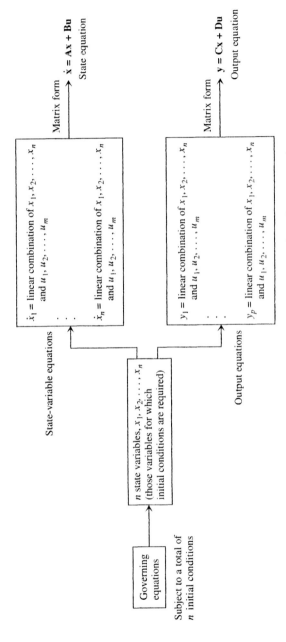

FIGURE 3.5 Procedure to find the state-space form from governing equations.

stitution of state variables, as follows:

$$m\dot{x}_2 + bx_2 + kx_1 = f(t)$$

Solving for $\dot{x}_2$, one obtains

$$\dot{x}_2 = \frac{1}{m}[-bx_2 - kx_1 + f(t)] \tag{3.16}$$

Equations (3.15) and (3.16) form the *state-variable equations* consisting of the time derivatives of the state variables and the algebraic functions of the state variables and system inputs; that is,

State-variable equations $\begin{cases} \dot{x}_1 = x_2 & (3.15) \\ \dot{x}_2 = \dfrac{1}{m}[-bx_2 - kx_1 + f(t)] & (3.16) \end{cases}$

Also, assuming linearity of the elements has caused these equations to be in the general form of Eq. (3.11). System output is, by assumption, the displacement of the block; hence

$$y = \underset{\text{displacement}}{x} \implies y = x_1 \tag{3.17}$$

which agrees with the general form of Eq. (3.12). Expressing the state-variable equations in matrix form yields the state equation

$$\begin{Bmatrix} \dot{x}_1 \\ \dot{x}_2 \end{Bmatrix} = \begin{bmatrix} 0 & 1 \\ -k/m & -b/m \end{bmatrix} \begin{Bmatrix} x_1 \\ x_2 \end{Bmatrix} + \begin{bmatrix} 0 \\ 1/m \end{bmatrix} f \tag{3.18}$$

so that in relation to Eq. (3.13a) we have

$$\dot{\mathbf{x}} = \mathbf{A}\mathbf{x} + \mathbf{B}u$$

where $\mathbf{x} = \begin{Bmatrix} x_1 \\ x_2 \end{Bmatrix}$, $\mathbf{A} = \begin{bmatrix} 0 & 1 \\ -k/m & -b/m \end{bmatrix}$, $\mathbf{B} = \begin{bmatrix} 0 \\ 1/m \end{bmatrix}$, $u = f(t)$

Rewriting Eq. (3.17) using vector notation yields the output equation

$$y = \begin{bmatrix} 1 & 0 \end{bmatrix} \begin{Bmatrix} x_1 \\ x_2 \end{Bmatrix} + 0 \cdot u$$

and in relation to Eq. (3.13b) we have

$$y = \mathbf{C}\mathbf{x} + \mathbf{D}u \quad \text{where} \quad \mathbf{C} = \begin{bmatrix} 1 & 0 \end{bmatrix}, \quad \mathbf{D} = 0 \tag{3.19}$$

Ultimately, a combination of Eqs. (3.18) and (3.19) defines the system's state-space form

$$\begin{cases} \dot{\mathbf{x}} = \mathbf{A}\mathbf{x} + \mathbf{B}u & \text{state equation} \\ y = \mathbf{C}\mathbf{x} + \mathbf{D}u & \text{output equation} \end{cases}$$

in which all vectors and matrices have been properly defined. Observe also that the two state variables are independent, because it is impossible to express displacement as an algebraic function of velocity. The same approach may be used to obtain the state-space representation of more complicated sets of differential equations. To that end, consider a two-degree-of-freedom electrical system.

EXAMPLE 3.8. Consider the electrical circuit in Fig. 3.6, in which the voltage $e(t)$ is the input, and q_1 and q_2 denote electric charges. The constant parameters R, L_1, L_2, C_1, and C_2 denote resistance, inductances, and capacitances, respectively. The system's governing equations are obtained through application of Kirchhoff's voltage law (see Chapter 5), as

$$\begin{cases} L_1\ddot{q}_1 + R(\dot{q}_1 - \dot{q}_2) + \dfrac{1}{C_1}q_1 = e & (3.20a) \\ L_2\ddot{q}_2 + R(\dot{q}_2 - \dot{q}_1) + \dfrac{1}{C_2}q_2 = 0 & (3.20b) \end{cases}$$

(a) By choosing a suitable set of state variables, obtain the state equation.
(b) Assuming that the system output is q_1, find the output equation. Repeat for the case in which the outputs are q_1 and $\dot{q}_1$.

Solution

(a) Because each differential equation in Eq. (3.20) is second-order, a total of four initial conditions are required; hence, there exist four state variables. Therefore, state variables should be chosen as

$$x_1 = q_1, \qquad x_2 = q_2, \qquad x_3 = \dot{q}_1, \qquad x_4 = \dot{q}_2$$

As a result, the state-variable equations are

$$\dot{x}_1 = x_3$$
$$\dot{x}_2 = x_4$$
$$\dot{x}_3 = \ddot{q}_1 = \frac{1}{L_1}\left[-R(x_3 - x_4) - \frac{1}{C_1}x_1 + e(t)\right] \qquad (3.21)$$
$$\dot{x}_4 = \ddot{q}_2 = \frac{1}{L_2}\left[-R(x_4 - x_3) - \frac{1}{C_2}x_2\right]$$

The third and fourth equations are obtained by substituting the state variables in Eqs. (3.20a) and (3.20b), and the first two are automatic relations be-

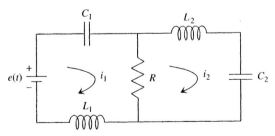

FIGURE 3.6 Two-loop electrical circuit.

tween the state variables, independent of the dynamics. The state equation is obtained by expressing Eq. (3.21) in matrix form, $\dot{\mathbf{x}} = \mathbf{Ax} + \mathbf{B}u$, as

$$\begin{Bmatrix} \dot{x}_1 \\ \dot{x}_2 \\ \dot{x}_3 \\ \dot{x}_4 \end{Bmatrix} = \begin{bmatrix} 0 & 0 & 1 & 0 \\ 0 & 0 & 0 & 1 \\ -1/(L_1 C_1) & 0 & -R/L_1 & R/L_1 \\ 0 & -1/(L_2 C_2) & R/L_2 & -R/L_2 \end{bmatrix} \begin{Bmatrix} x_1 \\ x_2 \\ x_3 \\ x_4 \end{Bmatrix} + \begin{bmatrix} 0 \\ 0 \\ 1/L_1 \\ 0 \end{bmatrix} e(t)$$

(b) By assumption, the output is q_1, that is,

$$y = \underset{\text{first charge}}{q_1} \quad \Longrightarrow \quad y = \underset{\text{first state}}{x_1}$$

and the output equation is $y = \mathbf{Cx} + \mathbf{D}u$ where $\mathbf{C} = [1 \ 0 \ 0 \ 0]$, and $D = 0$. However, for the case of two outputs, q_1 and $\dot{q}_1$, we have

$$\mathbf{y} = \underset{\substack{\text{first charge} \\ \text{and current}}}{\begin{bmatrix} q_1 \\ \dot{q}_1 \end{bmatrix}} \quad \Longrightarrow \quad \mathbf{y} = \underset{\substack{\text{first and} \\ \text{third states}}}{\begin{bmatrix} x_1 \\ x_3 \end{bmatrix}}$$

and the output equation is $\mathbf{y} = \mathbf{Cx} + \mathbf{D}u$, where

$$\mathbf{C} = \begin{bmatrix} 1 & 0 & 0 & 0 \\ 0 & 0 & 1 & 0 \end{bmatrix} \quad \text{and} \quad \mathbf{D} = 0$$

Nonuniqueness of State Variables

It was mentioned earlier that a chosen set of state variables is not a unique set. That is, a different set of variables may be selected so that the system model is represented by equations that resemble Eq. (3.13), known as state-space form. As a result, this latter form will be different from the one corresponding to the original set of state variables. However, both representations will contain the same information regarding the nature and behavior of the dynamic system. Consider a dynamic system for which a set of n state variables, $x_1, x_2, \ldots, x_n$, has been properly selected and for which the corresponding state-space form is

$$\begin{cases} \dot{\mathbf{x}} = \mathbf{Ax} + \mathbf{B}u \\ \mathbf{y} = \mathbf{Cx} + \mathbf{D}u \end{cases} \tag{3.13}$$

Consider the matrix transformation

$$\mathbf{x} = \mathbf{P}\tilde{\mathbf{x}} \tag{3.22}$$

where $\mathbf{P}$ is a nonsingular, $n \times n$ matrix with constant entries. Substitution of Eq. (3.22) into Eq. (3.13) yields

$$\begin{cases} \mathbf{P}\dot{\tilde{\mathbf{x}}} = \mathbf{AP}\tilde{\mathbf{x}} + \mathbf{B}u & (3.23a) \\ \mathbf{y} = \mathbf{CP}\tilde{\mathbf{x}} + \mathbf{D}u & (3.23b) \end{cases}$$

Premultiplication of Eq. (3.23a) by $\mathbf{P}^{-1}$ results in

$$\begin{cases} \dot{\tilde{\mathbf{x}}} = \mathbf{P}^{-1}\mathbf{A}\mathbf{P}\tilde{\mathbf{x}} + \mathbf{P}^{-1}\mathbf{B}\mathbf{u} \\ \mathbf{y} = \mathbf{C}\mathbf{P}\tilde{\mathbf{x}} + \mathbf{D}\mathbf{u} \end{cases} \quad \overset{\tilde{A}=P^{-1}AP,\ \tilde{B}=P^{-1}B,\ \tilde{C}=CP}{\Longrightarrow} \quad \begin{cases} \dot{\tilde{\mathbf{x}}} = \tilde{\mathbf{A}}\tilde{\mathbf{x}} + \tilde{\mathbf{B}}\mathbf{u} \\ \mathbf{y} = \tilde{\mathbf{C}}\tilde{\mathbf{x}} + \mathbf{D}\mathbf{u} \end{cases} \quad (3.24)$$

which is precisely in the general form of Eq. (3.13) and hence represents a state-space form corresponding to a set of new state variables, $\tilde{x}_1, \tilde{x}_2, \ldots, \tilde{x}_n$, components of the new state vector, $\tilde{\mathbf{x}}$. This supports the earlier claim that a set of state variables is indeed not unique. Furthermore, state-space representations defined by Eqs. (3.13) and (3.24) contain the same information about the dynamic system under consideration.

Decoupling

In most situations, the selected state variables, $x_1, x_2, \ldots, x_n$, are **coupled** through the elements of the state matrix $\mathbf{A}$. That is, $\mathbf{A}$ is a full matrix, and hence the corresponding state-variable equations cannot be treated independently. To circumvent this problem we may use a specific type of matrix transformation, $\mathbf{x} = \mathbf{P}\tilde{\mathbf{x}}$, in which $\mathbf{P}$ is the modal matrix associated with $\mathbf{A}$. As discussed in Chapter 2, assuming that $\mathbf{A}$ has a complete set of linearly independent eigenvectors, then these vectors constitute $\mathbf{P}$, making $\mathbf{P}$ nonsingular. Consequently, by Theorem 2.5, we have

$$\tilde{\mathbf{A}} = \mathbf{P}^{-1}\mathbf{A}\mathbf{P} = \Lambda$$

where Λ is a diagonal matrix whose entries along the main diagonal are exactly the eigenvalues of $\mathbf{A}$. Then the state equation corresponding to $\tilde{\mathbf{x}}$ takes the special form of

$$\dot{\tilde{\mathbf{x}}} = \mathbf{P}^{-1}\mathbf{A}\mathbf{P}\tilde{\mathbf{x}} + \mathbf{P}^{-1}\mathbf{B}\mathbf{u} = \Lambda\tilde{\mathbf{x}} + \mathbf{P}^{-1}\mathbf{B}\mathbf{u} \quad (3.25)$$

or, componentwise,

$$\begin{Bmatrix} \dot{\tilde{x}}_1 \\ \dot{\tilde{x}}_2 \\ \vdots \\ \dot{\tilde{x}}_n \end{Bmatrix} = \begin{bmatrix} \lambda_1 & & & \\ & \lambda_2 & & \\ & & \ddots & \\ & & & \lambda_n \end{bmatrix} \begin{Bmatrix} \tilde{x}_1 \\ \tilde{x}_2 \\ \vdots \\ \tilde{x}_n \end{Bmatrix} + [\mathbf{P}^{-1}\mathbf{B}][\mathbf{u}] \quad (3.26)$$

where $\lambda_1, \lambda_2, \ldots, \lambda_n$ are eigenvalues of $\mathbf{A}$. Then it is clear that each row of Eq. (3.26) is a first-order differential equation in one state variable, independent of the others, and can be treated separately. In this circumstance, we say that the coupled system has been **decoupled**.

EXAMPLE 3.9. Suppose the governing equation of a single-degree-of-freedom system is given by $\ddot{x} + 5\dot{x} + 4x = f(t)$, where x and f are the system output and input, respectively. Obtain two sets of state variables and their corresponding state-space forms.

Solution. Using the same reasoning as before, there exist two state variables, $x_1 = x$ and $x_2 = \dot{x}$. As a result, the state-space form is

CHAPTER 3: System Model Representation   123

$$\begin{cases} \dot{\mathbf{x}} = \mathbf{A}\mathbf{x} + \mathbf{B}u \\ y = \mathbf{C}\mathbf{x} + Du \end{cases}$$

where $\mathbf{x} = \begin{bmatrix} x_1 \\ x_2 \end{bmatrix}$, $\mathbf{A} = \begin{bmatrix} 0 & 1 \\ -4 & -5 \end{bmatrix}$, $\mathbf{B} = \begin{bmatrix} 0 \\ 1 \end{bmatrix}$, $\mathbf{C} = \begin{bmatrix} 1 & 0 \end{bmatrix}$,

$D = 0$, $u = f$

To obtain a second set of state variables, we first calculate the eigenvalues and the corresponding eigenvectors of the state matrix $\mathbf{A}$. These are readily seen to be

$$\lambda_1 = -4, \qquad \lambda_2 = -1$$

$$\mathbf{v}_1 = \begin{bmatrix} -1 \\ 4 \end{bmatrix}, \qquad \mathbf{v}_2 = \begin{bmatrix} 1 \\ -1 \end{bmatrix}$$

Recall from Chapter 2 that because eigenvalues of $\mathbf{A}$ are distinct, $\mathbf{v}_1$ and $\mathbf{v}_2$ are linearly independent. Form the modal matrix as

$$\mathbf{P} = [\mathbf{v}_1 \quad \mathbf{v}_2] = \begin{bmatrix} -1 & 1 \\ 4 & -1 \end{bmatrix}$$

which is nonsingular and has the property that

$$\mathbf{P}^{-1}\mathbf{A}\mathbf{P} = \Lambda = \begin{bmatrix} -4 & 0 \\ 0 & -1 \end{bmatrix}$$

If we now use the matrix transformation $\mathbf{x} = \mathbf{P}\tilde{\mathbf{x}}$, the resulting state equation is as in Eq. (3.25),

$$\dot{\tilde{\mathbf{x}}} = \Lambda\tilde{\mathbf{x}} + \mathbf{P}^{-1}\mathbf{B}u$$

$$\begin{Bmatrix} \dot{\tilde{x}}_1 \\ \dot{\tilde{x}}_2 \end{Bmatrix} = \begin{bmatrix} -4 & 0 \\ 0 & -1 \end{bmatrix} \begin{Bmatrix} \tilde{x}_1 \\ \tilde{x}_2 \end{Bmatrix} + \frac{1}{3}\begin{bmatrix} 1 & 1 \\ 4 & 1 \end{bmatrix} \begin{bmatrix} 0 \\ 1 \end{bmatrix} u$$

$$= \begin{bmatrix} -4 & 0 \\ 0 & -1 \end{bmatrix} \begin{Bmatrix} \tilde{x}_1 \\ \tilde{x}_2 \end{Bmatrix} + \begin{bmatrix} 1/3 \\ 1/3 \end{bmatrix} u$$

indicating the independence of the two first-order, differential equations in the new state variables $\tilde{x}_1$ and $\tilde{x}_2$. Similarly, the output equation can be expressed as

$$y = \mathbf{C}\mathbf{P}\tilde{\mathbf{x}}$$

$$= [1 \quad 0]\begin{bmatrix} -1 & 1 \\ 4 & -1 \end{bmatrix}\tilde{\mathbf{x}} = [-1 \quad 1]\tilde{\mathbf{x}}$$

In summary, the state-space form associated with the new state vector $\tilde{\mathbf{x}}$ is

$$\begin{cases} \dot{\tilde{\mathbf{x}}} = \begin{bmatrix} -4 & 0 \\ 0 & -1 \end{bmatrix}\tilde{\mathbf{x}} + \begin{bmatrix} 1/3 \\ 1/3 \end{bmatrix} u \\ y = [-1 \quad 1]\tilde{\mathbf{x}} \end{cases}$$

PROBLEM SET 3.2

For each given differential equation, choose a suitable set of state variables and find the resulting state-variable equations. Then, express in matrix form to obtain the state equation.

3.4. $2\ddot{x} + 3\dot{x} + x = f(t)$

3.5. $\dddot{x} + \ddot{x} + 2\dot{x} + x = 2f(t)$

3.6. Consider the RLC circuit shown in Fig. P3.6 for which the governing differential equation is

$$L\ddot{q} + R\dot{q} + \frac{1}{C}q = e(t)$$

Here, q denotes the electric charge and i represents the current, and they are related through $i = dq/dt$.
(a) By selecting a set of state variables, determine the state equation.
(b) Write the output equation if electric charge is the system output. Repeat for the case in which circuit current is the output.

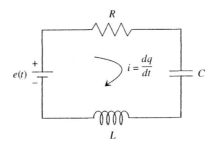

FIGURE P3.6 RLC circuit.

3.7. Consider the rotational mechanical system shown in Fig. P3.7 in which τ is an applied torque. The system is, as will be shown in Chapter 4, governed by

$$\begin{cases} J_1\ddot{\theta}_1 + K_1\theta_1 + K_2(\theta_1 - \theta_2) = \tau \\ J_2\ddot{\theta}_2 + K_3\theta_2 - K_2(\theta_1 - \theta_2) = 0 \end{cases}$$

(a) Determine the state-space form if the system output is θ_1.
(b) Repeat for the case in which both θ_1 and θ_2 are outputs.

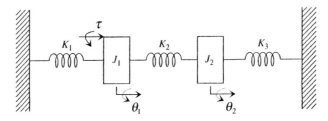

FIGURE P3.7 Rotational mechanical system.

3.8. The governing equations for a rotational mechanical system are

$$\begin{cases} J\ddot{\theta}_1 + K_1\theta_1 + K_2(\theta_1 - \theta_2) = \tau \\ K_3\theta_2 = K_2(\theta_1 - \theta_2) \end{cases}$$

J, K_1, K_2, and K_3 are constant physical parameters, θ_1 and θ_2 denote angular displacements, and τ is an applied torque. Note that the second equation merely provides an algebraic relation between θ_1 and θ_2, and is not a differential equation. Eliminate θ_2 from the two equations to obtain a single differential equation in θ_1 only. Choose a set of state variables and find the state equation. Assuming that θ_2 is the system output, write the output equation.

3.9. In Problem 3.8, eliminate θ_1 from the governing equations to find a single differential equation in θ_2 only. Assuming that θ_2 is the output, write the state-space representation for the system.

3.10. A certain dynamic system is governed by the third-order differential equation

$$\dddot{x} + 5\ddot{x} + 4\dot{x} = f(t)$$

(a) Through selection of a suitable set of state variables, determine the state-space form. Assume that x is the system output.
(b) Using a matrix transformation, $\mathbf{x} = \mathbf{P}\tilde{\mathbf{x}}$, construct a new state-space representation for this system, in which the state matrix is diagonal.

3.3 INPUT-OUTPUT (I/O) EQUATION

Another form in which a system model may be represented is an **input-output (I/O) equation,** which is a single differential equation in terms of the system input, system output, and their (time) derivatives

$$y^{(n)} + a_1 y^{(n-1)} + \cdots + a_{n-1}\dot{y} + a_n y = b_0 u^{(m)} + b_1 u^{(m-1)}$$
$$+ \cdots + b_{m-1}\dot{u} + b_m u, \quad m \leq n \quad (3.27)$$

In Eq. (3.27), a_i ($i = 1, 2, \ldots, n$) and b_k ($k = 0, 1, \ldots, m$) are constant coefficients (for a linear system), y is the system output, and u is the input. For a dynamic system involving many generalized coordinates, one often finds it extremely difficult or impossible to obtain the input-output equation directly from the governing equations. This is mainly because, in most cases, the generalized coordinates are *coupled* through the governing differential equations. This means that the derivatives of many of these coordinates may appear simultaneously in some of the differential equations, thus making it quite difficult to obtain a single differential equation in one coordinate via elimination of the unwanted variables. To circumvent this problem, an alternative approach using Laplace transformation is introduced.

Strategy

The idea is to take the Laplace transform of each differential equation in the system model, assuming zero initial conditions. Consequently, a set of algebraic equations

126   Dynamic Systems: Modeling and Analysis

in terms of the transform functions of the coordinates will be obtained. Then the unwanted variables may be eliminated through algebraic methods such as substitution or Cramer's rule (Chapter 2) to produce a single equation in terms of the Laplace transforms of the desired coordinate and input. Ultimately, this equation is transformed into time domain and interpreted as a differential equation in the form of Eq. (3.27).

EXAMPLE 3.10. Consider the mechanical system of Example 3.3 with the equation of motion

$$\ddot{x} + b\dot{x} + kx = f(t)$$

where the applied force $f(t)$ represents the system input. Obtain the input-output equation assuming that the system output is the displacement, $x(t)$, of the block, and assuming zero initial conditions.

Solution. By assumption, the output is $y = x$ and the input is $u = f$. Direct substitution of these into the equation of motion results in

$$\ddot{y} + b\dot{y} + ky = u$$

which is in the general form of Eq. (3.27) with

$$n = 2, \quad a_1 = b, \quad a_2 = k, \quad m = 0, \quad b_0 = 1$$

where m is the order of the right-hand side of Eq. (3.27). Therefore, the governing equation is already in the desired form, and no further analysis is needed.

EXAMPLE 3.11. Obtain the input-output equation for the mechanical system of Example 3.4, where the input and output are $f(t)$ and $x_1(t)$, respectively.

Solution. From previous results, the equations of motion are

$$\begin{cases} m_1\ddot{x}_1 + b_1\dot{x}_1 + k_1 x_1 - k_2(x_2 - x_1) - b_2(\dot{x}_2 - \dot{x}_1) = 0 & (3.8a) \\ m_2\ddot{x}_2 + b_2(\dot{x}_2 - \dot{x}_1) + k_2(x_2 - x_1) = f(t) & (3.8b) \end{cases}$$

The input-output equation must be a differential equation in terms of $f(t)$, $x_1(t)$, and their time derivatives. Taking the Laplace transform of Eqs. (3.8a and b) results in

$$\begin{cases} m_1 s^2 X_1(s) + b_1 s X_1(s) + k_1 X_1(s) - k_2[X_2(s) - X_1(s)] - b_2[s X_2(s) - s X_1(s)] = 0 \\ m_2 s^2 X_2(s) + b_2[s X_2(s) - s X_1(s)] + k_2[X_2(s) - X_1(s)] = F(s) \end{cases}$$

$$(3.28)$$

Collect like terms in Eq. (3.28) to obtain

$$\begin{cases} [m_1 s^2 + (b_1 + b_2)s + k_1 + k_2] X_1(s) - (b_2 s + k_2) X_2(s) = 0 \\ -(b_2 s + k_2) X_1(s) + (m_2 s^2 + b_2 s + k_2) X_2(s) = F(s) \end{cases}$$

or, in matrix form,

$$\begin{bmatrix} m_1 s^2 + (b_1 + b_2)s + k_1 + k_2 & -(b_2 s + k_2) \\ -(b_2 s + k_2) & m_2 s^2 + b_2 s + k_2 \end{bmatrix} \begin{Bmatrix} X_1(s) \\ X_2(s) \end{Bmatrix} = \begin{Bmatrix} 0 \\ F(s) \end{Bmatrix}$$

Because the output is x_1, directly solve the above for $X_1(s)$ via Cramer's rule, as follows

$$X_1(s) = \frac{\begin{vmatrix} 0 & -(b_2 s + k_2) \\ F(s) & m_2 s^2 + b_2 s + k_2 \end{vmatrix}}{\begin{vmatrix} m_1 s^2 + (b_1 + b_2)s + k_1 + k_2 & -(b_2 s + k_2) \\ -(b_2 s + k_2) & m_2 s^2 + b_2 s + k_2 \end{vmatrix}}$$

$$\Rightarrow \quad X_1(s) = \frac{(b_2 s + k_2)F(s)}{[m_1 s^2 + (b_1 + b_2)s + k_1 + k_2](m_2 s^2 + b_2 s + k_2) - (b_2 s + k_2)^2} \quad (3.29)$$

Algebraic manipulation of Eq. (3.29) yields

$$\{m_1 m_2 s^4 + [m_1 b_2 + m_2(b_1 + b_2)]s^3 + [m_1 k_2 + b_1 b_2 + (k_1 + k_2)m_2]s^2$$
$$+ (b_2 k_1 + b_1 k_2)s + k_1 k_2\} X_1(s) = (b_2 s + k_2)F(s)$$

In time domain, representing the system's output and input, this equation then reads

$$m_1 m_2 x_1^{(4)} + [m_1 b_2 + m_2(b_1 + b_2)]\dddot{x}_1 + [m_1 k_2 + b_1 b_2 + (k_1 + k_2)m_2]\ddot{x}_1 \quad (3.30)$$
$$+ (b_2 k_1 + b_1 k_2)\dot{x}_1 + k_1 k_2 x_1 = b_2 \dot{f} + k_2 f$$

which is the system's input-output equation. As expected, Eq. (3.30) is a differential equation relating input f, output x_1, and their derivatives, and is precisely in the general form of Eq. (3.27).

Observing the complexity of the differential equations here, and the fact that they are coupled, one can certainly appreciate the above technique as opposed to the elimination and substitution process. Direct elimination of x_2 and $\dot{x}_2$ would require a considerable amount of algebraic manipulation, which naturally increases with the number of variables involved in the system.

Remarks:

1. By now it is realized that an input-output equation provides a relation between one output and one input. Thus, when dealing with a multi-input/multi-output (MIMO) system, there exists an input-output equation corresponding to every pair of input and output. For instance, for a system with one input and two outputs, there exist two input-output equations.
2. A solution to the state equation

$$\begin{cases} \dot{\mathbf{x}} = \mathbf{A}\mathbf{x} + \mathbf{B}\mathbf{u} & \text{state equation} \\ \mathbf{x}(0) = \mathbf{x}_0 & \text{initial state} \end{cases}$$

can be found once there is complete knowledge of the **initial state-vector**, $\mathbf{x}_0$, i.e., the initial values of all state variables. On the other hand, to solve the input-output equation

$$y^{(n)} + a_1 y^{(n-1)} + \cdots + a_{n-1} \dot{y} + a_n y = b_0 u^{(m)} + b_1 u^{(m-1)} + \cdots + b_m u$$

one needs to have complete knowledge of $y(0), \dot{y}(0), \ldots, y^{(n-1)}(0)$, (i.e., initial values of the output) as well as the first $n-1$ derivatives of the output. However, from a practical standpoint, determination of these initial values can pose serious problems. Therefore, for higher-order systems, as in Example 3.11, solving the state equation may in fact require less computational effort as compared to the input-output equation.

PROBLEM SET 3.3

● **3.11.** Consider a dynamic system with input $f(t)$ and output x_1, whose state-variable equations are

$$\begin{cases} \dot{x}_1 = x_2 \\ \dot{x}_2 = \frac{1}{2}[-3x_2 - 2x_1 + f(t)] \end{cases}$$

Directly from these equations, determine the input-output equation.

3.12. In Problem 3.11, find the input-output equation via Laplace transformation of the state-variable equations, solving for $X_1(s)$, and interpreting the result in time domain. Assume zero initial conditions.

3.13. In Example 3.11, assume that x_2 is the output. Find the input-output equation.

3.14. Consider the rotational system studied in Problem 3.7. The equations of motion were given as

$$\begin{cases} J_1\ddot{\theta}_1 + K_1\theta_1 + K_2(\theta_1 - \theta_2) = \tau \\ J_2\ddot{\theta}_2 + K_3\theta_2 - K_2(\theta_1 - \theta_2) = 0 \end{cases}$$

Suppose that the output is θ_2 and that the system is subjected to zero initial conditions. Given specific parameter values $J_1 = 1 = J_2$ and $K_1 = K_2 = K_3 = K$, find the input-output equation.

3.4 TRANSFER FUNCTION

Once again, consider a linear, time-invariant (constant coefficients) system described by

$$y^{(n)} + a_1 y^{(n-1)} + \cdots + a_{n-1}\dot{y} + a_n y = b_0 u^{(m)} + b_1 u^{(m-1)} \\ + \cdots + b_{m-1}\dot{u} + b_m u, \quad m \le n \quad (3.27)$$

in which u and y denote the system input and output, respectively. Furthermore, assume that initial conditions are all zero; i.e., $u(0) = 0 = \cdots = u^{(m-1)}(0)$ and $y(0) = 0 = \cdots = y^{(n-1)}(0)$. Take Laplace transforms of both sides of Eq. (3.27) to obtain

$$(s^n + a_1 s^{n-1} + \cdots + a_{n-1}s + a_n)Y(s) = (b_0 s^m + b_1 s^{m-1} + \cdots + b_m)U(s) \quad (3.31)$$

Then, assuming zero initial conditions, the **transfer function** is defined as the ratio of the Laplace transform of the output and the Laplace transform of the input. From

Eq. (3.31), the transfer function is then determined to be in the form of a rational function,

$$G(s) = \frac{Y(s)}{U(s)} = \frac{b_0 s^m + b_1 s^{m-1} + \cdots + b_m}{s^n + a_1 s^{n-1} + \cdots + a_{n-1} s + a_n} \qquad (3.32)$$

Recall that each pair of system input and output corresponds to an input-output equation. The same is true here in the sense that corresponding to each pair of input and output, there exists a single transfer function. For instance, in Example 3.3, there is only one, and in Example 3.4, there may be one or more depending on the choice(s) of the system output. In general, for a MIMO system with p inputs and q outputs there are a total of pq transfer functions. Grouping these transfer functions in the form of a matrix yields the $q \times p$ **transfer function matrix** (or simply, **transfer matrix**).

Transfer functions can be used to determine the output of a system associated with a known input. From Eq. (3.32) it is readily seen that $Y(s) = G(s)U(s)$ in which $G(s)$ and $U(s)$ are both known. Taking the inverse Laplace of both sides, the output is

$$y(t) = L^{-1}\{Y(s)\} = L^{-1}\{G(s)U(s)\} \qquad (3.33)$$

The inverse Laplace transform on the right-hand side of Eq. (3.33) may then be determined using the techniques discussed in Chapter 1.

EXAMPLE 3.12. Consider the single-degree-of-freedom mechanical system studied in Example 3.3. Assume it to be subject to zero initial conditions. Suppose x is the output and $f(t)$ is the input. Determine the transfer function.

Solution. Taking the Laplace transform of the equation of motion, assuming zero initial conditions, results in

$$(ms^2 + bs + k)X(s) = F(s)$$

so that the system transfer function is obtained as

$$\frac{X(s)}{F(s)} = \frac{1}{ms^2 + bs + k}$$

EXAMPLE 3.13. In the system shown in Fig. 3.7, $x(t)$ and $y(t)$ denote the output and input, respectively. Given that the equation of motion is $m\ddot{x} + c\dot{x} + kx = c\dot{y} + ky$, find the system transfer function, assuming zero initial conditions.

Solution. Taking the Laplace transform of both sides of the equation of motion and collecting like terms, one obtains

$$(ms^2 + cs + k)X(s) = (cs + k)Y(s)$$

Consequently, the transfer function is

$$G(s) = \frac{X(s)}{Y(s)} = \frac{cs + k}{ms^2 + cs + k}$$

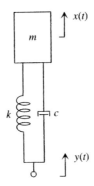

FIGURE 3.7 Single-degree-of-freedom system.

EXAMPLE 3.14. Consider the two-degree-of-freedom mechanical system in Fig. 3.8 subjected to zero initial conditions. The equations of motion are given as

$$\begin{cases} m_1\ddot{x}_1 + c\dot{x}_1 - c\dot{x}_2 + kx_1 - kx_2 = f_1(t) \\ m_2\ddot{x}_2 - c\dot{x}_1 + c\dot{x}_2 - kx_1 + kx_2 = f_2(t) \end{cases} \quad (3.34)$$

where x_1 and x_2 are system outputs, and f_1 and f_2 are system inputs. Determine the transfer function matrix.

Solution. Because there are two inputs and two outputs, there are four transfer functions, denoted by $G_{11}(s)$, $G_{12}(s)$, $G_{21}(s)$, and $G_{22}(s)$. Consequently, the transfer matrix is formed as

$$\mathbf{G}(s) = \begin{bmatrix} G_{11}(s) & G_{12}(s) \\ G_{21}(s) & G_{22}(s) \end{bmatrix}$$

More specifically, these transfer functions are defined as follows

$$G_{11}(s) = \left.\frac{X_1(s)}{F_1(s)}\right|_{F_2(s)=0}, \quad G_{12}(s) = \left.\frac{X_1(s)}{F_2(s)}\right|_{F_1(s)=0}$$

$$G_{21}(s) = \left.\frac{X_2(s)}{F_1(s)}\right|_{F_2(s)=0}, \quad G_{22}(s) = \left.\frac{X_2(s)}{F_2(s)}\right|_{F_1(s)=0} \quad (3.35)$$

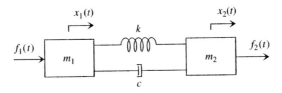

FIGURE 3.8 Two-degree-of-freedom mechanical system.

Express the equations of motion, Eq. (3.34), in second-order matrix form (Section 3.1), as

$$\begin{bmatrix} m_1 & 0 \\ 0 & m_2 \end{bmatrix} \begin{Bmatrix} \ddot{x}_1 \\ \ddot{x}_2 \end{Bmatrix} + \begin{bmatrix} c & -c \\ -c & c \end{bmatrix} \begin{Bmatrix} \dot{x}_1 \\ \dot{x}_2 \end{Bmatrix} + \begin{bmatrix} k & -k \\ -k & k \end{bmatrix} \begin{Bmatrix} x_1 \\ x_2 \end{Bmatrix} = \begin{Bmatrix} f_1 \\ f_2 \end{Bmatrix} \quad (3.36)$$

Taking the Laplace transform of both sides of Eq. (3.36) after setting initial conditions to zero, one obtains

$$\begin{bmatrix} m_1 s^2 + cs + k & -(cs + k) \\ -(cs + k) & m_2 s^2 + cs + k \end{bmatrix} \begin{Bmatrix} X_1(s) \\ X_2(s) \end{Bmatrix} = \begin{Bmatrix} F_1(s) \\ F_2(s) \end{Bmatrix} \quad (3.37)$$

Next, use Cramer's rule to solve for $X_1(s)$. This requires replacing the first column of the coefficient matrix by the vector on the right-hand side, and

$$X_1(s) = \frac{1}{\Delta(s)} \begin{vmatrix} F_1(s) & -(cs + k) \\ F_2(s) & m_2 s^2 + cs + k \end{vmatrix} = \frac{(m_2 s^2 + cs + k)F_1(s) + (cs + k)F_2(s)}{\Delta(s)}$$

$$= \frac{m_2 s^2 + cs + k}{\Delta(s)} F_1(s) + \frac{cs + k}{\Delta(s)} F_2(s)$$

where $\Delta(s)$ denotes the determinant of the 2×2 coefficient matrix in Eq. (3.37), and is defined as

$$\Delta(s) = \begin{vmatrix} m_1 s^2 + cs + k & -(cs + k) \\ -(cs + k) & m_2 s^2 + cs + k \end{vmatrix}$$

$$= (m_1 s^2 + cs + k)(m_2 s^2 + cs + k) - (cs + k)^2$$

Similarly, solve Eq. (3.37) for $X_2(s)$ using Cramer's rule. This time the second column of the coefficient matrix is replaced by the vector on the right-hand side, and

$$X_2(s) = \frac{1}{\Delta(s)} \begin{vmatrix} m_1 s^2 + cs + k & F_1(s) \\ -(cs + k) & F_2(s) \end{vmatrix} = \frac{(m_1 s^2 + cs + k)F_2(s) + (cs + k)F_1(s)}{\Delta(s)}$$

$$= \frac{m_1 s^2 + cs + k}{\Delta(s)} F_2(s) + \frac{cs + k}{\Delta(s)} F_1(s)$$

Subsequently, all four transfer functions, defined through the relations in Eq. (3.35), can be obtained as

$$G_{11}(s) = \left. \frac{X_1(s)}{F_1(s)} \right|_{F_2(s)=0} = \frac{m_2 s^2 + cs + k}{\Delta(s)}$$

$$= \frac{m_2 s^2 + cs + k}{m_1 m_2 s^4 + (m_1 + m_2)cs^3 + (m_1 + m_2)ks^2}$$

$$G_{12}(s) = \left. \frac{X_1(s)}{F_2(s)} \right|_{F_1(s)=0} = \frac{cs + k}{\Delta(s)} = \frac{cs + k}{m_1 m_2 s^4 + (m_1 + m_2)cs^3 + (m_1 + m_2)ks^2}$$

$$G_{21}(s) = \left. \frac{X_2(s)}{F_1(s)} \right|_{F_2(s)=0} = \frac{cs + k}{\Delta(s)} = \frac{cs + k}{m_1 m_2 s^4 + (m_1 + m_2)cs^3 + (m_1 + m_2)ks^2}$$

$$G_{22}(s) = \left.\frac{X_2(s)}{F_2(s)}\right|_{F_1(s)=0} = \frac{m_1 s^2 + cs + k}{\Delta(s)}$$

$$= \frac{m_1 s^2 + cs + k}{m_1 m_2 s^4 + (m_1 + m_2)cs^3 + (m_1 + m_2)ks^2}$$

Then, constitute the transfer matrix, $\mathbf{G}(s)$, as defined earlier.

EXAMPLE 3.15. Determine the transfer matrix for the electrical circuit of Example 3.8. Assume that q_1 and q_2 are system outputs, and that initial conditions are zero.

Solution. The governing equations of this system are given by Eq. (3.20) and may be expressed in the standard second-order matrix form as

$$\begin{bmatrix} L_1 & 0 \\ 0 & L_2 \end{bmatrix} \begin{Bmatrix} \ddot{q}_1 \\ \ddot{q}_2 \end{Bmatrix} + \begin{bmatrix} R & -R \\ -R & R \end{bmatrix} \begin{Bmatrix} \dot{q}_1 \\ \dot{q}_2 \end{Bmatrix} + \begin{bmatrix} 1/C_1 & 0 \\ 0 & 1/C_2 \end{bmatrix} \begin{Bmatrix} Q_1 \\ Q_2 \end{Bmatrix} = \begin{Bmatrix} e \\ 0 \end{Bmatrix}$$

Laplace transformation of the governing equations, taking into account zero initial conditions, yields

$$\begin{bmatrix} L_1 s^2 + Rs + 1/C_1 & -Rs \\ -Rs & L_2 s^2 + Rs + 1/C_2 \end{bmatrix} \begin{Bmatrix} Q_1(s) \\ Q_2(s) \end{Bmatrix} = \begin{Bmatrix} E(s) \\ 0 \end{Bmatrix}$$

Using Cramer's rule, solve for $Q_1(s)$ and $Q_2(s)$ separately to obtain

$$Q_1(s) = \frac{1}{\Delta(s)} \begin{vmatrix} E(s) & -Rs \\ 0 & L_2 s^2 + Rs + 1/C_2 \end{vmatrix} = \frac{L_2 s^2 + Rs + 1/C_2}{\Delta(s)} E(s) \implies$$

$$G_1(s) = \frac{Q_1(s)}{E(s)} = \frac{L_2 s^2 + Rs + 1/C_2}{\Delta(s)}$$

$$Q_2(s) = \frac{1}{\Delta(s)} \begin{vmatrix} L_1 s^2 + Rs + 1/C_1 & E(s) \\ -Rs & 0 \end{vmatrix} = \frac{Rs}{\Delta(s)} E(s) \implies$$

$$G_2(s) = \frac{Q_2(s)}{E(s)} = \frac{Rs}{\Delta(s)}$$

where

$$\Delta(s) = (L_1 s^2 + Rs + 1/C_1)(L_2 s^2 + Rs + 1/C_2) - R^2 s^2$$

Insert this into the expressions of $G_1(s)$ and $G_2(s)$ and simplify to obtain

$$G_1(s) = \frac{L_2 C_1 C_2 s^2 + RC_1 C_2 s + C_1}{L_1 L_2 C_1 C_2 s^4 + RC_1 C_2 (L_1 + L_2) s^3 + (L_1 C_1 + L_2 C_2) s^2 + R(C_1 + C_2)s + 1}$$

and

$$G_2(s) = \frac{RC_1 C_2 s}{L_1 L_2 C_1 C_2 s^4 + RC_1 C_2 (L_1 + L_2) s^3 + (L_1 C_1 + L_2 C_2) s^2 + R(C_1 + C_2)s + 1}$$

Consequently, the transfer matrix is defined as

$$\mathbf{G}(s) = \begin{Bmatrix} G_1(s) \\ G_2(s) \end{Bmatrix}$$

Relation Between State-Space Form and Transfer Function

Regardless of the type of representation for the mathematical model of a dynamic system, similar information about the system may be extracted. More specifically, given that the state-space form of a system model is available, its transfer function or transfer matrix can be determined using the state, input, output, and direct transmission matrices. To this end, we consider two separate cases: single input-single output (SISO) systems, and multiple input-multiple output (MIMO) systems.

Single Input-Single Output (SISO) Systems

Consider a dynamic system with a single input and a single output with a state-space representation

$$\begin{cases} \dot{\mathbf{x}} = \mathbf{A}\mathbf{x} + \mathbf{B}u \\ y = \mathbf{C}\mathbf{x} + Du \end{cases} \quad (3.38)$$

and a transfer function

$$G(s) = \frac{Y(s)}{U(s)} \quad (3.39)$$

Assume zero initial state; i.e., $\mathbf{x}(0) = \mathbf{0}_{n \times 1}$. Laplace transformation of state and output equations yields

$$\begin{cases} s\mathbf{X}(s) = \mathbf{A}\mathbf{X}(s) + \mathbf{B}U(s) \\ Y(s) = \mathbf{C}\mathbf{X}(s) + DU(s) \end{cases} \Longrightarrow \begin{cases} (s\mathbf{I} - \mathbf{A})\mathbf{X}(s) = \mathbf{B}U(s) \\ Y(s) = \mathbf{C}\mathbf{X}(s) + DU(s) \end{cases}$$

$$\Longrightarrow \begin{cases} \mathbf{X}(s) = (s\mathbf{I} - \mathbf{A})^{-1}\mathbf{B}U(s) & (3.40a) \\ Y(s) = \mathbf{C}\mathbf{X}(s) + DU(s) & (3.40b) \end{cases}$$

Substitution of $\mathbf{X}(s)$ from Eq. (3.40a) into Eq. (3.40b) results in

$$Y(s) = \mathbf{C}(s\mathbf{I} - \mathbf{A})^{-1}\mathbf{B}U(s) + DU(s) = [\mathbf{C}(s\mathbf{I} - \mathbf{A})^{-1}\mathbf{B} + D]U(s)$$

Thus, the transfer function defined by Eq. (3.39) may be expressed in terms of the state, input, output, and direct transmission matrices as

$$G(s) = \frac{Y(s)}{U(s)} = \mathbf{C}(s\mathbf{I} - \mathbf{A})^{-1}\mathbf{B} + D \quad (3.41)$$

Recall that

$$(s\mathbf{I} - \mathbf{A})^{-1} = \frac{1}{|s\mathbf{I} - \mathbf{A}|} \text{adj}(s\mathbf{I} - \mathbf{A})$$

Insert into Eq. (3.41) and simplify to obtain

$$G(s) = \frac{\mathbf{C}\,\text{adj}(s\mathbf{I} - \mathbf{A})\mathbf{B}}{|s\mathbf{I} - \mathbf{A}|} + D = \frac{\mathbf{C}\,\text{adj}(s\mathbf{I} - \mathbf{A})\mathbf{B} + |s\mathbf{I} - \mathbf{A}|D}{|s\mathbf{I} - \mathbf{A}|} = \frac{N(s)}{|s\mathbf{I} - \mathbf{A}|} \quad (3.42)$$

in which $N(s)$ is an nth-degree polynomial in s. From previous discussion in Chapter 1, poles of $G(s)$ are those values of s for which the denominator of $G(s)$ vanishes;

i.e., $|s\mathbf{I} - \mathbf{A}| = 0$. However, in Chapter 2 we learned that this equation is the characteristic equation associated with matrix $\mathbf{A}$, and hence, generates the eigenvalues of $\mathbf{A}$. Thus, the eigenvalues of the state matrix are identical to the poles of the system transfer function.

EXAMPLE 3.16. Determine the transfer function for the single-degree-of-freedom mechanical system of Example 3.3, using its state-space form.

Solution. The state-space form for this system was determined in Example 3.7 to be:

$$\begin{cases} \dot{\mathbf{x}} = \mathbf{A}\mathbf{x} + \mathbf{B}u \\ y = \mathbf{C}\mathbf{x} + \mathbf{D}u \end{cases}$$

where

$$\mathbf{x} = \begin{Bmatrix} x_1 \\ x_2 \end{Bmatrix}, \quad \mathbf{A} = \begin{bmatrix} 0 & 1 \\ -k/m & -b/m \end{bmatrix}, \quad \mathbf{B} = \begin{bmatrix} 0 \\ 1/m \end{bmatrix},$$

$$u = f(t), \quad \mathbf{C} = \begin{bmatrix} 1 & 0 \end{bmatrix}, \quad D = 0$$

Prior to substitution into Eq. (3.41), we note that

$$(s\mathbf{I} - \mathbf{A}) = \begin{bmatrix} s & -1 \\ k/m & s + (b/m) \end{bmatrix} \implies$$

$$(s\mathbf{I} - \mathbf{A})^{-1} = \frac{1}{s(s + b/m) + k/m} \begin{bmatrix} s + b/m & 1 \\ -k/m & s \end{bmatrix}$$

Consequently, Eq. (3.41) yields

$$G(s) = \begin{bmatrix} 1 & 0 \end{bmatrix} \frac{1}{s(s + b/m) + k/m} \begin{bmatrix} s + b/m & 1 \\ -k/m & s \end{bmatrix} \begin{bmatrix} 0 \\ 1/m \end{bmatrix}$$

$$= \frac{1}{s(s + b/m) + k/m} \begin{bmatrix} 1 & 0 \end{bmatrix} \begin{bmatrix} 1/m \\ s/m \end{bmatrix}$$

$$= \frac{1}{s(s + b/m) + k/m} \cdot \frac{1}{m} = \frac{1}{ms^2 + bs + k}$$

which is in agreement with the result of Example 3.12.

Multiple Input-Multiple Output (MIMO) Systems

The formulation leading to Eq. (3.41) was based on the assumption of a single input and a single output that caused $G(s)$ and D to be scalars (1×1). When the system has multiple inputs and outputs, $G(s)$ and D extend to the transfer matrix, $\mathbf{G}(s)$, and the direct transmission matrix, $\mathbf{D}$, respectively. In treating these systems, there exist two possible scenarios: (a) a specific transfer function, or a few selected transfer functions, are desired; or (b) the entire transfer matrix is sought. In either situation, what turns out to be a significant consideration is the adjustment of sizes of the matrices **B**, **C**, and **D** in Eq. (3.41). Proper modification of these matrices leads to the desired transfer function or transfer matrix. In what follows, both cases will be considered for a relatively simple example.

EXAMPLE 3.17. Suppose a dynamic system has the following governing equations

$$\begin{cases} \ddot{x}_1 + \dot{x}_1 - \dot{x}_2 + x_1 - x_2 = u_1 \\ \ddot{x}_2 - \dot{x}_1 + \dot{x}_2 - x_1 + x_2 = u_2 \end{cases} \quad (3.43)$$

where u_1 and u_2 denote the system inputs, and x_1 and x_2 represent the outputs. Determine the transfer function $X_1(s)/U_1(s)$, via a modification of Eq. (3.41).

Solution. This is clearly what was labeled as case (a) earlier. Proceeding as before, there are four state variables,

$$x_1 = x_1, \qquad x_2 = x_2, \qquad x_3 = \dot{x}_1, \qquad x_4 = \dot{x}_2$$

which result in the following state-variable equations

$$\begin{cases} \dot{x}_1 = x_3 \\ \dot{x}_2 = x_4 \\ \dot{x}_3 = -x_1 + x_2 - x_3 + x_4 + u_1 \\ \dot{x}_4 = x_1 - x_2 + x_3 - x_4 + u_2 \end{cases}$$

Consequently, the state equation is

$$\dot{\mathbf{x}} = \mathbf{A}\mathbf{x} + \mathbf{B}\mathbf{u}$$

where $\quad \mathbf{x} = \begin{Bmatrix} x_1 \\ x_2 \\ x_3 \\ x_4 \end{Bmatrix}, \quad \mathbf{A} = \begin{bmatrix} 0 & 0 & 1 & 0 \\ 0 & 0 & 0 & 1 \\ -1 & 1 & -1 & 1 \\ 1 & -1 & 1 & -1 \end{bmatrix},$

$$\mathbf{B} = \begin{bmatrix} 0 & 0 \\ 0 & 0 \\ 1 & 0 \\ 0 & 1 \end{bmatrix}, \quad \mathbf{u} = \begin{bmatrix} u_1 \\ u_2 \end{bmatrix}$$

Because $G_1(s) = X_1(s)/U_1(s)$ is the transfer function to be determined, x_1 must be chosen as the output and u_1 as the input. In other words, the problem at hand reduces to a single input-single output system, and may be treated as before. However, in doing so, certain matrices must be adjusted properly. Because u_1 is the input, the input matrix **B**, which was originally 4×2, reduces to a 4×1 matrix $\mathbf{B}_1$, which is the first column of **B**. This is because the elements of the first column of **B** correspond to u_1. Furthermore, since x_1 is the output, the output equation reads

$$y = \mathbf{C}_1 \mathbf{x} \quad \text{where} \quad \mathbf{C}_1 = [1 \ 0 \ 0 \ 0]$$

136 Dynamic Systems: Modeling and Analysis

With this information available, we now apply Eq. (3.41) as

$$G_1(s) = \frac{X_1(s)}{U_1(s)} = \mathbf{C}_1(s\mathbf{I} - \mathbf{A})^{-1}\mathbf{B}_1 + D \tag{3.44}$$

with $\mathbf{B}_1 = \begin{bmatrix} 0 \\ 0 \\ 1 \\ 0 \end{bmatrix}$ and $D = 0$

Calculate $(s\mathbf{I} - \mathbf{A})^{-1}$ and insert into Eq. (3.44) to find the desired transfer function

$$G_1(s) = \mathbf{C}_1(s\mathbf{I} - \mathbf{A})^{-1}\mathbf{B}_1 + D$$

$$= \begin{bmatrix} 1 & 0 & 0 & 0 \end{bmatrix} \frac{1}{s^2(s^2 + 2s + 2)}$$

$$\times \begin{bmatrix} s(s+1)^2 & s & s^2+s+1 & s+1 \\ s & s(s+1)^2 & s+1 & s^2+s+1 \\ -s^2 & s^2 & s(s^2+s+1) & s^2+s \\ s^2 & -s^2 & s^2+s & s(s^2+s+1) \end{bmatrix} \begin{bmatrix} 0 \\ 0 \\ 1 \\ 0 \end{bmatrix}$$

$$= \frac{1}{s^2(s^2+2s+2)} \begin{bmatrix} 1 & 0 & 0 & 0 \end{bmatrix} \begin{bmatrix} s^2+s+1 \\ s+1 \\ s(s^2+s+1) \\ s^2+s \end{bmatrix} = \frac{s^2+s+1}{s^2(s^2+2s+2)}$$

This result may be readily verified through Laplace transformation of the governing equations, Eq. (3.43), assuming zero initial conditions, and rearranging:

$$\begin{cases} (s^2+s+1)X_1(s) - (s+1)X_2(s) = U_1(s) \\ -(s+1)X_1(s) + (s^2+s+1)X_2(s) = U_2(s) \end{cases}$$

Solving for $X_1(s)$ via Cramer's rule, one obtains

$$X_1(s) = \frac{s^2+s+1}{s^2(s^2+2s+2)} U_1(s) + \frac{s+1}{s^2(s^2+2s+2)} U_2(s)$$

Setting $U_2(s) = 0$, we have

$$\frac{X_1(s)}{U_1(s)} = \frac{s^2+s+1}{s^2(s^2+2s+2)}$$

EXAMPLE 3.18. Referring to the system of Example 3.17, determine the transfer matrix $\mathbf{G}(s)$.

Solution. This problem falls under category (b), discussed earlier. We are seeking the transfer matrix associated with inputs u_1 and u_2, and outputs x_1 and x_2; hence, $\mathbf{G}(s)$ is 2×2. Then, matrices in Eq. (3.41) are adjusted accordingly to give

$$G(s) = C(sI - A)^{-1}B + D$$

$$= \begin{bmatrix} 1 & 0 & 0 & 0 \\ 0 & 1 & 0 & 0 \end{bmatrix} \frac{1}{s^2(s^2 + 2s + 2)}$$

$$\times \begin{bmatrix} s(s+1)^2 & s & s^2+s+1 & s+1 \\ s & s(s+1)^2 & s+1 & s^2+s+1 \\ -s^2 & s^2 & s(s^2+s+1) & s^2+s \\ s^2 & -s^2 & s^2+s & s(s^2+s+1) \end{bmatrix} \begin{bmatrix} 0 & 0 \\ 0 & 0 \\ 1 & 0 \\ 0 & 1 \end{bmatrix} + \begin{bmatrix} 0 & 0 \\ 0 & 0 \end{bmatrix}$$

$$= \frac{1}{s^2(s^2+2s+2)} \begin{bmatrix} 1 & 0 & 0 & 0 \\ 0 & 1 & 0 & 0 \end{bmatrix} \begin{bmatrix} s^2+s+1 & s+1 \\ s+1 & s^2+s+1 \\ s(s^2+s+1) & s^2+s \\ s^2+s & s(s^2+s+1) \end{bmatrix}$$

$$= \begin{bmatrix} \dfrac{s^2+s+1}{s^2(s^2+2s+2)} & \dfrac{s+1}{s^2(s^2+2s+2)} \\ \dfrac{s+1}{s^2(s^2+2s+2)} & \dfrac{s^2+s+1}{s^2(s^2+2s+2)} \end{bmatrix}$$

As expected, the (1, 1) entry of $G(s)$, corresponding to the input-output pair of u_1 and x_1, is in complete agreement with what was obtained in Example 3.17.

PROBLEM SET 3.4

3.15. The equation of motion of a rotational mechanical system is derived as

$$J\ddot{\theta}_o + B\dot{\theta}_o + K\theta_o = K\theta_i$$

where θ_i and θ_o denote angular displacements, and are system input and output, respectively. Parameters J, B, and K are constants. Assuming that the system is subjected to zero initial conditions, determine the transfer function $\Theta_o(s)/\Theta_i(s)$.

3.16. The governing equations of an electromechanical system can be shown to be (see Chapter 5)

$$\begin{cases} L\dfrac{di}{dt} + Ri + K_1\omega = v \\ J\dfrac{d\omega}{dt} + B\omega - K_2 i = 0 \end{cases}$$

Here, i and ω denote the current and the angular velocity, respectively, and are system outputs, and applied voltage v is the input. Parameters J, L, R, B, K_1, and K_2 are constants. Find the two possible transfer functions, represented by

$$G_1(s) = \frac{I(s)}{V(s)} \quad \text{and} \quad G_2(s) = \frac{\Omega(s)}{V(s)}$$

and determine the transfer matrix $G(s)$.

3.17. Express the governing equations of the system in Problem 3.16 in the form of a state equation. Use the state, input, and output matrices to find the transfer function when
(a) current i is the output, and
(b) angular velocity ω is the output.

How should your results compare with individual transfer functions obtained in Problem 3.16?

3.18. In Problem 3.17, determine the transfer matrix for the case in which the system has two outputs, i and ω. How should this compare with the transfer matrix of Problem 3.16?

3.19. Consider the system of Problem 3.15. Using a suitable set of state variables, express the equation of motion in the form of state equation. Determine the transfer function via the state, input, and output matrices. How does this expression compare with what was obtained in Problem 3.15?

3.5 STATE-SPACE REPRESENTATION FROM THE INPUT-OUTPUT EQUATION

In this section, a general procedure will be introduced to represent a dynamic system's input-output equation in its state-space form. The strategy directly depends on whether time derivatives of the input are present or absent in the input-output equation. In the event that the time derivatives of the input u are absent in the input-output equation, as in Example 3.10, the procedure becomes more simple and easier to implement. In the general case, when derivatives of u are present in the input-output equation, as in Eq. (3.30) of Example 3.11, the formulation tends to be more complicated.

Suppose the input u and output y of a dynamic system are related through the input-output equation, as

$$y^{(n)} + a_1 y^{(n-1)} + \cdots + a_{n-1}\dot{y} + a_n y = b_0 u^{(n)} + b_1 u^{(n-1)} + \cdots + b_n u \quad (3.45)$$

Then, taking the Laplace transformation of this equation and assuming zero initial conditions, the corresponding transfer function is obtained as

$$\frac{Y(s)}{U(s)} = \frac{b_0 s^n + b_1 s^{n-1} + \cdots + b_n}{s^n + a_1 s^{n-1} + \cdots + a_n} \quad (3.46)$$

Rewrite the transfer function, defining $W(s)$, as

$$\frac{Y(s)}{U(s)} = \frac{Y(s)}{W(s)} \cdot \frac{W(s)}{U(s)} = \underbrace{(b_0 s^n + b_1 s^{n-1} + \cdots + b_n)}_{Y(s)/W(s)} \underbrace{\left[\frac{1}{s^n + a_1 s^{n-1} + \cdots + a_n}\right]}_{W(s)/U(s)}$$

(3.47)

Interpretation of the newly constructed transfer functions in the time domain yields

$$\frac{Y(s)}{W(s)} = b_0 s^n + b_1 s^{n-1} + \cdots + b_n \implies y(t) = b_0 w^{(n)} + b_1 w^{(n-1)} + \cdots + b_n w \quad (3.48)$$

and

$$\frac{W(s)}{U(s)} = \frac{1}{s^n + a_1 s^{n-1} + \cdots + a_n} \implies w^{(n)} + a_1 w^{(n-1)} + \cdots + a_n w = u \quad (3.49)$$

Equation (3.49) represents an nth-order differential equation and hence, n initial conditions are required for a complete solution; that is, $w(0), \dot{w}(0), \ldots, w^{(n-1)}(0)$. Following the discussion in Section 3.2, the state variables are then chosen as

$$\begin{cases} x_1 = w \\ x_2 = \dot{w} \\ \vdots \\ x_n = w^{(n-1)} \end{cases} \quad (3.50)$$

The corresponding n state-variable equations may then be obtained as before. The first $n-1$ of these equations are merely automatic relations between the state variables, and the last one is generated using Eq. (3.49).

$$\begin{cases} \dot{x}_1 = x_2 \\ \dot{x}_2 = x_3 \\ \vdots \\ \dot{x}_n = w^{(n)} = -a_n x_1 - a_{n-1} x_2 - \cdots - a_1 x_n + u \end{cases} \quad (3.51)$$

Therefore, expressing the state-variable equations, Eq. (3.51), in matrix form, the *state equation* is given as

$$\dot{\mathbf{x}} = \mathbf{A}\mathbf{x} + \mathbf{B}u$$

where
$$\mathbf{x} = \begin{bmatrix} x_1 \\ x_2 \\ \vdots \\ x_n \end{bmatrix}, \quad \mathbf{A} = \begin{bmatrix} 0 & 1 & 0 & \cdots & 0 \\ 0 & 0 & 1 & \cdots & 0 \\ \vdots & \vdots & & \vdots & \vdots \\ 0 & 0 & \cdots & 0 & 1 \\ -a_n & -a_{n-1} & \cdots & -a_2 & -a_1 \end{bmatrix}, \quad \mathbf{B} = \begin{bmatrix} 0 \\ 0 \\ \vdots \\ 0 \\ 1 \end{bmatrix} \quad (3.52)$$

in which the state matrix $\mathbf{A}$ is referred to as the **lower companion matrix.** The system output y is given by Eq. (3.48), which can be expressed in terms of the state variables as

$$\begin{aligned} y &= b_0 w^{(n)} + b_1 w^{(n-1)} + \cdots + b_n w \\ &= b_0 \dot{x}_n + b_1 x_n + \cdots + b_n x_1 \end{aligned} \quad (3.53)$$

Substituting the last equation in Eq. (3.51) for $\dot{x}_n$ in Eq. (3.53), we have

$$y = b_0(-a_n x_1 - a_{n-1} x_2 - \cdots - a_1 x_n + u) + b_1 x_n + \cdots + b_n x_1$$

Rearrange and collect like terms to obtain

$$y = (-b_0 a_n + b_n) x_1 + (-b_0 a_{n-1} + b_{n-1}) x_2 + \cdots + (-b_0 a_1 + b_1) x_n + b_0 u \quad (3.54)$$

As a result, rewriting Eq. (3.54) using the matrix notation, the *output equation* is given as

$$y = \mathbf{C}\mathbf{x} + \mathbf{D}u$$

where $\quad \mathbf{C} = \begin{bmatrix} -b_0 a_n + b_n & -b_0 a_{n-1} + b_{n-1} & \cdots & -b_0 a_1 + b_1 \end{bmatrix}, \quad \mathbf{D} = b_0$
(3.55)

Equations (3.52) and (3.55) constitute the system's state-space form.

Remark: It is also common practice to define the state variables different from those in Eq. (3.50). In the event that they are selected as $x_1 = w^{(n-1)}$, $x_2 = w^{(n-2)}, \ldots, x_{n-1} = \dot{w}$, $x_n = w$, the corresponding state matrix takes the form

$$\begin{bmatrix} -a_1 & -a_2 & \cdots & \cdots & -a_{n-1} & -a_n \\ 1 & 0 & \cdots & \cdots & 0 & 0 \\ \vdots & \vdots & & & \vdots & \vdots \\ 0 & 0 & \cdots & 1 & 0 & 0 \\ 0 & 0 & \cdots & 0 & 1 & 0 \end{bmatrix}$$

and is known as the **upper companion matrix**. The obvious computational advantage of the lower and upper companion matrices lies in their special structure. Matrices of these types, containing several zero entries, are known as **sparse matrices**.

Special case

As mentioned earlier, if the time derivatives of input u are absent, the formulation simplifies substantially. To this end, consider a specific form of the input-output equation, Eq. (3.45), as

$$y^{(n)} + a_1 y^{(n-1)} + \cdots + a_{n-1} \dot{y} + a_n y = bu \qquad (3.56)$$

In this case, the state-space form can be directly obtained. Following similar reasoning as before, introduce a set of n state variables, as

$$x_1 = y, \quad x_2 = \dot{y}, \ldots, \quad x_n = y^{(n-1)}$$

The first $n - 1$ state-variable equations provide the relations between the state variables, and the last one is a direct result of Eq. (3.56), that is,

$$\begin{cases} \dot{x}_1 = x_2 \\ \dot{x}_2 = x_3 \\ \vdots \\ \dot{x}_{n-1} = x_n \\ \dot{x}_n = -a_1 x_n - \cdots - a_{n-1} x_2 - a_n x_1 + bu \end{cases} \qquad (3.57)$$

Thus, the state equation is given by

$$\dot{\mathbf{x}} = \mathbf{A}\mathbf{x} + \mathbf{B}u$$

where $\mathbf{x} = \begin{bmatrix} x_1 \\ x_2 \\ \vdots \\ x_n \end{bmatrix}$, $\mathbf{A} = \begin{bmatrix} 0 & 1 & 0 & \cdots & 0 \\ 0 & 0 & 1 & \cdots & 0 \\ \vdots & \vdots & & & \vdots \\ 0 & 0 & \cdots & 0 & 1 \\ -a_n & -a_{n-1} & \cdots & & -a_1 \end{bmatrix}$, $\mathbf{B} = \begin{bmatrix} 0 \\ 0 \\ \vdots \\ 0 \\ b \end{bmatrix}$ (3.58)

The system output is simply y, which turns out to be the first state variable selected; that is, $y = x_1$. Therefore, the output equation is given by

$$y = \mathbf{Cx} + Du \quad \text{where} \quad \mathbf{C} = [1 \ 0 \ \cdots \ 0], \quad D = 0 \quad (3.59)$$

Equations (3.58) and (3.59) define the system's state-space form.

EXAMPLE 3.19. A dynamic system is described by its transfer function, as

$$\frac{Y(s)}{U(s)} = \frac{1}{s^2 + 2s + 1}$$

Determine the state-space form.

Solution. Manipulation of the transfer function results in

$$\ddot{y} + 2\dot{y} + y = u$$

which is the system's input-output equation and agrees with the formulation of Eq. (3.56). Comparison reveals that

$$n = 2, \quad a_1 = 2, \quad a_2 = 1, \quad b = 1$$

Hence, defining the state variables, and the resulting state vector as

$$\mathbf{x} = \begin{bmatrix} x_1 = y \\ x_2 = \dot{y} \end{bmatrix}$$

the state-space form is

$$\begin{cases} \dot{\mathbf{x}} = \begin{bmatrix} 0 & 1 \\ -1 & -2 \end{bmatrix} \mathbf{x} + \begin{bmatrix} 0 \\ 1 \end{bmatrix} u \\ y = [1 \ 0] \mathbf{x} \end{cases}$$

EXAMPLE 3.20. Obtain the state-space representation for the input-output equation below

$$\dddot{y} + 4\ddot{y} + 2\dot{y} + 3y = 2\dot{u} + u \quad (3.60)$$

Solution. This is in the general form of Eq. (3.45). Following the general procedure outlined above, the system's transfer function, assuming zero initial conditions, is

$$\frac{Y(s)}{U(s)} = \frac{2s + 1}{s^3 + 4s^2 + 2s + 3} = \underbrace{(2s + 1)}_{Y(s)/W(s)} \underbrace{\left[\frac{1}{s^3 + 4s^2 + 2s + 3} \right]}_{W(s)/U(s)}$$

and in time domain,

$$\frac{Y(s)}{W(s)} = 2s + 1 \implies y = 2\dot{w} + w \quad \text{and} \quad \frac{W(s)}{U(s)} = \frac{1}{s^3 + 4s^2 + 2s + 3}$$

$$\implies \dddot{w} + 4\ddot{w} + 2\dot{w} + 3w = u$$

Defining the state variables and the corresponding state vector as

$$\mathbf{x} = \begin{bmatrix} x_1 = w \\ x_2 = \dot{w} \\ x_3 = \ddot{w} \end{bmatrix}$$

the state-variable equations are determined and expressed in matrix form to give the state equation, as

$$\begin{cases} \dot{x}_1 = x_2 \\ \dot{x}_2 = x_3 \\ \dot{x}_3 = -3x_1 - 2x_2 - 4x_3 + u \end{cases} \implies \dot{\mathbf{x}} = \begin{bmatrix} 0 & 1 & 0 \\ 0 & 0 & 1 \\ -3 & -2 & -4 \end{bmatrix} \mathbf{x} + \begin{bmatrix} 0 \\ 0 \\ 1 \end{bmatrix} u$$

The output equation is

$$y = 2\dot{w} + w = 2x_2 + x_1 \implies y = [1 \ 2 \ 0]\mathbf{x}$$

Of course, once the general procedure is established, the intermediate steps need not be taken every time. For instance, for the current problem the state-space form could have been obtained directly via the formulation listed in Eqs. (3.52) and (3.55). Compare the input-output equation at hand, Eq. (3.60), with the general form, Eq. (3.45) with $n = 3$, and identify all constants, as follows:

$$\dddot{y} + \underset{\underset{4}{\downarrow}}{a_1}\ddot{y} + \underset{\underset{2}{\downarrow}}{a_2}\dot{y} + \underset{\underset{3}{\downarrow}}{a_3}y = \underset{\underset{0}{\downarrow}}{b_0}\dddot{u} + \underset{\underset{0}{\downarrow}}{b_1}\ddot{u} + \underset{\underset{2}{\downarrow}}{b_2}\dot{u} + \underset{\underset{1}{\downarrow}}{b_3}u$$

Substitution of the identified constants into Eqs. (3.52) and (3.55) yields the state-space form, as

$$\begin{cases} \dot{\mathbf{x}} = \begin{bmatrix} 0 & 1 & 0 \\ 0 & 0 & 1 \\ -3 & -2 & -4 \end{bmatrix} \mathbf{x} + \begin{bmatrix} 0 \\ 0 \\ 1 \end{bmatrix} u \\ y = [1 \ 2 \ 0]\mathbf{x} + 0 \cdot u \end{cases}$$

which completely agrees with the earlier result.

EXAMPLE 3.21. Determine the state-space representation of the input-output equation

$$\ddot{y} + 2\dot{y} + y = \ddot{u} + 3\dot{u} + 2u$$

Solution. We proceed as in the previous example, omitting all intermediate steps. That is, directly compare the input-output equation with Eq. (3.45) with $n = 2$, and identify all constants involved; that is,

$$\ddot{y} + \underset{\underset{2}{\downarrow}}{a_1}\dot{y} + \underset{\underset{1}{\downarrow}}{a_2}y = \underset{\underset{1}{\downarrow}}{b_0}\ddot{u} + \underset{\underset{3}{\downarrow}}{b_1}\dot{u} + \underset{\underset{2}{\downarrow}}{b_2}u$$

Insertion of these numerical values into Eqs. (3.52) and (3.55) yields the state-space form,

$$\begin{cases} \dot{\mathbf{x}} = \begin{bmatrix} 0 & 1 \\ -1 & -2 \end{bmatrix} \mathbf{x} + \begin{bmatrix} 0 \\ 1 \end{bmatrix} u \\ y = [1 \ \ 1] \mathbf{x} + u \end{cases}$$

PROBLEM SET 3.5

3.20. The transfer function of a dynamic system is given as

$$\frac{Y(s)}{U(s)} = \frac{2}{s(s^2 + s + 2)}$$

(a) Determine the input-output equation and, subsequently, find the state-space form.
(b) Using the state, input, and output matrices obtained in part (a), find the transfer function. Is this in agreement with the transfer function provided originally?

3.21. Repeat Problem 3.20 for a system with transfer function

$$\frac{Y(s)}{U(s)} = \frac{s+1}{s(s+3)}$$

3.22. The input-output equation for the two-degree-of-freedom mechanical system of Example 3.11 was shown to be in the form

$$m_1 m_2 x_1^{(4)} + [m_1 b_2 + m_2(b_1 + b_2)] \dddot{x}_1$$
$$+ [m_1 k_2 + b_1 b_2 + (k_1 + k_2) m_2] \ddot{x}_1$$
$$+ (b_2 k_1 + b_1 k_2) \dot{x}_1 + k_1 k_2 x_1 = b_2 \dot{f} + k_2 f$$

where x_1 and f denote the displacement of the first block and the applied force, respectively, and are regarded as the system's output and input. Determine the state-space form.

3.23. The input-output equation for a two-tank liquid-level system (see Chapter 6) is given by

$$2\ddot{h} + 5\dot{h} + h = 2\dot{q} + 2q$$

in which h and q represent the liquid level and the inflow rate, respectively, and are considered output and input. Find the state-space form.

3.24. A certain dynamic system is governed by the following equations

$$\begin{cases} 2\ddot{x}_1 + 2\dot{x}_1 + x_1 - x_2 = f(t) \\ 3x_2 - x_1 = 0 \end{cases}$$

where f and x_1 are the input and output, respectively.
(a) Write the input-output equation.
(b) Determine the state-space representation.

144 Dynamic Systems: Modeling and Analysis

3.25. Repeat Problem 3.24 for a dynamic system with the following governing equations
$$\begin{cases} 2\ddot{x}_1 + 2\dot{x}_1 + x_1 - x_2 = f(t) \\ \dot{x}_2 + 3x_2 - x_1 = 0 \end{cases}$$
where f and x_1 are the input and output, respectively.

3.26. Repeat Problem 3.25 for the case in which x_2 is the output and f is the input.

3.6 LINEARIZATION

Often in systems analysis, one encounters nonlinear elements in the differential equations describing a system's model. Seeking a solution for such systems in closed-form is at times a formidable task, and very difficult to achieve. One possible way to circumvent the difficulties associated with nonlinearities is to solve the system numerically via effective techniques such as the fourth-order Runge-Kutta method. However, what this section is concerned with is the derivation of a linear model approximating a nonlinear system. The objective is to obtain a linear model, associated with certain inputs and initial conditions, whose response will be in agreement with that of the nonlinear system, within a reasonable degree of accuracy. The idea is based on **Taylor series expansion** of the nonlinear terms about a point known as an **operating point** (or **equilibrium point**). Then, within a *small neighborhood* of this point, where the variables undergo small deviations relative to the coordinates of the operating point, each nonlinear term is approximated by the linear terms in the Taylor series expansion. For now, we assume that the operating point is known; later, a procedure will be introduced to determine this point. First, we consider the graphical interpretation of the linearization procedure.

Graphical Interpretation

Consider a nonlinear function f of an independent variable x and let $P : (\bar{x}, \bar{f})$ be an operating point, as shown in Fig. 3.9. Suppose $A : (x, f)$ denotes a typical point on the graph of $f(x)$ and that l_1 is the line connecting P and A. Define

$$\Delta x(t) = x(t) - \bar{x} \quad \text{and} \quad \Delta f(t) = f(t) - \bar{f} \qquad (3.61)$$

where, in the discussion of dynamic systems, parameter t represents time. Δx and Δf are known as **increment variables** associated with x and f, respectively. Note that variables with overbars denote the coordinates of the operating point P, and thus are regarded as fixed constants. Also note that Δx and Δf are time-dependent, as are x and f. Then, recall from calculus that when A is located in a small neighborhood of P (that is, Δx and Δf are relatively small) then the slope of l_1 is approximated by the slope of the line l_2, tangent to the curve at point P. Denoting this slope by m, we then have

$$m = \left. \frac{df}{dx} \right|_{x=\bar{x}}$$

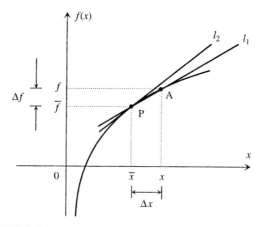

FIGURE 3.9 Nonlinear function and an operating point.

meaning that the derivative function, df/dx, is evaluated at the operating point. As a result, the equation of the tangent line l_2 is given by

$$f - \bar{f} = m(x - \bar{x}) \implies \Delta f = m\Delta x \tag{3.62}$$

Therefore, relative to a new coordinate system with axes Δx and Δf, Eq. (3.62) describes the equation of a straight line with slope m that passes through the new origin. It is readily seen that this new origin is indeed the operating point P, as shown in Fig. 3.10.

The degree of accuracy of the linear approximation is dependent on how $f(x)$ behaves in a small neighborhood of the operating point. In the event that $f(x)$ does not exhibit any radical behavior in this neighborhood (that is, it is smooth) then it may be approximated by a linear model, described by $\Delta f = m\Delta x$.

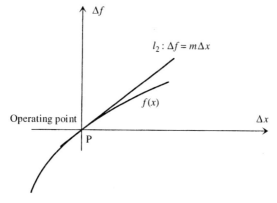

FIGURE 3.10 Linear approximation of a nonlinear function relative to a new coordinate system.

Taylor Series Expansion

The geometrical interpretation of the linearization process may now be accompanied by its analytical counterpart. For our purposes, we consider two types of nonlinear functions: functions of one variable and functions of two variables. For each class of functions, Taylor series expansion will be employed, of which the linear terms will be retained for linear approximation.

Functions of one variable

Suppose $y = f(x)$ represents a nonlinear term in the governing equation of a dynamic system and $(\bar{x}, \bar{f})$ denotes a prescribed operating point. It is then clear that $\bar{y} = f(\bar{x}) = \bar{f}$. The Taylor series expansion of the nonlinear function $f(x)$ about the operating point, assuming that the derivatives of $f(x)$ exist, is given as

$$f(x) = \overbrace{f(\bar{x}) + f'(x)|_{x=\bar{x}}(x - \bar{x})}^{\text{linear terms}} + \frac{1}{2!}f''(x)|_{x=\bar{x}}(x - \bar{x})^2 + \cdots \quad (3.63)$$

In the event that $(x - \bar{x})$ is small, all higher-order terms involving powers of $(x - \bar{x})$, higher than one, are then considered negligible relative to the first two terms in Eq. (3.63). Thus, a linear approximation of $f(x)$ is

$$f(x) \approx f(\bar{x}) + f'(x)|_{x=\bar{x}}(x - \bar{x})$$

or, alternatively, using the notation in Eq. (3.61),

$$y \approx \bar{y} + f'(x)|_{x=\bar{x}}\Delta x \quad \text{where} \quad \Delta x = x - \bar{x} \quad (3.64)$$

EXAMPLE 3.22. Linearize the nonlinear function defined by $y = x^2$ about the operating point $(1, 1)$. Investigate the accuracy of the linearized model for $x = 0.8$ and $x = 0.9$.

Solution. By assumption, $\bar{x} = 1$ and $\bar{y} = 1$. The nonlinear function is a function of one variable, x, and from Eq. (3.64), we have

$$y \approx \bar{y} + f'(x)|_{x=\bar{x}}\Delta x \quad \Longrightarrow \quad y \approx \bar{y} + [2x]_{x=1}\Delta x \quad (3.65)$$
$$\Longrightarrow \quad y_{\text{approx}} \approx 1 + 2\Delta x$$

This describes the linear approximation of the original function in a small neighborhood of the operating point. Because the graph of $y = x^2$ is a smooth curve, at least in a close vicinity of P, then the degree of accuracy of the linear model solely depends on how far x is located from P. For instance, if $x = 0.8$, as in Fig. 3.11, then by Eq. (3.65), we have

$$y_{\text{approx}} = 1 + 2\Delta x = 1 + 2(0.8 - 1) = 0.60$$

while

$$y_{\text{actual}} = (0.8)^2 = 0.64$$

which shows a 6 percent relative error. On the other hand, when $x = 0.9$, the numerical results are

$$y_{\text{approx}} = 1 + 2\Delta x = 1 + 2(0.9 - 1) = 0.80$$

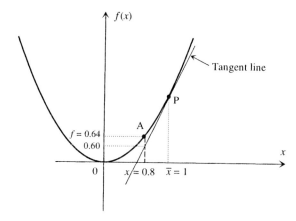

FIGURE 3.11 Approximation of a nonlinear curve about the operating point.

while
$$y_{actual} = (0.9)^2 = 0.81$$
indicating that the relative error has been reduced to 1.1 percent.

Functions of two variables

Suppose $w = f(x, y)$ represents a nonlinear function and $(\bar{x}, \bar{y}, \bar{f})$ denotes an operating point. It is then clear that $\bar{w} = f(\bar{x}, \bar{y})$. Assuming that all partial derivatives of $f(x, y)$ exist, its Taylor series expansion about the operating point is defined as

$$f(x, y) = f(\bar{x}, \bar{y}) + \overbrace{\left.\frac{\partial f}{\partial x}\right|_{(\bar{x},\bar{y})} (x - \bar{x}) + \left.\frac{\partial f}{\partial y}\right|_{(\bar{x},\bar{y})} (y - \bar{y})}^{\text{linear terms}} + \frac{1}{2!} \left.\frac{\partial^2 f}{\partial x^2}\right|_{(\bar{x},\bar{y})} (x - \bar{x})^2$$

$$+ \left.\frac{\partial^2 f}{\partial x \partial y}\right|_{(\bar{x},\bar{y})} (x - \bar{x})(y - \bar{y}) + \frac{1}{2!} \left.\frac{\partial^2 f}{\partial y^2}\right|_{(\bar{x},\bar{y})} (y - \bar{y})^2 + \cdots$$

(3.66)

In the event that $(x - \bar{x})$ and $(y - \bar{y})$ are both relatively small, all higher-order terms involving $(x - \bar{x})$ and $(y - \bar{y})$ are then considered negligible relative to the first three terms in Eq. (3.66). Thus, a linear approximation of $f(x, y)$ is given by

$$f(x, y) \cong f(\bar{x}, \bar{y}) + \left.\frac{\partial f}{\partial x}\right|_{(\bar{x},\bar{y})} (x - \bar{x}) + \left.\frac{\partial f}{\partial y}\right|_{(\bar{x},\bar{y})} (y - \bar{y})$$

$$\implies w \cong \bar{w} + \left.\frac{\partial f}{\partial x}\right|_{(\bar{x},\bar{y})} \Delta x + \left.\frac{\partial f}{\partial y}\right|_{(\bar{x},\bar{y})} \Delta y$$

where
$$\Delta x = x - \bar{x}, \qquad \Delta y = y - \bar{y} \qquad (3.67)$$

EXAMPLE 3.23. Consider a nonlinear function of two variables $w = f(x, y) = xy^2$ and an operating point $(\bar{x}, \bar{y}, \bar{w}) = (2, 1, 2)$. Determine the linear approximation of w about the operating point.

Solution. In a small neighborhood of the operating point, the linear approximation of the original function is obtained via Eq. (3.67), as

$$w_{approx} = 2 + [y^2]_{(2,1)} \Delta x + [2xy]_{(2,1)} \Delta y$$
$$\Longrightarrow w_{approx} = 2 + 1\Delta x + 4\Delta y$$
$$\longrightarrow w_{approx} = \Delta x + 4\Delta y + 2 \qquad (3.68)$$

Once again, the accuracy of the approximation is completely dependent on how far x and y are chosen relative to the operating point. To illustrate this, suppose $x = 1.5$ and $y = 0.7$, so that

$$w_{approx} = \Delta x + 4\Delta y + 2 = (1.5 - 2) + 4(0.7 - 1) + 2 = 0.3$$

while
$$w_{actual} = (1.5)(0.7)^2 = 0.735$$

exhibiting a substantial error of 57 percent. However, the accuracy is considerably improved if $x = 1.5$ and $y = 0.9$

$$w_{approx} = 1.10 \quad \text{while} \quad w_{actual} = (1.5)(0.9)^2 = 1.215$$

with an associated relative error of 9 percent. Major factors contributing to such a significant error reduction are that $(y - \bar{y})$ has been made smaller and that the original function is proportional to y^2.

EXAMPLE 3.24. Consider the single-tank, liquid-level system shown in Fig. 3.12. Assume that the outflow rate q_0 and the liquid level h obey the nonlinear relation $q_0 = k\sqrt{h}$, where k is some proportionality constant. Furthermore, let $\bar{h}$ and $q_i = \bar{q} = q_0$ denote the liquid level and the flow rate at steady state, respectively. Determine a linear relation between q_0 and h if they undergo small deviations from their respective steady-state values.

Solution. The operating point is defined by the steady-state values of h and q_0, i.e., $(\bar{h}, \bar{q})$. Expanding q_0 in a Taylor series about this operating point, retaining the linear terms only, one obtains

$$q_0 \cong q_0(\bar{h}) + \left[\frac{d}{dh}q_0\right]_{h=\bar{h}} \cdot (h - \bar{h})$$

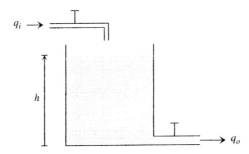

FIGURE 3.12 Single-tank, liquid-level system.

where

$$\frac{d}{dh}q_0 = \frac{d}{dh}k\sqrt{h} = k\frac{1}{2\sqrt{h}} \quad \text{and} \quad q_0(\bar{h}) = \bar{q}$$

Substitution into the Taylor series above, yields

$$q_0 \cong \bar{q} + \frac{k}{2\sqrt{\bar{h}}} \cdot (h - \bar{h})$$

Noting that the nonlinear relation $q_0 = k\sqrt{h}$ is also valid at steady state (i.e., $\bar{q} = k\sqrt{\bar{h}}$) we have $k = \bar{q}/\sqrt{\bar{h}}$. Insert this into the last equation to obtain

$$q_0 \cong \bar{q} + \frac{\bar{q}}{2\bar{h}} \cdot (h - \bar{h})$$

which indeed represents a linear relation between the liquid level and the outflow rate near steady state.

Determination of the Operating Point

So far, as in the last two examples, complete information about operating points has been provided. In several instances, however, the operating point is not directly given and needs to be properly determined. In doing so, a few important facts need to be taken into account: (1) an operating point refers to some type of an equilibrium condition for the dynamic system, (2) each variable is replaced with the corresponding coordinate of the operating point, a constant, and (3) system input, $u(t)$, will assume a constant value of $\bar{u}$, which normally represents its average value.

EXAMPLE 3.25. Consider the mechanical system shown in Fig. 3.13 involving a nonlinear spring, where $f_s(x) = x|x|$ denotes the nonlinear spring force, as shown in Fig. 3.14. $u(t)$ represents the applied force (input) and is assumed to be the unit-step function. The system is subjected to initial conditions $x(0) = 0$ and $\dot{x}(0) = 1$.

The equation of motion is given by

$$\ddot{x} + \dot{x} + \underbrace{x|x|}_{f_s(x)} = u(t) \tag{3.69}$$

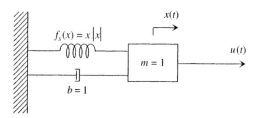

FIGURE 3.13 Mechanical system with a nonlinear spring.

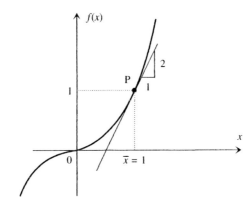

FIGURE 3.14 Linearization of a nonlinear spring force.

Determine an operating point and obtain a *linear model* approximating the nonlinear mechanical system about this point. Identify the initial conditions associated with the linear model.

Solution. Equation (3.69) clearly contains a nonlinear term, $x|x|$, which must be linearized in a small neighborhood of an operating point. The operating point, however, is not completely available and must be determined while the above-mentioned facts, labeled (1), (2), and (3), are taken into consideration. To this end, in Eq. (3.69), substitute $x = \bar{x}$ and $u = \bar{u} = 1$, as the average value of the unit-step function is 1. Then, we have

$$\underbrace{\ddot{\bar{x}}}_{0} + \underbrace{\dot{\bar{x}}}_{0} + \bar{x}|\bar{x}| = 1 \quad \Longrightarrow \quad \bar{x}|\bar{x}| = 1 \qquad (3.70)$$

$$\text{because } \bar{x} \text{ is constant}$$

Because of the presence of the absolute value, two cases must be considered as follows:

Case 1. Assume $\bar{x} > 0$. This implies that $|\bar{x}| = \bar{x}$, and Eq. (3.70) becomes $\bar{x}^2 = 1$. Solution yields

$$\bar{x} = \pm 1 \quad \underset{\text{by assumption, } \bar{x}>0}{\Longrightarrow} \quad \bar{x} = 1$$

Case 2. Assume $\bar{x} < 0$. This implies that $|\bar{x}| = -\bar{x}$, and Eq. (3.70) becomes $-\bar{x}^2 = 1$. This yields

$$\bar{x}^2 = -1$$

so no real solution exists. Therefore, *Case 1* provides the only valid solution, $\bar{x} = 1$. Consider the nonlinear term $f_s(x) = x|x|$. Its complete analytical description, and its derivative with respect to x, can be given as

$$f_s(x) = x|x| = \begin{cases} x^2 & \text{if } x \geq 0 \\ -x^2 & \text{if } x < 0 \end{cases} \implies f_s'(x) = \begin{cases} 2x & \text{if } x \geq 0 \\ -2x & \text{if } x < 0 \end{cases} \quad (3.71)$$

$f_s(x)$ is linearized about the operating point, via Eq. (3.64), as follows:

$$f_s(x) \cong f_s(\bar{x}) + f_s'(x)|_{x=\bar{x}} \Delta x \quad (3.72)$$

Because $\bar{x} = 1 > 0$, the first segments in the descriptions of $f_s(x)$ and $f_s'(x)$ in Eq. (3.71) will apply. Hence, we have $f_s(\bar{x}) = f_s(1) = 1$ and $f_s'(x)|_{x=1} = [2x]_{x=1} = 2$. The latter is, of course, the slope of the tangent line to the graph of $f_s(x)$ at (1, 1), illustrated in Fig. 3.14. Therefore, Eq. (3.72) reduces to

$$f_s(x) \cong 1 + 2\Delta x \quad (3.73)$$

which is clearly a linear expression. This completes the linearization process of the only nonlinear term in the equation of motion.

Next, in order to obtain a linear model that approximates the original nonlinear system, the following steps need to be undertaken:

- Recall that $\Delta x = x - \bar{x}$ so that $x = \bar{x} + \Delta x = 1 + \Delta x$. Substitute for x in the original nonlinear model, Eq. (3.69), taking into account that $\bar{x}$ is a constant; i.e., $\dot{\bar{x}} = \ddot{\bar{x}} = 0$.
- In Eq. (3.69), substitute $u(t) = \bar{u} + \Delta u(t) = 1 + \Delta u(t)$ with $\Delta u(t) = 0$; because u is the unit-step and $\bar{u} = 1$, there are zero deviations. Also replace the nonlinear term by its linear approximation, given by Eq. (3.73), to obtain

$$\frac{d^2}{dt^2}(1 + \Delta x) + \frac{d}{dt}(1 + \Delta x) + \underbrace{[1 + 2\Delta x]}_{\text{linear approximation}} = 1 + \underbrace{\Delta u(t)}_{0}$$

Simplification yields

$$\Delta\ddot{x} + \Delta\dot{x} + 1 + 2\Delta x = 1 \implies \Delta\ddot{x} + \Delta\dot{x} + 2\Delta x = 0 \quad (3.74)$$

Equation (3.74), a linear second-order differential equation in Δx with constant coefficients, is a linear model that approximates the original system. Notice that the new variable for the linear model is the increment variable Δx, in place of x. The input to the linear model is the increment variable corresponding to $u(t)$, which is zero in this case.

- Finally, to obtain the complete description of the linear model, a new set of initial conditions should be specified in relation to Δx. To this end, recall that $\Delta x(t) = x(t) - \bar{x}$. Evaluation at $t = 0$ results in

$$\Delta x(0) = \underbrace{x(0)}_{0} - \underbrace{\bar{x}}_{1} = -1$$

and differentiation yields

$$\Delta\dot{x}(t) = \dot{x}(t) - \dot{\bar{x}} = \dot{x}(t) \stackrel{\text{at } t=0}{\implies} \Delta\dot{x}(0) = \dot{x}(0) = 1$$

Hence, the linear model is completely described by

$$\Delta\ddot{x} + \Delta\dot{x} + 2\Delta x = 0, \qquad \Delta x(0) = -1, \qquad \Delta\dot{x}(0) = 1$$

For convenience, we may use $\tilde{x}$ in place of Δx so that the linear model is rewritten as

$$\ddot{\tilde{x}} + \dot{\tilde{x}} + 2\tilde{x} = 0, \qquad \tilde{x}(0) = -1, \qquad \dot{\tilde{x}}(0) = 1$$

Note that because $\tilde{x}$ denotes deviations of x relative to the operating point, (i.e., $x = \bar{x} + \tilde{x}$), the solution to the linearized model does not approximate the solution to the nonlinear system.

Procedure: The steps undertaken in Example 3.25 suggest a general procedure that must be followed. This may be outlined as a four-step procedure:

1. To determine the operating point, replace the dependent variable(s) such as x and y by $\bar{x}$ and $\bar{y}$, and the input u by its average value $\bar{u}$, if specified. Solve the resulting (nonlinear) algebraic equation(s) for constants such as $\bar{x}$ and $\bar{y}$.
   * In the event that the input is composed of two or more functions, set the time-varying portion(s) equal to zero. For instance, if $u(t) = k + g(t)$, with $k =$ constant, set $g(t) = 0$ so that $\bar{u} = k$.
2. Linearize the nonlinear term(s) about the operating point via Taylor series expansion.
3. In the original nonlinear model, replace the dependent variable(s) such as x and y by $\bar{x} + \Delta x$ and $\bar{y} + \Delta y$, the input u by $\bar{u} + \Delta u$, and the nonlinear term(s) by their linear approximations.
4. Properly adjust the initial conditions.

EXAMPLE 3.26. Assume that the nonlinear state-variable equations for a system are given as

$$\begin{cases} \dot{x}_1 = x_2 - 1 \\ \dot{x}_2 = (x_1 - 1)^3 x_2 + 1 \end{cases} \tag{3.75}$$

subject to initial conditions $x_1(0) = 1$ and $x_2(0) = 1$. Obtain a linear model.

Solution. The operating point (not specified) is obtained by substituting $x_1 = \bar{x}_1$ and $x_2 = \bar{x}_2$ into the original equations, and taking into account that $\dot{\bar{x}}_1 = 0 = \dot{\bar{x}}_2$. Note that in both equations, the respective inputs are merely constants that serve as their respective average values.

Step 1.

$$\begin{cases} 0 = \bar{x}_2 - 1 \\ 0 = (\bar{x}_1 - 1)^3 \bar{x}_2 + 1 \end{cases} \Longrightarrow \begin{cases} \bar{x}_2 = 1 \\ (\bar{x}_1 - 1)^3 \bar{x}_2 + 1 = 0 \end{cases} \tag{3.76}$$

Solving Eq. (3.76) defines the operating point, $(\bar{x}_1, \bar{x}_2) = (0, 1)$.

Step 2. Next, let the nonlinear term in the state-variable equations be denoted by $f(x_1, x_2) = (x_1 - 1)^3 x_2$. Linearize via Eq. (3.66), as

$$f(x_1, x_2) \cong f(\bar{x}_1, \bar{x}_2) + \frac{\partial f}{\partial x_1}\bigg|_{(0,1)} (x_1 - \bar{x}_1) + \frac{\partial f}{\partial x_2}\bigg|_{(0,1)} (x_2 - \bar{x}_2)$$

$$= -1 + \left[3(x_1 - 1)^2 x_2\right]_{(0,1)} \Delta x_1 + \left[(x_1 - 1)^3\right]_{(0,1)} \Delta x_2 \qquad (3.77)$$

$$= -1 + 3\Delta x_1 - \Delta x_2$$

Step 3. Substitute the linearized term given by Eq. (3.77) into the original system, Eq. (3.75). Also, insert $x_1 = \bar{x}_1 + \Delta x_1 = \Delta x_1$ and $x_2 = \bar{x}_2 + \Delta x_2 = 1 + \Delta x_2$. Note that because the inputs were constants to begin with, their respective increment variables are simply zero. Subsequently, we have

$$\begin{cases} \Delta \dot{x}_1 = (1 + \Delta x_2) - 1 \\ \dfrac{d}{dt}(1 + \Delta x_2) = \underbrace{-1 + 3\Delta x_1 - \Delta x_2}_{\text{linearized term}} + 1 \end{cases} \xrightarrow{\text{simplify}} \begin{cases} \Delta \dot{x}_1 = \Delta x_2 \\ \Delta \dot{x}_2 = 3\Delta x_1 - \Delta x_2 \end{cases} \qquad (3.78)$$

Step 4. As was done in the previous example, adjust the initial conditions according to the change of variables, as

$$\Delta x_1(0) = 1 \quad \text{and} \quad \Delta x_2(0) = 0 \qquad (3.79)$$

The system in Eq. (3.78), together with the initial conditions given by Eq. (3.79), describes the linear model completely. As before, rewrite Eqs. (3.78) and (3.79) as

$$\begin{cases} \dot{\tilde{x}}_1 = \tilde{x}_2 \\ \dot{\tilde{x}}_2 = 3\tilde{x}_1 - \tilde{x}_2 \end{cases}, \quad \tilde{x}_1(0) = 1, \quad \tilde{x}_2(0) = 0$$

EXAMPLE 3.27. As an example of case (*) in Step 1 of the linearization procedure, consider a nonlinear system described by

$$\begin{cases} \dot{x}_1 = x_2 - 1 \\ \dot{x}_2 = (x_1 - 1)^3 x_2 + \underbrace{1 + \sin t}_{u(t)} \end{cases} \qquad (3.80)$$

which is the system studied in Example 3.26 with the exception that the input in the second equation is now in the form $u(t) = 1 + \sin t$. Assuming initial conditions $x_1(0) = 1$ and $x_2(0) = 1$, determine an operating point and the corresponding linear model.

Solution

Step 1. To determine the operating point, substitute $x_1 = \bar{x}_1$ and $x_2 = \bar{x}_2$ into the original system, Eq. (3.80). Furthermore, according to (*), the time-varying portion of the input must be set to zero (i.e., $\sin t = 0$) to obtain the average value of $u(t) = 1$. This implies that

$$u(t) = \underbrace{1}_{\bar{u}} + \underbrace{\sin t}_{\Delta u} = \bar{u} + \Delta u$$

Consequently, the resulting nonlinear algebraic system is identical to that in Eq. (3.76), and hence $(\bar{x}_1, \bar{x}_2) = (0, 1)$.

Step 2. Because the operating point has remained unchanged and the nonlinear term is as before, the result on linear approximation from the previous example still holds; that is,

$$f(x_1, x_2) \cong -1 + 3\Delta x_1 - \Delta x_2$$

Step 3. Substitute this linear approximation, as well as

$$x_1 = \bar{x}_1 + \Delta x_1 = \Delta x_1,$$
$$x_2 = \bar{x}_2 + \Delta x_2 = 1 + \Delta x_2,$$
$$u(t) = \bar{u} + \Delta u = 1 + \sin t$$

into the original system, Eq. (3.80), to obtain the linear model, as

$$\begin{cases} \dot{\Delta x}_1 = (1 + \Delta x_2) - 1 \\ \dfrac{d}{dt}(1 + \Delta x_2) = \underbrace{-1 + 3\Delta x_1 - \Delta x_2}_{\text{linearized term}} + 1 + \sin t \end{cases}$$

$$\overset{\text{simplify}}{\Longrightarrow} \begin{cases} \dot{\Delta x}_1 = \Delta x_2 \\ \dot{\Delta x}_2 = 3\Delta x_1 - \Delta x_2 + \sin t \end{cases} \quad (3.81)$$

Step 4. Because the operating point, as well as the initial conditions for the nonlinear system, are the same as in the previous example, it is readily seen that the initial conditions corresponding to the linear model are once again given by $\Delta x_1(0) = 1$ and $\Delta x_2(0) = 0$. More conveniently, the linear model is given by

$$\begin{cases} \dot{\tilde{x}}_1 = \tilde{x}_2 \\ \dot{\tilde{x}}_2 = 3\tilde{x}_1 - \tilde{x}_2 + \sin t \end{cases}, \quad \tilde{x}_1(0) = 1, \quad \tilde{x}_2(0) = 0$$

Numerical Solution of a Nonlinear Model

Up to this point, for a dynamic system containing one or more nonlinear elements, we have managed to determine an operating point and subsequently find the linear model that approximates it. Furthermore, we have established that such approximation is reasonably accurate as long as the variables are within a small neighborhood of the operating point. To demonstrate the effectiveness of the linearization procedure, an effort is now made to (1) solve the original nonlinear model numerically, (2) solve the linear model exactly, and (3) present both in graphical forms for comparison. While task (2) may be accomplished via analytical methods, such as Laplace transformation, task (1) requires employment of a suitable numerical technique. The most commonly used such numerical method is the **fourth-order Runge-Kutta method** (RK4). Note that task (2), solving the linear model, can also be accomplished via application of the Runge-Kutta method. However, in many situations, especially with time-invariant coefficients, closed-form solutions can be obtained via analytical methods, such as the Laplace transform and undetermined coefficients.

Fourth-Order Runge-Kutta Method

A single first-order differential equation

Consider a single first-order ordinary differential equation subjected to a prescribed initial condition, as

$$\dot{y} = f(t, y), \quad a \le t \le b, \quad y(a) = \alpha \quad (3.82)$$

Partition the time interval $[a, b]$ into N equally spaced divisions, each with a length of $h = (b - a)/N$, so that

$$t_0 = a, \quad t_1 = a + h, \ldots, t_N = a + Nh$$

are the mesh points. Start with $y(t_0)$; i.e., α. In order to evaluate y at the next mesh point, t_1, four function evaluations of $f(t, y)$ must be performed. Subsequently, $y(t_1)$ is available and serves as the initial condition for the next subinterval, $[t_1, t_2]$, to evaluate $y(t_2)$. This process continues until y has been evaluated at all mesh points. This method is referred to as fourth-order, because the *global truncation error* is of the order $O(h^4)$. The basic structure of the algorithm is as follows:

ALGORITHM 3.1. Fourth-order Runge-Kutta method (RK4) to solve a single first-order differential equation

- Initialize:

    $y_0 = \alpha$

- Perform four different function evaluations at each time step: for $i = 0, 1, 2, \ldots, N - 1$, do

    $k_1 = hf(t_i, y_i)$
    $k_2 = hf(t_i + \tfrac{1}{2}h, y_i + \tfrac{1}{2}k_1)$
    $k_3 = hf(t_i + \tfrac{1}{2}h, y_i + \tfrac{1}{2}k_2)$
    $k_4 = hf(t_{i+1}, y_i + k_3)$

- Evaluate y at the next mesh point:

    $y_{i+1} = y_i + \tfrac{1}{6}[k_1 + 2k_2 + 2k_3 + k_4]$

- Continue....

EXAMPLE 3.28. Solve the following initial-value problem via the fourth-order Runge-Kutta method:

$$\dot{x} + x^3 = 1, \qquad x(0) = 0, \qquad 0 \leq t \leq 3.5 \tag{3.83}$$

Solution. Prior to implementation of the numerical method, Eq. (3.83) must be expressed in the general form of Eq. (3.82), i.e.,

$$\dot{x} = \underbrace{1 - x^3}_{f(t,x)}, \qquad x(0) = 0, \qquad 0 \leq t \leq 3.5$$

Note that for this problem, $f(t, x)$ is actually $f(x)$, because it is not an explicit function of t, but rather an implicit one. A modified version of Algorithm 3.1, suited

TABLE 3.1 Modification of Algorithm 3.1 to solve Eq. (3.83)

- Define increment size and the initial value:
  $h = 0.25$
  $N = 3.5/h$
  $x(1) = 0$

- Set initial time:
  $t(1) = 0$

- Perform function evaluations:
  for $j = 1 : N - 1$, do
  $k1 = h * \text{func1}(t(j), x(j))$
  $k2 = h * \text{func1}(t(j) + h/2, x(j) + 0.5 * k1)$
  $k3 = h * \text{func1}(t(j) + h/2, x(j) + 0.5 * k2)$
  $k4 = h * \text{func1}(t(j) + h, x(j) + k3)$

- Update x and t:
  $x(j + 1) = x(j) + (k1 + 2 * k2 + 2 * k3 + k4)/6$
  $t(j + 1) = t(j) + h$

- Continue . . . .

- User-defined function:
  $f1 = \text{func1}(t, x)$
  $f1 = 1 - x^3$
  return

TABLE 3.2 Solutions to Eq. (3.83) using RK4

Step i	t_i	x_i	Step i	t_i	x_i
1	0.00	0.0000	9	2.00	0.9894
2	0.25	0.2490	10	2.25	0.9949
3	0.50	0.4852	11	2.50	0.9976
4	0.75	0.6828	12	2.75	0.9989
5	1.00	0.8229	13	3.00	0.9995
6	1.25	0.9080	14	3.25	0.9997
7	1.50	0.9543	15	3.50	0.9999
8	1.75	0.9778			

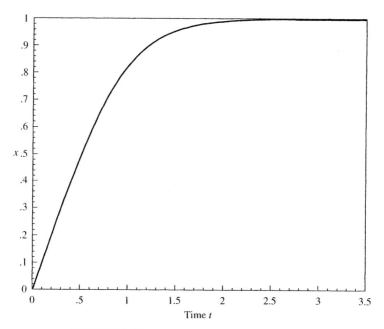

FIGURE 3.15 Solution to Eq. (3.83) via RK4.

for the problem at hand, is shown in Table 3.1 and is used to calculate values of x at different mesh points. The numerical results are presented in tabulated form, Table 3.2, and in graphical form, Fig. 3.15.

A system of first-order differential equations

The Runge-Kutta method may be generalized to solve a system of first-order differential equations. To that end, consider the following system subjected to known initial conditions:

$$\begin{cases} \dot{y}_1 = f_1(t, y_1, y_2, \ldots, y_n) \\ \dot{y}_2 = f_2(t, y_1, y_2, \ldots, y_n) \\ \vdots \\ \dot{y}_n = f_n(t, y_1, y_2, \ldots, y_n) \end{cases}, \quad \begin{cases} y_1(0) = \alpha_1 \\ y_2(0) = \alpha_2 \\ \vdots \\ y_n(0) = \alpha_n \end{cases} \quad (3.84)$$

Partition $[a, b]$ as before, into N equally spaced divisions. The algorithm, an extension of the previous case, is given below.

ALGORITHM 3.2. Fourth-order Runge-Kutta method (RK4) to solve a system of first-order differential equations

- Initialize:
  $t = a$
  $y_1 = \alpha_1, y_2 = \alpha_2, \ldots, y_n = \alpha_n$

- For each f_j, perform four function evaluations:

    for $i = 1, 2, \ldots, N$, do

        for $j = 1, 2, \ldots, n$, do

            $k_{1,j} = h f_j(t, y_1, y_2, \ldots, y_n)$

        for $j = 1, 2, \ldots, n$, do

            $k_{2,j} = h f_j\left(t + \tfrac{1}{2}h,\ y_1 + \tfrac{1}{2}k_{1,1},\ y_2 + \tfrac{1}{2}k_{1,2}, \ldots, y_n + \tfrac{1}{2}k_{1,n}\right)$

        for $j = 1, 2, \ldots, n$, do

            $k_{3,j} = h f_j\left(t + \tfrac{1}{2}h,\ y_1 + \tfrac{1}{2}k_{2,1},\ y_2 + \tfrac{1}{2}k_{2,2}, \ldots, y_n + \tfrac{1}{2}k_{2,n}\right)$

        for $j = 1, 2, \ldots, n$, do

            $k_{4,j} = h f_j\left(t + h,\ y_1 + k_{3,1},\ y_2 + k_{3,2}, \ldots, y_n + k_{3,n}\right)$

- Update values of y at the next mesh point:

    for $j = 1, 2, \ldots, n$, do

        $y_j = y_j + \tfrac{1}{6}\left[k_{1,j} + 2k_{2,j} + 2k_{3,j} + j_{4,j}\right]$

- Increment t:

    $t = a + ih$

- Continue....

EXAMPLE 3.29. Solve the following second-order differential equation via RK4, subject to the given initial conditions.

$$\ddot{x} + \dot{x} + x^3 = 1, \qquad x(0) = 0, \qquad \dot{x}(0) = -1$$

Solution. Note that the Runge-Kutta method is not applicable to this equation in its present form. Through introduction of a set of state variables, the resulting state-variable equations may be expressed in the general form of Eq. (3.84). Let $x_1 = x$ and $x_2 = \dot{x}$ be the state variables. Then,

$$\begin{cases} \dot{x}_1 = x_2 \\ \dot{x}_2 = 1 - x_1^3 - x_2 \end{cases} \tag{3.85}$$

A modified version of Algorithm 3.2 is shown in Table 3.3 and is used to calculate values of x_1 and x_2 at each mesh point. The numerical results are presented in Table 3.4 and plotted in Fig. 3.16.

EXAMPLE 3.30. For the mechanical system in Example 3.25, present the response of the nonlinear system and that of the linear model, graphically, and compare.

Solution. The equation of motion of the nonlinear system, and the initial conditions, are

$$\ddot{x} + \dot{x} + x|x| = u(t), \qquad x(0) = 0, \qquad \dot{x}(0) = 1 \tag{3.86}$$

TABLE 3.3 Modification of Algorithm 3.2 to solve Eq. (3.85)

- Define increment size and the initial value:
  $h = 0.10$

  $N = \dfrac{14}{h}$

  $x1(1) = 0$
  $x2(1) = -1$

- Set initial time:
  $t(1) = 0$

- Perform function evaluations:
  for $j = 1 : N - 1$, do
  $k(1, 1) = h * x2(j)$
  $k(1, 2) = h * \text{func2}(x1(j), x2(j))$
  $k(2, 1) = h * (x2(j) + 0.5 * k(1, 2))$
  $k(2, 2) = h * \text{func2}(x1(j) + 0.5 * k(1, 1), x2(j) + 0.5 * k(1, 2))$
  $k(3, 1) = h * (x2(j) + 0.5 * k(2, 2))$
  $k(3, 2) = h * \text{func2}(x1(j) + 0.5 * k(2, 1), x2(j) + 0.5 * k(2, 2))$
  $k(4, 1) = h * (x2(j) + k(3, 2))$
  $k(4, 2) = h * \text{func2}(x1(j) + k(3, 1), x2(j) + k(3, 2))$

- Update x and t:
  $x1(j + 1) = x1(j) + (k(1, 1) + 2 * k(2, 1) + 2 * k(3, 1) + k(4, 1))/6$
  $x2(j + 1) = x2(j) + (k(1, 2) + 2 * k(2, 2) + 2 * k(3, 2) + k(4, 2))/6$
  $t(j + 1) = t(j) + h$

- Continue . . . .

- User-defined function:
  $f2 = \text{func2}(x, y)$
  $f2 = 1 - x^3 - y$
  return

TABLE 3.4 Solutions to Eq. (3.85) using RK4

Step i	t_i	$x_{1,i}$	Step i	t_i	$x_{1,i}$
1	0.00	0.0000	50	4.90	0.9472
2	0.10	−0.0903	60	5.90	0.9177
3	0.20	−0.1625	80	7.90	1.0266
10	0.90	−0.2832	100	9.90	0.9912
15	1.40	−0.0972	120	11.90	1.0027
20	1.90	0.2129	130	12.90	0.9979
31	3.00	1.0289	141	14.00	0.9995
40	3.90	1.2107			

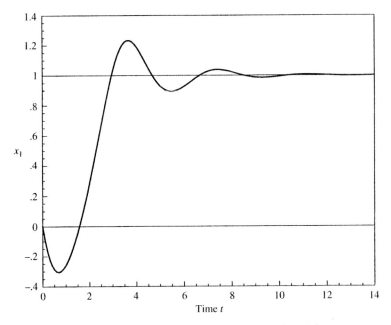

FIGURE 3.16 Solution x_1 to Eq. (3.85) via RK4.

where u was assumed to be the unit-step. Solution of this equation via the Runge-Kutta method requires that it be expressed as a system of first-order differential equations. To that end, we introduce state variables $x_1 = x$ and $x_2 = \dot{x}$ so that the state-variable equations are

Nonlinear system
$$\begin{cases} \dot{x}_1 = x_2 \\ \dot{x}_2 = -x_2 - x_1 |x_1| \end{cases}, \quad x_1(0) = 0, \quad x_2(0) = 1 \tag{3.87}$$

which is precisely in the general form of Eq. (3.84). On the other hand, recall that the linear model was obtained earlier as

Linear model
$$\ddot{\tilde{x}} + \dot{\tilde{x}} + 2\tilde{x} = 0, \quad \tilde{x}(0) = -1, \quad \dot{\tilde{x}}(0) = 1 \tag{3.88}$$

Whereas the solution, $\tilde{x}$, to Eq. (3.88) is not compatible to x_1 in Eq. (3.87), $\tilde{x} + \bar{x}$ is. The fourth-order Runge-Kutta method is employed to solve the nonlinear system in Eq. (3.86). The linear system in Eq. (3.88) may be solved using either the Laplace transform or the characteristic equation. The results, in plotted forms, are presented in Fig. 3.17.

EXAMPLE 3.31. Consider a pendulum mounted on a moving cart, as shown in Fig. 3.18. The friction at the hinge joint is included in the model. For a given set of parameter values, the equations of motion can be derived as

$$\begin{cases} \ddot{\theta} + \ddot{x}\cos\theta + 9.8\sin\theta + 0.5\dot{\theta} = 0 & (3.89a) \\ \ddot{x} + 0.5\ddot{\theta}\cos\theta - 0.5\dot{\theta}^2\sin\theta + 0.5\dot{x} + 0.5x = 0.5u(t) & (3.89b) \end{cases}$$

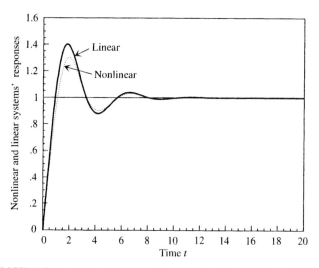

FIGURE 3.17 Dotted line: nonlinear model (via RK4); solid line: linear model (via Laplace transform).

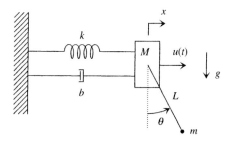

FIGURE 3.18 Pendulum mounted on a moving cart.

The system is driven by an applied force u, and is subjected to zero initial conditions. Assume that the angular displacement θ is small. Determine an operating point and the corresponding linear model.

Solution

Step 1. Let $(\bar{\theta}, \bar{x})$ denote the operating point. Substituting $\theta = \bar{\theta}$, $x = \bar{x}$, and $\bar{u} = 0$ into Eq. (3.89) yields

$$\begin{cases} 9.8 \sin \bar{\theta} = 0 \\ 0.5\bar{x} = 0 \end{cases} \implies \bar{\theta} = 0, \bar{x} = 0$$

Step 2. In Eq. (3.89), note that the nonlinear terms are those involving $\sin \theta$ and $\cos \theta$. To linearize these terms, expand each in a Taylor series about the operating

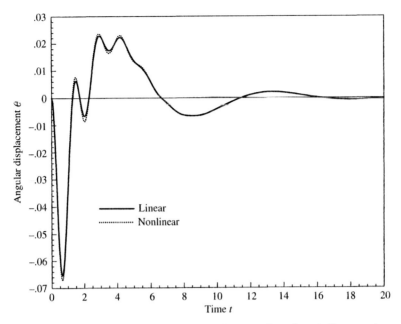

FIGURE 3.19 Angular displacement θ versus time, for nonlinear and linear models.

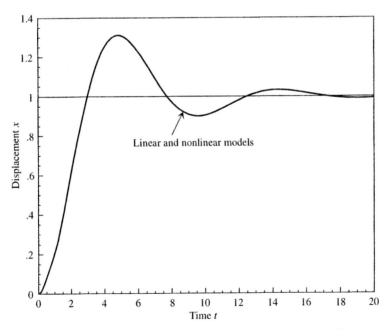

FIGURE 3.20 Displacement x versus time, for linear and nonlinear models.

point, as

$$\sin\theta \approx \sin\bar{\theta} + \left[\frac{d}{d\theta}\sin\theta\right]_{\theta=\bar{\theta}} \cdot \Delta\theta = \Delta\theta \quad (3.90)$$

$$\cos\theta \approx \cos\bar{\theta} + \left[\frac{d}{d\theta}\cos\theta\right]_{\theta=\bar{\theta}} \cdot \Delta\theta = 1 \quad (3.91)$$

In summary, for small θ, $\sin\theta \approx \theta$, $\cos\theta \approx 1$, and $\theta\dot{\theta}^2 \approx 0$.

Step 3. Substitute the linearized terms, Eqs. (3.90) and (3.91), as well as

$$\theta = \bar{\theta} + \Delta\theta = \Delta\theta, \quad x = \bar{x} + \Delta x = \Delta x, \quad u = \bar{u} + \Delta u = \Delta u$$

into Eq. (3.89) to obtain

$$\begin{cases} \ddot{\Delta\theta} + \ddot{\Delta x} + 9.8\,\Delta\theta + 0.5\,\dot{\Delta\theta} = 0 \\ \ddot{\Delta x} + 0.5\,\ddot{\Delta\theta} + 0.5\,\dot{\Delta x} + 0.5\,\Delta x = 0.5\,\Delta u \end{cases} \quad (3.92)$$

Note that the third term in Eq. (3.89b) has vanished because upon insertion of $\sin\theta \cong \theta$, it read $-0.5\theta\dot{\theta}^2$, which is approximately zero by step 2 above. Eq. (3.92) represents the linear model that approximates the original system of Eq. (3.89). Furthermore, initial conditions for the linear model remain zero. As always, renaming the variables for convenience, the linear model may be expressed as

$$\begin{cases} \ddot{\tilde{\theta}} + \ddot{\tilde{x}} + 9.8\tilde{\theta} + 0.5\dot{\tilde{\theta}} = 0 \\ \ddot{\tilde{x}} + 0.5\ddot{\tilde{\theta}} + 0.5\dot{\tilde{x}} + 0.5\tilde{x} = 0.5\tilde{u} \end{cases} \quad (3.93)$$

We now investigate the accuracy of the outcome of this linearization procedure. To this end, let the system input $u(t)$ be a unit-step function. The nonlinear system, Eq. (3.89), and its linear model, Eq. (3.93), are both subjected to zero initial conditions. Solutions of both systems are presented graphically in Figs. 3.19 and 3.20. Figure 3.19 shows the angular displacement θ for both systems versus time, indicating a slight difference between the two, while Fig. 3.20 points out that time variations of x in both cases are practically the same.

PROBLEM SET 3.6

3.27. Linearize $y = x^3$ about the operating point (1, 1). Investigate the accuracy of the linearized function for (a) $x = 0.75$, (b) $x = 0.85$, and (c) $x = 0.95$. In each case, calculate the relative error.

3.28. Linearize $y = x|x|$ about the operating point (2, 4). Investigate the accuracy of the linearized function for (a) $x = 1.5$, (b) $x = 1.9$, and (c) $x = 2.2$. In each case, calculate the relative error.

3.29. Linearize $f(x, y) = x^3 y$ about the operating point $(1, 2, 2)$. Investigate the accuracy of the linearized function for (a) $(x, y) = (0.9, 2.1)$, and (b) $(x, y) = (0.8, 2.1)$.

3.30. The mathematical model of a mechanical system is given by

$$\ddot{x} + \dot{x}^3 + 3x = u(t), \qquad x(0) = x_0, \qquad \dot{x}(0) = v_0$$

where x_0 and v_0 denote prescribed initial conditions, and $\dot{x}^3$ models a certain nonlinear damping. The input $u(t)$ is assumed to have an average value of 3. Determine the operating point. Then, obtain the linear model of the nonlinear system about the operating point and identify the initial conditions associated with it. Hint: Treat the nonlinear element $\dot{x}^3$ as a function of $\dot{x}$, and write the corresponding Taylor series expansion.

3.31. The equation of motion of a mechanical system involving a nonlinear spring is given as

$$\ddot{x} + \dot{x} + x|x| = 1 + M \sin \omega t$$

and is subject to initial conditions $x(0) = 0$ and $\dot{x}(0) = 1$. Find the operating point and obtain the linear model about this point. Evaluate the initial conditions for the linear model.

3.32. The governing equation of a dynamic system is given by

$$\ddot{x} + 2\dot{x} + x|x| = 1 + \sin 2t$$

and is subjected to initial conditions $x(0) = 1$ and $\dot{x}(0) = 0$.
(a) Find the operating point and the linearized model.
(b) Solve the nonlinear model via the fourth-order Runge-Kutta method in the time interval $[0, 2.8]$ with increments of $h = 0.08$.
(c) Solve the linear model analytically via the method of undetermined coefficients. Present your results in tabulated form and also graphically.

3.33. A nonlinear dynamic system is described by

$$\begin{cases} \dot{x}_1 = -x_1 |x_1| - x_2 - 5 + \sin t \\ \dot{x}_2 = x_1 - x_2 - 3 \end{cases}, \qquad x_1(0) = 2, \qquad x_2(0) = -2$$

Find the operating point and, subsequently, the linearized model and the corresponding initial conditions.

SUMMARY

The *configuration form* of an n-degree-of-freedom system, governed by second-order differential equations, is

$$\begin{cases} \ddot{q}_1 = f_1(q_1, q_2, \ldots, q_n, \dot{q}_1, \dot{q}_2, \ldots, \dot{q}_n, t) \\ \ddot{q}_2 = f_2(q_1, q_2, \ldots, q_n, \dot{q}_1, \dot{q}_2, \ldots, \dot{q}_n, t) \\ \quad \vdots \\ \ddot{q}_n = f_n(q_1, q_2, \ldots, q_n, \dot{q}_1, \dot{q}_2, \ldots, \dot{q}_n, t) \end{cases}$$

subject to initial conditions

$$\underbrace{q_1(0), \ldots, q_n(0)}_{\text{Initial generalized coordinates}}$$

$$\underbrace{\dot{q}_1(0), \ldots, \dot{q}_n(0)}_{\text{Initial generalized velocities}}$$

The *second-order matrix form* for an $n \times n$ system of second-order differential equations is

$$\mathbf{m\ddot{x} + c\dot{x} + kx = f}$$

where $\mathbf{m}$ = mass matrix ($n \times n$),
$\mathbf{c}$ = damping matrix ($n \times n$),
$\mathbf{k}$ = stiffness matrix ($n \times n$)
$\mathbf{x}$ = configuration vector ($n \times 1$) = vector of generalized coordinates,
$\mathbf{f}$ = vector of external forces ($n \times 1$)

A set of *state variables* is the smallest set of independent variables that completely describes the state of the system. The number of state variables is equal to the total number of initial conditions that are required for a complete description of the system. The state variables are chosen as those variables for which initial conditions are required. A set of state variables is *not unique*.

The *state-space form* of the model of a system (n state variables, m inputs, and p outputs) is expressed as

$$\begin{cases} \mathbf{\dot{x} = Ax + Bu} & \text{state equation} \\ \mathbf{y = Cx + Du} & \text{output equation} \end{cases}$$

where

$$\mathbf{x} = \begin{bmatrix} x_1 \\ x_2 \\ \cdot \\ \cdot \\ x_n \end{bmatrix}_{n \times 1} = \text{state vector}, \quad \mathbf{u} = \begin{bmatrix} u_1 \\ u_2 \\ \cdot \\ \cdot \\ u_m \end{bmatrix}_{m \times 1} = \text{input vector},$$

$$\mathbf{y} = \begin{bmatrix} y_1 \\ y_2 \\ \cdot \\ \cdot \\ y_p \end{bmatrix}_{p \times 1} = \text{output vector},$$

$$\mathbf{A} = \begin{bmatrix} a_{11} & a_{12} & \cdot & \cdot & a_{1n} \\ a_{21} & a_{22} & \cdot & \cdot & a_{2n} \\ \cdot & \cdot & & & \cdot \\ \cdot & \cdot & & & \\ a_{n1} & a_{n2} & \cdot & \cdot & a_{nn} \end{bmatrix}_{n \times n} = \text{state matrix},$$

$$\mathbf{B} = \begin{bmatrix} b_{11} & b_{12} & \cdot & \cdot & b_{1m} \\ b_{21} & b_{22} & \cdot & \cdot & b_{2m} \\ \cdot & \cdot & & & \cdot \\ \cdot & \cdot & & & \cdot \\ b_{n1} & b_{n2} & \cdot & \cdot & b_{nm} \end{bmatrix}_{n \times m} = \text{input matrix}$$

$$\mathbf{C} = \begin{bmatrix} c_{11} & c_{12} & \cdot & \cdot & c_{1n} \\ c_{21} & c_{22} & \cdot & \cdot & c_{2n} \\ \cdot & \cdot & & & \cdot \\ \cdot & \cdot & & & \cdot \\ c_{p1} & c_{p2} & \cdot & \cdot & c_{pn} \end{bmatrix}_{p \times n} = \text{output matrix,}$$

$$\mathbf{D} = \begin{bmatrix} d_{11} & d_{12} & \cdot & \cdot & d_{1m} \\ d_{21} & d_{22} & \cdot & \cdot & d_{2m} \\ \cdot & \cdot & & & \cdot \\ \cdot & \cdot & & & \cdot \\ d_{p1} & d_{p2} & \cdot & \cdot & d_{pm} \end{bmatrix}_{p \times m} = \text{direct transmission matrix}$$

A coupled system may be *decoupled* via matrix $\mathbf{P}$, composed of the linearly independent eigenvectors of the state matrix. If $\mathbf{A}$ is the state matrix in the original coupled system, then $\tilde{\mathbf{A}} = \mathbf{P}^{-1}\mathbf{A}\mathbf{P} = \Lambda$ is the state matrix for the decoupled system, where Λ is diagonal and consists of the eigenvalues of matrix $\mathbf{A}$ along its main diagonal.

For a SISO system, the *transfer function* is defined as the ratio of the Laplace transforms of the output and the input, assuming zero initial conditions. The poles of the transfer function are the eigenvalues of the state matrix. For a MIMO system, the *transfer matrix* is a matrix whose entries are the transfer functions between each possible pair of input and output.

Given that the input-output equation of a system is in the general form

$$y^{(n)} + a_1 y^{(n-1)} + \cdots + a_{n-1}\dot{y} + a_n y = b_0 u^{(n)} + b_1 u^{(n-1)} + \cdots + b_n u$$

then the state-space form is expressed as

$$\dot{\mathbf{x}} = \mathbf{A}\mathbf{x} + \mathbf{B}u$$

where

$$\mathbf{x} = \begin{bmatrix} x_1 \\ x_2 \\ \cdot \\ \cdot \\ x_n \end{bmatrix}, \quad \mathbf{A} = \begin{bmatrix} 0 & 1 & 0 & \cdot & \cdot & 0 \\ 0 & 0 & 1 & \cdot & \cdot & 0 \\ \cdot & & & & & \cdot \\ \cdot & & & & & \cdot \\ 0 & 0 & \cdot & \cdot & 0 & 1 \\ -a_n & -a_{n-1} & \cdot & \cdot & & -a_1 \end{bmatrix}, \quad \mathbf{B} = \begin{bmatrix} 0 \\ 0 \\ \cdot \\ \cdot \\ 0 \\ 1 \end{bmatrix},$$

$$y = \mathbf{C}\mathbf{x} + \mathbf{D}u$$

where $\mathbf{C} = [-b_0 a_n + b_n \quad -b_0 a_{n-1} + b_{n-1} \cdots -b_0 a_1 + b_1]$, $D = b_0$

where $\mathbf{A}$ is known as the **lower-companion matrix**.

Linearization of a nonlinear system may be outlined as a four-step procedure:

1. To determine the *operating point*, replace the dependent variable(s) such as x and y by $\bar{x}$ and $\bar{y}$, and the input u by its average value $\bar{u}$, if specified. Solve the resulting (nonlinear) algebraic equation(s) for constants such as $\bar{x}$ and $\bar{y}$.
* In the event that the input is composed of two or more functions, set the time-varying portion(s) equal to zero. For instance, if $u(t) = k + g(t)$, with $k = $ constant, set $g(t) = 0$ so that $\bar{u} = k$.
2. Linearize the nonlinear term(s) about the operating point via Taylor series expansion.
3. In the original nonlinear model, replace the dependent variable(s) such as x and y by $\bar{x} + \Delta x$ and $\bar{y} + \Delta y$, the input u by $\bar{u} + \Delta u$, and the nonlinear term(s) by their linear approximations.
4. Properly adjust the initial conditions.

The **fourth-order Runge-Kutta** method is used to determine a numerical solution to a system of nonlinear, first-order differential equations, in general. If the system is in the form

$$\dot{y} = f(t, y), \qquad a \leq t \leq b, \qquad y(a) = \alpha$$

then the interval $[a, b]$ is divided into N equally spaced divisions, each with a length of $h = (b - a)/N$. In each subinterval, four function evaluations are performed to approximate the value of the solution at the right-end point.

PROBLEMS

3.34. The mathematical model for an electrical circuit is governed by a second-order ordinary differential equation

$$\ddot{q} + \dot{q} + 2q = e(t)$$

where $q(t)$ and $e(t)$ denote the electric charge and the applied voltage, respectively. Express the governing equation in *state-space form* (state equation + output equation) assuming the output is $q(t)$.

3.35. Suppose the equations describing a certain system are

$$\begin{cases} \ddot{x}_1 + 2\dot{x}_1 + x_1 - x_2 = f(t) \\ x_1 - x_2 = 2x_2 \end{cases}$$

where $f(t)$ represents an applied force and is the input.
(a) Assuming that x_1 is the system output, obtain the state-space representation.
(b) Determine the input-output equation.

3.36. Consider the following system of differential equations in which the input is $u(t)$ and the output vector consists of x_1 and x_2.

$$\begin{cases} \ddot{x}_1 + \dot{x}_1 + 3x_1 - 4x_2 = 0 \\ \dot{x}_2 = x_1 - x_2 + u(t) \end{cases}$$

(a) By choosing suitable state variables, obtain the state-space representation.
(b) Find two input-output equations, one for each of the outputs x_1 and x_2.
(c) Assuming zero initial conditions, obtain the transfer functions $X_1(s)/U(s)$ and $X_2(s)/U(s)$.

Obtain the state-space representations corresponding to the input-output equations in Problems 3.37, 3.38, and 3.39.

3.37. $\dddot{y} + 3\ddot{y} + 4\dot{y} + 5y = \ddot{u} + 3\dot{u} + u$

3.38. $\ddot{y} + 2\dot{y} + 3y = 2\ddot{u} + \dot{u} + 3u$

3.39. $\dddot{y} + \ddot{y} + 2\dot{y} + 3y = 2u$

3.40. The governing equations for a second-order nonlinear system are

$$\begin{cases} \dot{x}_1 = -2x_1 + x_2^3 \\ \dot{x}_2 = x_1 + 4 + t \end{cases}$$

where x_1 and x_2 represent the state variables.
(a) Determine the operating point, denoted by $(\bar{x}_1, \bar{x}_2)$.
(b) Through linearization of any nonlinear terms in the original equations about the operating point, obtain the linear model.

3.41. A mechanical system containing a *nonlinear spring* is subject to an applied force $f(t)$. The equation of motion for this system is given as

$$m\ddot{x} + b\dot{x} + \underbrace{x^3}_{\text{nonlinear spring force}} = f(t)$$

(a) Assuming that the operating conditions are associated with the case when $\bar{f} = 0$, find the operating point.
(b) Obtain the linear model for this system.

3.42. Consider the two-degree-of-freedom system in Fig. P3.42 with equations of motion

$$\begin{cases} m_1\ddot{x}_1 + b_1\dot{x}_1 + k_1 x_1 - k_2(x_2 - x_1) - b_2(\dot{x}_2 - \dot{x}_1) = f_1(t) \\ m_2\ddot{x}_2 + b_2(\dot{x}_2 - \dot{x}_1) + k_2(x_2 - x_1) = f_2(t) \end{cases}$$

Assume zero initial conditions. System inputs are f_1 and f_2, and outputs are x_1 and x_2. Determine the transfer function matrix.

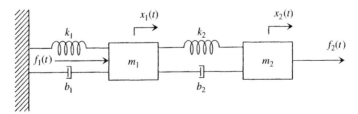

FIGURE P3.42

3.43. Consider the mechanical system in Fig. P3.43. The equations of motion for this system are given as

$$\begin{cases} m_1\ddot{x}_1 + c\dot{x}_1 - c\dot{x}_2 + kx_1 - kx_2 = f(t) \\ m_2\ddot{x}_2 - c\dot{x}_1 + c\dot{x}_2 - kx_1 + kx_2 = 0 \end{cases}$$

Assuming zero initial conditions, obtain the two transfer functions $X_1(s)/F(s)$ and $X_2(s)/F(s)$. Subsequently, form the transfer function matrix.

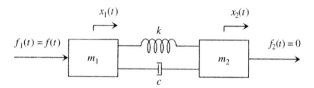

FIGURE P3.43

3.44. Consider the three-degree-of-freedom mechanical system shown in Fig. P3.44. The equations of motion are (see Chapter 4)

$$\begin{cases} m_1\ddot{x}_1 - b_2\dot{x}_2 + (b_1 + b_2)\dot{x}_1 + (k_1 + k_2)x_1 - k_2x_2 = f_1(t) \\ m_2\ddot{x}_2 - b_2\dot{x}_1 + (b_3 + b_2)\dot{x}_2 - b_3\dot{x}_3 - k_2x_1 + (k_3 + k_2)x_2 - k_3x_3 = f_2(t) \\ m_3\ddot{x}_3 + b_3\dot{x}_3 - b_3\dot{x}_2 + k_3x_3 - k_3x_2 = f_3(t) \end{cases}$$

(a) Express the equations of motion in second-order matrix form. What do you notice about the nature of mass, stiffness, and damping matrices?
(b) Obtain the state equation by choosing a suitable set of state variables.

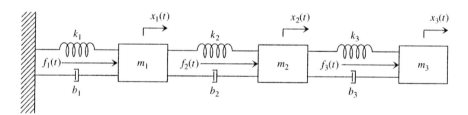

FIGURE P3.44

3.45. The input-output equation for the RLC circuit in Fig. P3.45 is defined as

$$L\frac{d^2i}{dt^2} + R\frac{di}{dt} + \frac{1}{C}i = \dot{e}(t)$$

in which current $i(t)$ and applied voltage $e(t)$ represent the output and input, respectively.
(a) Assuming zero initial conditions, obtain the system's transfer function.
(b) Express the input-output equation in state-space form.

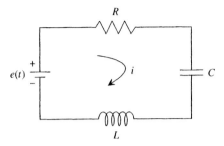

FIGURE P3.45

3.46. Consider the system shown in Fig. P3.46 in which applied forces $f_1(t)$ and $f_2(t)$ are the inputs.
   (a) Determine the state variables, and subsequently the state equation.
   (b) Assuming that the system output is x_1, obtain the output equation. Repeat for the case when the outputs are x_1 and $\dot{x}_1$.

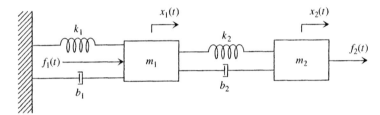

FIGURE P3.46 Two-degree-of-freedom mechanical system.

3.47. Consider the two-tank liquid-level system shown below (see Chapter 6). Assuming that q_i and q_o represent system input and output, respectively, the transfer function is given in the form

$$\frac{Q_o(s)}{Q_i(s)} = \frac{1}{R_1C_1R_2C_2s^2 + (R_2C_1 + R_2C_2 + R_1C_1)s + 1}$$

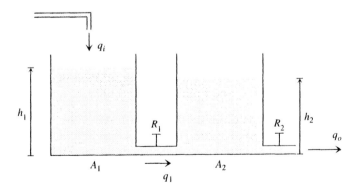

FIGURE P3.47 Two-tank liquid-level system.

where C_1, C_2, R_1, and R_2 denote capacitances of the two tanks, and resistances of the two valves, respectively.
(a) Determine the input-output equation.
(b) By selecting suitable state variables, express the I/O equation in state-space form.

3.48. In the (undamped) mechanical system in Fig. P3.48, displacements x_1 and x_2 are system outputs, and the applied force $f(t)$ is the input. Provided that the equations of motion are as follows, determine the transfer matrix.

$$\begin{bmatrix} m_1 & 0 \\ 0 & m_2 \end{bmatrix} \begin{Bmatrix} \ddot{x}_1 \\ \ddot{x}_2 \end{Bmatrix} + \begin{bmatrix} k_1 + k_2 & -k_2 \\ -k_2 & k_2 \end{bmatrix} \begin{Bmatrix} x_1 \\ x_2 \end{Bmatrix} = \begin{Bmatrix} 0 \\ f \end{Bmatrix}$$

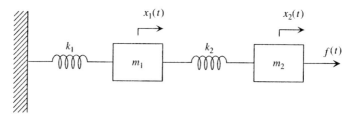

FIGURE P3.48 Two-degree-of-freedom, undamped mechanical system.

CHAPTER 4

Mechanical Systems

The fundamental dynamics and modeling of mechanical systems are treated in this chapter. Mechanical systems are either in translational or rotational motion or both. Mechanical elements include mass element, spring element, and damper element, translational and rotational. The concepts of equivalence, degrees of freedom, and constraints are discussed. Two important types of constraints are holonomic and nonholonomic constraints. Newton's laws are used for translational systems, whereas the moment equations are for rotational systems. They are used together for modeling of combined systems of translational and rotational. Gear-train systems are also included.

4.1 INTRODUCTION

The objective of an engineering analysis of a mechanical system is prediction of its behavior. Since real systems are usually quite complicated when viewed in detail, an "exact" analysis of any system is often impossible. Thus, simplifying assumptions must be made to reduce the system to an idealized version whose behavior approximates that of the real system. The process by which a physical system is simplified to obtain a mathematically tractable situation is called **modeling.** The resulting simplified version of the real system is called the **mathematical model,** or simply the **model,** of the system.

In this chapter, we shall concentrate on the modeling of mechanical systems. These systems can be divided into two categories: translational and rotational. Certain systems may be purely translational or rotational, whereas others may be hybrid (mixed).

4.2 MECHANICAL ELEMENTS

To model a mechanical system, we must first understand the elements of the system. There are three types of mechanical elements: mass, spring, and damper. In this section, the *elemental equations* of mechanical elements are presented.

Mass Element

Translational mass

Let us consider a particle traveling at (translational) velocity **v**. The word *translational* is placed in parentheses, in this case, because it is often omitted for brevity. The motion is under the influence of an externally applied force **f**, and the linear momentum of the particle is **p**. The force is equal to the time rate of change of the momentum, in vector form, as

$$\mathbf{f} = \frac{d\mathbf{p}}{dt} \tag{4.1}$$

This fundamental relation holds true for both newtonian mechanics and the special theory of relativity.

In newtonian mechanics, the linear momentum is given as

$$\mathbf{p} = m\mathbf{v} \tag{4.2a}$$

where m is the mass of the particle. However, in the special theory of relativity, the linear momentum is expressed as

$$\mathbf{p} = \frac{m\mathbf{v}}{\sqrt{1 - (v/c)^2}} \tag{4.2b}$$

where m = rest mass of the particle
v = "speed" of the particle, or the magnitude of the velocity vector **v**
c = speed of light in vacuum, 3×10^8 m/s

In the normal operation of most machines, mass is assumed to be constant where the theory of relativity can be ignored. Thus, Newton's second law of motion, for newtonian mechanics, is expressed as

$$\mathbf{f} = \frac{d}{dt}(m\mathbf{v}) = m\frac{d\mathbf{v}}{dt} = m\mathbf{a} \tag{4.3}$$

where **a** is the (translational) acceleration vector.

Note that the acceleration vector **a** is *absolute* (i.e., measured with respect to an inertial reference frame). Although the earth is not truly an inertial reference, it is common practice to use the ground as a reference for motion. The error introduced by this practice is insignificant.

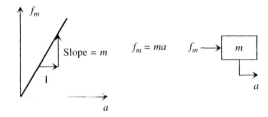

FIGURE 4.1 Translational motion: linear relation among the force f_m, mass m, and acceleration a.

The acceleration vector **a**, velocity vector **v**, and position vector **r** are all related by the time derivatives,

$$\mathbf{a} = \frac{d\mathbf{v}}{dt} = \frac{d^2\mathbf{r}}{dt^2} \qquad (4.4a)$$

Introducing the simple "dot notation" for time derivatives, we have

$$\mathbf{a} = \dot{\mathbf{v}} = \ddot{\mathbf{r}} \qquad (4.4b)$$

Figure 4.1 shows a linear relationship between the force and acceleration, where the mass m is the proportional constant. The subscript m on the force, f_m, is used to denote that the force is associated with the mass element.

The Right-Hand Rule (RHR)

In contrast to translational motion, rotational motions may be difficult for some people to work with. The right-hand rule (RHR), as shown in Fig. 4.2, helps to alleviate this problem. The rotational vector points in the direction of the thumb, and the sense of rotation follows the curve of the four fingers. A set of two arrow types is used to show rotational vectors. The direction of the rotational vector (primary) is shown by a *solid* line-and-arrow, and the sense of rotation (secondary) is by a *broken* curve-and-arrow.

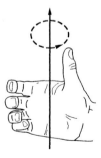

FIGURE 4.2 The right-hand rule for rotational motions: the direction of the angular vector points in the thumb direction.

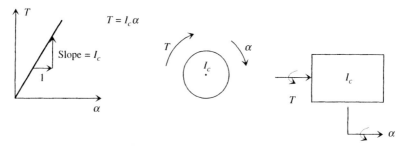

FIGURE 4.3 Rotational motion: linear relation among the torque T, centroidal mass moment of inertia I_c, and angular acceleration α.

Rotational mass

The centroidal **mass moment of inertia,** I_c, about the longitudinal axis of a shaft (or a disk) is given as shown in Fig. 4.3, where T is an externally applied torque. Note that the angular acceleration α is *absolute* and the mass moment of inertia is used in dynamics, whereas **area moment of inertia** is used in strength of materials (or mechanics of materials).

The angular acceleration vector $\boldsymbol{\alpha}$ is defined as

$$\boldsymbol{\alpha} = \frac{d\boldsymbol{\omega}}{dt} = \dot{\boldsymbol{\omega}} \tag{4.5}$$

where $\boldsymbol{\omega}$ = angular velocity vector.

In scalar form, we have

$$\alpha = \frac{d\omega}{dt} = \frac{d^2\theta}{dt^2} \tag{4.6}$$

$$\alpha = \dot{\omega} = \ddot{\theta} \tag{4.7}$$

where ω = angular speed
θ = angular displacement

Spring Element

Translational spring

The stiffness k of a **translational spring** is given as shown in Fig. 4.4, where f_k is an externally applied force. A linear spring is assumed, so k is constant. The spring is also assumed massless, or of negligible mass; therefore, the forces at both ends of the spring are equal in magnitude but opposite in direction. This result can be shown by using Newton's second law with zero mass.

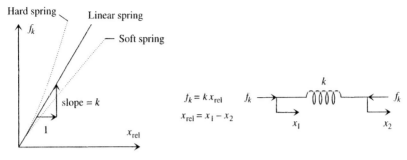

FIGURE 4.4 Translational motion: linear relation among the spring force f_k, stiffness k, and relative displacement x_{rel}. Assume $x_1 > x_2 > 0$.

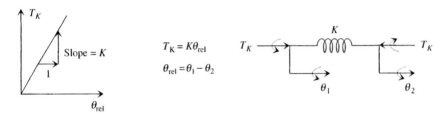

FIGURE 4.5 Rotational motion: linear relation among the torque T_K, torsional stiffness K, and relative angular displacement θ_{rel}. Assume $\theta_1 > \theta_2 > 0$.

Rotational spring

The stiffness K of a torsional spring is given as shown in Fig. 4.5, where T_K is an externally applied torque. Note that the capital letter K is used for torsional stiffness whereas the lowercase k indicates **translational stiffness.**

Damper Element

There are three types of damping in engineering mechanics:

- Viscous damping (with fluid)
- Coulomb damping (dry friction)
- Structural damping (hysteresis damping)

Only viscous damping and Coulomb damping are treated here. Viscous damping is the simplest type of damping and is our focus in this text, whereas structural damping is beyond the scope of the text.

Viscous damping

The friction generated between two surfaces separated by a liquid is called **viscous damping.** A **viscous damper** is also called a **dashpot.** Figure 4.6 shows a

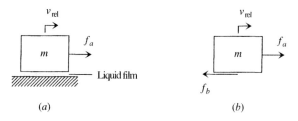

FIGURE 4.6 Viscous damping: (a) physical system, (b) FBD.

block of mass m sliding on a horizontal surface, where the two surfaces are separated by a liquid film. A force f_a is applied to the block in the direction shown. To focus on the horizontal motion, only the horizontal force components are shown on the free-body diagram (FBD); the vertical force components are ignored. The damping force, f_b, opposes the motion and depends on the nature of the fluid flow between the two surfaces. The exact friction force for viscous damping is complex, thus for modeling in system dynamics, we use a linear relationship for viscous damper. The magnitude of the damping force is given as

$$f_b = bv_{\text{rel}} \qquad v_{\text{rel}} > 0 \qquad (4.8)$$

Note that this damping is called **positive damping,** because it dissipates the energy of the system. This kind of damping helps to stabilize the system. On the other hand, the damping associated with self-excited vibrations is called **negative damping** because it adds energy into the system and thus tends to make the system unstable.

Coulomb damping

The friction generated between two dry surfaces is called Coulomb damping or dry friction. To illustrate the concept, let us consider a block of mass m in contact with a horizontal surface and under the influence of gravity and the applied force f_a (Fig. 4.7). The notation is given as

f_a = applied force
f_f = (general) friction force
f_k = kinetic friction force
f_s = static friction force
$(f_s)_{\max}$ = maximum static friction force
N = normal force

Although the system is dynamic, the friction force is static if the relative velocity between the two contacting surfaces is zero. The (general) friction force thus becomes the static friction force, $f_f \rightarrow f_s$. The value of f_s ranges from zero to the maximum value:

$$0 \leq f_s \leq [(f_s)_{\max} = \mu_s N] \qquad \text{if } v_{\text{rel}} = 0 \qquad (4.9)$$

At the transition, when sliding impends, the static friction force attains the maximum value.

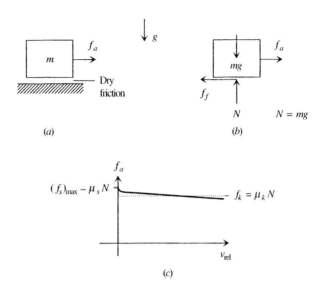

FIGURE 4.7 Coulomb damping: (*a*) physical system, (*b*) FBD, (*c*) friction force.

As soon as sliding occurs, the (general) friction force becomes the kinetic (sliding) friction force, $f_f \rightarrow f_k$. The kinetic friction force starts from a value below $(f_s)_{max}$ and then continues to decrease as the relative velocity, v_{rel}, increases. This characteristic explains certain phenomena in self-excited vibrations, such as the vibration of a violin's strings. However, for most applications, we assume that the kinetic friction force is nearly constant:

$$f_k \approx \mu_k N \quad \text{if } v_{rel} > 0 \qquad (4.10)$$

where $\mu_k < \mu_s$.

Translational viscous damper

The viscous damping coefficient b of a translational damper is given as shown in Fig. 4.8, where f_b is an externally applied force. Linear damper is assumed, so b is constant. The damper is also assumed massless, or of negligible mass; therefore,

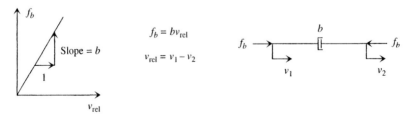

FIGURE 4.8 Translational motion: linear relation among the damping force f_b, viscous damping coefficient b, and relative velocity v_{rel}. Assume $v_1 > v_2 > 0$.

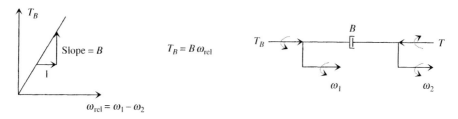

FIGURE 4.9 Rotational motion: linear relation among the damping torque T_B, torsion viscous damping coefficient B, and relative angular velocity ω_{rel}. Assume $\omega_1 > \omega_2 > 0$.

the forces at both ends of the damper are equal in magnitude but opposite in direction. This result can be shown, again, by Newton's second law with zero mass.

A Note on Notation. The symbol c is often (or traditionally) used to denote the viscous damping coefficient. But in certain applications, it may be confused with the center of mass (or mass center) C; therefore, both b and c will be used interchangeably in this book.

Rotational viscous damper

The viscous damping coefficient B of a torsional damper is given as shown in Fig. 4.9, where T_B is an externally applied torque. Note that the capital letter B is used for torsional damper whereas the lowercase b is for translational damper. This notation helps to distinguish between the two types of damper.

4.3 EQUIVALENCE

The concept of equivalence is convenient in many applications, but it must be used with care. For example, a system of springs may be represented by an equivalent spring. The springs may be in parallel, in series, or mixed (hybrid). The derivation, or proof, of the equivalent stiffness for a system of springs in parallel is relatively simple, whereas that for springs in series is relatively more complex.

THEOREM 4.1. If there are n springs in *parallel*, then the equivalent stiffness k_{eq} is equal to the sum of all the individual spring stiffnesses k_i:

$$k_{eq} = k_1 + k_2 + \cdots + k_n \qquad (4.11)$$

As an example, consider a system of two springs, k_1 and k_2, in *parallel* as shown in Fig. 4.10. The system of two springs in parallel is equivalent to a single spring whose stiffness is

$$k_{eq} = k_1 + k_2$$

Proof. From Fig. 4.10,

$$f = f_1 + f_2$$
$$k_{eq}x = k_1 x + k_2 x = (k_1 + k_2)x$$
$$k_{eq} = k_1 + k_2$$

180 Dynamic Systems: Modeling and Analysis

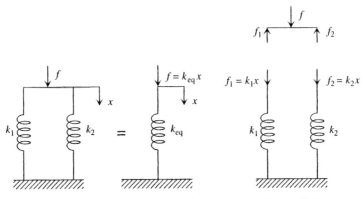

FIGURE 4.10 Equivalence for two springs in parallel.

THEOREM 4.2. If there are n springs in *series*, then the reciprocal of the equivalent stiffness k_{eq} is equal to the sum of all the reciprocals of the individual spring stiffnesses k_i:

$$\frac{1}{k_{eq}} = \frac{1}{k_1} + \frac{1}{k_2} + \cdots + \frac{1}{k_n} \tag{4.12}$$

As an example, consider a system of two springs, k_1 and k_2, in *series* as shown in Fig. 4.11. The equivalent stiffness of the system is

$$k_{eq} = \frac{k_1 k_2}{k_1 + k_2}$$

Proof. From Fig. 4.11, the total deformation δ is given by

$$\delta = \delta_1 + \delta_2$$

$$\frac{f}{k_{eq}} = \frac{f_1}{k_1} + \frac{f_2}{k_2}$$

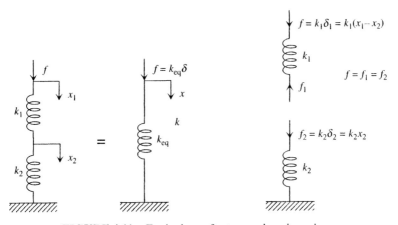

FIGURE 4.11 Equivalence for two springs in series.

But $f = f_1 = f_2$, thus

$$\frac{1}{k_{eq}} = \frac{1}{k_1} + \frac{1}{k_2}$$

$$k_{eq} = \frac{k_1 k_2}{k_1 + k_2}$$

For a system of damping elements, the equivalent damper can be shown using the same logic and similar steps. However, it is not the same with a system of mass elements. The reason is that spring and damper elements are two-terminal elements, whereas mass element is a single-terminal element.

4.4 DEGREES OF FREEDOM

An important concept in the description of dynamic systems is **degree of freedom** (**DOF**) or **number of degrees of freedom** (**#DOF**).

DEFINITION 4.1. The **number of degrees of freedom** of a dynamic system is defined as the number of *independent* generalized coordinates that specify the configuration of the system.

Generalized coordinates need not be restricted only to the actual coordinates of position. The positional coordinates are physical *coordinates*. On the other hand, a generalized coordinate could be anything, e.g., positional coordinate, translational displacement, rotational displacement, pressure, voltage, or current. Generalized coordinates need not be of the same type. The generalized coordinates of a mechanical system can be a mixed set of translational displacements and rotational displacements.

Consider the mechanical system shown in Fig. 4.12. The mass m moves in a horizontal direction, and x is the displacement measured from the static equilibrium position of the mass. Thus, the displacement x is the generalized coordinate. If the origin of the coordinate system is at the static equilibrium position, then x is also the positional coordinate.

As an example of the mixed type of generalized coordinates, consider the pendulum system shown in Fig. 4.13. The cart of mass M moves in a horizontal direction, and x is the displacement measured from the static equilibrium position of the cart. Attached to the cart is a pendulum subsystem consisting of a *massless rod* of length L and a concentrated mass (or point mass) of mass m at the

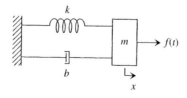

FIGURE 4.12 A mechanical system showing displacement as generalized coordinate.

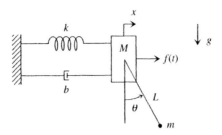

FIGURE 4.13 A mechanical system showing mixed types of generalized coordinates.

rod tip. For this system, the two independent generalized coordinates are x and θ. They are independent because if x is fixed, θ can vary independently. Similarly, if θ is fixed, x can have a range of values.

Definition 4.1 is a simple definition of number of degrees of freedom, which is a useful concept for obtaining a mathematical model (differential equation) of the dynamic system. The following definition, Definition 4.2, is another definition of degrees of freedom, one that involves constraints.

DEFINITION 4.2. The number of degrees of freedom is defined as the number of *dependent* generalized coordinates, n, that specify the configuration of the system, *minus* the number of *independent* equations of constraint, m.

$$\#\mathrm{DOF} = n - m \qquad (4.13)$$

In Definition 4.2 the number of degrees of freedom of a dynamic system is determined by the number of *dependent* generalized coordinates and the number of constraint equations. If n coordinates are used to define the configuration of a system, and if there are m independent equations of constraint relating these coordinates, then there are $(n - m)$ degrees of freedom.

EXAMPLE 4.1. Consider the classical simple pendulum in planar motion on the vertical xy-plane (Fig. 4.14). The pendulum system consists of a massless rod of constant length L that is attached to a point mass m at the end. Determine the constraint equation and the number of degrees of freedom.

Solution. The configuration of the system is specified by the coordinates of the point mass, either by rectangular coordinates (x, y) or polar coordinates (L, θ). If we

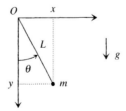

FIGURE 4.14 A simple pendulum.

choose x and y as the coordinates ($n = 2$), then we must use the following constraint ($m = 1$)

$$x^2 + y^2 = L^2 = \text{constant}$$

The system has only one degree of freedom: $n - m = 2 - 1 = 1$.

For a dynamic system, if the purpose is to obtain the differential equations, then the analyst should select a set of independent coordinates so that constraint equations are not needed. Simpler is better here. For the pendulum system of Example 4.1, if we select θ as the only coordinate ($n = 1$), then no constraint equation is needed ($m = 0$). The number of degrees of freedom does not change: $n - m = 1$. Note that for this system, the two polar coordinates (L, θ) are reduced to the single coordinate θ because the length L is constant.

Rolling Constraints

Consider a circular disk rolling on a horizontal plane along a straight line (Fig. 4.15). The constraint is that the disk rolls without slipping. It is desired to obtain the relations between the translational and corresponding angular variables. Without slipping, or with the **no-slip condition**, the length AB must be the same as the arc length $A'B$,

$$\overline{AB} = \widehat{A'B} \qquad (4.14)$$

With this constraint, the relation between the (translational) displacement x and the angular displacement θ is

$$x = r\theta \qquad (4.15)$$

This result yields other relations for translational and angular velocities and accelerations, as

$$\dot{x} = r\dot{\theta} \quad \text{or} \quad v = r\omega \qquad (4.16)$$

$$\ddot{x} = r\ddot{\theta} \quad \text{or} \quad a = r\alpha \qquad (4.17)$$

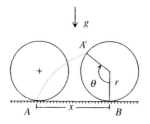

FIGURE 4.15 A rolling disk on a horizontal plane, along a straight line, without slip.

EXAMPLE 4.2. Consider a circular cylinder rolling on an inclined plane without slipping (Fig. 4.16). Determine the relation between the translational acceleration of mass C and the angular acceleration α.

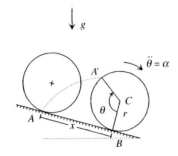

FIGURE 4.16 A rolling disk on an inclined plane with the no-slip condition.

Solution. From the constraint, the no-slip condition, we have

$$x = r\theta$$

Thus,

$$\ddot{x} = r\ddot{\theta} \quad \text{or} \quad a_c = r\alpha$$

Holonomic Constraints and Nonholonomic Constraints

There are two types of constraints: **holonomic constraints** and **nonholonomic constraints**. Systems with holonomic constraints are relatively easy to deal with, whereas those with nonholonomic constraints are more difficult. Fortunately, many engineering systems that we often encounter contain holonomic constraints. For these systems, the number of degrees of freedom is the same as the number of independent generalized coordinates, which is also the same as the number of independent differential equations.

Holonomic constraints

Holonomic constraints are equations expressed in terms of the coordinates and possibly time. The classical simple pendulum in planar motion of Example 4.1 is an excellent example of holonomic constraints.

For a system with holonomic constraints, we can always find a set of *independent* generalized coordinates. This number of independent generalized coordinates equals the number of degrees of freedom. For the pendulum of Example 4.1, it is best to select θ as the independent generalized coordinate because the system has only one degree of freedom.

EXAMPLE 4.3. Consider the mechanical system shown in Fig. 4.17, which is constrained to move in a straight line. Determine the independent coordinates, the number of degrees of freedom, and the number of independent differential equations.

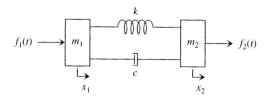

FIGURE 4.17 Two-mass system.

Solution. The configuration of the system is specified by the displacements of the two masses, in rectangular coordinates. Thus, we have two independent generalized coordinates: x_1 and x_2. The system is constrained in such a way that the two masses can only move along a straight line; this type of constraint is holonomic. Therefore, the number of degrees of freedom is the same as the number of independent generalized coordinates. The number of degrees of freedom (#DOF) is 2, thus, the number of differential equations (DEs) is also 2.

Nonholonomic constraints

Unlike holonomic constraints, nonholonomic constraints must be expressed in terms of the *differentials* of the coordinates and possibly time. Consider a circular coin of radius R that rolls without slipping on a horizontal xy-plane (Fig. 4.18). The plane of the coin is vertical, that is, perpendicular to the horizontal xy-plane, and its axis of rotation goes through the center of mass C and is parallel to the xy-plane. The path is on the xy-plane, and the projected coordinate of the center of mass C on the xy-plane coincides with the contact point A. This is an example of nonholonomic constraints:

$$dx - R\sin\phi\, d\theta = 0 \qquad (4.18)$$
$$dy - R\cos\phi\, d\theta = 0 \qquad (4.19)$$

where (x, y) = coordinate of the contact point A
(x, y, r) = coordinate of the center of mass C
θ = angular displacement of the coin about the axis through its center and perpendicular to its plane
ϕ = angle between the plane of the coin and the yz-plane.

For a system containing nonholonomic constraints, the number of required *dependent* generalized coordinates is greater than the number of degrees of freedom. Since the nonholonomic constraints are *not* integrable, we cannot integrate the constraints to obtain a set of *independent* generalized coordinates. Recall that the number of independent generalized coordinates is equal to the number of degrees of freedom.

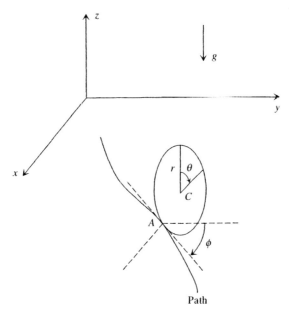

FIGURE 4.18 A rolling coin as an example of nonholonomic constraint.

The rolling coin example has four dependent generalized coordinates—x and y (location) and ϕ and α (orientation)—but it has only two degrees of freedom. We can vary only two coordinates independently. For example, we can change ϕ and α independently, but these changes will result in changes in x and y. This problem can be solved by using the required dependent generalized coordinates together with the nonholonomic constraints.

PROBLEM SET 4.1–4.4

4.1. Consider a system of two springs of stiffnesses k_1 and k_2 in parallel. Find the equivalent spring stiffness k_{eq}.

4.2. Consider a system of three springs of stiffnesses k_1, k_2, and k_3 in parallel. Find the equivalent spring stiffness k_{eq}.

4.3. Consider a system of two springs of stiffnesses k_1 and k_2 in series. Find the equivalent spring stiffness k_{eq}.

4.4. Consider a system of three springs of stiffnesses k_1, k_2, and k_3 in series. Find the equivalent spring stiffness k_{eq}.

4.5. Consider a hybrid (mixed) system of three springs as shown in Fig. P4.5. The two springs k_1 and k_2 are in series, and are both in parallel with the spring k_3. Find the equivalent spring stiffness k_{eq}.

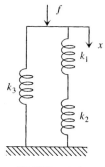

FIGURE P4.5

4.6. Consider a system of two dampers of viscous damping coefficients b_1 and b_2 in parallel. Find the equivalent viscous damping coefficient b_{eq}.

4.7. Consider a system of three dampers of viscous damping coefficients b_1, b_2, and b_3 in parallel. Find the equivalent viscous damping coefficient b_{eq}.

4.8. Consider a system of two dampers of viscous damping coefficients b_1 and b_2 in series. Find the equivalent viscous damping coefficient b_{eq}.

4.9. Consider a system of three viscous damping coefficients b_1, b_2, and b_3 in series. Find the equivalent viscous damping coefficient b_{eq}.

4.10. Consider a system of two rigid masses m_1 and m_2 in series. Find the equivalent mass m_{eq}.

4.11. What about a system of two rigid masses m_1 and m_2 in parallel? Can you find an equivalent mass m_{eq}?

4.12. For each of the following dynamic systems, determine the number of degrees of freedom, the independent generalized coordinates, and the number of independent differential equations:
(a) A point mass in space
(b) A point mass on a plane
(c) A rigid body in space
(d) A single pendulum pivoted at one end and constrained in a vertical plane

4.13. Indicate the type(s) of constraint for Problem 4.12.

4.14. Throw a coin on a flat surface such that it rolls. Indicate the type of constraint.

4.15. Consider the hybrid (mixed) system of three springs shown in Fig. P4.15. The system is constrained to move in a vertical plane and in a vertical direction. Assume that the concept of equivalence is not used. Determine the number of degrees of freedom, the independent generalized coordinates, and hence the number of independent differential equations.

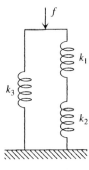

FIGURE P4.15

4.16. Consider the mechanical systems shown in Fig. P4.16. The system is constrained to move on a horizontal plane and along a straight line. Determine the number of degrees of freedom, the independent generalized coordinates, and hence the number of independent differential equations.

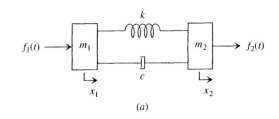

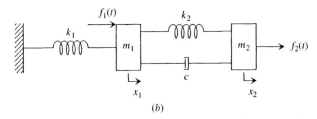

FIGURE P4.16

4.17. A circular coin of radius R and mass m rolling down, without slipping, on an inclined plane whose angle of inclination is ϕ, is shown in Fig. P4.17.
  (a) Express the centroidal mass moment of inertia of the coin in terms of the given parameters.
  (b) Determine the equation of constraint. Hence, deduce the number of degrees of freedom, the independent generalized coordinate, and the number of independent differential equations.

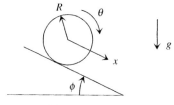

FIGURE P4.17

4.18. Consider a simple pendulum pivoted at point O as shown in Fig. P4.18. The pendulum consists of a uniform rod of mass m and length L and a point mass (concentrated mass) at the rod tip. The system is constrained to move in a vertical plane.
   (a) In terms of the given parameters, express the mass-moment of inertia about point O.
   (b) Determine the number of degree(s) of freedom, the independent generalized coordinate(s), and the number of independent differential equation(s).

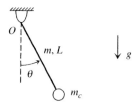

FIGURE P4.18

4.19 Consider the inverted pendulum system shown in Fig. P4.19. The system consists of a cart of mass M and a uniform rod of mass m and length L where the concentrated mass m_c is attached at the tip of the rod. The rod is pivoted on the cart. The applied force is f, and the coefficient of viscous damping on the cart is c. Assume that the system is constrained to move in a vertical plane and the cart rolls without slipping on a horizontal line. Select the independent generalized coordinates, and hence determine the number of degrees of freedom and the number of independent differential equations.

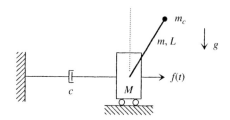

FIGURE P4.19

4.20. Consider the pendulum system shown in Fig. P4.20. The system consists of a cart of mass M, and a uniform and massless rod of mass m and length L. The rod is pivoted on the cart. The applied force is f, and the coefficient of viscous damping on the cart is

c. Assume that the system is constrained to move in a vertical plane and the cart rolls without slipping on a horizontal line. Select the independent generalized coordinates, and hence determine the number of degrees of freedom and the number of independent differential equations.

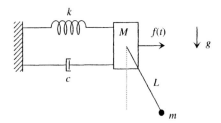

FIGURE P4.20

4.5 TRANSLATIONAL SYSTEMS

Newton's Second Law

In many textbooks, *Newton's second law* is often written casually in vector form as

$$\sum \mathbf{F} = m\mathbf{a} \qquad (4.20)$$

and in rectangular-coordinate component form as

$$\sum F_x = ma_x \qquad (4.21)$$

$$\sum F_y = ma_y \qquad (4.22)$$

Newton's second law as written appears simple, but how to apply the law correctly is another story. The law, as written, is only good for conceptual discussion or development. The sign convention is not shown, although this is where students have difficulty in solving problems. As a result, many students do not fully understand Newton's second law, and they tend to misuse it. This section presents Newton's second law with notation in detail in hopes that it will help to alleviate the problem.

For a system of n rigid bodies subject to N externally applied forces, the most general vector form of Newton's second law can be shown as

$$\sum_{i=1}^{N} \mathbf{F}_i = m\mathbf{a}_c = \sum_{j=1}^{n} m_j \mathbf{a}_{cj} = m_1 \mathbf{a}_{c1} + m_2 \mathbf{a}_{c2} + \cdots + m_n \mathbf{a}_{cn} \qquad (4.23)$$

where

$$m = \sum_{j=1}^{n} m_j \qquad (4.24)$$

$\mathbf{F}_i$ = ith externally applied physical force
$\mathbf{a}_c$ = absolute acceleration of the mass center of the system of rigid bodies
$\mathbf{a}_{cj}$ = absolute acceleration of the mass center of the jth rigid body

In component form for x- and y-directions, Newton's second law is given as

$$\xrightarrow{+} \sum_{i=1}^{N} F_{ix} = ma_{cx} = \sum_{j=1}^{n} m_j a_{cjx} = m_1 a_{c1x} + m_2 a_{c2x} + \cdots + m_n a_{cnx} \xrightarrow{+} \quad (4.25)$$

$$+\uparrow \sum_{i=1}^{N} F_{iy} = ma_{cy} = \sum_{j=1}^{n} m_j a_{cjy} = m_1 a_{c1y} + m_2 a_{c2y} + \cdots + m_n a_{cny} \uparrow + \quad (4.26)$$

where a rectangular coordinate system (cartesian coordinate system) is used with x horizontal and positive to the right and y vertical and positive up. The component form is the one to be used for problem solving, whereas the vector form is mainly for theory development.

Note that in applying Newton's second law, the correct point in the formulation is the *center of mass* C, not the *center of gravity* G. This is why the subscript c, not G, is chosen. However, in many engineering applications, the gravity field may be considered uniform, so the center of mass and the center of gravity coincide, $C \equiv G$, even though they are different by definition.

For simplicity and for quick writing, some of the subscripts or summation indices may be dropped. For a system of n rigid bodies,

$$\sum \mathbf{F} = m\mathbf{a}_c = \sum m_j \mathbf{a}_{cj} \quad (4.27)$$

$$\xrightarrow{+} \sum F_x = ma_{cx} = \sum m_j a_{cjx} \xrightarrow{+} \quad (4.28)$$

$$+\uparrow \sum F_y = ma_{cy} = \sum m_j a_{cjy} \uparrow + \quad (4.29)$$

For a single rigid body,

$$\sum \mathbf{F} = m\mathbf{a}_c \quad (4.30)$$

$$\xrightarrow{+} \sum F_x = ma_{cx} \xrightarrow{+} \quad (4.31)$$

$$+\uparrow \sum F_y = ma_{cy} \uparrow + \quad (4.32)$$

For a concentrated mass (also called a point mass), or a single particle, the subscript c is dropped since there is no distinction between the center of mass C and any other points of the mass:

$$\sum \mathbf{F} = m\mathbf{a} \quad (4.33)$$

$$\xrightarrow{+} \sum F_x = ma_x \xrightarrow{+} \quad (4.34)$$

$$+\uparrow \sum F_y = ma_y \uparrow + \quad (4.35)$$

192 Dynamic Systems: Modeling and Analysis

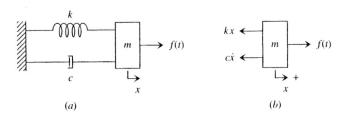

FIGURE 4.19 A mass-spring-damper system: (*a*) physical system, (*b*) FBD.

Free-Body Diagram and Sign Convention

Drawing correct free-body diagrams is the most important step in analyzing mechanical systems. For analyzing mechanical systems from the force/moment approach (as opposed to the energy approach), an incorrect free-body diagram (FBD) certainly will lead to incorrect mathematical models or differential equations of motion.

Let us consider a simple system consisting of a block of mass m, a spring of stiffness k, and a viscous damper of viscous damping coefficient b. Figure 4.19 shows the physical system and the FBD. It may be described in short as a system of mass m, spring k, and viscous damper b. For simplicity hereafter, unless stated otherwise, we will refer to this as the mass-spring-damper system.

The positive direction is the direction shown by the arrow next to the displacement. This is the sign convention. Thus, the positive direction is to the right as shown by the arrow accompanying displacement x (Fig. 4.19*a*). Placing a positive sign, +, next to x (an optional step) may be helpful to remind the reader that this is the positive direction (Fig. 4.19*b*). With this convention, although only the positive displacement x is shown, it implies that both the velocity $\dot{x}$ and acceleration $\ddot{x}$ are also positive to the right.

The FBD should be constructed using the sign convention established above. In addition, only the *magnitudes* (positive) of the forces are shown on the FBDs. The physical directions of the forces are taken care of by the arrows. For example, the restoring spring force with positive magnitude kx points to the left when positive displacement x occurs (x points to the right). Similarly, the damping force (opposing the motion) with positive magnitude $c\dot{x}$ points to the left since positive velocity $\dot{x}$ points to the right.

Modeling of a System with Viscous Damping

Let us take the simple system shown in Fig. 4.19 again as an example, which includes the FBD, assuming $x > 0$. Thus the spring is in *tension*. We derive the differential equation of motion using Newton's laws. Note that if we assume x negative, $x < 0$, then the spring k would be in *compression*. The directions of the spring force and damping force must be reversed on the FBD; however, the resulting differential equation will be the same. This is left as an exercise to the reader.

Applying Newton's second law, in the x-direction,

$$\xrightarrow{+} \sum F_x = ma_{cx} \xrightarrow{+}$$

$$\underbrace{f(t) - c\dot{x} - kx}_{\text{physical forces}} = \underbrace{m}_{\text{mass}} \underbrace{\ddot{x}}_{\text{acceleration}} \qquad (4.36)$$

Rearranging the equation in the standard input-output differential equation form,

$$m\ddot{x} + c\dot{x} + kx = f(t) \qquad (4.37)$$

Dividing the above equation by m, we have

$$\ddot{x} + \frac{c}{m}\dot{x} + \frac{k}{m}x = \frac{f(t)}{m}$$

Thus, the differential equation is expressed in terms of the natural frequency ω_n and the damping ratio ζ, as

$$\ddot{x} + 2\zeta\omega_n\dot{x} + \omega_n^2 x = \frac{f(t)}{m}$$

where
$$\omega_n = \sqrt{\frac{k}{m}} \qquad \zeta = \frac{c}{2m\omega_n} = \frac{c}{2\sqrt{km}}$$

Notice that the parameter m still appears on the right-hand side of the equation for the mass-spring-damper system.

Modeling of a System with Coulomb Damping

Consider the simple system with a Coulomb damper as shown in Fig. 4.20. The block of mass m, which is attached to a spring of stiffness k, slides on a dry surface. The motion is along a horizontal direction and under the influence of gravity, and the horizontal force applied to the mass is f. The FBD is included for different conditions on the displacement and velocity. It is desired to obtain the differential equation of motion.

Applying Newton's second law, in the x-direction,

$$\xrightarrow{+} \sum F_x = ma_{cx} \xrightarrow{+}$$

In the case of positive displacement and velocity, the equation is

$$f(t) - kx - \mu_k N = m\ddot{x}, \qquad x > 0, \qquad \dot{x} > 0 \qquad (4.38)$$

which becomes, after rearranging,

$$m\ddot{x} + kx + \mu_k N = f(t), \qquad x > 0, \qquad \dot{x} > 0 \qquad (4.39)$$

If the displacement and velocity are negative, the equation is

$$f(t) + k(-x) + \mu_k N = m\ddot{x} \qquad (4.40)$$

194 Dynamic Systems: Modeling and Analysis

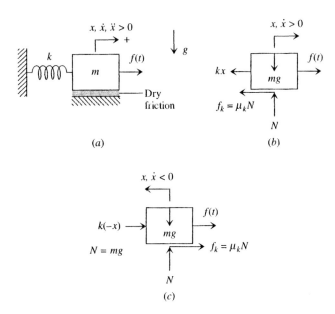

FIGURE 4.20 System with Coulomb damping: (*a*) physical system, (*b*) FBD with *positive* displacement and velocity, (*c*) FBD with *negative* displacement and velocity.

or

$$m\ddot{x} + kx - \mu_k N = f(t), \quad x < 0, \quad \dot{x} < 0 \quad (4.41)$$

Thus, different conditions on velocity yield different results in the differential equation. If we introduce the *signum* function

$$\text{sgn}(\dot{x}) = \frac{\dot{x}}{|\dot{x}|} = \begin{cases} +1 & \text{if } \dot{x} > 0 \\ -1 & \text{if } \dot{x} < 0 \end{cases} \quad (4.42)$$

then we may represent the system by a single equation

$$m\ddot{x} + kx + \mu_k N \frac{\dot{x}}{|\dot{x}|} = f(t) \quad \text{for all } \dot{x} \quad (4.43)$$

which is valid for all velocity. After introducing $N = mg$, the equation becomes

$$m\ddot{x} + kx + \mu_k mg \frac{\dot{x}}{|\dot{x}|} = f(t) \quad \text{for all } \dot{x} \quad (4.44)$$

Since the signum function of $\dot{x}$ is not linearizable about $\dot{x} = 0$, the preceding equation is not linearizable. Thus, obtaining a closed-form analytical solution for this problem is a great challenge. Recall that the Laplace transform and other familiar methods can only be used for *linear* differential equations, or linearized versions. In practice, however, obtaining the response plots is fairly simple via computer simulation using a graphical software package such as SystemBuild or Easy 5.

It is interesting to see that the kinetic friction force may be represented in different ways, as shown in Fig. 4.21. These types of formulations may be advantageous

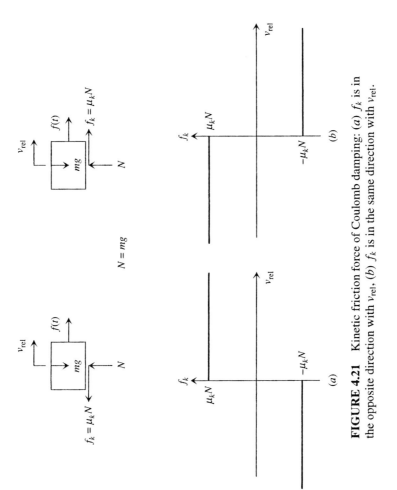

FIGURE 4.21 Kinetic friction force of Coulomb damping: (*a*) f_k is in the opposite direction with v_{rel}, (*b*) f_k is in the same direction with v_{rel}.

for programming because they are algebraic as opposed to physical. The kinetic friction force f_k is shown opposing the motion in Fig. 4.21a, whereas it is shown in the same direction with the motion in Fig. 4.21b. Either way is correct, but Fig. 4.21a may be preferred.

D'Alembert's Principle

D'Alembert's principle can be manipulated from Newton's second law as

$$\underbrace{\sum \mathbf{F}}_{\text{physical forces}} + \underbrace{(-m\mathbf{a}_c)}_{\text{fictitious force}} = \mathbf{0} \tag{4.45}$$

or simply

$$\sum \mathbf{F} - m\mathbf{a}_c = \mathbf{0} \tag{4.46}$$

This equation is known as **D'Alembert's principle**. The equation has $\sum \mathbf{F}$ as the sum of all the physical forces and $-m\mathbf{a}_c$ as the **inertial force**, also called the **fictitious force, reversed effective force, apparent force,** and other terms. The term $-m\mathbf{a}_c$, with the minus sign, causes some confusion. This equation may be attractive to some people because it looks like a classical static force balance. Many students, however, have difficulty in distinguishing between a physical force and a nonphysical force (fictitious force).

From a pedagogical point of view, beginning students should *not* use D'Alembert's principle. It is recommended to use Newton's second law instead. However, if one insists on using D'Alembert's principle, care must be taken so that the inertial force is considered correctly.

The following is a simple example that shows how the differential equation of motion may be derived using two different methods: Newton's second law (preferred) and D'Alembert's principle. To use the methods correctly, the free-body diagram must be constructed accurately.

EXAMPLE 4.4. Reconsider the simple system shown in Fig. 4.19a. Draw a free-body diagram, and then derive the differential equation of motion using D'Alembert's principle.

Solution. For D'Alembert's principle, a free-body diagram of the mass m is shown in Fig. 4.22. Note that the fictitious force $(-m\ddot{x})$ must be shown by a *broken line* to distinguish it from other physical forces. Applying D'Alembert's principle, in the x-direction,

$$\xrightarrow{+} \sum F_x + (-ma_{cx}) = 0 \tag{1}$$

$$\underbrace{f(t) - c\dot{x} - kx}_{\text{physical forces}} + \underbrace{(-m\ddot{x})}_{\text{fictitious force}} = 0 \tag{2}$$

or simply

$$f(t) - c\dot{x} - kx - m\ddot{x} = 0 \tag{3}$$

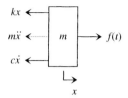

FIGURE 4.22 Free-body diagram for D'Alembert's principle.

After rearranging, the equation becomes

$$m\ddot{x} + c\dot{x} + kx = f(t)$$

which is identical to the previous result, as expected.

EXAMPLE 4.5. Consider the simple pendulum shown in Fig. 4.23. The string is of length L, which is massless (or negligible mass) and inextensible. The bob is of mass m, a point mass (zero or negligible dimension). The friction is negligible. For each of the two methods, draw a free-body diagram and then derive the differential equation of motion using

(a) Newton's second law
(b) D'Alembert's principle

Solution

Method 1: Newton's second law (preferred). For Newton's second law, the free-body diagram of the mass m is shown in Fig. 4.24, where T is tension. Applying Newton's second law in the x-direction, which is perpendicular to the length of the string,

$$\overset{+}{\nearrow} \sum F_x = ma_{cx} \overset{+}{\nearrow}$$

$$\underbrace{-mg\sin\theta}_{\text{physical force}} = \underbrace{m}_{\text{mass}} \underbrace{L\ddot{\theta}}_{\text{acceleration}}$$

Rearranging the equation in the standard input-output differential equation form,

$$mL\ddot{\theta} + mg\sin\theta = 0$$

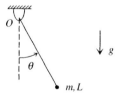

FIGURE 4.23 A simple pendulum system.

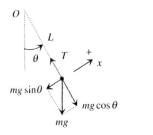

FIGURE 4.24 FBD for Newton's second law.

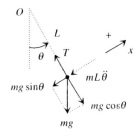

FIGURE 4.25 FBD for D'Alembert's principle.

or

$$\ddot{\theta} + \frac{g}{L}\sin\theta = 0$$

Method 2: D'Alembert's principle. For D'Alembert's principle, a free-body diagram of the mass m is shown in Fig. 4.25. Notice that the fictitious force $(-mL\ddot{\theta})$ is shown by a broken line. Applying D'Alembert's principle, in the x-direction,

$$\overset{+}{\nearrow}\sum F_x + (-ma_{cx}) = 0$$

$$\underbrace{-mg\sin\theta}_{\text{physical force}} + \underbrace{(-mL\ddot{\theta})}_{\text{fictitious force}} = 0$$

or simply

$$mL\ddot{\theta} + mg\sin\theta = 0$$

which is the same as obtained previously.

Gravity and Differential Equation of Spring-Mass Systems

In this section, we will show that the gravity term does not enter into the governing differential equation if the displacement is measured from the static equilibrium.

Consider the mass-spring system shown in Fig. 4.26, where x and y are measured from the static equilibrium position and the undeformed position, respectively. Also, x_{st} is the static deformation of the spring. Applying Newton's second law, in the y-direction,

$$+\downarrow \sum F_y = ma_{cy} \downarrow +$$
$$-ky + mg = m\ddot{y} \qquad (4.47)$$
$$m\ddot{y} + ky = mg$$

Now, for x,

$$+\downarrow \sum F_x = ma_{cx} \downarrow +$$
$$-k(x_{st} + x) + mg = m\ddot{x}$$

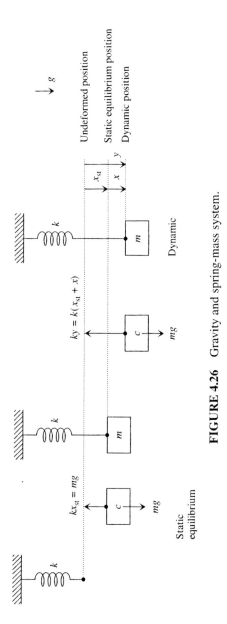

FIGURE 4.26 Gravity and spring-mass system.

Since the static equilibrium condition gives

$$kx_{st} = mg$$

the differential equation becomes

$$m\ddot{x} + kx = 0 \qquad (4.48)$$

As we have seen, the weight mg and the static spring force kx_{st} cancel each other out, resulting in a simpler equation.

In summary, the gravity term, g or mg, does not enter into the governing differential equation of the spring-mass system if the displacement is measured from the static equilibrium position.

Systems with Displacement Input

There are different types of inputs for mechanical systems. The two types of inputs are *generalized forces* and *generalized displacements*. By generalized forces we mean that the inputs may be forces or moments, and by generalized displacements we imply that the inputs may be velocity or acceleration, translational or rotational. Systems with generalized force inputs may be more familiar to the reader than generalized displacement inputs.

An example of a system with displacement input is an automobile suspension system. Figure 4.27 shows a very simple model, which includes the FBD for $y > x$. The system consists of a mass m connected by a damper b and a spring k, in parallel. (The elasticity and damping of the tires are not modeled.) The disturbance input $y(t)$ represents the displacement of the tire due to the surface of the road. This input is a function of the road surface and the speed of the vehicle.

Applying Newton's second law, in the y-direction,

$$+\uparrow \sum F_x = ma_{cx} \uparrow +$$

$$k(y - x) + b(\dot{y} - \dot{x}) = m\ddot{x} \qquad (4.49)$$

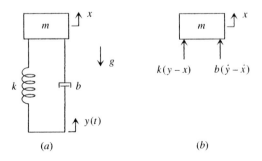

FIGURE 4.27 A simple system with displacement input: (*a*) system, (*b*) FBD ($y > x$).

Thus, the differential equation is

$$m\ddot{x} + b\dot{x} + kx = b\dot{y} + ky \tag{4.50}$$

or in terms of the natural frequency and damping ratio,

$$\ddot{x} + 2\zeta\omega_n\dot{x} + \omega_n^2 x = 2\zeta\omega_n\dot{y} + \omega_n^2 y \tag{4.51}$$

Note that at the point where the displacement input $y(t)$ is applied, a force, exerted onto the system by the road surface, does exist. This force is given as

$$f(t) = k(y - x) + b(\dot{y} - \dot{x}) \tag{4.52}$$

After solving for $x(t)$, the force $f(t)$ is determined.

Transfer Functions and State-Space Representation for SDOF Systems

Different representations of mathematical models have been treated extensively in previous chapters. The transfer function (TF) of a system relates the input to the output in the Laplace transform domain as

$$G(s) = \frac{\text{Output}(s)}{\text{Input}(s)} \tag{4.53}$$

where Output$(s) = L\{\text{output}(t)\}$
Input$(s) = L\{\text{input}(t)\}$

The following examples will show these representations for certain fundamental mechanical systems.

EXAMPLE 4.6. For the system shown in Fig. 4.19, the differential equation is

$$m\ddot{x} + c\dot{x} + kx = f(t)$$

(a) Determine the transfer functions $X(s)/F(s)$ and $V(s)/F(s)$.
(b) Determine the state-space representation, assuming the output is the displacement x and the state variables are

$$\begin{Bmatrix} x_1 \\ x_2 \end{Bmatrix} = \begin{Bmatrix} x \\ \dot{x} \end{Bmatrix}$$

Solution

(a) The transfer function can be obtained easily once the differential equation is determined, as shown previously. Taking the Laplace transform of both sides of the preceding equation with all the initial conditions set to zero,

$$\left(ms^2 + cs + k\right)X(s) = F(s)$$

Thus the transfer function relating the input $f(t)$ to the output $x(t)$ is

$$G_x(s) = \frac{X(s)}{F(s)} = \frac{1}{ms^2 + cs + k}$$

Since $v(t) = \dot{x}(t)$, the transfer function relating the input $f(t)$ to the output $v(t)$ is

$$G_v(s) = \frac{V(s)}{F(s)} = \frac{sX(s)}{F(s)} = \frac{s}{ms^2 + cs + k}$$

(b) A state-space representation is readily obtained from the differential equation, as shown previously. The state variables x_1 and x_2, the input u, and the output y are defined, respectively, as

$$\mathbf{x} = \underbrace{\begin{Bmatrix} x_1 \\ x_2 \end{Bmatrix}}_{\text{mathematical}} = \underbrace{\begin{Bmatrix} x \text{ (displacement)} \\ \dot{x} \text{ (velocity)} \end{Bmatrix}}_{\text{physical}}, \qquad \underbrace{u}_{\text{mathematical}} = \underbrace{f \text{ (force)}}_{\text{physical}},$$

$$\underbrace{y}_{\text{mathematical}} = \underbrace{x \text{ (measured by a sensor)}}_{\text{physical}}$$

where $\mathbf{x}$ is the state vector. (The word *vector* is used in the linear algebra sense.) Thus, the differential equation is rearranged as

$$\ddot{x} = -\frac{k}{m}x - \frac{c}{m}\dot{x} + \frac{1}{m}f(t)$$

or

$$\dot{x}_2 = -\frac{k}{m}x_1 - \frac{c}{m}x_2 + \frac{1}{m}u$$

Finally, the state-space representation (state equation and output equation) is given as

$$\begin{Bmatrix} \dot{x}_1 \\ \dot{x}_2 \end{Bmatrix} = \begin{bmatrix} 0 & 1 \\ -k/m & -c/m \end{bmatrix} \begin{Bmatrix} x_1 \\ x_2 \end{Bmatrix} + \begin{Bmatrix} 0 \\ 1/m \end{Bmatrix} u \qquad \text{(state equation)}$$

$$y = \begin{bmatrix} 1 & 0 \end{bmatrix} \begin{Bmatrix} x_1 \\ x_2 \end{Bmatrix} + 0 \cdot u \qquad \text{(output equation)}$$

Two-Degree-of-Freedom (TDOF) Systems

A mechanical system that requires two independent coordinates to specify its configuration is called a two-degree-of-freedom (TDOF) system. TDOF is a special class of multiple-degree-of-freedom (MDOF) systems.

EXAMPLE 4.7. Consider the TDOF system shown in Fig. 4.28.

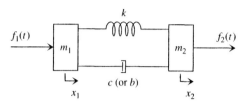

FIGURE 4.28 A TDOF system.

(a) Draw the necessary free-body diagrams and derive the differential equations
(b) Determine the state-space representation, assuming the output is the displacement x_1 and the state variables are

$$\begin{Bmatrix} x_1 \\ x_2 \\ x_3 \\ x_4 \end{Bmatrix} = \begin{Bmatrix} x_1 \\ x_2 \\ \dot{x}_1 \\ \dot{x}_2 \end{Bmatrix}$$

Solution

(a) Assume

$$x_1 > x_2 > 0 \quad \text{(spring is in } compression\text{)}$$

This condition implies

$$\dot{x}_1 > \dot{x}_2 > 0$$

The FBD of the system is shown in Fig. 4.29. Applying Newton's second law for the masses m_1 and m_2, respectively,

$$\overset{+}{\rightarrow} \sum F_x = ma_{cx} \overset{+}{\rightarrow}$$

$$f_1(t) - c(\dot{x}_1 - \dot{x}_2) - k(x_1 - x_2) = m_1 \ddot{x}_1$$
$$f_2(t) + c(\dot{x}_1 - \dot{x}_2) + k(x_1 - x_2) = m_2 \ddot{x}_2$$

Rearranging the equations into the standard input-output form,

$$m_1 \ddot{x}_1 + c\dot{x}_1 - c\dot{x}_2 + kx_1 - kx_2 = f_1(t)$$
$$m_2 \ddot{x}_2 - c\dot{x}_1 + c\dot{x}_2 - kx_1 + kx_2 = f_2(t)$$

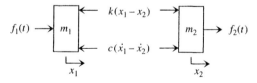

FIGURE 4.29 FBD.

Expressing the equations into the standard *second-order matrix form*,

$$\begin{bmatrix} m_1 & 0 \\ 0 & m_2 \end{bmatrix} \begin{Bmatrix} \ddot{x}_1 \\ \ddot{x}_2 \end{Bmatrix} + \begin{bmatrix} c & -c \\ -c & c \end{bmatrix} \begin{Bmatrix} \dot{x}_1 \\ \dot{x}_2 \end{Bmatrix} + \begin{bmatrix} k & -k \\ -k & k \end{bmatrix} \begin{Bmatrix} x_1 \\ x_2 \end{Bmatrix} = \begin{Bmatrix} f_1 \\ f_2 \end{Bmatrix}$$

(b) Starting with the differential equations of motion, the state vector **x**, the input u_1 and u_2, and the output y (measured by a sensor) are defined, respectively, as

$$\underbrace{\begin{Bmatrix} x_1 \\ x_2 \\ x_3 \\ x_4 \end{Bmatrix}}_{\text{mathematical}} = \underbrace{\begin{Bmatrix} x_1 \\ x_2 \\ \dot{x}_1 \\ \dot{x}_2 \end{Bmatrix}}_{\text{physical}}, \qquad u_1 = f_1(t), \qquad u_2 = f_2(t), \qquad y = x_1$$

Then

$$\ddot{x}_1 = -\frac{k}{m_1}x_1 + \frac{k}{m_1}x_2 - \frac{c}{m_1}\dot{x}_1 + \frac{c}{m_1}\dot{x}_2 + \frac{1}{m_1}f_1(t)$$

$$\ddot{x}_2 = \frac{k}{m_2}x_1 - \frac{k}{m_2}x_2 + \frac{c}{m_2}\dot{x}_1 - \frac{c}{m_2}\dot{x}_2 + \frac{1}{m_2}f_2(t)$$

or

$$\dot{x}_3 = -\frac{k}{m_1}x_1 + \frac{k}{m_1}x_2 - \frac{c}{m_1}x_3 + \frac{c}{m_1}x_4 + \frac{1}{m_1}u_1$$

$$\dot{x}_4 = \frac{k}{m_2}x_1 - \frac{k}{m_2}x_2 + \frac{c}{m_2}x_3 - \frac{c}{m_2}x_4 + \frac{1}{m_2}u_2$$

where $\dot{x}_3 = \ddot{x}_1$ and $\dot{x}_4 = \ddot{x}_2$. Thus, the state-space representation is

$$\begin{Bmatrix} \dot{x}_1 \\ \dot{x}_2 \\ \dot{x}_3 \\ \dot{x}_4 \end{Bmatrix} = \begin{bmatrix} 0 & 0 & 1 & 0 \\ 0 & 0 & 0 & 1 \\ -k/m_1 & k/m_1 & -c/m_1 & c/m_1 \\ k/m_2 & -k/m_2 & c/m_2 & -c/m_2 \end{bmatrix} \begin{Bmatrix} x_1 \\ x_2 \\ x_3 \\ x_4 \end{Bmatrix}$$

$$+ \begin{Bmatrix} 0 \\ 0 \\ 1/m_1 \\ 0 \end{Bmatrix} u_1 + \begin{Bmatrix} 0 \\ 0 \\ 0 \\ 1/m_2 \end{Bmatrix} u_2 \qquad \text{(state equation)}$$

$$y = \begin{bmatrix} 1 & 0 & 0 & 0 \end{bmatrix} \begin{Bmatrix} x_1 \\ x_2 \\ x_3 \\ x_4 \end{Bmatrix} + 0 \cdot u_1 + 0 \cdot u_2 \qquad \text{(output equation)}$$

where the state equation is a set of four first-order differential equations.

EXAMPLE 4.8. Consider another two-DOF system (Fig. 4.30), but assume

$$x_2 > x_1 > 0 \qquad \text{(all springs are in } tension\text{)} \tag{a}$$

Draw the necessary FBD, and then derive the differential equations.

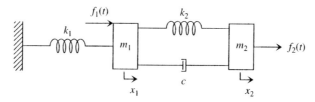

FIGURE 4.30   Another TDOF system.

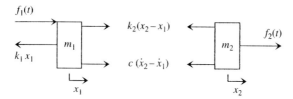

FIGURE 4.31   FBD.

Solution. This is another way of assuming the condition on the springs (see Example 4.7). The FBD of the system is shown in Fig. 4.31. Applying Newton's second law for the masses m_1 and m_2, respectively,

$$\xrightarrow{+} \sum F_x = ma_{cx} \xrightarrow{+}$$

$$f_1(t) - k_1 x_1 + c(\dot{x}_2 - \dot{x}_1) + k_2(x_2 - x_1) = m_1 \ddot{x}_1$$

$$f_2(t) - c(\dot{x}_2 - \dot{x}_1) - k_2(x_2 - x_1) = m_2 \ddot{x}_2$$

Rearranging the equations into the standard input-output differential equation form,

$$m_1 \ddot{x}_1 + c\dot{x}_1 - c\dot{x}_2 + (k_1 - k_2)x_1 - k_2 x_2 = f_1(t)$$

$$m_2 \ddot{x}_2 - c\dot{x}_1 + c\dot{x}_2 - k_2 x_1 + k_2 x_2 = f_2(t)$$

Expressing the equations into the standard second-order matrix form,

$$\begin{bmatrix} m_1 & 0 \\ 0 & m_2 \end{bmatrix} \begin{Bmatrix} \ddot{x}_1 \\ \ddot{x}_2 \end{Bmatrix} + \begin{bmatrix} c & -c \\ -c & c \end{bmatrix} \begin{Bmatrix} \dot{x}_1 \\ \dot{x}_2 \end{Bmatrix} + \begin{bmatrix} k_1 + k_2 & -k_2 \\ -k_2 & k_2 \end{bmatrix} \begin{Bmatrix} x_1 \\ x_2 \end{Bmatrix} = \begin{Bmatrix} f_1 \\ f_2 \end{Bmatrix}$$

Systems with Massless Junctions

A system of massless junction is a system of springs and dampers without any masses. Deriving the differential equations for such a system is similar to deriving them for a system with masses. Simply let the masses be zero at the junctions.

As an example, let us consider the system of massless junctions shown in Fig. 4.32. The FBDs are included where we assume $y > x_1 > x_2 > 0$ (all the springs are in compression). Applying Newton's second law for the two massless junctions,

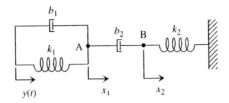

FIGURE 4.32 A TDOF system with displacement input; FBD ($y > x_1 > x_2 > 0$).

respectively,

$$\xrightarrow{+} \sum F_x = ma_{cx} = 0$$

A: $\quad k_1(y - x_1) + b_1(\dot{y} - \dot{x}_1) - b_2(\dot{x}_1 - \dot{x}_2) = 0$

B: $\quad b_2(\dot{x}_1 - \dot{x}_2) - k_2 x_2 = 0$

After rearranging, the equations become

$$(b_1 + b_2)\dot{x}_1 - b_2\dot{x}_2 + k_1 x_1 = b_1\dot{y} + k_1 y$$

$$-b_2\dot{x}_1 + b_2\dot{x}_2 + k_2 x_2 = 0$$

or in the standard second-order matrix form,

$$\begin{bmatrix} b_1 + b_2 & -b_2 \\ -b_2 & b_2 \end{bmatrix} \begin{Bmatrix} \dot{x}_1 \\ \dot{x}_2 \end{Bmatrix} + \begin{bmatrix} k_1 & 0 \\ 0 & k_2 \end{bmatrix} \begin{Bmatrix} x_1 \\ x_2 \end{Bmatrix} = \begin{Bmatrix} b_1\dot{y} + k_1 y \\ 0 \end{Bmatrix}$$

Skeleton Approach

The skeleton approach is a simple method developed by the authors for obtaining the governing differential equations for MDOF systems. The method is simple and quick, which is particularly useful with mass-spring-damper systems of high order. Although there are many techniques for obtaining the governing differential equations, these methods require a significant amount of time for higher-order systems. With the skeleton approach, in contrast, the effort required is nearly independent of the size of the system.

Background

The differential equations of mass-spring-damper systems are given in the general and compact second-order matrix form as

$$\mathbf{m\ddot{x} + c\dot{x} + kx = f} \qquad (4.54)$$

CHAPTER 4: Mechanical Systems   207

For TDOF systems, as an example, we have

$$\mathbf{x} = \begin{Bmatrix} x_1 \\ x_2 \end{Bmatrix}, \quad \dot{\mathbf{x}} = \begin{Bmatrix} \dot{x}_1 \\ \dot{x}_2 \end{Bmatrix}, \quad \ddot{\mathbf{x}} = \begin{Bmatrix} \ddot{x}_1 \\ \ddot{x}_2 \end{Bmatrix} \quad (4.55)$$

$$\mathbf{m} = \begin{bmatrix} m_{11} & m_{12} \\ m_{21} & m_{22} \end{bmatrix}, \quad \mathbf{c} = \begin{bmatrix} c_{11} & c_{12} \\ c_{21} & c_{22} \end{bmatrix}, \quad \mathbf{k} = \begin{bmatrix} k_{11} & k_{12} \\ k_{21} & k_{22} \end{bmatrix}, \quad \mathbf{f} = \begin{Bmatrix} f_1 \\ f_2 \end{Bmatrix} \quad (4.56)$$

A check on the correctness of the differential equations of dynamic systems is always useful. The following are the necessary conditions for correct results. The following results are for stable and *pure* systems:

- All the elements on the main diagonals are nonnegative (either positive or zero).
- All the mass, damping, and stiffness matrices are symmetric (with respect to the main diagonals).
- The off-diagonal elements of both the damping and stiffness matrices are nonpositive (either negative or zero).
- The off-diagonal elements of the mass matrix are nonnegative.

A pure system is defined as strictly of one type, e.g., mechanical systems with purely translational or rotational motions. Liquid-level systems are also dynamic and pure systems.

Method

To illustrate the method, let us consider the TDOF system shown in Fig. 4.28.

Step 1. Make a *skeleton* or a frame in the second-order matrix form. For TDOF systems, we have

$$\underbrace{\begin{bmatrix} \cdots & \cdots \\ \cdots & \cdots \end{bmatrix}}_{2 \times 2} \underbrace{\begin{Bmatrix} \ddot{x}_1 \\ \ddot{x}_2 \end{Bmatrix}}_{2 \times 1} + \underbrace{\begin{bmatrix} \cdots & \cdots \\ \cdots & \cdots \end{bmatrix}}_{2 \times 2} \underbrace{\begin{Bmatrix} \dot{x}_1 \\ \dot{x}_2 \end{Bmatrix}}_{2 \times 1} + \underbrace{\begin{bmatrix} \cdots & \cdots \\ \cdots & \cdots \end{bmatrix}}_{2 \times 2} \underbrace{\begin{Bmatrix} x_1 \\ x_2 \end{Bmatrix}}_{2 \times 1} = \underbrace{\begin{Bmatrix} \cdots \\ \cdots \end{Bmatrix}}_{2 \times 1} \quad (4.57)$$

It is useful to note the sizes of the matrices, so one can check for matrix compatibility.

Step 2. Add the matrix elements into the frame to obtain the final equations. These are the elements of the mass, damping, and stiffness matrices, and vectoral forces.

Mass matrix. We need to determine the elements of the mass matrix:

$$\begin{matrix} m_{11} & m_{12} \\ m_{21} & m_{22} \end{matrix}$$

Diagonal elements. Place a finger at x_1 or $\ddot{x}_1$: the mass at x_1 is m_1, thus

$$m_{11} = m_1$$

Place a finger at x_2 or $\ddot{x}_2$: the mass at x_2 is m_2, thus

$$m_{22} = m_2$$

Off-diagonal elements. Place a left finger at x_1 or $\ddot{x}_1$ and a right finger at x_2 or $\ddot{x}_2$: there is no mass connecting x_1 and x_2, thus the coupling terms are zero,

$$m_{12} = m_{21} = 0$$

We now place these elements into the mass matrix of the equation,

$$\begin{bmatrix} m_1 & 0 \\ 0 & m_2 \end{bmatrix} \begin{Bmatrix} \ddot{x}_1 \\ \ddot{x}_2 \end{Bmatrix} + \begin{bmatrix} \cdots & \cdots \\ \cdots & \cdots \end{bmatrix} \begin{Bmatrix} \dot{x}_1 \\ \dot{x}_2 \end{Bmatrix} + \begin{bmatrix} \cdots & \cdots \\ \cdots & \cdots \end{bmatrix} \begin{Bmatrix} x_1 \\ x_2 \end{Bmatrix} = \begin{Bmatrix} \cdots \\ \cdots \end{Bmatrix} \quad (4.58)$$

Damping matrix. Similarly, we need to determine the elements of the damping matrix:

$$\begin{matrix} c_{11} & c_{12} \\ c_{21} & c_{22} \end{matrix}$$

Diagonal elements. Place a finger at x_1 or $\dot{x}_1$: the damper at x_1 is c, thus

$$c_{11} = c$$

Place a finger at x_2 or $\dot{x}_2$: the damper at x_2 is b, thus

$$c_{22} = c$$

Off-diagonal elements. Place a left finger at x_1 or $\dot{x}_1$ and a right finger at x_2 or $\dot{x}_2$: the damper connecting x_1 and x_2 is c. For the consistent set of displacements as shown (Fig. 4.28), the off-diagonal elements of the damping and stiffness matrices are negative (or zero). Thus the coupling terms are

$$c_{12} = c_{21} = -c$$

Recall that the damping matrix is symmetric with respect to the main diagonal.

We now place these elements into the damping matrix of the equation,

$$\begin{bmatrix} m_1 & 0 \\ 0 & m_2 \end{bmatrix} \begin{Bmatrix} \ddot{x}_1 \\ \ddot{x}_2 \end{Bmatrix} + \begin{bmatrix} c & -c \\ -c & c \end{bmatrix} \begin{Bmatrix} \dot{x}_1 \\ \dot{x}_2 \end{Bmatrix} + \begin{bmatrix} \cdots & \cdots \\ \cdots & \cdots \end{bmatrix} \begin{Bmatrix} x_1 \\ x_2 \end{Bmatrix} = \begin{Bmatrix} \cdots \\ \cdots \end{Bmatrix} \quad (4.59)$$

Stiffness matrix. In a similar manner, for the elements of the stiffness matrix, we obtain

$$k_{11} = k$$
$$k_{22} = k$$
$$k_{12} = k_{21} = -k$$

Inserting these elements into the damping matrix of the equation,

$$\begin{bmatrix} m_1 & 0 \\ 0 & m_2 \end{bmatrix} \begin{Bmatrix} \ddot{x}_1 \\ \ddot{x}_2 \end{Bmatrix} + \begin{bmatrix} c & -c \\ -c & c \end{bmatrix} \begin{Bmatrix} \dot{x}_1 \\ \dot{x}_2 \end{Bmatrix} + \begin{bmatrix} k & -k \\ -k & k \end{bmatrix} \begin{Bmatrix} x_1 \\ x_2 \end{Bmatrix} = \begin{Bmatrix} \cdots \\ \cdots \end{Bmatrix} \quad (4.60)$$

Force matrix. Place a finger at x_1: the externally applied force at x_1 is $f_1(t)$, thus

$$f_1 = f_1(t)$$

Place a finger at x_2: the externally applied force at x_2 is $f_2(t)$, thus

$$f_2 = f_2(t)$$

Finally, we complete the matrix equation,

$$\begin{bmatrix} m_1 & 0 \\ 0 & m_2 \end{bmatrix} \begin{Bmatrix} \ddot{x}_1 \\ \ddot{x}_2 \end{Bmatrix} + \begin{bmatrix} c & -c \\ -c & c \end{bmatrix} \begin{Bmatrix} \dot{x}_1 \\ \dot{x}_2 \end{Bmatrix} + \begin{bmatrix} k & -k \\ -k & k \end{bmatrix} \begin{Bmatrix} x_1 \\ x_2 \end{Bmatrix} = \begin{Bmatrix} f_1(t) \\ f_2(t) \end{Bmatrix} \quad (4.61)$$

EXAMPLE 4.9. Using the skeleton approach, determine the differential equations of the three-DOF system shown in Fig. 4.33. This is an example of a MDOF system of higher order.

Solution.

Step 1. Make a *skeleton* or a frame in the second-order matrix form.

$$\begin{bmatrix} \cdots & \cdots & \cdots \\ \cdots & \cdots & \cdots \\ \cdots & \cdots & \cdots \end{bmatrix} \begin{Bmatrix} \ddot{x}_1 \\ \ddot{x}_2 \\ \ddot{x}_3 \end{Bmatrix} + \begin{bmatrix} \cdots & \cdots & \cdots \\ \cdots & \cdots & \cdots \\ \cdots & \cdots & \cdots \end{bmatrix} \begin{Bmatrix} \dot{x}_1 \\ \dot{x}_2 \\ \dot{x}_3 \end{Bmatrix} + \begin{bmatrix} \cdots & \cdots & \cdots \\ \cdots & \cdots & \cdots \\ \cdots & \cdots & \cdots \end{bmatrix} \begin{Bmatrix} x_1 \\ x_2 \\ x_3 \end{Bmatrix} = \begin{Bmatrix} \cdots \\ \cdots \\ \cdots \end{Bmatrix}$$

Step 2. Add the matrix elements into the frame to obtain the final equations. Following the procedure, we obtain

$$\begin{bmatrix} m_1 & 0 & 0 \\ 0 & m_2 & 0 \\ 0 & 0 & m_3 \end{bmatrix} \begin{Bmatrix} \ddot{x}_1 \\ \ddot{x}_2 \\ \ddot{x}_3 \end{Bmatrix} + \begin{bmatrix} 0 & 0 & 0 \\ 0 & c & -c \\ 0 & -c & c \end{bmatrix} \begin{Bmatrix} \dot{x}_1 \\ \dot{x}_2 \\ \dot{x}_3 \end{Bmatrix}$$

$$+ \begin{bmatrix} k_1 + k_2 & -k_2 & 0 \\ -k_2 & k_2 + k_3 & -k_3 \\ 0 & -k_3 & k_3 \end{bmatrix} \begin{Bmatrix} x_1 \\ x_2 \\ x_3 \end{Bmatrix} = \begin{Bmatrix} k_1 y \\ 0 \\ f(t) \end{Bmatrix}$$

A check is in order here. This is a *stable* and *pure* dynamic system (purely mechanical and translational). All the elements on the main diagonals of the mass, damping, and stiffness matrices are nonnegative. In addition, all the mass, damping, and stiffness matrices are symmetric. The elements on the off-diagonals of both the damping and stiffness matrices are nonpositive. Notice that the off-diagonal elements of the mass matrix are nonnegative (zero in this case). Also, $k_1 y$ of the force column matrix has dimension of force.

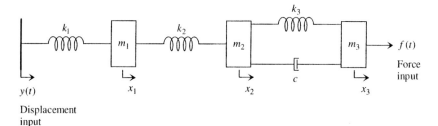

FIGURE 4.33 A three-DOF system.

PROBLEM SET 4.5

4.21. A simple mechanical system of mass, spring, and damper is shown in Fig. P4.21. Draw the necessary free-body diagram and derive the differential equation.

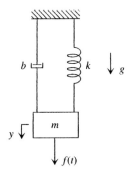

FIGURE P4.21

4.22. Using the differential equation obtained from Problem 4.21, determine the transfer function.

4.23. Using the differential equation obtained from Problem 4.21, obtain the state-space representation for two different cases of output: (a) displacement, (b) velocity.

4.24. Consider the hybrid (mixed) system of three massless springs shown in Fig. P4.24. Draw the necessary free-body diagram and derive the differential equation.

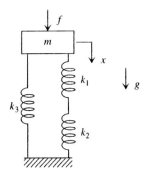

FIGURE P4.24

4.25. Using the differential equation obtained from Problem 4.24, determine the transfer function.

4.26. Using the differential equation obtained from Problem 4.24, obtain the state-space representation for two different cases of output: (a) displacement, (b) velocity.

4.27. Consider the translational system shown in Fig. P4.27. Assume that the system is constrained to move in a vertical plane and in a horizontal direction. (a) Draw the necessary free-body diagrams and derive the differential equations, then (b) put the equations in the second-order matrix form. (c) Obtain a state-space representation where the output is the displacement x_1.

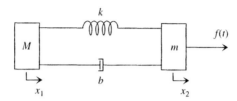

FIGURE P4.27

4.28. Reconsider the translational system shown in Problem 4.27. Without any derivation, provide the differential equations directly in the second-order matrix form. Use the skeleton approach.

4.29. Using the differential equation obtained from Problem 4.27, determine all the transfer functions.

4.30. Using the differential equations obtained from Problem 4.27, determine a state-space representation when the outputs are the displacements x_1 and x_2.

4.31. Consider the translational system shown in Fig. P4.31. Assume that the system is constrained to move in a vertical plane and in a horizontal direction. (a) Draw the necessary free-body diagrams and derive the differential equations, then (b) put the equations in the second-order matrix form.

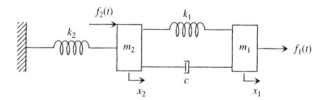

FIGURE P4.31

4.32. Reconsider the translational system shown in Problem 4.31. Without any derivation, provide the differential equations directly in the second-order matrix form. Use the skeleton approach.

4.33. Using the differential equations obtained from Problem 4.31, determine all the transfer functions.

4.34. Using the differential equations obtained from Problem 4.31, determine the state-space representation when the outputs are (a) the displacements x_1 and x_2, (b) the velocities.

4.35. Consider the translational system shown in Fig. P4.35. Assume that the system is constrained to move in a vertical plane and in a horizontal direction. (a) Draw the necessary free-body diagrams and derive the differential equations, then (b) put the equations in the second-order matrix form.

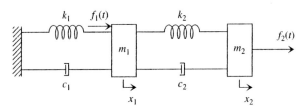

FIGURE P4.35

4.36. Reconsider the translational system shown in Problem 4.35. Without any derivation, provide the differential equations directly in the second-order matrix form. Use the skeleton approach.

4.37. Using the differential equations obtained from Problem 4.35, determine all the transfer functions.

4.38. Using the differential equations obtained from Problem 4.35, determine the state-space representation when the outputs are the displacements x_1 and x_2.

4.39. Consider the translational system shown in Fig. P4.39. Assume that the system is constrained to move in a vertical plane and in a horizontal direction. (a) Draw the necessary free-body diagrams and derive the differential equations, then (b) put the equations in the second-order matrix form.

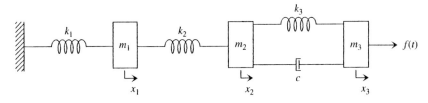

FIGURE P4.39

4.40. Reconsider the translational system shown in Problem 4.39. Without any derivation, provide the differential equations directly in the second-order matrix form. Use the skeleton approach.

CHAPTER 4: Mechanical Systems 213

4.6 ROTATIONAL SYSTEMS

In this section we will cover the general moment equation and mathematical modeling. The moment equation is applicable to systems of particles and rigid bodies in three dimensions (3D) and two dimensions (2D). However, with mathematical modeling, we are mainly concerned with 2D or planar motion.

When a rigid body moves arbitrarily in 3D space, the axis of rotation keeps changing. This creates a great challenge to the analyst. Because the modeling of rigid bodies in 3D is rather complex, it is beyond the scope of this text.

In many engineering applications, the motion of rigid bodies is primarily 2D. We, therefore, will focus on the modeling of rigid bodies only in planar motion. However, we will include certain fundamentals of rigid bodies in 3D to help the interested reader in further studies.

The Moment Equation

The general moment equation is introduced from the most general to the most specific. A system of particles will be discussed, then a rigid body in three dimensions (3D), and finally a rigid body in two dimensions (2D). When a system of particles is connected rigidly, i.e., the distances among the particles are fixed (unchanged), the system of particles becomes a rigid body. When a rigid body is constrained to rotate about a fixed axis, the 3D problem becomes 2D. This is the logical development.

A system of particles

For a *system of particles* in an arbitrary motion, the *general moment equation*, after using Newton's second law, *can be shown* as

$$\sum \mathbf{M}_p = \dot{\mathbf{H}}_p + m\mathbf{r}_{c/p} \times \mathbf{a}_p \qquad (4.62)$$

where P = an arbitrary, accelerating point
 C = *center of mass*, or *mass center*, of the system of particles
 $\sum \mathbf{M}_p$ = sum of all externally applied moments about point P
 $\mathbf{H}_p$ = angular momentum vector about point P
 $\dot{\mathbf{H}}_p$ = time-rate-of-change of H_p
 m = total mass of the system of particles
 $\mathbf{r}_{c/p}$ = position vector of the mass center C with respect to point P
 $\mathbf{a}_p$ = acceleration vector of point P

Special cases

1. If point P is the same as the *fixed* point O, $P \equiv O$, then $\mathbf{a}_p \equiv \mathbf{a}_o = \mathbf{0}$. Point O is fixed in space so it has zero acceleration, thus the general moment equation reduces to

$$\sum \mathbf{M}_o = \dot{\mathbf{H}}_o \qquad (4.63)$$

2. If point P is the same as the mass center C, $P \equiv C$, then $\mathbf{r}_{c/p} \equiv \mathbf{r}_{c/c} = \mathbf{0}$. Thus the general moment equation reduces to

$$\sum \mathbf{M}_c = \dot{\mathbf{H}}_c \qquad (4.64)$$

3. If $\mathbf{r}_{c/p} \| \mathbf{a}_p$ or $\mathbf{r}_{c/s} \| \mathbf{a}_s$, that is, point P is the same as a special point, point S, where the position vector of the mass center C with respect to point P, $\mathbf{r}_{c/p}$, is parallel with the acceleration vector of point P, then $\mathbf{r}_{c/p} \times \mathbf{a}_p \equiv \mathbf{r}_{c/s} \times \mathbf{a}_s = \mathbf{0}$. Similarly, the general moment equation reduces to

$$\sum \mathbf{M}_s = \dot{\mathbf{H}}_s \qquad (4.65)$$

Notes

- We define $\sum \mathbf{M}_p = \dot{\mathbf{H}}_p + m \mathbf{r}_{c/p} \times \mathbf{a}_p$ as the *general moment equation*, not as Newton's second law. Some people imprecisely call the special cases of this equation, i.e., $\sum \mathbf{M}_c = \dot{\mathbf{H}}_c$ or $\sum \mathbf{M}_o = \dot{\mathbf{H}}_o$, as Newton's second law. Strictly speaking, Newton's second law, $\sum \mathbf{F} = m \mathbf{a}_c$, is simply *used* in certain steps of the development of the final result, which is the general moment equation.
- An alternate form of the moment equation can be shown as

$$\sum \mathbf{M}_p = \dot{\mathbf{H}}_c + m \mathbf{r}_{c/p} \times \mathbf{a}_c \qquad (4.66)$$

where $\dot{\mathbf{H}}_c$ = time rate of change of the angular momentum vector about the mass center C
$\mathbf{a}_c$ = acceleration vector of the mass center C

However, in many applications the vector $\mathbf{a}_c$ is more difficult to obtain than $\mathbf{a}_p$.

Rigid body in 3D

For a *rigid body* in an arbitrary motion in three-dimensional (3D) space, the *general moment equation* is also given by Eq. (4.62). This expression is the same as that for a system of particles, because a rigid body is a special case of a system of particles where the particles are rigidly fixed. The distance between any two particles of the body is constant or fixed. In general, it is difficult to obtain $\dot{\mathbf{H}}_p$ for complex systems. For a rigid body, the angular momentum, mass moment of inertia, and angular velocity are related in matrix form as

$$\mathbf{H}_p = \mathbf{I}_p \boldsymbol{\omega} \qquad (4.67)$$

or

$$\{H\}_p = [I]_p \{\omega\} \qquad (4.68)$$

or

$$\begin{Bmatrix} H_x \\ H_y \\ H_z \end{Bmatrix}_p = \begin{bmatrix} I_{xx} & I_{xy} & I_{xz} \\ I_{yx} & I_{yy} & I_{yz} \\ I_{zx} & I_{zy} & I_{zz} \end{bmatrix}_p \begin{Bmatrix} \omega_x \\ \omega_y \\ \omega_z \end{Bmatrix} \qquad (4.69)$$

where $\mathbf{H_p}$ = angular momentum column matrix (3×1) of the rigid body about point P
$\mathbf{I}_p$ = mass moment of inertia square matrix (3 × 3) of the rigid body about point P
$\boldsymbol{\omega}$ = angular velocity column matrix (3 × 1) of the rigid body
$(H_x)_p, (H_y)_p, (H_z)_p$ = three cartesian components of the angular momentum vector $\mathbf{H}_p$
$\omega_x, \omega_y, \omega_z$ = three cartesian components of the angular velocity vector $\boldsymbol{\omega}$

In general, when a nonsymmetrical rigid body is in motion, the time derivatives of its nine component I's ($I_{xx}, I_{xy}, I_{xz}; I_{yx}, I_{yy}, I_{yz}; I_{zx}, I_{zy}, I_{zz}$) are nonzero if xyz is fixed to the rigid body. This body-coordinate frame xyz moves in the inertial reference frame XYZ. Thus, it is difficult to obtain $\dot{\mathbf{H}}_p$ for general motion.

This fact makes the study of the dynamics of a rigid body in general 3D a challenging task. However, if the rigid body is in 2D, or planar motion, then the complexity is reduced significantly.

Rigid Body in 2D

For a rigid body in planar motion, or 2D, the moment equation is given by

$$\sum \mathbf{M}_p = I_p \boldsymbol{\alpha} + m\mathbf{r}_{c/p} \times \mathbf{a}_p \tag{4.70}$$

where I_p = moment of inertia about the arbitrary (accelerating) point P fixed in the body
$\boldsymbol{\alpha}$ = angular acceleration vector of the rigid body

The *parallel-axis theorem* gives

$$I_p = I_c + md^2 \tag{4.71}$$

where I_c = centroidal mass moment of inertia
d = distance from the mass center C to the point P

For a rigid body that contains n separated masses, the general expression for the combined moment of inertia about an arbitrary point P is

$$I_p = \sum_{j=1}^{n} I_{cj} + m_j d_j^2 = (I_{c1} + m_1 d_1^2) + (I_{c2} + m_2 d_2^2) + \cdots + (I_{cn} + m_n d_n^2) \tag{4.72}$$

Note that *only* for a rigid body in planar motion, the mass moment of inertia I_p is a scalar quantity and the time rate of change of angular momentum reduces to

$$\dot{\mathbf{H}}_p = I_p \boldsymbol{\alpha} \tag{4.73}$$

For a rigid body in general 3D, the mass moment of inertia I_p is a second-order tensor, which is much more complex.

Summary

1. If point O (fixed in space *and* fixed in the body) exists, then use

$$\sum \mathbf{M}_o = I_o \boldsymbol{\alpha}$$

2. Otherwise, use

$$\sum \mathbf{M}_c = I_c \alpha$$

3. Or use the general moment equation

$$\sum \mathbf{M}_p = I_p \alpha + m \mathbf{r}_{c/p} \times \mathbf{a}_p$$

4. If there is a special point S such that $\mathbf{r}_{c/s} \| \mathbf{a}_s$, then use

$$\sum \mathbf{M}_s = I_s \alpha$$

As an example of a special point, let us consider the rolling, without slipping, of a circular disk of mass m and radius r (Fig. 4.34). The motion is on a horizontal plane along a straight line.

Point O (fixed in space *and* fixed in the body) does *not* exist, thus we *cannot* use $\sum \mathbf{M}_o = I_o \alpha$. Also, it is inconvenient to take the sum of moments about the center of mass C, using $\sum \mathbf{M}_C = I_C \alpha$, because we have to deal with f_s and N. We would have to consider them first, then would try to eliminate them later. For this problem, the best formula to use is the general moment equation,

$$\sum \mathbf{M}_p = I_p \alpha + m \mathbf{r}_{c/p} \times \mathbf{a}_p \qquad \text{where } I_p = I_c + md^2 \qquad (4.74)$$

With P replaced by IC,

$$\sum \mathbf{M}_{IC} = I_{IC} \alpha + m \mathbf{r}_{c/IC} \times \mathbf{a}_{IC} \qquad \text{where } I_{IC} = I_c + md^2 \qquad (4.75)$$

The contact point is called **instantaneous center of rotation (IC)**. The point IC has zero velocity, $v_{IC} = 0$, but its acceleration is nonzero. The magnitude of the acceleration is $a_{IC} = r\dot{\theta}^2$ and the direction is from IC to the **center of geometry** of the disk. With the coordinate system shown, the acceleration vector is

$$\mathbf{a}_{IC} = r\dot{\theta}^2 \mathbf{j} \qquad (4.76)$$

Note that, for a uniform circular disk, the center of geometry coincides with the center of mass. Therefore,

$$\mathbf{r}_{c/IC} \times \mathbf{a}_{IC} = \mathbf{0} \qquad (4.77)$$

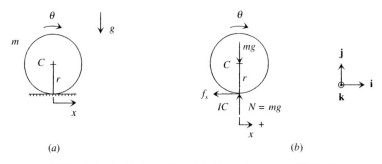

FIGURE 4.34 Rolling disk: (*a*) physical system, (*b*) FBD.

thus

$$\sum \mathbf{M}_{IC} = I_{IC}\alpha \quad \text{where} \quad I_{IC} = I_c + md^2 \quad (4.78)$$

Angular Momentum and Moments of Inertia

The angular momentum vector of a system of n particles rotating about a reference point P is defined as

$$\mathbf{H}_p = \sum_{i=1}^{n} \mathbf{r}_{i/p} \times m_i \mathbf{v}_{i/p} \quad (4.79)$$

where $\mathbf{r}_{i/p}$ = position vector of the ith particle with respect to point P
  m_i = mass of the ith particle
  $\mathbf{v}_{i/p}$ = velocity vector of the ith particle with respect to point P

From the fundamentals of kinematics of rotating frames, if $\mathbf{A}$ is an arbitrary vector in a frame rotating with angular velocity $\boldsymbol{\omega}$, then

$$\frac{d\mathbf{A}}{dt} = \left(\frac{\partial \mathbf{A}}{\partial t}\right)_{\text{rel}} + \boldsymbol{\omega} \times \mathbf{A} \quad (4.80)$$

or

$$\dot{\mathbf{A}} = \left(\dot{\mathbf{A}}\right)_{\text{rel}} + \boldsymbol{\omega} \times \mathbf{A} \quad (4.81)$$

Thus,

$$v_{i/p} = \dot{\mathbf{r}}_{i/p} = (\dot{\mathbf{r}}_{i/p})_{\text{rel}} + \boldsymbol{\omega} \times \mathbf{r}_{i/p} \quad (4.82)$$

For a rigid body, $(\dot{\mathbf{r}}_{i/p})_{\text{rel}} = 0$ and the ith particle becomes the ith point, so the subscripts may be dropped for brevity. The velocity of the ith point with respect to point P becomes

$$\mathbf{v} = \dot{\mathbf{r}} = \left(\dot{\mathbf{r}}\right)_{\text{rel}} + \boldsymbol{\omega} \times \mathbf{r} = \boldsymbol{\omega} \times \mathbf{r} \quad (4.83)$$

since $\left(\dot{\mathbf{r}}\right)_{\text{rel}} = 0$. In addition, the summation sign in the expression of the angular momentum becomes the integral, thus

$$\mathbf{H}_p = \int_V \mathbf{r} \times \mathbf{v}\rho \, dV = \int_V \mathbf{r} \times \boldsymbol{\omega} \times \mathbf{r}\rho \, dV \quad (4.84)$$

The velocity vector of the ith point with respect to point P may be given as

$$\mathbf{r} = x\mathbf{i} + y\mathbf{j} + z\mathbf{k} \quad (4.85)$$

and the angular velocity vector is

$$\boldsymbol{\omega} = \omega_x \mathbf{i} + \omega_y \mathbf{j} + \omega_z \mathbf{k} \quad (4.86)$$

Performing the cross product,

$$\boldsymbol{\omega} \times \mathbf{r} = (\omega_x \mathbf{i} + \omega_y \mathbf{j} + \omega_z \mathbf{k}) \times (x\mathbf{i} + y\mathbf{j} + z\mathbf{k}) \quad (4.87)$$

or

$$\boldsymbol{\omega} \times \mathbf{r} = \begin{vmatrix} \mathbf{i} & \mathbf{j} & \mathbf{k} \\ \omega_x & \omega_y & \omega_z \\ x & y & z \end{vmatrix} = (z\omega_y - y\omega_z)\mathbf{i} + (x\omega_z - z\omega_x)\mathbf{j} + (y\omega_x - x\omega_y)\mathbf{k} \tag{4.88}$$

and

$$\mathbf{r} \times (\boldsymbol{\omega} \times \mathbf{r}) = \begin{vmatrix} \mathbf{i} & \mathbf{j} & \mathbf{k} \\ x & y & z \\ z\omega_y - y\omega_z & x\omega_z - z\omega_x & y\omega_x - x\omega_y \end{vmatrix} \tag{4.89}$$

$$= [y(y\omega_x - x\omega_y) - z(x\omega_z - z\omega_x)]\mathbf{i}$$
$$+ [z(z\omega_y - y\omega_z) - x(y\omega_x - x\omega_y)]\mathbf{j}$$
$$+ [x(x\omega_z - z\omega_x) - y(z\omega_y - y\omega_z)]\mathbf{k}$$
$$= [(y^2 + z^2)\omega_x - xy\omega_y - xz\omega_z]\mathbf{i}$$
$$+ [-yx\omega_x + (z^2 + x^2)\omega_y - yz\omega_z]\mathbf{j} \tag{4.90}$$
$$+ [-zx\omega_x - zy\omega_y + (x^2 + y^2)\omega_z]\mathbf{k}$$

Thus,

$$\mathbf{H}_p = \int_V \mathbf{r} \times (\boldsymbol{\omega} \times \mathbf{r}) \rho \, dV \tag{4.91}$$

$$= \left[\int_V (y^2 + z^2) \rho \, dV \omega_x - \int_V xy \rho \, dV \omega_y - \int_V xz \rho \, dV \omega_z \right] \mathbf{i}$$
$$+ \left[-\int_V yx \rho \, dV \omega_x + \int_V (z^2 + x^2) \rho \, dV \omega_y - \int_V yz \rho \, dV \omega_z \right] \mathbf{j}$$
$$+ \left[-\int_V zx \rho \, dV \omega_x - \int_V zy \rho \, dV \omega_y + \int_V (x^2 + y^2) \rho \, dV \omega_z \right] \mathbf{k}$$
$$\tag{4.92}$$

The moments of inertia about point P may be defined as

$$(I_{xx})_P = \int_V (y^2 + z^2) \rho \, dV \tag{4.93a}$$

$$(I_{yy})_P = \int_V (z^2 + x^2) \rho \, dV \tag{4.93b}$$

$$(I_{zz})_P = \int_V (x^2 + y^2) \rho \, dV \tag{4.93c}$$

$$(I_{xy})_P = (I_{yx})_P = -\int_V xy \rho \, dV \tag{4.93d}$$

$$(I_{yz})_p = (I_{zy})_p = -\int_V yz\rho\, dV \qquad (4.93e)$$

$$(I_{zx})_p = (I_{xz})_p = -\int_V zx\rho\, dV \qquad (4.93f)$$

To avoid any confusion between the two different types of moments of inertia in mechanics, we use *mass moments of inertia* for dynamics and *area moments of inertia* for mechanics of materials.

The angular momentum vector about point P is thus given as

$$\mathbf{H}_p = (H_x)_p \mathbf{i} + (H_y)_p \mathbf{j} + (H_z)_p \mathbf{k} \qquad (4.94)$$

where

$$(H_x)_p = (I_{xx})_p \omega_x + (I_{xy})_p \omega_y + (I_{xz})_p \omega_z \qquad (4.95a)$$

$$(H_y)_p = (I_{yx})_p \omega_x + (I_{yy})_p \omega_y + (I_{yz})_p \omega_z \qquad (4.95b)$$

$$(H_z)_p = (I_{zx})_p \omega_x + (I_{zy})_p \omega_y + (I_{zz})_p \omega_z \qquad (4.95c)$$

In terms of matrices,

$$\mathbf{H}_p = \mathbf{I}_p \boldsymbol{\omega} \qquad (4.96)$$

TABLE 4.1 Moments of inertia of two common rigid bodies

Uniform thin rod	Uniform circular cylinder
$(I_{xx})_c = (I_{yy})_c = \frac{1}{12}mL^2$	$(I_{xx})_c = (I_{yy})_c = \dfrac{m}{12}(3r^2 + L^2)$
$(I_{zz})_c = (I_{zz})_o \cong 0$	$(I_{zz})_c = (I_{zz}) = \frac{1}{2}mr^2$
$(I_{xx})_o = (I_{yy})_o = \frac{1}{3}mL^2$	$(I_{xx})_o = (I_{yy})_o = \dfrac{m}{12}(3r^2 + 4L^2)$

or

$$\begin{Bmatrix} H_x \\ H_y \\ H_z \end{Bmatrix}_p = \begin{bmatrix} I_{xx} & I_{xy} & I_{xz} \\ I_{yx} & I_{yy} & I_{yz} \\ I_{zx} & I_{zy} & I_{zz} \end{bmatrix}_p \begin{Bmatrix} \omega_x \\ \omega_y \\ \omega_z \end{Bmatrix} \qquad (4.97)$$

The moments of inertia of common uniform rigid bodies are given in Table 4.1. Notice that the word *thin* implies that the radius is negligible, $r \cong 0$. In fact, a uniform thin rod is simply a special case of a uniform circular cylinder.

Notice that for a *uniform circular disk* of mass m and radius r, the moment of inertia about an axis through its mass center C and perpendicular to the plane of the disk is

$$I_c = \tfrac{1}{2}mr^2 \qquad (4.98)$$

This is a direct result from the properties of a uniform circular cylinder. For brevity, this moment of inertia is often referred to as the *centroidal mass moment of inertia* in planar motion (2D).

Modeling of Rigid Bodies in Planar Motion

In this section we cover the mathematical modeling of rigid bodies in planar motion. First, we look at single-degree-of-freedom (SDOF) systems, then we continue with two-degree-of-freedom (TDOF) systems. We then consider other multiple-degree-of-freedom (MDOF) systems of higher order.

Single degree-of-freedom systems

One of the most fundamental systems in rotation is the simple pendulum with inextensible string (Fig. 4.23). The string is of length L, which is massless (or negligible mass) and inextensible. The bob is of mass m, which is a point mass (zero or negligible dimension). The friction is negligible.

This system serves as an excellent example of SDOF systems. Thus, we start with this simple system first, then move on to relatively more complex cases.

Previously, this simple pendulum has been treated using two methods of summing the forces: Newton's second law and D'Alembert's principle. We now use the moment equation (sum of the moments) as another method to show that the same result can be obtained.

Because this pendulum system consists of a point mass and a string, either one of the methods is applicable. However, when the string is replaced by a rod, the moment equation is superior. Thus, in general, for systems involving rotational motion, it may be more efficient to use the moment equation.

EXAMPLE 4.10. Reconsider the simple pendulum shown in Fig. 4.23, and derive the differential equation using the moment equation.

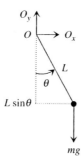

FIGURE 4.35 FBD for the moment equation.

Solution. The free-body diagram for the moment equation is shown in Fig. 4.35. Applying the moment equation about the fixed point O,

$$+\circlearrowleft \sum M_o = I_o \alpha \circlearrowleft +$$

$$-\underbrace{L\sin\theta}_{\text{moment arm}} \cdot \underbrace{mg}_{\text{force}} = \left(0 + mL^2\right)\ddot{\theta}$$

Rearranging the equation in the standard input-output differential equation form,

$$mL^2\ddot{\theta} + mgL\sin\theta = 0$$

or

$$\ddot{\theta} + \frac{g}{L}\sin\theta = 0$$

which is the same result obtained previously.

EXAMPLE 4.11. A pendulum of a thin uniform rod of mass m and length L is shown in Fig. 4.36 together with its FBD. The friction is assumed negligible.

(a) Derive the governing nonlinear differential equation.
(b) Linearize the equation, and then obtain the natural frequency in terms of given parameters.

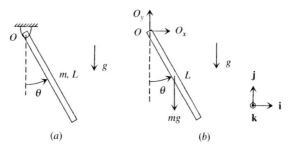

FIGURE 4.36 A pendulum system: (a) physical system, (b) FBD.

Solution

(a) Applying the moment equation about point O, in z (or **k** direction),

$$\sum (M_z)_o = I_o \alpha \qquad \text{where } I_o \equiv (I_{zz})_o, \qquad \alpha \equiv \alpha_z$$

Although this expression is precise, the following, with the sign convention, is simpler and useful. For the given angle θ,

$$+ \circlearrowleft \sum M_o = I_o \alpha \circlearrowleft +$$

Thus

$$-\frac{L}{2} \sin\theta \, mg = I_o \ddot{\theta}$$

or

$$I_o \ddot{\theta} + mg \frac{L}{2} \sin\theta = 0$$

(b) *Linearization:* For small motions, $\theta \ll 1$ rad, from Example 3.31 of Section 3.6, we have

$$\sin\theta \approx \theta$$

The linearized differential equation is

$$\ddot{\theta} + \omega_n^2 \theta = 0$$

where

$$\omega_n = \sqrt{\frac{mgL}{2I_o}}$$

From the parallel-axis theorem, we have

$$I_o = I_c + md^2 = I_c + m\left(\frac{L}{2}\right)^2$$

Since the rod is uniform,

$$I_c = \frac{1}{12} mL^2, \qquad I_o = \frac{1}{3} mL^2$$

EXAMPLE 4.12. Consider the pendulum system shown in Fig. 4.37 together with its FBD. The system consists of a uniform thin rod of mass M and length L with a concentrated mass m attached at the tip. The friction at the joint is modeled as a damper with *coefficient of torsional viscous damping B*.

(a) Derive the governing nonlinear differential equation.
(b) Linearize the equation, then obtain the natural frequency and damping ratio in terms of given parameters.

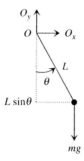

FIGURE 4.35 FBD for the moment equation.

Solution. The free-body diagram for the moment equation is shown in Fig. 4.35. Applying the moment equation about the fixed point O,

$$+\circlearrowright \sum M_o = I_o \alpha \circlearrowright +$$

$$-\underbrace{L \sin \theta}_{\text{moment arm}} \cdot \underbrace{mg}_{\text{force}} = (0 + mL^2)\ddot{\theta}$$

Rearranging the equation in the standard input-output differential equation form,

$$mL^2 \ddot{\theta} + mgL \sin \theta = 0$$

or

$$\ddot{\theta} + \frac{g}{L} \sin \theta = 0$$

which is the same result obtained previously.

EXAMPLE 4.11. A pendulum of a thin uniform rod of mass m and length L is shown in Fig. 4.36 together with its FBD. The friction is assumed negligible.

(*a*) Derive the governing nonlinear differential equation.
(*b*) Linearize the equation, and then obtain the natural frequency in terms of given parameters.

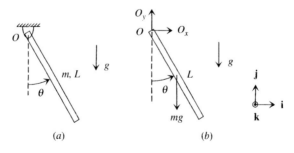

FIGURE 4.36 A pendulum system: (*a*) physical system, (*b*) FBD.

Solution

(a) Applying the moment equation about point O, in z (or **k** direction),

$$\sum (M_z)_o = I_o \alpha \quad \text{where } I_o = (I_{zz})_o, \quad \alpha = \alpha_z$$

Although this expression is precise, the following, with the sign convention, is simpler and useful. For the given angle θ,

$$+\circlearrowleft \sum M_o = I_o \alpha \circlearrowleft +$$

Thus

$$-\frac{L}{2} \sin\theta\, mg = I_o \ddot{\theta}$$

or

$$I_o \ddot{\theta} + mg \frac{L}{2} \sin\theta = 0$$

(b) *Linearization:* For small motions, $\theta \ll 1$ rad, from Example 3.31 of Section 3.6, we have

$$\sin\theta \approx \theta$$

The linearized differential equation is

$$\ddot{\theta} + \omega_n^2 \theta = 0$$

where

$$\omega_n = \sqrt{\frac{mgL}{2I_o}}$$

From the parallel-axis theorem, we have

$$I_o = I_c + md^2 = I_c + m\left(\frac{L}{2}\right)^2$$

Since the rod is uniform,

$$I_c = \frac{1}{12} mL^2, \quad I_o = \frac{1}{3} mL^2$$

EXAMPLE 4.12. Consider the pendulum system shown in Fig. 4.37 together with its FBD. The system consists of a uniform thin rod of mass M and length L with a concentrated mass m attached at the tip. The friction at the joint is modeled as a damper with *coefficient of torsional viscous damping B*.

(a) Derive the governing nonlinear differential equation.
(b) Linearize the equation, then obtain the natural frequency and damping ratio in terms of given parameters.

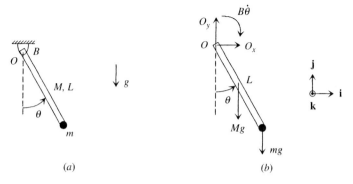

FIGURE 4.37 A pendulum system: (a) physical system, (b) FBD.

Solution

(a) The moment equation for a rigid body about a fixed point O, is

$$+\circlearrowleft \sum M_o = I_o \alpha \circlearrowleft +$$

$$-\frac{L}{2}\sin\theta \cdot Mg - L\sin\theta \cdot mg - B\dot\theta = I_o\ddot\theta$$

$$I_o\ddot\theta + B\dot\theta + \left(\frac{M}{2} + m\right)gL\sin\theta = 0$$

where

$$I_o = \sum_{j=1}^{n=2} I_{cj} + m_j d_j^2 = I_{c1} + m_1 d_1^2 + I_{c2} + m_2 d_2^2$$

$$I_o = \left[\frac{1}{12}ML^2 + M\left(\frac{L}{2}\right)^2\right] + \left[0 + mL^2\right] = \frac{1}{3}ML^2 + mL^2 = \left(\frac{M}{3} + m\right)L^2$$

(b) For small motions,

$$\theta \ll 1 \text{ rad}, \qquad \sin\theta \approx \theta$$

$$I_o\ddot\theta + B\dot\theta + \left(\frac{M}{2} + m\right)gL\theta = 0$$

or

$$\ddot\theta + 2\zeta\omega_n\dot\theta + \omega_n^2\theta = 0$$

where

$$\omega_n = \sqrt{\frac{(M/2 + m)gL}{I_o}} = \sqrt{\frac{(M/2 + m)gL}{(M/3 + m)L^2}} = \sqrt{\frac{3(M + 2m)g}{2(M + 3m)L}}$$

$$\zeta = \frac{B}{2\omega_n I_o} = \frac{B}{\sqrt{(2/3)(M + 2m)(M + 3m)gL^3}}$$

224 Dynamic Systems: Modeling and Analysis

since

$$2\zeta\omega_n = \frac{B}{I_o}$$

EXAMPLE 4.13. Consider the rotational system shown in Fig. 4.38 together with its FBD. The circular disk of mass m and radius r rolls, without slip, on a horizontal plane along a straight line. A translational spring of stiffness k is attached at the mass center C of the disk.

(a) Derive the differential equation.
(b) Obtain the natural frequency in terms of given parameters.

Solution

(a) For this problem, the best formula is

$$\sum M_{IC} = I_{IC}\alpha \quad \text{where } I_{IC} = I_c + md^2$$

Applying

$$+\curvearrowleft \sum M_{IC} = I_{IC}\alpha\curvearrowleft +$$

$$-r \cdot kx = I_{IC}\ddot{\theta}$$

From the no-slip constraint

$$x = r\theta$$

the equation becomes

$$I_{IC}\ddot{\theta} + kr^2\theta = 0$$

where

$$I_{IC} = I_c + md^2$$

$$I_{IC} = \tfrac{1}{2}mr^2 + mr^2 = \tfrac{3}{2}mr^2$$

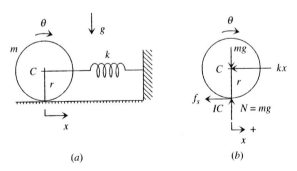

(a) (b)

FIGURE 4.38 Rolling disk: (a) physical system, (b) FBD.

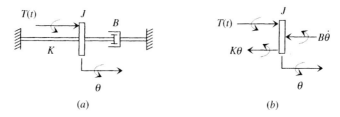

FIGURE 4.39 A SDOF torsional system: (*a*) physical system, (*b*) FBD.

(*b*) The equation can be rewritten as

$$\ddot{\theta} + \omega_n^2 \theta = 0$$

where

$$\omega_n = \sqrt{\frac{kr^2}{I_{IC}}} = \sqrt{\frac{kr^2}{(3/2)mr^2}} = \sqrt{\frac{2k}{3m}}$$

EXAMPLE 4.14. Consider the single-degree-of-freedom (SDOF) torsional system shown in Fig. 4.39 together with its FBD. The system consists of a shaft of torsional stiffness K, a disk of mass-moment of inertia J, and a torsional damper B. Derive the differential equation.

Solution. Applying the moment equation about the mass center along the longitudinal axis,

$$+ \circlearrowright \sum M_c = I_c \alpha \circlearrowright +$$

This sign convention is simpler and useful for the given angle θ. Thus,

$$T(t) - K\theta - B\dot{\theta} = J\ddot{\theta}$$

The differential equation in the input-output form is

$$J\ddot{\theta} + B\dot{\theta} + K\theta = T(t)$$

Two-degree-of-freedom systems

Two-degree-of-freedom (TDOF) systems are important in many engineering applications. To obtain the essential dynamic characteristics, certain complex systems may be modeled as TDOF. As an illustrative example, a TDOF torsional system follows.

EXAMPLE 4.15. Consider the TDOF torsional system shown in Fig. 4.40. The system consists of a shaft of torsional stiffness K, two disks of polar moments of inertia J_1 and J_2, and a torsional damper B. Draw the necessary free-body diagrams and derive the differential equations. Then express the equations in the second-order matrix form.

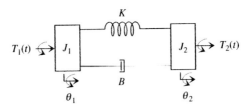

FIGURE 4.40 A TDOF torsional system.

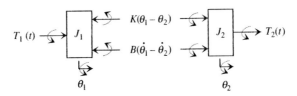

FIGURE 4.41 FBD.

Solution. Assuming $\theta_1 > \theta_2 > 0$, the free-body diagrams of the disks are shown in Fig. 4.41. Applying the moment equation about the mass centers along the longitudinal axis,

$$\curvearrowright + \sum M_c = I_c \alpha \curvearrowright +$$

Disk 1: $\quad T_1(t) - K(\theta_1 - \theta_2) - B(\dot\theta_1 - \dot\theta_2) = J_1 \ddot\theta_1$

Disk 2: $\quad T_2(t) - K(\theta_1 - \theta_2) + B(\dot\theta_1 - \dot\theta_2) = J_2 \ddot\theta_2$

The differential equations of the system are expressed in the standard input-output form as

$$J_1 \ddot\theta_1 + B\dot\theta_1 - B\dot\theta_2 + K\theta_1 - K\theta_2 = T_1(t)$$
$$J_2 \ddot\theta_2 - B\dot\theta_1 + B\dot\theta_2 - K\theta_1 + K\theta_2 = T_2(t)$$

and in the second-order matrix form as

$$\begin{bmatrix} J_1 & 0 \\ 0 & J_2 \end{bmatrix} \begin{Bmatrix} \ddot\theta_1 \\ \ddot\theta_2 \end{Bmatrix} + \begin{bmatrix} B & -B \\ -B & B \end{bmatrix} \begin{Bmatrix} \dot\theta_1 \\ \dot\theta_2 \end{Bmatrix} + \begin{bmatrix} K & -K \\ -K & K \end{bmatrix} \begin{Bmatrix} \theta_1 \\ \theta_2 \end{Bmatrix} = \begin{Bmatrix} T_1(t) \\ T_2(t) \end{Bmatrix}$$

Note that this torsional system is analogous to the corresponding TDOF translational system (Fig. 4.28).

4.7 MIXED SYSTEMS: TRANSLATIONAL AND ROTATIONAL

Mixed translational and rotational systems do not require any additional formulas. Although a few different methods can be used to model these systems, getting the model in an efficient and systematic manner may be a challenge. This section provides an elegant way of obtaining the governing differential equations using the FBD-and-force/moment approach. The method uses both the moment equation and Newton's second law. The rotational equation (the θ-equation) is obtained first, by using the moment equation; the translational equation, (the x-equation) is determined next by using the force equation (Newton's second law). At the end, the differential equations are combined to form the system equations of the mixed (hybrid) system.

For each of the two equations, two diagrams are helpful: an FBD showing all the relevant forces and moments, and a kinematics diagram showing the position and acceleration vectors. For the rotational equation, consider only a rigid body in rotational motion, whereas for the translational equation, consider the whole system.

For a rigid body with a single mass, the general moment equation for the rotational motion is given as

$$\sum \mathbf{M}_p = I_p \alpha + m\mathbf{r}_{c/p} \times \mathbf{a}_p$$

However, for a rigid body with multiple masses, the general moment equation becomes

$$\sum_{i=1}^{N} \mathbf{M}_{pi} = \left(\sum_{j=1}^{n} I_{pj}\right)\alpha + \sum_{j=1}^{n} \left(m_j \mathbf{r}_{cj/p} \times \mathbf{a}_p\right)$$

As an example, for a two-mass rigid body ($n = 2$) and three external forces causing moments ($N = 3$), the general moment equation is

$$\sum_{i=1}^{N=3} \mathbf{M}_{pi} = \left(\sum_{j=1}^{n=2} I_{pj}\right)\alpha + \sum_{j=1}^{n=2} \left(m_j \mathbf{r}_{cj/p} \times \mathbf{a}_p\right)$$

For the translational motion, Newton's second law is applied to the *entire* system. In the x-direction, for example, we have

$$\xrightarrow{+} \sum_{i=1}^{N=3} F_{ix} = ma_{cx} = \sum_{j=1}^{n=2} m_j a_{cjx} = m_1 a_{c1x} + m_2 a_{c2x} \xrightarrow{+}$$

For this situation, only the horizontal force components need to be shown (the vertical force components are ignored) on the FBD because we are only concerned about the horizontal motion.

Note that one of the most fundamental relations in kinematics, relating two acceleration vectors in this case, is given as

$$\mathbf{a}_{cj} = \mathbf{a}_p + \mathbf{a}_{cj/p}$$

where $\mathbf{a}_{cj}$ = acceleration of the jth center of mass (of the jth rigid body)
$\mathbf{a}_p$ = acceleration of the arbitrary point P
$\mathbf{a}_{cj/p}$ = acceleration of the jth center of mass with respect to the point P

Thus, the x-component of the vector $\mathbf{a}_{cj}$ is

$$a_{cjx} = a_{px} + (a_{cj/p})_x$$

228 Dynamic Systems: Modeling and Analysis

Also, from calculus, the cross product of the two vectors $\mathbf{r}_{c/p}$ and $\mathbf{a}_p$ is

$$\mathbf{r}_{c/p} \times \mathbf{a}_p = |\mathbf{r}_{c/p}||\mathbf{a}_p|\sin\phi$$

where ϕ is the angle between the two vectors.

Illustrative examples follow.

EXAMPLE 4.16. The pulley system shown in Fig. 4.42 consists of a pulley of inertia I_c and radius r, a block of mass m, and a translational spring of stiffness k. The pulley rotates about its fixed center of mass, in planar motion. The inertia I_c is the centroidal mass moment of inertia about the rotation axis. Derive the differential equation.

Solution. Free-body diagrams of the system in general motion and at static equilibrium are shown in Fig. 4.43. In this figure, δ is the static deformation of the spring. The displacement x is measured from the static equilibrium position (when the spring has deformation δ). The geometric constraint is

$$\ddot{x} = r\ddot{\theta} \qquad (a)$$

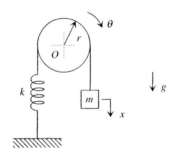

FIGURE 4.42 A pulley system.

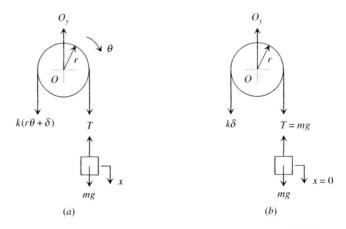

FIGURE 4.43 FBD: (a) general motion, (b) static equilibrium.

At static equilibrium

$$\sum M_o = 0$$

thus the static spring force is

$$k\delta = mg \quad (b)$$

Applying the moment equation and Newton's second law for general motion,

$$+\curvearrowright \sum M_o = I_o \alpha \curvearrowright +$$

$$rT - rk(r\theta + \delta) = I_c \ddot{\theta} \quad (c)$$

$$+\downarrow \sum F_x = ma_{cx} \downarrow +$$

$$-T + mg = m\ddot{x} \quad (d)$$

Substituting Eq. (a) into (d),

$$-T + mg = mr\ddot{\theta} \quad (e)$$

The unknown T may now be eliminated by multiplying Eq. (e) by r, and then the result is added to Eq. (c),

$$-rk(r\theta + \delta) - rmg = \left(I_c + mr^2\right)\ddot{\theta} \quad (f)$$

After introducing the static equilibrium condition, $mg = k\delta$, into the preceding equation, the differential equation is obtained as

$$\left(I_c + mr^2\right)\ddot{\theta} + kr^2\theta = 0 \quad (g)$$

EXAMPLE 4.17. Consider the pendulum system shown in Fig. 4.44. The cart of mass M moves in a horizontal direction, and x is the displacement measured from the static equilibrium position of the cart. Attached to the cart is a pendulum subsystem consisting of a massless rod (or inextensible string) of length L and a concentrated mass (or point mass) of mass m at the rod tip.

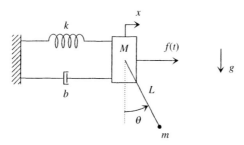

FIGURE 4.44 A pendulum system with mixed translational and rotational motions.

(a) Draw the necessary free-body diagrams and *derive* the general nonlinear differential equations.
(b) Linearize the equations for small motions, then express the equations in the second-order matrix form.

Solution

(a) This is a mixed two-degree-of-freedom system: rotational and translational. The rotational equation, θ-equation, will be obtained first. The FBD and kinematics of the rigid-body rod are shown in Fig. 4.45.

The general moment equation for a single-mass rigid body is

$$\sum \mathbf{M}_p = I_p \alpha + m \mathbf{r}_{c/p} \times \mathbf{a}_p$$

The cross product of the two vectors is

$$\mathbf{r}_{c/p} \times \mathbf{a}_p = |\mathbf{r}_{c/p}||\mathbf{a}_p| \sin \phi = L\ddot{x} \sin(90° - \theta) = L\ddot{x} \cos \theta$$

From the parallel-axis theorem, $I_p = mL^2$. Thus,

$$-L \sin \theta \, mg \mathbf{k} = mL^2 \ddot{\theta} \mathbf{k} + mL\ddot{x} \cos \theta \mathbf{k}$$

After canceling the unit vector, $\mathbf{k}$, the θ-equation is given as

$$-L \sin \theta \, mg = mL^2 \ddot{\theta} + mL\ddot{x} \cos \theta \tag{a}$$

The translational equation (the x-equation) is obtained next. The free-body and kinematics diagrams of the entire system consisting of the rod and the cart are shown in Fig. 4.46. On the free-body diagram, only the horizontal force components are shown (the vertical force components are ignored) because we are only concerned about the horizontal motion. On the kinematics diagram, $L\dot{\theta}^2$ and $L\ddot{\theta}$ are the centripetal and tangential accelerations of the rod tip, respectively.

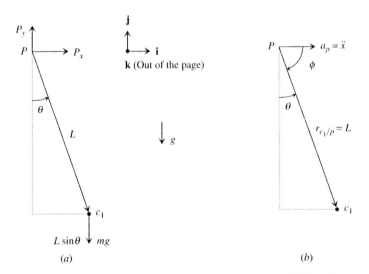

FIGURE 4.45 Diagrams for the moment equation: (a) FBD with all the forces, (b) kinematics with position and acceleration vectors.

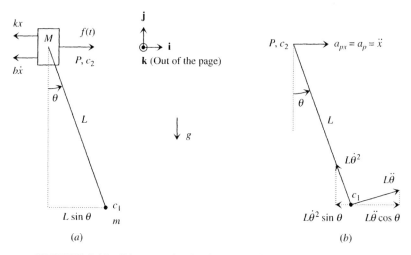

FIGURE 4.46 Diagrams for the force equation: (*a*) FBD with only the horizontal force components, (*b*) kinematics with position and acceleration vectors.

Applying Newton's second law for the system of rod and cart in the *x*-direction,

$$\xrightarrow{+} \sum_{j=1}^{3} F_{jx} = ma_{cx} = \sum_{i=1}^{2} m_i a_{cix} = m_1 a_{c1x} + m_2 a_{c2x} \xrightarrow{+}$$

Note that the fundamental relation in kinematics relating two acceleration vectors is given as

$$\mathbf{a}_{ci} = \mathbf{a}_p + \mathbf{a}_{ci/p}$$

Thus, the *x*-component of the vector $\mathbf{a}_{ci}$ is

$$a_{cix} = a_{px} + (a_{ci/p})_x$$

After applying Newton's second law, the *x*-equation is given as

$$f(t) - kx - b\dot{x} = m\underbrace{\left(\ddot{x} + L\ddot{\theta}\cos\theta - L\dot{\theta}^2\sin\theta\right)}_{a_{c1x}} + M\underbrace{\ddot{x}}_{a_{c2x}} \quad (b)$$

The system equations, Eqs. (*a*) and (*b*), are rearranged in the standard input-output form as

$$mL^2\ddot{\theta} + mL\ddot{x}\cos\theta + mgL\sin\theta = 0$$
$$(M + m)\ddot{x} + m\left(L\ddot{\theta}\cos\theta - L\dot{\theta}^2\sin\theta\right) + b\dot{x} + kx = f(t) \quad (c)$$

(*b*) Linearization: For small motions,

$$\theta \ll 1 \text{ rad}, \quad \cos\theta \approx 1, \quad \sin\theta \approx \theta, \quad \dot{\theta}^2 \approx 0$$

The linearized system equations are

$$mL^2\ddot{\theta} + mL\ddot{x} + mgL\theta = 0$$
$$(M + m)\ddot{x} + ml\ddot{\theta} + b\dot{x} + kx = f(t)$$
(d)

The differential equations are expressed in the second-order matrix form, with the θ-equation on top, as

$$\begin{bmatrix} mL^2 & mL \\ mL & M+m \end{bmatrix} \begin{Bmatrix} \ddot{\theta} \\ \ddot{x} \end{Bmatrix} + \begin{bmatrix} 0 & 0 \\ 0 & b \end{bmatrix} \begin{Bmatrix} \dot{\theta} \\ \dot{x} \end{Bmatrix} + \begin{bmatrix} mgL & 0 \\ 0 & k \end{bmatrix} \begin{Bmatrix} \theta \\ x \end{Bmatrix} = \begin{Bmatrix} 0 \\ f(t) \end{Bmatrix} \quad (e)$$

Because this system is stable, the elements on all the main diagonals are nonnegative. In contrast with the inverted pendulum, this system has the positive term mgL in the stiffness matrix.

EXAMPLE 4.18. Consider the inverted pendulum system shown in Fig. 4.47. The system consists of a cart of mass M, and a uniform rod of mass m and length L, where the concentrated mass m_c is attached at the rod tip. The rod is pivoted on the cart. The applied force is f, and the coefficient of viscous damping on the cart is c. Assume that the system is constrained to move in a vertical plane and that the cart rolls without slipping on a horizontal line.

(a) Draw the necessary free-body diagrams and *derive* the general nonlinear differential equations.
(b) Linearize the equations for small motions; then express the equations in the second-order matrix form.

Solution

(a) This is a mixed two-degree-of-freedom system: rotational and translational. The rotational equation is obtained first. The FBD and kinematics diagrams of the rigid-body rod are shown in Fig. 4.48. For convenience let $l = L/2$.
The general moment equation for a single-mass rigid body is

$$\sum \mathbf{M}_p = I_p \boldsymbol{\alpha} + m\mathbf{r}_{c/p} \times \mathbf{a}_p$$

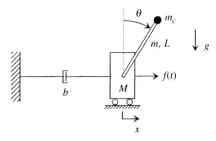

FIGURE 4.47 An inverted pendulum system.

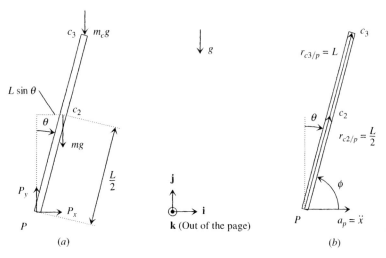

FIGURE 4.48 Diagrams for the moment equation: (a) FBD with all the forces, (b) kinematics with position and acceleration vectors.

However, for a multiple-mass rigid body the general moment equation becomes

$$\sum_{j=1}^{N} \mathbf{M}_{pj} = \left(\sum_{i=1}^{n} I_{pi}\right)\alpha + \sum_{i=1}^{n} \left(m_i \mathbf{r}_{ci/p} \times \mathbf{a}_p\right)$$

For this two-mass rigid body ($n = 2$) and two external forces causing moments ($N = 2$), the general moment equation is

$$\sum_{j=1}^{N=2} \mathbf{M}_{pj} = \left(\sum_{i=1}^{n=2} I_{pi}\right)\alpha + \sum_{i=1}^{n=2} \left(m_i \mathbf{r}_{ci/p} \times \mathbf{a}_p\right)$$

From the kinematics diagram, the cross product is

$$\mathbf{r}_{ci/p} \times \mathbf{a}_p = |\mathbf{r}_{ci/p}||\mathbf{a}_p|\sin\phi = |\mathbf{r}_{ci/p}|\ddot{x}\cos\theta$$

Thus,

$$\frac{L}{2}\sin\theta mg(-\hat{\mathbf{k}}) + L\sin\theta m_c g(-\hat{\mathbf{k}})$$
$$= I_p\ddot{\theta}(-\hat{\mathbf{k}}) + \left(m\frac{L}{2}\ddot{x}\cos\theta + m_c L\ddot{x}\cos\theta\right)(-\hat{\mathbf{k}})$$

After canceling the unit vector, $\hat{\mathbf{k}}$, the θ-equation is given as

$$\frac{L}{2}\sin\theta mg + L\sin\theta m_c g = I_p\ddot{\theta} + m\frac{L}{2}\ddot{x}\cos\theta + m_c L\ddot{x}\cos\theta \qquad (a)$$

where the mass moment of inertia is

$$I_p = \sum_{i=1}^{n=2} I_{ci} + m_i d_i^2 = I_{c1} + m_1 d_1^2 + I_{c2} + m_2 d_2^2$$

$$I_p = \left[\frac{1}{12}mL^2 + m\left(\frac{L}{2}\right)^2\right] + \left[0 + m_c L^2\right] = \frac{1}{3}mL^2 + m_c L^2 = \left(\frac{m}{3} + m_c\right)L^2$$

The translational equation is obtained next. The free-body and kinematics diagrams of the entire system consisting of the rod and the cart are shown in Fig. 4.49. Only the horizontal components of the forces are shown in the free-body diagram, and only the acceleration components of the cart and the rod are shown in the kinematics diagram. The acceleration components of the concentrated mass are similar (not shown).

Applying Newton's second law for a system of rod and cart, in x-direction,

$$\xrightarrow{+} \sum_{j=1}^{N=2} F_{jx} = ma_{cx} = \sum_{i=1}^{n=3} m_i a_{cix} = m_1 a_{c1x} = m_2 a_{c2x} = m_3 a_{c3x} \xrightarrow{+}$$

Note that the fundamental relation in kinematics relating two acceleration vectors is given as

$$\mathbf{a}_{ci} = \mathbf{a}_p + \mathbf{a}_{ci/p}$$

Thus, the x-component of the vector $\vec{a}_{ci}$ is

$$a_{cix} = a_{px} + \left(a_{ci/p}\right)_x$$

Thus,

$$f(t) - b\dot{x} = M\ddot{x} + m\left(\ddot{x} + \frac{L}{2}\ddot{\theta}\cos\theta - \frac{L}{2}\dot{\theta}^2\sin\theta\right) \quad (b)$$
$$+ m_c(\ddot{x} + L\ddot{\theta}\cos\theta - L\dot{\theta}^2\sin\theta)$$

The system equations, Eqs. (a) and (b), are rearranged in the standard input-output form as

$$I_p \ddot{\theta} + \left(\frac{m}{2} + m_c\right)L\ddot{x}\cos\theta - \left(\frac{m}{2} + m_c\right)gL\sin\theta = 0$$

$$(M + m + m_c)\ddot{x} + \left(\frac{m}{2} + m_c\right)L\ddot{\theta}\cos\theta - \left(\frac{m}{2} + m_c\right)L\dot{\theta}^2\sin\theta + b\dot{x} = f(t)$$

For brevity, if we define

$$m_t = M + m + m_c$$

$$m_o = \frac{m}{2} + m_c$$

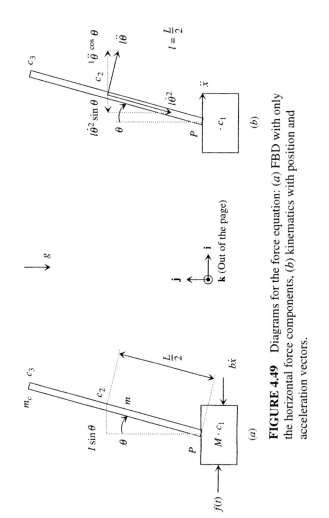

FIGURE 4.49 Diagrams for the force equation: (*a*) FBD with only the horizontal force components, (*b*) kinematics with position and acceleration vectors.

then the system equations become

$$I_p\ddot{\theta} + m_o L\ddot{x}\cos\theta - m_o gL\sin\theta = 0 \qquad (c)$$
$$m_t\ddot{x} + m_o L\ddot{\theta}\cos\theta - m_o L\dot{\theta}^2\sin\theta + b\dot{x} = f(t)$$

(b) Linearization: For small motions,

$$\theta \ll 1 \text{ rad}, \qquad \cos\theta \approx 1, \qquad \sin\theta \approx \theta, \qquad \dot{\theta}^2 \approx 0$$

The linearized system equations are

$$I_p\ddot{\theta} + m_o L\ddot{x} - m_o gL\theta = 0 \qquad (d)$$
$$m_o L\ddot{\theta} + m_t\ddot{x} + b\dot{x} = f(t)$$

Notice that the inverted pendulum system is *unstable*; thus the negative term $-m_o gL$ appears in the main diagonal of the stiffness matrix.

The system equations are given in the second-order matrix form as

$$\begin{bmatrix} I_p & m_o L \\ m_o L & m_t \end{bmatrix}\begin{Bmatrix} \ddot{\theta} \\ \ddot{x} \end{Bmatrix} + \begin{bmatrix} 0 & 0 \\ 0 & b \end{bmatrix}\begin{Bmatrix} \dot{\theta} \\ \dot{x} \end{Bmatrix} + \begin{bmatrix} -m_o gL & 0 \\ 0 & 0 \end{bmatrix}\begin{Bmatrix} \theta \\ x \end{Bmatrix} = \begin{Bmatrix} 0 \\ f(t) \end{Bmatrix} \qquad (e)$$

PROBLEM SET 4.6–4.7

4.41 A circular coin of radius R and mass m rolling down, without slipping, on an inclined plane whose angle of inclination is ϕ, is shown in Fig. P4.41. (a) Express the centroidal mass moment of inertia of the coin in terms of the given parameters; (b) Draw the necessary free-body diagrams and derive the differential equation.

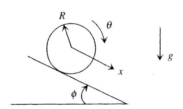

FIGURE P4.41

4.42. Consider the simple pendulum pivoted at point O as shown in Fig. P4.42. The pendulum consists of a uniform rod of mass m and length L and a point mass (concentrated

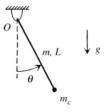

FIGURE P4.42

mass) at the rod tip. The system is constrained to move in a vertical plane. Draw the necessary free-body diagram and derive the general nonlinear differential equation, and then obtain the linearized differential equation and the natural frequency.

4.43. Using the differential equation obtained from Problem 4.41, obtain the state-space representation for two different cases of output: (a) displacement, (b) velocity.

4.44. Consider the torsional system shown in Fig. P4.44. Assume that the system is constrained to rotate about its fixed longitudinal axis. The mass moments of inertia of disks 1 and 2 are J_1 and J_2, respectively. The torsional stiffnesses are K_1 and K_2, and the torsional damper is C. (a) Draw the necessary free-body diagrams and *derive* the differential equations, and then (b) put the equations in the second-order matrix form.

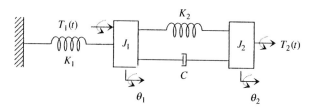

FIGURE P4.44

4.45. Reconsider the system shown in Problem 4.44. Without any derivation, *provide* the differential equations in the second-order matrix form. Use the skeleton approach.

4.46. Using the differential equations obtained from Problem 4.44, determine all the transfer functions.

4.47. Using the differential equations obtained from Problem 4.44, determine the state-space representation when the outputs are the displacements θ_1 and θ_2.

4.48. Consider the inverted pendulum system shown in Fig. P4.48. The system consists of a cart of mass M and a uniform rod of mass m and length L. The rod is pivoted on the cart. The applied force is f, and the coefficient of viscous damping on the cart is b.

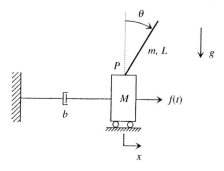

FIGURE P4.48

Assume that the system is constrained to move in a vertical plane and the cart rolls without slipping on a horizontal line. (a) Draw the necessary free-body diagrams and *derive* the general nonlinear differential equations, (b) linearize the equations for small motions, and (c) express the equations in the second-order matrix form.

4.49. Consider the pendulum shown in Fig. P4.49. The differential equations are given as

$$\begin{bmatrix} I_p & m_o L \\ m_o L & m_t \end{bmatrix} \begin{Bmatrix} \ddot{\theta} \\ \ddot{x} \end{Bmatrix} + \begin{bmatrix} 0 & 0 \\ 0 & b \end{bmatrix} \begin{Bmatrix} \dot{\theta} \\ \dot{x} \end{Bmatrix} + \begin{bmatrix} -m_o g L & 0 \\ 0 & 0 \end{bmatrix} \begin{Bmatrix} \theta \\ x \end{Bmatrix} = \begin{Bmatrix} 0 \\ u \end{Bmatrix}$$

where

$$m_t = M + m + m_c, \qquad m_o = \frac{m}{2} + m_c, \qquad I_p = \left(\frac{m}{3} + m_c\right)L^2.$$

Determine the transfer functions and transfer matrix.

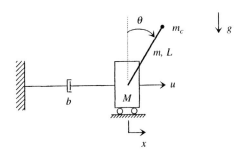

FIGURE P4.49

4.50. Using the differential equations obtained from Problem 4.49, determine the state-space representation when the outputs are the displacements θ and x.

4.51. Determine the transfer functions and transfer matrix of a pendulum system whose differential equations are given as

$$\begin{bmatrix} mL^2 & mL \\ mL & m_t \end{bmatrix} \begin{Bmatrix} \ddot{\theta} \\ \ddot{x} \end{Bmatrix} + \begin{bmatrix} 0 & 0 \\ 0 & c \end{bmatrix} \begin{Bmatrix} \dot{\theta} \\ \dot{x} \end{Bmatrix} + \begin{bmatrix} mgL & 0 \\ 0 & 0 \end{bmatrix} \begin{Bmatrix} \theta \\ x \end{Bmatrix} = \begin{Bmatrix} 0 \\ f(t) \end{Bmatrix}$$

where $m_t = M + m$.

4.8 GEAR-TRAIN SYSTEMS

Gear-train systems are important in many engineering applications. To obtain the mathematical model of a gear-train system, the following will be used: the moment equation, Newton's laws, and free-body diagrams. Although the fundamental laws are still applied, the application requires special consideration for an intuitive understanding.

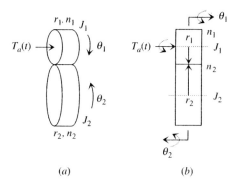

FIGURE 4.50 Diagrams of a gear-train system: (*a*) perspective, (*b*) for analysis.

A simple one-degree-of-freedom gear-train system is shown in Fig. 4.50. The system consists of two gears of polar moments of inertia J_1 and J_2 and radii r_1 and r_2, respectively. The numbers of gear teeth on gear 1 and gear 2 are n_1 and n_2, respectively. The applied torque on gear 1 is $T_a(t)$. The gears are assumed rigid, and they are meshed without backlash. A mathematical model in the form of a differential equation of motion is desired. Figure 4.50*b* is preferred since it shows clear directions of the torque and angular displacements, without any ambiguity. A clear sign convention is a must for derivation and problem solving.

Free-body diagrams for rotational gear-train systems must be shown with care. Figure 4.51, a free-body diagram, shows a clear picture of the physics (or engineering) of the gear system. Either Fig. 4.51*a* or Fig. 4.51*b* or both may be used to show the details: the action-reaction force F, the input (applied) torque, and the angular

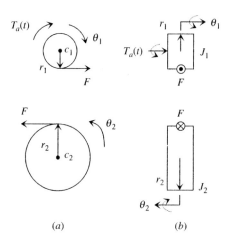

FIGURE 4.51 Free-body diagrams of the gear-train system: (*a*) axial view, (*b*) side view.

240   Dynamic Systems: Modeling and Analysis

(rotational) displacements θ_1 and θ_2. The torque T_a applied onto gear 1 causes an action force F_1 on gear 2. As a result of Newton's *third* law (action-reaction law), gear 2 experiences a reaction force F_2. These two forces are equal in magnitude, F, but opposite in direction. If the gears are assumed rigid and have no backlash, the geometric constraint is

$$r_1\theta_1 = r_2\theta_2 \tag{4.99a}$$

Thus
$$\theta_2 = \frac{r_1}{r_2}\theta_1, \qquad \ddot{\theta}_2 = \frac{r_1}{r_2}\ddot{\theta}_1 \tag{4.99b}$$

The gear-radii ratio is given in terms of the gear-teeth ratio as

$$\frac{r_1}{r_2} = \frac{n_1}{n_2} = N \tag{4.100}$$

The gear-teeth ratio may be used if desired, but care must be taken since N is defined differently among authors.

As illustrative examples, the differential equation of a simple gear-train system will be considered first, then a relatively more complicated one follows.

EXAMPLE 4.19. For the gear-train system shown in Fig. 4.50, obtain the differential equation.

Solution. The free-body diagram was shown in Fig. 4.51. For convenience, the constraint is given here

$$r_1\theta_1 = r_2\theta_2 \tag{a}$$

$$\theta_2 = \frac{r_1}{r_2}\theta_1, \qquad \ddot{\theta}_2 = \frac{r_1}{r_2}\ddot{\theta}_1 \tag{b}$$

Applying the moment equation about the center of mass C,

Gear 1: $+ \circlearrowright \sum M_c = I_c\alpha \circlearrowright +$ (c)

$$T_a(t) - r_1 F = J_1\ddot{\theta}_1 \tag{d}$$

Gear 2: $+ \circlearrowleft \sum M_c = I_c\alpha \circlearrowleft +$

$$r_2 F = J_2\ddot{\theta}_2 \tag{e}$$

Because this is a single-degree-of-freedom system, the mathematical model (differential equation) must be expressed in only one angular displacement. If the equation is expressed in terms of θ_1, then θ_2 must be eliminated. Introducing the geometric constraint, Eq. (*a*), into Eq. (*e*),

$$r_2 F = J_2 \frac{r_1}{r_2}\ddot{\theta}_1 \tag{f}$$

The force F can be eliminated in many different ways, but the following may be the most systematic, and thus least prone to mistake. Multiplying Eq. (*f*) by r_1/r_2 and then adding up the result to Eq. (d), the system equation is

$$\left[J_1 + J_2\left(\frac{r_1}{r_2}\right)^2\right]\ddot{\theta}_1 = T_a(t)$$

EXAMPLE 4.20. A three-degree-of-freedom gear-train system is shown in Fig. 4.52, which consists of four gears of moments of inertia J_1, J_2, J_3, and J_4. Gears 1 and 2 are connected by a relatively long shaft, and similarly with gears 3 and 4. Gears 2 and 3 are meshed and their radii are r_2 and r_3, respectively. The applied torque and load torque are $T_a(t)$ and $T_L(t)$ on gear 1 and gear 4, respectively. The gears are assumed rigid and have no backlash. Obtain the differential equations.

Solution. The shafts are relatively long and thus are assumed to be flexible. A physical model with flexibility represented by torsional springs is shown in Fig. 4.53. The free-body diagrams of the gears are shown Fig. 4.54. Since the gears are assumed rigid and have no backlash, the geometric constraint is

$$r_2\theta_2 = r_3\theta_3 \qquad (a)$$

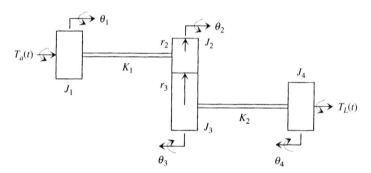

FIGURE 4.52 A gear-train system.

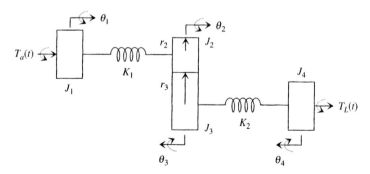

FIGURE 4.53 A physical model.

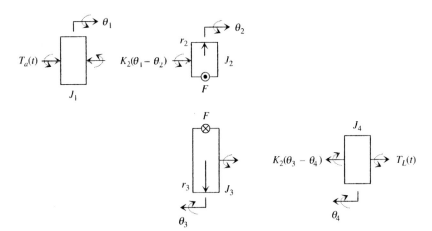

FIGURE 4.54 Free-body diagrams of the gear-train system: side view.

Thus
$$\theta_3 = \frac{r_2}{r_3}\theta_2, \quad \ddot\theta_3 = \frac{r_2}{r_3}\ddot\theta_2 \qquad (b)$$

Applying the moment equation about the mass center C,

Gear 1:
$$T_a(t) - K_1(\theta_1 - \theta_2) = J_1\ddot\theta_1 \qquad (c)$$

Gear 2:
$$K_1(\theta_1 - \theta_2) - r_2 F = J_2\ddot\theta_2 \qquad (d)$$

Gear 3:
$$r_2 F - K_2(\theta_3 - \theta_4) = J_3\ddot\theta_3 \qquad (e)$$

Gear 4:
$$-T_L(t) + K_2(\theta_3 - \theta_4) = J_4\ddot\theta_4 \qquad (f)$$

Introducing the geometric constraint, Eq. (b), into Eqs. (e) and (f),

$$r_3 F - K_2\left(\frac{r_2}{r_3}\theta_2 - \theta_4\right) = J_3\frac{r_2}{r_3}\ddot\theta_2 \qquad (g)$$

$$-T_L(t) + K_2\left(\frac{r_2}{r_3}\theta_2 - \theta_4\right) = J_4\ddot\theta_4 \qquad (h)$$

Multiplying Eq. (g) by r_2/r_3 and then adding the result to Eq. (d), we obtain

$$K_1(\theta_1 - \theta_2) - \frac{r_2}{r_3}K_2\left(\frac{r_2}{r_3}\theta_2 - \theta_4\right) = \left[J_2 + J_3\left(\frac{r_2}{r_3}\right)^2\right]\ddot\theta_2 \qquad (i)$$

Rearranging the three system equations,

$$J_1\ddot{\theta}_1 + K\theta_1 - K\theta_2 = T_a(t) \qquad (j)$$

$$\left[J_2 + J_3\left(\frac{r_2}{r_3}\right)^2\right]\ddot{\theta}_2 - K_1\theta_1 + \left[K_1 + K_2\left(\frac{r_2}{r_3}\right)^2\right]\theta_2 - \frac{r_2}{r_3}K_2\theta_4 = 0 \qquad (k)$$

$$J_4\ddot{\theta}_4 - K_2\frac{r_2}{r_3}\theta_2 + K_2\theta_4 = -T_L(t) \qquad (l)$$

For multi-degree-of-freedom systems, it is convenient to express the system equations in standard matrix form. The differential equations are expressed in the standard second-order matrix form as

$$\begin{bmatrix} J_1 & 0 & 0 \\ 0 & J_2 + J_3(r_2/r_3)^2 & 0 \\ 0 & 0 & J_4 \end{bmatrix}\begin{Bmatrix}\ddot{\theta}_1 \\ \ddot{\theta}_2 \\ \ddot{\theta}_4\end{Bmatrix} +$$

$$\begin{bmatrix} K_1 & -K_1 & 0 \\ -K_1 & K_1 + K_2(r_2/r_3)^2 & -(r_2/r_3)K_2 \\ 0 & -(r_2/r_3)K_2 & K_2 \end{bmatrix}\begin{Bmatrix}\theta_1 \\ \theta_2 \\ \theta_4\end{Bmatrix} = \begin{Bmatrix} T_a(t) \\ 0 \\ -T_L(t) \end{Bmatrix} \qquad (m)$$

PROBLEM SET 4.8

4.52. Consider the simple one-degree-of-freedom gear-train system shown in Fig. P4.52. The system consists of two gears of polar moments of inertia J_1 and J_2 and radii r_1 and r_2, respectively. The numbers of gear teeth on gear 1 and gear 2 are n_1 and n_2, respectively. The applied torque on gear 1 is $T_a(t)$. The gears are assumed rigid, and they are meshed without backlash. Determine a mathematical model in the form of a differential equation of motion in θ_2.

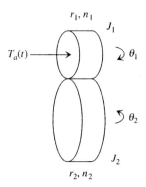

FIGURE P4.52

4.53. A three-degree-of-freedom gear-train system is shown in Fig. P4.53. The system consists of four gears of polar moments of inertia J_1, J_2, J_3, and J_4. The shaft connecting gears 1 and 2 is relatively short, so it can be approximated as rigid, or infinitely stiff, with

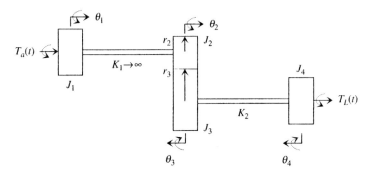

FIGURE P4.53

stiffness $K_1 \to \infty$. On the other hand, the shaft connecting gears 3 and 4 is relatively long; thus it can be approximated as flexible with stiffness K_2. Gears 2 and 3 are meshed and their radii are r_2 and r_3, respectively. The applied torque and load torque are $T_a(t)$ and $T_L(t)$ on gear 1 and gear 4, respectively. The gears are assumed to be rigid and have no backlash. Obtain the differential equations in the second-order matrix form.

4.54. Repeat Problem 4.53, but assume the shaft connecting gears 1 and 2 is relatively long (K_1 is finite) and that connecting gears 3 and 4 is relatively short (K_2 is approximately infinite).

4.9 LAGRANGE'S EQUATIONS

Lagrange's equations are expressed in one of the principal forms as

$$\frac{d}{dt}\left(\frac{\partial T}{\partial \dot{q}_i}\right) - \frac{\partial T}{\partial q_i} + \frac{\partial V}{\partial q_i} = Q_{nci}, \qquad i = 1, 2, \ldots, n \qquad (4.101)$$

where T = kinetic energy
 V = potential energy
 Q_{nci} = nonconservative generalized forces
 q_i = independent generalized coordinates
 n = total number of independent generalized coordinates

The kinetic energy, in general, is a function of the generalized coordinates q_i, the generalized velocities $\dot{q}_i$, and the time t,

$$T = T(q_i; \dot{q}_i; t) = T(q_1, q_2, \ldots, q_n; \dot{q}_1, \dot{q}_2, \ldots, \dot{q}_n; t) \qquad (4.102)$$

The potential energy V, in general, is the sum of the elastic potential energy V_e and the gravitational potential energy V_g,

$$V = V_e + V_g \qquad (4.103)$$

In contrast to kinetic energy, the potential energy in general is a function of the generalized coordinates q_i and the time t, not a function of the generalized velocities $\dot{q}_i$

$$V = V(q_i; t) = V(q_1, q_2, \ldots, q_n; t) \qquad (4.104)$$

One of the standard forms of Lagrange's equations is given as

$$\frac{d}{dt}\left(\frac{\partial L}{\partial \dot{q}_i}\right) - \frac{\partial L}{\partial q_i} = Q_{nci}, \quad i = 1, 2, \ldots, n \quad (4.105)$$

where the **Lagrangian** *function* L is defined as

$$L = T - V \quad (4.106)$$

This can be seen by substituting Eq. (4.106) into (4.101),

$$\frac{d}{dt}\left(\frac{\partial (T-V)}{\partial \dot{q}_i}\right) - \frac{\partial (T-V)}{\partial q_i} = Q_{nci}, \quad i = 1, 2, \ldots, n \quad (4.107)$$

Since V is not a function of $\dot{q}_i$, the equation becomes the same as Eq. (4.101). It is purely a personal choice when one decides which form of Lagrange's equations, Eq. (4.101) or (4.105), to work with.

Kinetic Energy

The kinetic energy of a particle is given as

$$T = \tfrac{1}{2} m v^2 \quad (4.108)$$

A particle has negligible size or dimension, thus no distinction is made between its center of mass and any other points of the particle.

The kinetic energy of a system of n particles is given as

$$T = \sum_{i=1}^{n} \frac{1}{2} m_i v_i^2 \quad (4.109)$$

or

$$T = \frac{1}{2} m v_c^2 + \sum_{i=1}^{n} \frac{1}{2} m_i v_{i/c}^2 \quad (4.110)$$

where $v_c^2 = \mathbf{v}_c \cdot \mathbf{v}_c$
  v_c = magnitude of the velocity (speed) of the center of mass c of the system (of n particles)
  m = total mass
  m_i = mass of the ith particle
  $v_{i/c}$ = speed of the ith particle with respect to c

The kinetic energy of the system can also be expressed using an arbitrary point P as a reference, as

$$T = \frac{1}{2} m v_p^2 + \sum_{i=1}^{n} \frac{1}{2} m_i v_{i/p}^2 + m \mathbf{v}_p \cdot \mathbf{v}_{c/p} \quad (4.111)$$

This expression may be used because, in certain applications, it is more difficult to compute v_c^2.

The kinetic energy of a rigid body in two dimensions (2D) is given as

$$T = \tfrac{1}{2}mv_c^2 + \tfrac{1}{2}I_c\omega^2 \tag{4.112}$$

where v_c = velocity of the center of mass c
I_c = mass moment of inertia about the center of mass
ω = angular velocity of the rigid body rotating about the fixed-direction axis

If we can find a fixed point O (fixed in the rigid body and fixed in space), then the kinetic energy is given in a simpler form

$$T = \tfrac{1}{2}I_o\omega^2 \tag{4.113}$$

where I_o is the mass moment of inertia about the fixed point O.

The kinetic energy of a system of rigid bodies in two dimensions (2D) is given as

$$T = \sum_{i=1}^{n} \tfrac{1}{2}m_i v_{ci}^2 + \tfrac{1}{2}I_{ci}\omega_i^2 \tag{4.114}$$

and the kinetic energy of a rigid body in three dimensions (3D) is given as

$$T = \tfrac{1}{2}mv_c^2 + \sum_{i=1}^{n} \tfrac{1}{2}m_i v_{i/c}^2 \tag{4.115}$$

The term $v_{i/c}^2$ may be expressed in terms of angular velocity $\boldsymbol{\omega}$ and centroidal mass moment of inertia $\mathbf{I_c}$. After some work, it can be shown that

$$T = \tfrac{1}{2}mv_c^2 + \tfrac{1}{2}\boldsymbol{\omega}^T \mathbf{I}_c \boldsymbol{\omega} \tag{4.116}$$

or

$$T = \tfrac{1}{2}mv_c^2 + \tfrac{1}{2}\begin{bmatrix}\omega_x & \omega_y & \omega_z\end{bmatrix}\begin{bmatrix} I_{xx} & I_{xy} & I_{xz} \\ I_{yx} & I_{yy} & I_{yz} \\ I_{zx} & I_{zy} & I_{zz}\end{bmatrix}\begin{Bmatrix}\omega_x \\ \omega_y \\ \omega_z\end{Bmatrix} \tag{4.117}$$

Potential Energy

As stated earlier, the potential energy, in general, is the sum of the elastic potential energy and the gravitational potential energy. The elastic potential energy of an elastic element is given as

$$V_e = \tfrac{1}{2}k\delta^2 \tag{4.118}$$

where δ is the elastic deformation from the undeformed configuration. Thus, for a system of elastic elements,

$$V_e = \sum_{i=1}^{n} \tfrac{1}{2}k_i\delta_i^2 \tag{4.119}$$

The gravitational potential energy of a mass is given as

$$V_g = mgh \tag{4.120}$$

where h is the height measured from a datum (reference) to the mass center c. This datum is defined (or selected) only for the gravitational potential energy V_g. For a system of rigid masses, we have

$$V_g = \sum_{i=1}^{n} m_i g h_i \qquad (4.121)$$

Nonconservative Forces

Nonconservative forces are forces that are *not* derivable from a potential function. Examples of nonconservative forces are frictional forces and externally applied forces. The forces that are derivable from a potential function are already included in the potential energy V, which is part of the Lagrangian function L.

The generalized forces are given by virtual work. The virtual work δW of the generalized Q_i in arbitrary virtual displacements δq_i is

$$\delta W = \sum_{i=1}^{n} Q_i \delta q_i = Q_1 \delta q_1 + Q_2 \delta q_2 + \cdots + Q_n \delta q_n \qquad (4.122)$$

To determine Q_j (for example, $j = 1$), we simply obtain δW first, then let all virtual displacements δq_i, except δq_j, equal zero; thus

$$Q_j = \frac{\delta W}{\delta q_j} \qquad \delta q_i = 0, \qquad j \neq i \qquad (4.123)$$

Thus, the nonconservative forces Q_{nci} are given by nonconservative virtual work δW_{nc} as

$$\delta W_{nc} = \sum_{i=1}^{n} Q_{nci} \delta q_i = Q_{nc1} \delta q_1 + Q_{nc2} \delta q_2 + \cdots + Q_{ncn} \delta q_n \qquad (4.124)$$

and Q_{ncj} is given as

$$Q_{ncj} = \frac{\delta W_{nc}}{\delta q_j} \qquad \delta q_i = 0, \qquad j \neq i \qquad (4.125)$$

EXAMPLE 4.21. Using Lagrange's equations, obtain the differential equation of the system shown in Fig. 4.19, shown again in Fig. 4.55.

Solution. This system has only one independent generalized coordinate, namely $q = x$. The kinetic energy is

$$T = \tfrac{1}{2} m v_c^2 = \tfrac{1}{2} m \dot{x}^2$$

The potential energy is

$$V = V_e = \tfrac{1}{2} k x^2 \qquad (V_g = 0)$$

For this system, the two nonconservative forces exerted on the mass are the applied force $f(t)$ and the damping force $c\dot{x}$. The applied force acts in the positive direction

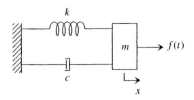

FIGURE 4.55 A mass-spring-damper system.

of the virtual displacement δx (as shown), whereas the damping force is in the negative direction. Thus, the nonconservative virtual work, which is done on the system, is given as

$$\delta W_{nc} = [f(t) - c\dot{x}]\delta x$$

Since

$$\delta W_{nc} = Q_{nc}\delta x$$

the nonconservative generalized force is obtained as

$$Q_{nc} = \frac{\delta W_{nc}}{\delta x} = f(t) - c\dot{x}$$

The general Lagrange's equations become

$$\frac{d}{dt}\left(\frac{\partial T}{\partial \dot{x}}\right) - \frac{\partial T}{\partial x} + \frac{\partial V}{\partial x} = Q_{nc}$$

where $\delta q = \delta x$, $\delta \dot{q} = \delta \dot{x}$. We have

$$\frac{\partial T}{\partial \dot{x}} = \frac{\partial}{\partial \dot{x}}\left(\frac{1}{2}m\dot{x}^2\right) = m\dot{x}$$

$$\frac{d}{dt}\left(\frac{\partial T}{\partial \dot{x}}\right) = m\ddot{x}$$

$$\frac{\partial T}{\partial x} = \frac{\partial}{\partial x}\left(\frac{1}{2}m\dot{x}^2\right) = 0$$

$$\frac{\partial V}{\partial x} = \frac{\partial}{\partial x}\left(\frac{1}{2}kx^2\right) = kx$$

As a result, we have

$$m\ddot{x} + kx = f(t) - c\dot{x}$$

Finally, the differential equation of motion is given in the standard input-output form as

$$m\ddot{x} + c\dot{x} + kx = f(t)$$

which is the same result as obtained previously using Newton's second law and the free-body diagram.

EXAMPLE 4.22. Repeat Example 4.22 but use Eq.(4.111), via the Lagrangian function, instead.

Solution. The Lagrangian function is
$$L = T - V = \tfrac{1}{2}m\dot{x}^2 - \tfrac{1}{2}kx^2$$
The Lagrange's equation becomes
$$\frac{d}{dt}\left(\frac{\partial L}{\partial \dot{x}}\right) - \frac{\partial L}{\partial x} = Q_{nc}$$
where
$$Q_{nc} = f(t) - c\dot{x}$$
We have
$$\frac{\partial L}{\partial \dot{x}} = \frac{\partial}{\partial \dot{x}}\left(\frac{1}{2}m\dot{x}^2 - \frac{1}{2}kx^2\right) = m\dot{x}$$
$$\frac{d}{dt}\left(\frac{\partial L}{\partial \dot{x}}\right) = m\ddot{x}$$
$$\frac{\partial L}{\partial x} = \frac{\partial}{\partial x}\left(\frac{1}{2}m\dot{x}^2 - \frac{1}{2}kx^2\right) = -kx$$
Thus, the differential equation is obtained as
$$m\ddot{x} + c\dot{x} + kx = f(t)$$
which agrees with the previous result.

EXAMPLE 4.23. Consider the system shown in Fig. 4.56. The displacements y and x are measured from the undeformed position (free length) and the static equilibrium position, respectively. Also, x_{st} is the static deformation of the spring. Use Lagrange's equations to obtain the differential equation in:

(a) y
(b) x

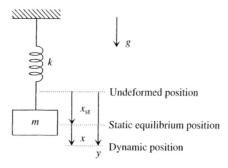

FIGURE 4.56 A simple translational mechanical system.

Solution

(a) The datum (reference) for V_g for y is shown in Fig. 4.57. This system has only one independent generalized coordinate, namely $q = y$. The kinetic energy is

$$T = \tfrac{1}{2}mv_c^2 = \tfrac{1}{2}m\dot{y}^2$$

The potential energy is

$$V = V_e + V_g = \tfrac{1}{2}k\delta^2 + mgh = \tfrac{1}{2}ky^2 + mg(-y)$$

The nonconservative virtual work δW_{nc} is zero:

$$\delta W_{nc} = 0$$

The Lagrange's equation becomes

$$\frac{d}{dt}\left(\frac{\partial T}{\partial \dot{y}}\right) - \frac{\partial T}{\partial y} + \frac{\partial V}{\partial y} = 0$$

We have

$$\frac{\partial T}{\partial \dot{y}} = \frac{\partial}{\partial \dot{y}}\left(\tfrac{1}{2}m\dot{y}^2\right) = m\dot{y} \quad\Longrightarrow\quad \frac{d}{dt}\left(\frac{\partial T}{\partial \dot{y}}\right) = m\ddot{y}$$

$$\frac{\partial T}{\partial y} = \frac{\partial}{\partial y}\left(\tfrac{1}{2}m\dot{y}^2\right) = 0$$

$$\frac{\partial V}{\partial y} = \frac{\partial}{\partial y}\left(\tfrac{1}{2}ky^2 - mgy\right) = ky - mg$$

Thus,

$$m\ddot{y} + ky - mg = 0$$

The differential equation is

$$m\ddot{y} + ky = mg$$

(b) The datum (reference) for V_g for x is shown in Fig. 4.58. The kinetic energy is

$$T = \tfrac{1}{2}mv_c^2 = \tfrac{1}{2}m\dot{x}^2$$

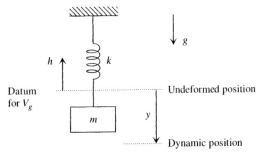

FIGURE 4.57 Datum for V_g for y.

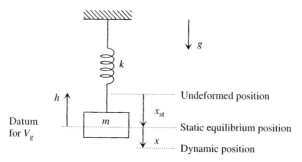

FIGURE 4.58 Datum for V_g for x.

The potential energy is

$$V = V_e + V_g = \tfrac{1}{2}k\delta^2 + mgh = \tfrac{1}{2}k(x_{st} + x)^2 + mg(-x)$$

The nonconservative virtual work δW_{nc} is zero:

$$\delta W_{nc} = 0$$

The Lagrange's equation becomes

$$\frac{d}{dt}\left(\frac{\partial T}{\partial \dot{x}}\right) - \frac{\partial T}{\partial x} + \frac{\partial V}{\partial x} = 0$$

We have

$$\frac{\partial T}{\partial \dot{x}} = \frac{\partial}{\partial \dot{x}}\left(\frac{1}{2}m\dot{x}^2\right) = m\dot{x} \implies \frac{\partial T}{\partial \dot{x}} = m\ddot{x}$$

$$\frac{\partial T}{\partial x} = \frac{\partial}{\partial x}\left(\frac{1}{2}m\dot{x}^2\right) = 0$$

$$\frac{\partial V}{\partial x} = \frac{\partial}{\partial x}\left[\frac{1}{2}k(x_{st} + x)^2 - mgx\right]$$

$$= k(x_{st} + x)(1) - mg = kx_{st} - mg + kx = kx$$

because at static equilibrium $kx_{st} = mg$. Finally, the differential equation is

$$m\ddot{x} + kx = 0$$

where the term mg does not enter into the equation. We can conclude that it is simpler to represent the mathematical model in terms of x, which is the displacement measured from the static equilibrium position. Thus, gravity may be ignored for this type of mass-spring system.

EXAMPLE 4.24. Using Lagrange's equations, obtain the differential equation of the system shown in Fig. 4.30.

Solution. This system has only two independent generalized coordinates, namely $q_1 = x_1$, $q_2 = x_2$. The kinetic energy is

$$T = \sum_{i=1}^{n=2} \frac{1}{2} m_i v_{ci}^2 = \frac{1}{2} m_1 \dot{x}_1^2 + \frac{1}{2} m_2 \dot{x}_2^2$$

The potential energy is

$$V = V_e = \sum_{i=1}^{n=2} \frac{1}{2} k_i \delta_i^2 = \frac{1}{2} k_1 x_1^2 + \frac{1}{2} k_2 (x_1 - x_2)^2 \qquad (V_g = 0)$$

The virtual work done by the nonconservative forces through virtual displacements is given as

$$\delta W = \sum_{i=1}^{n=2} Q_{nci} \delta q_i = Q_{nc1} \delta q_1 + Q_{nc2} \delta q_2$$

$$= [f_1(t) - c(\dot{x}_1 - \dot{x}_2)] \delta x_1 + [f_2(t) - c(\dot{x}_2 - \dot{x}_1)] \delta x_2$$

When $i = 1$, $q_1 = x_1$, we have

$$\frac{d}{dt}\left(\frac{\partial T}{\partial \dot{x}_1}\right) - \frac{\partial T}{\partial x_1} + \frac{\partial V}{\partial x_1} = Q_{nc1}$$

Since $\delta q_1 = \delta x_1$, $\delta \dot{q}_1 = \delta \dot{x}_1$, we have

$$\frac{d}{dt}\left(\frac{\partial T}{\partial \dot{x}_1}\right) = \frac{d}{dt}\left[\frac{\partial T}{\partial \dot{x}_1}\right] = \frac{\partial}{\partial \dot{x}_1}\left(\frac{1}{2}m_1\dot{x}_1^2 + \frac{1}{2}m_2\dot{x}_2^2\right) = m_1\dot{x}_1 \Big] = m_1\ddot{x}_1$$

$$\frac{\partial T}{\partial x_1} = \frac{\partial}{\partial x_1}\left(\frac{1}{2}m_1\dot{x}_1^2 + \frac{1}{2}m_2\dot{x}_2^2\right) = 0$$

$$\frac{\partial V}{\partial x_1} = \frac{\partial}{\partial x_1}\left(\frac{1}{2}k_1 x_1^2 + \frac{1}{2}k_2(x_1 - x_2)^2\right) = k_1 x_1 + k_2(x_1 - x_2)(1)$$

$$Q_{nc1} = \frac{\delta W_{nc}}{\delta x_1}\bigg|_{\delta x_2 = 0} = f_1(t) - c(\dot{x}_1 - \dot{x}_2)$$

Thus,

$$m_1 \ddot{x}_1 + (k_1 + k_2) x_1 + k_2 x_2 = f_1(t) - c(\dot{x}_1 - \dot{x}_2)$$

The differential equation is

$$m_1 \ddot{x}_1 + c\dot{x}_1 - c\dot{x}_2 + (k_1 + k_2) x_1 + k_2 x_2 = f_1(t)$$

When $i = 2$, $q_2 = x_2$, we have

$$\frac{d}{dt}\left(\frac{\partial T}{\partial \dot{x}_2}\right) - \frac{\partial T}{\partial x_2} + \frac{\partial V}{\partial x_2} = Q_{nc2}$$

Since $\delta q_2 = \delta x_2$, $\delta \dot{q}_2 = \delta \dot{x}_2$, we have

$$\frac{d}{dt}\left(\frac{\partial T}{\partial \dot{x}_2}\right) = \frac{d}{dt}\left[\frac{\partial T}{\partial \dot{x}_2}\right] = \frac{\partial}{\partial \dot{x}_2}\left(\frac{1}{2}m_1\dot{x}_1^2 + \frac{1}{2}m_2\dot{x}_2^2\right) = m_2\dot{x}_2 = m_2\ddot{x}_2$$

$$\frac{\partial T}{\partial x_2} = \frac{\partial}{\partial x_2}\left(\frac{1}{2}m_1\dot{x}_1^2 + \frac{1}{2}m_2\dot{x}_2^2\right) = 0$$

$$\frac{\partial V}{\partial x_2} = \frac{\partial}{\partial x_2}\left(\frac{1}{2}k_1 x_1^2 + \frac{1}{2}k_2(x_1 - x_2)^2\right) = k_2(x_1 - x_2)(-1)$$

$$Q_{nc2} = \left.\frac{\delta W_{nc}}{\delta x_2}\right|_{\delta x_1 = 0} = f_2(t) - c(\dot{x}_2 - \dot{x}_1)$$

Thus,

$$m_2\ddot{x}_2 - k_2 x_1 + k_2 x_2 = f_2(t) - c(\dot{x}_2 - \dot{x}_1)$$

The differential equation is

$$m_2\ddot{x}_2 - c\dot{x}_1 + c\dot{x}_2 - k_2 x_1 + k_2 x_2 = f_2(t)$$

Expressing the equations into the standard second-order matrix form,

$$\begin{bmatrix} m_1 & 0 \\ 0 & m_2 \end{bmatrix}\begin{Bmatrix} \ddot{x}_1 \\ \ddot{x}_2 \end{Bmatrix} + \begin{bmatrix} c & -c \\ -c & c \end{bmatrix}\begin{Bmatrix} \dot{x}_1 \\ \dot{x}_2 \end{Bmatrix} + \begin{bmatrix} k_1 + k_2 & -k_2 \\ -k_2 & k_2 \end{bmatrix}\begin{Bmatrix} x_1 \\ x_2 \end{Bmatrix} = \begin{Bmatrix} f_1 \\ f_2 \end{Bmatrix}$$

which is the same, as expected, as that obtained by using Newton's second law and free-body diagrams.

EXAMPLE 4.25. Using Lagrange's equations, obtain the differential equation of the system shown in Fig. 4.36.

Solution. The datum (reference) for V_g is shown in Fig. 4.59. This system has a fixed point O (fixed in the rigid body and fixed in space). Also, it has only one independent generalized coordinate, namely $q = \theta$. The kinetic energy is

$$T = \tfrac{1}{2}I_o\omega^2 = \tfrac{1}{2}\left(\tfrac{1}{3}mL^2\right)\dot{\theta}^2$$

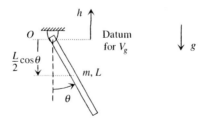

FIGURE 4.59 Datum for V_g.

The potential energy is

$$V = V_g = mgh = mg\left(-\frac{L}{2}\cos\theta\right) \quad (V_e = 0)$$

The nonconservative virtual work δW_{nc} is zero, so the nonconservative generalized force is also zero,

$$\delta W_{nc} = Q_{nc} = 0$$

The Lagrange's equations become

$$\frac{d}{dt}\left(\frac{\partial T}{\partial \dot\theta}\right) - \frac{\partial T}{\partial \theta} + \frac{\partial V}{\partial \theta} = 0$$

Since $\delta q = \delta\theta$, $\delta\dot q = \delta\dot\theta$, we have

$$\frac{d}{dt}\left(\frac{\partial T}{\partial \dot\theta}\right) = \frac{d}{dt}\left[\frac{\partial T}{\partial \dot\theta}\right] = \frac{\partial}{\partial \dot\theta}\left(\frac{1}{2}\frac{1}{3}mL^2\dot\theta^2\right) = \frac{1}{3}mL^2\dot\theta\bigg] = \frac{1}{3}mL^2\ddot\theta$$

$$\frac{\partial T}{\partial \theta} = \frac{\partial}{\partial \theta}\left(\frac{1}{2}\frac{1}{3}mL^2\dot\theta^2\right) = 0$$

$$\frac{\partial V}{\partial \theta} = \frac{\partial}{\partial \theta}\left(-mg\frac{L}{2}\cos\theta\right) = mg\frac{L}{2}\sin\theta$$

Thus, the differential equation is

$$\frac{1}{3}mL^2\ddot\theta + mg\frac{L}{2}\sin\theta = 0$$

which is the same result obtained previously, as expected.

EXAMPLE 4.26. Consider the double pendulum shown in Fig. 4.60. The two point masses m_1 and m_2 are attached to two rigid links of lengths L_1 and L_2. The links are assumed massless because their masses are small compared to the point masses. The hinged joints constrain the system in planar motion. The friction is negligible, so the system can be assumed frictionless. Using Lagrange's equations, obtain the differential equations of motion.

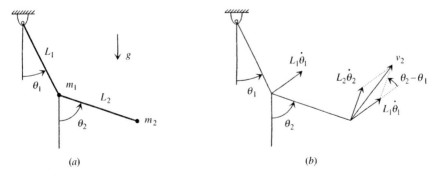

FIGURE 4.60 Double pendulum: (*a*) system, (*b*) velocity kinematics.

CHAPTER 4: Mechanical Systems   255

Solution. The kinetic energy of the system is

$$T = \sum_{i=1}^{2} \frac{1}{2} m_i v_{ci}^2 + \frac{1}{2} I_{ci} \omega_i^2 = \frac{1}{2} m_1 v_1^2 + \frac{1}{2} m_2 v_2^2$$

Notice that for point mass, the centroidal mass moment of inertia is zero, and the subscript c denoting the mass center is dropped. From the velocity kinematics, we have

$$v_1 = L_1 \dot\theta_1$$
$$v_2^2 = (L_1 \dot\theta_1)^2 + (L_2 \dot\theta_2)^2 + 2 L_1 \dot\theta_1 \cdot L_2 \dot\theta_2 \cdot \cos(\theta_2 - \theta_1)$$

where the law of cosines (Appendix C) has been applied. Thus,

$$T = \tfrac{1}{2} m_1 (L_1 \dot\theta_1)^2 + \tfrac{1}{2} m_2 [L_1^2 \dot\theta_1^2 + L_2^2 \dot\theta_2^2 + 2 L_1 L_2 \dot\theta_1 \dot\theta_2 \cos(\theta_2 - \theta_1)]$$
$$T = \tfrac{1}{2}(m_1 + m_2) L_1^2 \dot\theta_1^2 + \tfrac{1}{2} m_2 L_2^2 \dot\theta_2^2 + m_2 L_1 L_2 \dot\theta_1 \dot\theta_2 \cos(\theta_2 - \theta_1)$$

The potential energy is

$$V = V_g = \sum_{i=1}^{2} m_i g h_i = m_1 g h_1 + m_2 g h_2 \qquad (V_e = 0)$$

With the datum for V_g as shown, we have

$$h_1 = -L_1 \cos\theta_1 \qquad h_2 = -(L_1 \cos\theta_1 + L_2 \cos\theta_2)$$

Thus,

$$V = -(m_1 + m_2) g L_1 \cos\theta_1 - m_2 g L_2 \cos\theta_2$$

Applying the Lagrange's equations,

$$\frac{d}{dt}\left(\frac{\partial T}{\partial \dot q_i}\right) - \frac{\partial T}{\partial q_i} + \frac{\partial V}{\partial q_i} = Q_{nci} = 0, \qquad i = 1, 2$$

Because the nonconservative virtual work δ_{nc} is zero, the nonconservative generalized forces are also zero:

$$\delta W_{nc} = Q_{nci} = 0$$

For $i = 1$, $q_1 = \theta_1$, $Q_{nc1} = 0$,

$$\frac{d}{dt}\left[\frac{\partial L}{\partial \dot\theta_1}\right] = (m_1 + m_2) L_1^2 \dot\theta_1 + m_2 L_1 L_2 \dot\theta_2 \cos(\theta_2 - \theta_1)\Big] =$$

$$(m_1 + m_2) L_1^2 \ddot\theta_1 + m_2 L_1 L_2 \ddot\theta_2 \cos(\theta_2 - \theta_1) - m_2 L_1 L_2 \dot\theta_2 \sin(\theta_2 - \theta_1)(\dot\theta_2 - \dot\theta_1)$$

$$\frac{\partial T}{\partial \theta_1} = -m_2 L_1 L_2 \dot\theta_1 \dot\theta_2 \sin(\theta_2 - \theta_1)(-1)$$

$$\frac{\partial V}{\partial \theta_1} = (m_1 + m_2) g L_1 \sin\theta_1$$

Adding the results, the θ_1-equation is given as

$$(m_1 + m_2)L_1^2\ddot{\theta}_1 + m_2L_1L_2\ddot{\theta}_2 \cos(\theta_2 - \theta_1)$$
$$- m_2L_1L_2\dot{\theta}_2^2 \sin(\theta_2 - \theta_1) + (m_1 + m_2)gL_1 \sin\theta_1 = 0$$

For $i = 2$, $q_2 = \theta_2$, $Q_{nc2} = 0$,

$$\frac{d}{dt}\left[\frac{\partial L}{\partial \dot{\theta}_2}\right] = m_2L_2^2\dot{\theta}_2 + m_2L_1L_2\dot{\theta}_1 \cos(\theta_2 - \theta_1)$$
$$= m_2L_2^2\ddot{\theta}_2 + m_2L_1L_2\ddot{\theta}_1 \cos(\theta_2 - \theta_1) - m_2L_1L_2\dot{\theta}_1 \sin(\theta_2 - \theta_1)(\dot{\theta}_2 - \dot{\theta}_1)$$

$$\frac{\partial T}{\partial \theta_2} = -m_2L_1L_2\dot{\theta}_1\dot{\theta}_2 \sin(\theta_2 - \theta_1)(1)$$

$$\frac{\partial V}{\partial \theta_2} = m_2gL_2 \sin\theta_2$$

Similarly, after adding the results, the θ_2-equation is given as

$$m_2L_2^2\ddot{\theta}_2 + m_2L_1L_2\ddot{\theta}_1 \cos(\theta_2 - \theta_1) + m_2L_1L_2\dot{\theta}_1^2 \sin(\theta_2 - \theta_1) + m_2gL_2 \sin\theta_2 = 0$$

As a check, the differential equations are expressed in the second-order matrix form, after linearization for small motions, as

$$\begin{bmatrix}(m_1+m_2)L_1^2 & m_2L_1L_2 \\ m_2L_1L_2 & m_2L_2^2\end{bmatrix}\begin{Bmatrix}\ddot{\theta}_1 \\ \ddot{\theta}_2\end{Bmatrix} + \begin{bmatrix}(m_1+m_2)gL_1 & 0 \\ 0 & m_2gL_2\end{bmatrix}\begin{Bmatrix}\theta_1 \\ \theta_2\end{Bmatrix} = \begin{Bmatrix}0 \\ 0\end{Bmatrix}$$

Since the system is stable and the sign convention is consistent, the following results are obtained as expected:

- The main-diagonal elements of both the mass and stiffness matrices are nonnegative.
- Both the mass and stiffness matrices are symmetrical (with respect to the main diagonals).

In addition, since the sign convention is consistent, we also have

- The off-diagonal elements of the stiffness matrix are nonpositive.
- The off-diagonal elements of the mass matrix are nonnegative.

EXAMPLE 4.27. Reconsider the pendulum system in Example 4.18, as shown in Fig. 4.61. Using Lagrange's equations, derive the differential equations of motion.

Solution. The kinetic energy of the system is

$$T = \sum_{i=1}^{2}\frac{1}{2}m_iv_{ci}^2 = \frac{1}{2}m_1v_1^2 + \frac{1}{2}m_2v_2^2 \quad (I_{ci} = 0)$$

We have

$$m_1 = M \qquad v_1 = \dot{x}$$
$$m_2 = m \qquad v_2^2 = \dot{x}^2 + (L\dot{\theta})^2 + 2\dot{x}L\dot{\theta}\cos\theta$$

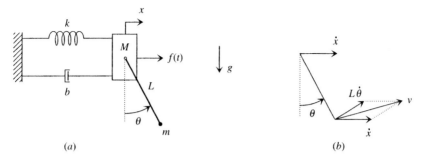

FIGURE 4.61 Pendulum system: (*a*) system (repeated), (*b*) velocity analysis.

Thus,
$$T = \tfrac{1}{2}M\dot{x}^2 + \tfrac{1}{2}m[\dot{x}^2 + L^2\dot{\theta}^2 + 2\dot{x}L\dot{\theta}\cos\theta]$$
$$T = \tfrac{1}{2}(M+m)\dot{x}^2 + \tfrac{1}{2}mL^2\dot{\theta}^2 + m\dot{x}L\dot{\theta}\cos\theta$$

The potential energy is
$$V = V_e + V_g = \tfrac{1}{2}kx^2 - mgL\cos\theta$$

where the datum for V_g is chosen as shown. The nonconservative virtual work is
$$\delta W = \sum_{i=1}^{n=2} Q_{nci}\delta q_i = Q_{nc1}\delta q_1 + Q_{nc2}\delta q_2 = [f(t) - b\dot{x}]\delta x$$

For $i = 1$, $q_1 = x$, $Q_{nc1} = f(t) - b\dot{x}$,
$$\frac{d}{dt}\left[\frac{\partial T}{\partial \dot{x}}\right] = M\dot{x} + m\dot{x} + mL\dot{\theta}\cos\theta\bigg] = (M+m)\ddot{x} + mL\ddot{\theta}\cos\theta - mL\dot{\theta}^2\sin\theta$$

$$\frac{\partial T}{\partial x} = 0$$

$$\frac{\partial V}{\partial x} = kx$$

Thus, the x-equation is given as
$$(M+m)\ddot{x} + b\dot{x} + mL\ddot{\theta}\cos\theta - mL\dot{\theta}^2\sin\theta + kx = f(t)$$

For $i = 2$, $q_2 = \theta$, $Q_{nc2} = 0$,
$$\frac{d}{dt}\left[\frac{\partial T}{\partial \dot{\theta}}\right] = mL^2\dot{\theta} + m\dot{x}L\cos\theta\bigg] = mL^2\ddot{\theta} + m\ddot{x}L\cos\theta - m\dot{x}L\sin\theta \cdot \dot{\theta}$$

$$\frac{\partial T}{\partial \theta} = -m\dot{x}L\dot{\theta}\sin\theta$$

$$\frac{\partial V}{\partial \theta} = mgL\sin\theta$$

Thus, the θ-equation is

$$mL^2\ddot{\theta} + m\ddot{x}L\cos\theta + mgL\sin\theta = 0$$

The system equations, the x-equation and the θ-equation, are the same as expected (Example 4.17.)

As a variation to the matrix equation shown in Example 4.17, the linearized differential equations may be given as

$$\begin{bmatrix} M+m & mL \\ mL & mL^2 \end{bmatrix} \begin{Bmatrix} \ddot{x} \\ \ddot{\theta} \end{Bmatrix} + \begin{bmatrix} b & 0 \\ 0 & 0 \end{bmatrix} \begin{Bmatrix} \dot{x} \\ \dot{\theta} \end{Bmatrix} + \begin{bmatrix} k & 0 \\ 0 & mgL \end{bmatrix} \begin{Bmatrix} x \\ \theta \end{Bmatrix} = \begin{Bmatrix} f(t) \\ 0 \end{Bmatrix}$$

where the x-equation is on top.

PROBLEM SET 4.9

4.55. Reconsider the simple mechanical system of mass, spring, and damper shown in Fig. P4.55. Using Lagrange's equations, derive the differential equation.

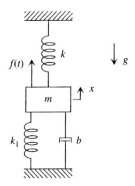

FIGURE P4.55

4.56. Reconsider the translational system shown in Fig. P4.27 of Problem 4.27. Using Lagrange's equations, derive the differential equations.

4.57. Using Lagrange's equations, derive the differential equations of the system shown in Fig. P4.57.

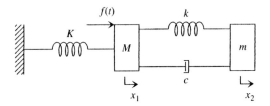

FIGURE P4.57

4.58. Reconsider the translational system shown in Fig. P4.35 of Problem 4.35. Using Lagrange's equations, derive the differential equations.

4.59. Reconsider the translational system shown in Fig. P4.39 of Problem 4.39. Using Lagrange's equations, derive the differential equations.

4.60. Reconsider the rotational system shown in Fig. P4.34 of Example 4.14. Using Lagrange's equation, derive the differential equation.

4.61. Consider the double pendulum shown in Fig. P4.61. The two point masses of equal mass m are attached to two rigid links of equal length L. The links are assumed massless since their masses are small compared to the point masses. The hinged joints constrain the system in planar motion. The friction is negligible, so the system can be assumed frictionless.
(a) Using Lagrange's equations, derive the general nonlinear differential equations of motion.
(b) Obtain the linearized differential equations for small motions, and then express the results in the second-order matrix form.

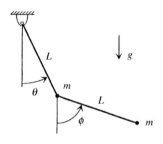

FIGURE P4.61

4.62. Consider the double pendulum shown in Fig. P4.62. The system consists of two identical uniform rods of mass m and length L. The hinged joints constrain the system in planar motion. The friction is negligible, so the system can be assumed frictionless. Obtain the general nonlinear differential equations of motion.

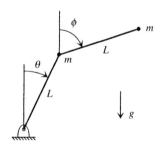

FIGURE P4.62

4.63. Consider the double pendulum shown in Fig. P4.63. The system consists of two uniform rods of masses m_1 and m_2 and lengths L_1 and L_2. The hinged joints constrain the system in planar motion. The friction is negligible, so the system can be assumed frictionless.
(a) Using Lagrange's equations, derive the general nonlinear differential equations of motion.
(b) Obtain the linearized differential equations for small motions, and then express the results in the second-order matrix form.

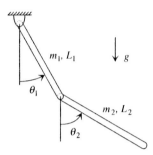

FIGURE P4.63

4.64. Consider the system shown in Fig. P4.64. The rod is uniform of mass m and length L.
(a) Using Lagrange's equations, derive the general nonlinear differential equations of motion.
(b) Obtain the linearized differential equations for small motions, and then express the results in the second-order matrix form.

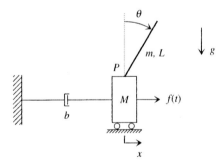

FIGURE P4.64

4.65. Consider the system shown in Fig. P4.65. The rod is uniform of mass m and length L. A point mass m_p is attached at the tip of the rod.
(a) Using Lagrange's equations, derive the general nonlinear differential equations of motion.
(b) Obtain the linearized differential equations for small motions, and then express the results in the second-order matrix form.

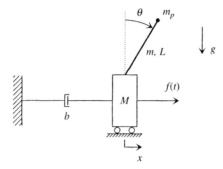

FIGURE P4.65

SUMMARY

The following are essential in the modeling of mechanical systems: Newton's second law is for translational systems, whereas the moment equations are for rotational systems. Both Newton's second law and the moment equations are used together for combined (mixed or hybrid) translational and rotational systems. Both the free-body diagram and acceleration analysis play a vital role in this fundamental method.

The most general vector form of Newton's second law, for a *system* of N rigid bodies (or particles), is given as

$$\sum_{i=1}^{N} \mathbf{F}_i = m\mathbf{a}_c = \sum_{j=1}^{n} m_j \mathbf{a}_{cj} = m_1 \mathbf{a}_{c1} + m_2 \mathbf{a}_{c2} + \cdots + m_n \mathbf{a}_{cn}$$

where

$$m = \sum_{j=1}^{n} m_j$$

$$\xrightarrow{+} \sum_{i=1}^{N} F_{ix} = ma_{cx} = \sum_{j=1}^{n} m_j a_{cjx} = m_1 a_{c1x} + m_2 a_{c2x} + \cdots + m_n a_{cnx} \xrightarrow{+}$$

$$+\uparrow \sum_{i=1}^{N} F_{iy} = ma_{cy} = \sum_{j=1}^{n} m_j a_{cjy} = m_1 a_{c1y} + m_2 a_{c2y} + \cdots + m_n a_{cny} \uparrow +$$

of which the positive x and y are horizontal to the right and vertically up, respectively, in a rectangular (cartesian) coordinate system.

For the type of mass-spring system shown in Fig. 4.62, if the displacement is measured from the static equilibrium, the gravity term mg does not enter into the governing differential equation. For a *single* rigid body in planar motion, which is subjected to N applied resulting moments, the moment equation about an arbitrary point P fixed in the body is given as

$$\sum_{i}^{N} \mathbf{M}_{pi} = I_p \alpha + m\mathbf{r}_{c/p} \times \mathbf{a}_p$$

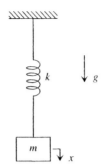

FIGURE 4.62 Gravity and mass-spring system.

where the *parallel-axis theorem* gives
$$I_p = I_c + md^2$$
For special points such as a point O that is fixed in space *and* fixed in the body or the mass center c, the moment equation reduces to
$$\sum M_o = I_o \alpha$$
$$\sum M_c = I_c \alpha$$

The general moment equation for a general rigid body containing n separated masses, in planar motion, is expressed as
$$\sum_{i=1}^{N} \mathbf{M}_{pi} = \left(\sum_{j=1}^{n} I_{pj} \right) \alpha + \sum_{j=1}^{n} \left(m_j \mathbf{r}_{cj/p} \times \mathbf{a}_p \right)$$
where
$$I_p = \sum_{j=1}^{n} I_{cj} + m_j d_j^2 = (I_{c1} + m_1 d_1^2) + (I_{c2} + m_2 d_2^2) + \cdots + (I_{cn} + m_n d_n^2)$$

In planar motion, the centroidal mass moment of inertia of a uniform circular disk of mass m and radius r is
$$I_c = \tfrac{1}{2} mr^2$$
and that of a uniform rod of mass m and radius L is
$$I_c = \tfrac{1}{12} mL^2$$

As a simple yet effective method, the skeleton approach can be used to obtain the governing differential equations for MDOF systems. The equations can be obtained quickly using this method, which is particularly useful for spring-mass-damper systems of higher order.

Lagrange's equations are another method of obtaining the governing differential equations. In contrast to the force-moment approach with Newton's second law and the moment equation, Lagrange's equations are an energy method, which does not

PROBLEMS

4.66. Using Lagrange's equations, derive the differential equations for the system shown in Fig. P4.44 (Problem 4.44).

4.67. Obtain the differential equations, in the second-order matrix form, for the system shown in Fig. P4.67.

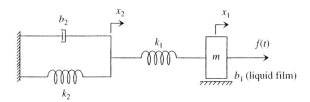

FIGURE P4.67

4.68. Obtain the differential equations, in the second-order matrix form, for the system shown in Fig. P4.68.

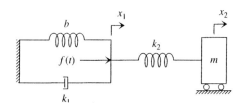

FIGURE P4.68

4.69. Obtain the differential equations, in the second-order matrix form, for the system shown in Fig. P4.69.

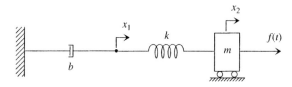

FIGURE P4.69

4.70. Consider the system shown in Fig. P4.70. Determine
(a) The differential equations in the second-order matrix form
(b) The transfer function $X_3(s)/F(s)$.

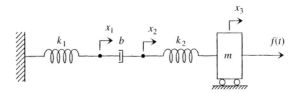

FIGURE P4.70

4.71. Consider the system shown in Fig. P4.71 where $y(t)$ and $f(t)$ are displacement input and force input, respectively. Obtain the differential equations in the second-order matrix form.

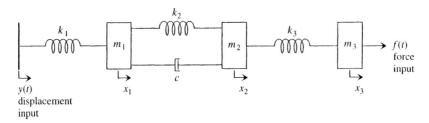

FIGURE P4.71

CHAPTER 5

Electrical, Electronic, and Electromechanical Systems

Electrical, electronic, and electromechanical systems are essential components of many dynamic systems. Kirchhoff's laws are important in electrical systems. Kirchhoff's voltage law (KVL) is used for the loop method, whereas Kirchhoff's current law (KCL) is used for the node method. Operational amplifiers (op-amps) are versatile electronic components whose voltage gains are extremely high, typically of 10^5 to 10^6. Last, but probably most important, electromechanical systems are combinations of electrical and mechanical parts. Motors and sensors are examples of such systems.

5.1 INTRODUCTION

Electrical systems, electronic systems, and electromechanical systems are essential subsystems of many complex and modern dynamic systems. A computer's disk drive and an automobile's cruise control serve as examples.

In this chapter, the fundamentals and modeling of electrical, electronic, and electromechanical systems are presented. Simple operational amplifiers (op-amps) are discussed as representatives of electronic systems. Because the concept of impedance is sometimes useful, electrical impedance and mechanical impedance are also covered. Examples are given to illustrate the principles and methods.

5.2 ELECTRICAL SYSTEMS

Electrical systems, or electrical circuits, play a critical role in dynamic systems. For example, a disk drive requires certain electrical circuits to function. The terms *electrical system* and *electrical circuits* are used interchangeably here. Electrical circuits have become essential in modern life.

In this section, we will cover the theory and applications of modeling electrical systems. Topics include electrical elements, Kirchhoff's voltage law (KVL), and Kirchhoff's current law (KCL).

Passive Electrical Elements

To model an electrical system, we must know its electrical elements. There are three primary types of passive electrical elements: resistor, inductor, and capacitor. The *elemental relations* of *voltages* are as follows:

Resistor

$$v_R = R i_R \quad (5.1a)$$

Inductor

$$v_L = L \frac{di_L}{dt} \quad (5.1b)$$

Capacitor

$$v_C = \frac{1}{C} \int i_C \, dt \quad (5.1c)$$

where R, L, and C are the resistance, inductance, and capacitance, respectively. For convenience, the elemental relations for *currents* are as follows:

Resistor $\qquad i_R = \dfrac{v_R}{R} \qquad (5.2a)$

Inductor $\qquad i_L = \dfrac{1}{L} \int v_L \, dt \qquad (5.2b)$

Capacitor $\qquad i_C = C \dfrac{dv_c}{dt} \qquad (5.2c)$

Kirchhoff's Laws

The two most useful physical laws for modeling electrical systems are Kirchhoff's laws. To apply these two laws effectively, a sign convention should be employed.

KIRCHHOFF'S CURRENT LAW (KCL), NODE METHOD. The algebraic sum of all currents entering (*in*) a circuit **node** is zero:

$$\sum_k (i_k)_{in} = 0 \tag{5.3}$$

Examples are shown in Fig. 5.1. We have

Figure 5.1*a*: $\quad i_1 + i_2 + i_3 = 0$ (5.4*a*)

Figure 5.1*b*: $\quad -(i_1 + i_2 + i_3) = 0$ (5.4*b*)

Figure 5.1*c*: $\quad i_1 + i_2 - i_3 = 0$ (5.4*c*)

KIRCHHOFF'S VOLTAGE LAW (KVL), LOOP METHOD. The sum of all voltage drops around a circuit **loop** is zero:

$$\sum v_{drop} = 0 \tag{5.5}$$

Although the KVL can also be stated with voltage gain as

$$\sum v_{gain} = 0 \tag{5.6}$$

it is simply inconvenient to apply in practice.

As an example, let us consider the electrical circuit shown in Fig. 5.2, which shows the recommended sign convention for the loop method. Within each element, the current flows from the high-voltage terminal to the low-voltage end. Across each element, the voltage drop has the same direction as the current flow. These arrow directions are logical for both current flow and voltage drops. For this loop we have

$$v_L + v_R + v_C - v_a = 0 \tag{5.7}$$

As with mechanical systems, a sign convention is needed for the modeling of electrical systems. Without a clear sign convention, it is difficult to keep track of currents and voltages. For node analysis, we need to show whether the currents enter or leave the node. Either one may be used depending on the choice of the analyst. For loop analysis, the voltage drops should be used because it is more convenient

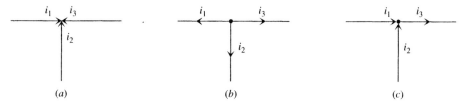

FIGURE 5.1 Sign convention for the node method: (*a*) all currents entering the node, (*b*) all currents leaving the node, (*c*) mixed.

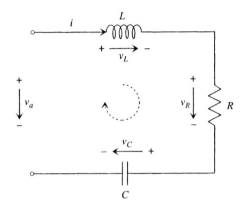

FIGURE 5.2 Sign convention for the loop method.

than the voltage gains. The following examples show the recommended useful sign convention.

EXAMPLE 5.1. Consider the simple RLC electrical system shown in Fig. 5.3. The system consists of a resistor of resistance R, an inductor of inductance L, a capacitor of capacitance C, and an applied (input) voltage v_a. Determine

(a) The differential equation
(b) The natural frequency and damping ratio
(c) The transfer function $V_c(s)/V_a(s)$, which relates the input voltage $v_a(t)$ to the voltage drop across the capacitor, $v_c(t)$

Solution

(a) The system is governed by one differential equation, because it has one input, $v_a(t)$, and one output, $i(t)$, as the independent current. Thus, we will apply KVL and the loop method. The sign convention is shown in Fig. 5.2. Applying KVL,

$$\sum v_{\text{drop}} = 0$$

$$v_L + v_R + v_C - v_a = 0$$

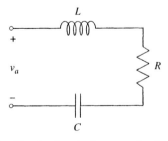

FIGURE 5.3 Simple electrical system for the loop method.

Introducing the elemental relations, $v_L = L(di_L/dt)$, $v_R = Ri_R$, $v_C = \frac{1}{C}\int i_C\, dt$, into the equation, and dropping the subscripts (because all the currents are the same), $i_L = i_R = i_C = i$.

$$L\frac{di}{dt} + Ri + \frac{1}{C}\int_0^t i\, dt = v_a$$

Notice that the above equation is not a differential equation; rather, it is an integro-differential (or integral-differential) equation.

Taking the time derivative of both sides of the equation to eliminate the integral term, we obtain

$$L\frac{d^2i}{dt^2} + R\frac{di}{dt} + \frac{1}{C}i = \frac{dv_a}{dt}$$

(b) For the purpose of getting the natural frequency and damping ratio, the equation is divided by L as

$$\frac{d^2i}{dt^2} + \frac{R}{L}\frac{di}{dt} + \frac{1}{LC}i = \frac{1}{L}\frac{dv_a}{dt}$$

Thus,

$$\omega_n = \frac{1}{\sqrt{LC}}$$

$$\zeta = \frac{R}{L2\omega_n} = \frac{1}{2}\sqrt{\frac{R^2C}{L}}$$

A check of dimension (or units) proved to be useful here. From Appendix A, the units of the inductance L, (electrical) capacitance C, and (electrical) resistance R are given as

$$[R] = \Omega = V/A$$
$$[L] = H = V \cdot s/A$$
$$[C] = F = A \cdot s/V$$

where the brackets represent "unit of" or "dimension of." Thus,

$$\frac{1}{[LC]} = \frac{1}{s^2} \qquad \frac{[R^2][C]}{[L]} = 1$$

Because rad is dimensionless, ω_n has a dimension of rad/s, and ζ is dimensionless.

(c) Taking the Laplace transform of the preceding differential equation and letting all the initial conditions be zero, the resulting transformed algebraic equation is

$$\left(s^2 + \frac{R}{L}s + \frac{1}{LC}\right)I(s) = \frac{s}{L}V_a(s)$$

The transfer function is

$$\frac{I(s)}{V_a(s)} = \frac{s}{L[s^2 + (R/L)s + 1/(LC)]}$$

Cross-multiplying,

$$(LCs^2 + RCs + 1)I(s) = CsV_a(s)$$

The fundamental relation for a capacitor is

$$v_C = \frac{1}{C}\int i_C\, dt$$

so the transfer function relating the input voltage to the voltage drop across the capacitor is

$$\frac{V_c(s)}{V_a(s)} = \frac{1}{Cs}\frac{I(s)}{V_a(s)} = \frac{1}{LC[s^2 + (R/L)s + 1/(LC)]}$$

EXAMPLE 5.2. Consider the electrical system shown in Fig. 5.4. Determine

(a) The differential equation, natural frequency, and damping ratio
(b) The transfer function $V_L(s)/V_a(s)$, which relates the input voltage $v_a(t)$ to the voltage measured across the inductor L, $v_L(t)$
(c) The transfer function $V_R(s)/V_a(s)$, which relates the input voltage $v_a(t)$ to the voltage measured across the resistor R, $v_R(t)$

Solution

(a) The system has one input, $v_a(t)$, and only one output, $v_1(t)$, as the independent voltage. We need only one node for one differential equation. This node, node 1, involves all the elements of the circuit. The node and the directions of the current flow are shown in Fig. 5.5. In general, we can either use KCL (node method) or KVL (loop method). For this problem we will use KCL because it is more convenient (one node for the governing differential equation).

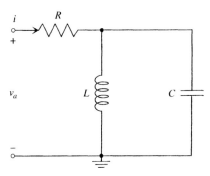

FIGURE 5.4 Electrical circuit for the node method.

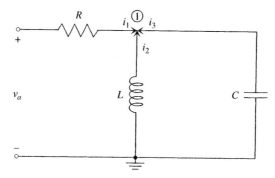

FIGURE 5.5 Sign convention for the node method.

At node 1,

$$\sum i_{in} = 0$$

$$i_1 + i_2 + i_3 = 0$$

$$\frac{v_a - v_1}{R} + \frac{1}{L}\int (0 - v_1)\,dt + C\frac{d}{dt}(0 - v_1) = 0$$

Differentiating the above equation with respect to time, we obtain

$$C\ddot{v}_1 + \frac{1}{R}\dot{v}_1 + \frac{1}{L}v_1 = \frac{1}{R}\dot{v}_a$$

or

$$\ddot{v}_1 + \frac{1}{RC}\dot{v}_1 + \frac{1}{LC}v_1 = \frac{1}{RC}\dot{v}_a$$

where

$$\omega_n = \frac{1}{\sqrt{LC}}$$

$$\zeta = \frac{1}{RC2\omega_n} = \frac{1}{2}\sqrt{\frac{L}{R^2 C}}$$

A check in dimension (or units) is in order here: ω_n has a dimension of rad/s and ζ is dimensionless.

(b) The transfer function is given as

$$\frac{V_L(s)}{V_a(s)} = \frac{V_1(s)}{V_a(s)} = \frac{s}{RC[s^2 + 1/(RC)s + 1/(LC)]}$$

(c) Because

$$v_R(t) = v_a(t) - v_1(t)$$

we have

$$\frac{V_R(s)}{V_a(s)} = 1 - \frac{V_1(s)}{V_a(s)} = 1 - \frac{s}{RC[s^2 + 1/(RC)s + 1/(LC)]}$$

EXAMPLE 5.3. Consider the electrical system shown in Fig. 5.6. Determine

(a) The system differential equations
(b) The transfer functions: $V_{L1}(s)/V_a(s)$, $V_c(s)/V_a(s)$, $I_1(s)/V_a(s)$

Solution

(a) The nodes including the current flows (going into the nodes) are shown in Fig. 5.7. The system has one input, $v_a(t)$, and two outputs, $v_1(t)$ and $v_2(t)$, as the independent voltages.

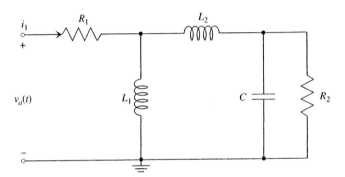

FIGURE 5.6 Electrical circuit for the node method.

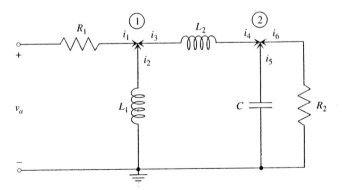

FIGURE 5.7 Circuit modeled with the node method.

CHAPTER 5: Electrical, Electronic, and Electromechanical Systems   273

For this problem we will use KCL for two nodes, nodes 1 and 2, to obtain the two differential equations for the system. At node 1,

$$\sum i_{in} = 0$$

$$i_1 + i_2 + i_3 = 0$$

$$\frac{v_a - v_1}{R_1} + \frac{1}{L_1}\int(0 - v_1)dt + \frac{1}{L_2}\int(v_2 - v_1)dt = 0$$

Differentiating the above equation with respect to time,

$$\frac{\dot{v}_a}{R_1} - \frac{\dot{v}_1}{R_1} - \frac{v_1}{L_1} + \frac{v_2}{L_2} - \frac{v_1}{L_2} = 0$$

Thus, the first differential equation in standard input-output form is

$$\frac{\dot{v}_1}{R_1} + \left(\frac{1}{L_1} + \frac{1}{L_2}\right)v_1 - \frac{v_2}{L_2} = \frac{\dot{v}_a}{R_1}$$

At node 2,

$$\sum i_{in} = 0$$

$$i_4 + i_5 + i_6 = 0$$

$$\frac{1}{L_2}\int(v_1 - v_2)dt + C\frac{d}{dt}(0 - v_2) + \frac{0 - v_2}{R_2} = 0$$

Differentiating the above equation with respect to time,

$$\frac{v_1 - v_2}{L_2} - C\ddot{v}_2 - \frac{\dot{v}_2}{R_2} = 0$$

Thus, the second differential equation in standard input-output form is

$$C\ddot{v}_2 + \frac{\dot{v}_2}{R_2} - \frac{v_1}{L_2} + \frac{v_2}{L_2} = 0$$

For checking and organizational purposes, the system equations are expressed in the second-order matrix form as

$$\begin{bmatrix} 0 & 0 \\ 0 & C \end{bmatrix}\begin{Bmatrix} \ddot{v}_1 \\ \ddot{v}_2 \end{Bmatrix} + \begin{bmatrix} 1/R_1 & 0 \\ 0 & 1/R_2 \end{bmatrix}\begin{Bmatrix} \dot{v}_1 \\ \dot{v}_2 \end{Bmatrix} + \begin{bmatrix} 1/L_1 + 1/L_2 & -1/L_2 \\ -1/L_2 & 1/L_2 \end{bmatrix}\begin{Bmatrix} v_1 \\ v_2 \end{Bmatrix} = \begin{Bmatrix} \dot{v}_a/R_1 \\ 0 \end{Bmatrix}$$

After taking the Laplace transform, with zero initial conditions, the terms are collected as

$$\begin{bmatrix} s/R_1 + 1/L_1 + 1/L_2 & -1/L_2 \\ -1/L_2 & Cs^2 + s/R_2 + 1/L_2 \end{bmatrix}\begin{Bmatrix} V_1(s) \\ V_2(s) \end{Bmatrix} = \begin{Bmatrix} sV_a(s)/R_1 \\ 0 \end{Bmatrix}$$

To ease the algebra, we need to eliminate the denominator terms. Multiplying the top and bottom of the preceding equation by $R_1L_1L_2$ and R_2L_2, respectively, we obtain

$$\begin{bmatrix} L_1L_2s + R_1(L_1+L_2) & -R_1L_1 \\ -R_2 & (R_2Cs+1)L_2s + R_2 \end{bmatrix} \begin{Bmatrix} V_1(s) \\ V_2(s) \end{Bmatrix} = \begin{Bmatrix} L_1L_2sV_a(s) \\ 0 \end{Bmatrix}$$

Applying Cramer's rule,

$$V_1(s) = \frac{1}{\Delta(s)} \begin{vmatrix} L_1L_2sV_a(s) & -R_1L \\ 0 & (R_2Cs+1)L_2s + R_2 \end{vmatrix}$$

$$= \frac{[(R_2Cs+1)L_2s + R_2]L_1L_2sV_a(s)}{\Delta(s)}$$

$$V_2(s) = \frac{1}{\Delta(s)} \begin{vmatrix} L_1L_2s + R_1(L_1+L_2) & L_1L_2sV_a(s) \\ -R_2 & 0 \end{vmatrix} = \frac{R_2L_1L_2sV_a(s)}{\Delta(s)}$$

where the characteristic polynomial is given as

$$\Delta(s) = [L_1L_2s + R_1(L_1+L_2)][(R_2Cs+1)L_2s + R_2] - R_1R_2L_1$$

or after expanding and collecting terms,

$$\Delta(s) = L_1L_2R_2Cs^3 + [L_1L_2 + R_1(L_1+L_2)R_2C]s^2$$
$$+ [R_1(L_1+L_2) + R_2L_1]s + R_1R_2$$

Thus, we have the transfer functions

$$\frac{V_1(s)}{V_a(s)} = \frac{[(R_2Cs+1)L_2s + R_2]L_1L_2s}{\Delta(s)}$$

$$\frac{V_2(s)}{V_a(s)} = \frac{R_2L_1L_2s}{\Delta(s)}$$

The transfer function $I_1(s)V_a(s)$ is obtained as

$$I_1(s) = \frac{V_a(s) - V_1(s)}{R_1} = \frac{V_a(s)}{R_1}\left[1 - \frac{V_1(s)}{V_a(s)}\right]$$

After a simple algebra, the transfer function is given as

$$\frac{I_1(s)}{V_a(s)} =$$

$$\frac{(L_1+L_2)R_2Cs^2 + (L_1+L_2)s + R_2}{L_1L_2R_2Cs^3 + [L_1L_2 + R_1(L_1+L_2)R_2C]s^2 + [R_1(L_1+L_2) + R_2L_1]s + R_1R_2}$$

where the numerator and the denominator are expressed as polynomials in s.

Remarks

- The denominator of the transfer function, or characteristic polynomial, serves as a check that the system is third order. This result is expected because the first and second differential equations of the system are first and second orders, respectively.

- The system differential equations are coupled; thus, they must be solved simultaneously as shown. Cramer's rule is systematic and less prone to mistakes; therefore, it is better than the substitution method.

5.3 ELECTRONIC SYSTEMS: OPERATIONAL AMPLIFIERS

An operational amplifier, abbreviated as **op-amp**, is an *electronic* amplifier with single output and very high voltage gain (typically 10^5 to 10^6). Op-amps are available in hundreds of different types and have many uses in electronic circuits. Because op-amps can only handle very small current, their use is limited to low-current applications.

Consider the schematic diagram of an op-amp shown in Fig. 5.8. This particular inverted configuration is common in practice. The op-amp equation is

$$v_o = K(v_+ - v_-) \tag{5.8}$$

where the gain K is a large positive number. Because v_o is small and finite and K is very large, we must have

$$v_+ - v_- \approx 0 \tag{5.9a}$$

or

$$v_+ \approx v_- \tag{5.9b}$$

This equation may be considered as the **op-amp equation.**

EXAMPLE 5.4. Consider the op-amp circuit shown in Fig. 5.9. Determine

(a) The governing differential equation
(b) The time constant
(c) The transfer function $V_o(s)/V_a(s)$

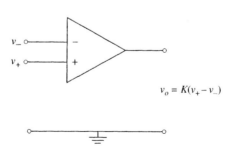

FIGURE 5.8 Operational amplifier in the inverted configuration; $K = 10^5$ to 10^6.

276 Dynamic Systems: Modeling and Analysis

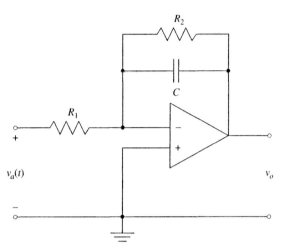

FIGURE 5.9 Op-amp circuit.

Solution

(a) The nodes and the current flows are shown in Fig. 5.10. The system has one input, $v_a(t)$, and two outputs, v_1 and v_o, as the independent voltages. The system has only one significant node: node 1. Applying KCL at node 1,

$$\sum i_{in} = 0$$

$$i_1 + i_2 = 0$$

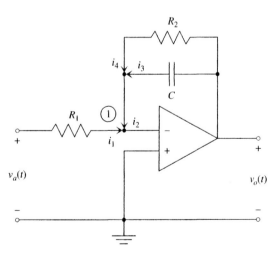

FIGURE 5.10 Op-amp circuit.

Because $i_2 = i_3 + i_4$, we have
$$i_1 + i_3 + i_4 = 0$$
$$\frac{v_a - v_1}{R_1} + C\frac{d}{dt}(v_o - v_1) + \frac{v_o - v_1}{R_2} = 0$$
From the op-amp equation, we have $v_+ \approx v_-$. Because $v_+ = v_{ground} = 0$,
$$v_- = v_1 \approx 0$$
Thus the differential equation becomes
$$C\dot{v}_o + \frac{v_o}{R_2} = -\frac{v_a}{R_1}$$

(b) Multiplying both sides of the preceding equation by R_2,
$$R_2 C \dot{v}_o + v_o = -\frac{R_2}{R_1} v_a$$
where the time constant τ is given readily as
$$\tau = R_2 C$$

(c) After taking the Laplace transforms, and setting all the initial conditions to zero, the terms are collected as
$$(R_2 C s + 1)V_o(s) = -\frac{R_2}{R_1} V_a(s)$$
Thus, the transfer function is
$$\frac{V_o(s)}{V_a(s)} = -\frac{R_2/R_1}{R_2 C s + 1}$$
Notice that after v_1 is eliminated by using the op-amp equation, the system is governed by only one differential equation in v_o.

PROBLEM SET 5.1–5.3

5.1. Consider the electrical system shown in Fig. P5.1, with the applied voltage v_a as the input. Determine the differential equation.

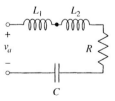

FIGURE P5.1

5.2. Using the differential equation obtained from Problem 5.1, obtain the state-space representation assuming the output is the current.

5.3. Consider the electrical system shown in Fig. P5.3, with the applied voltage v_a as the input. Determine
(a) the differential equation

and the following transfer functions:

(b) $I_L(s)/V_a(s)$
(c) $V_L(s)/V_a(s)$
(d) $V_C(s)/V_a(s)$

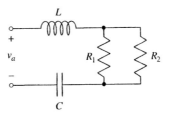

FIGURE P5.3

5.4. Consider the electrical system shown in Fig. P5.4, with the applied voltage v_a as the input.
(a) Determine the total number of independent differential equations.
(b) Derive the differential equations.
(c) Express the differential equations in the second-order matrix form.
(d) Obtain the transfer function relating the input voltage to the output voltage across the capacitor, $V_C(s)/V_a(s)$.

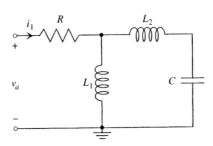

FIGURE P5.4

5.5. Consider the electrical system shown in Fig. P5.5 with the applied voltage v_i as the input.
(a) Derive the differential equations.
(b) Determine all the transfer functions relating the input voltage to the output voltages.

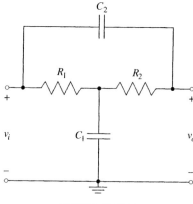

FIGURE P5.5

5.6. Consider the electrical system shown in Fig. P5.6, with the applied voltage v_a as the input.
  (a) Obtain the differential equation.
  (b) Determine the transfer function relating the input voltage v_a to the output voltage v_o.

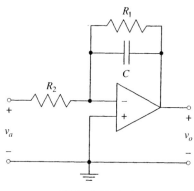

FIGURE P5.6

5.4 ELECTROMECHANICAL SYSTEMS

DC (direct current) motors are essential in control systems. In general, from the analysis, supported by experiment, we can develop a function of the torque in terms of voltage and speed. From the curves, linearized relations can be obtained.

Two primary types of DC motors are **armature-controlled DC motors** and **field-controlled DC motors.** In this section, we shall obtain mathematical models of these machines. Of the two types, armature-controlled DC motors are more popular.

Elemental Relations of Electromechanical Systems

In electromechanical systems, the electrical and mechanical parts are coupled. The mechanical motion of the rotor relative to the stator affects the electromotive-force (emf) voltage developed within the motor. There are two critical elemental relations for electromechanical systems. In general, the back emf voltage e_b across a DC motor is proportional to the angular speed of the motor rotor (the rotor of the motor),

$$e_b = K_e \omega = K_e \dot{\theta} \qquad (5.10)$$

and, similarly, the torque developed by the motor is proportional to the current

$$T = K_t i \qquad (5.11)$$

where e_b = back emf voltage
θ = angular displacement of the rotor (of the motor), $\dot{\theta} = \omega$
T = torque applied to the rotor (developed by the motor)
K_e = emf constant of the motor, V/Krpm (V = volt, 1 Krpm = 1000 rpm)
K_t = torque constant of the motor, in.-oz$_f$/A (A = ampere)

The two constants K_e and K_t may appear different, but they are actually the same. They are often given with different numerical values in engineering units. For convenience, the values of these two constants are both listed in the specifications sheet of a motor. However, they are exactly the same, having the same numerical value if expressed in a consistent set of units system such as MKS (m, kg, and s) or CGS (cm, g, and s). It is analogous to saying that 1 in. = 2.54 cm. Knowing one constant, we can obtain the other by using the following conversion factor

$$1 \frac{\text{in.-oz}_f}{A} = 0.741 \frac{V}{\text{Krpm}} \qquad (5.12)$$

Armature-Controlled DC Motors

Consider the armature-controlled DC motor shown in Fig. 5.11. The electromechanical system consists of two parts: electrical and mechanical. We will treat this prob-

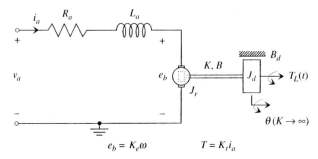

FIGURE 5.11 System with an armature-controlled DC motor.

lem with two different models for the shaft. The shaft is assumed massless for both cases. In the first case, as a simple model, the shaft is assumed to be rigid. However, in the second case, which is more complex, the shaft is modeled as a torsional spring and a torsional viscous damper in parallel to represent a physical model of a flexible and damped shaft. In this system,

B = coefficient of torsional viscous damping of the shaft
B_d = coefficient of torsional viscous damping associated with the disk
i_a = armature current
i_f = field current (not shown, constant)
J_d = mass moment of inertia of the disk
J_r = mass moment of inertia of the rotor (of the motor)
K = torsional stiffness of the shaft
L_a = armature inductance
R_a = armature resistance
v_a = armature voltage (input)

A simple model: rigid shaft

Because the shaft is rigid ($K \to \infty$) and massless, we can lump the rotor, the shaft, and the disk together as a single rigid body that rotates at the same speed, ω. The combined mass moment of inertia is

$$J = J_r + J_d \tag{5.13}$$

The system has one input, v_a, and two outputs, i_a and θ, as the independent variables (see Fig. 5.11). Thus, the system is governed by two independent differential equations.

For the electrical circuit, we apply the KVL method (loop analysis),

$$\sum v_{\text{drop}} = 0 \tag{5.14}$$

$$v_{Ra} + v_{La} + e_b - v_a = 0$$

Applying the elemental electrical relations, we have

$$R_a i_a + L_a \frac{di_a}{dt} + e_b = v_a \tag{5.15}$$

Introducing the relation $e_b = K_e \omega = K_e \dot{\theta}$,

$$L_a \frac{di_a}{dt} + R_a i_a + K_e \dot{\theta} = v_a \quad (i_a\text{-equation}) \tag{5.16}$$

For the mechanical part, the free-body diagram is shown in Fig. 5.12. Applying the moment equation about the center of mass along the longitudinal axis,

$$+\circlearrowright \sum M_c = I_c \alpha \circlearrowright +$$

$$T + T_L - B_d \dot{\theta} = J \ddot{\theta} \tag{5.17}$$

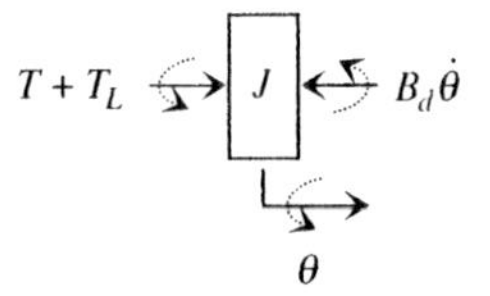

FIGURE 5.12 Free-body diagram.

Introducing the motor torque relation

$$T = K_t i_a$$

and rearranging the differential equation in the standard input-output form,

$$J\ddot{\theta} + B_d\dot{\theta} - K_t i_a = T_L \quad (\theta\text{-equation}) \tag{5.18}$$

Note that the current is *not* the input to the *system* although it is the input to the mechanical part.

Finally, the system is represented by the system equations as

$$J\ddot{\theta} + B_d\dot{\theta} - K_t i_a = T_L(t)$$

$$L_a\frac{di_a}{dt} + R_a i_a + K_e\dot{\theta} = v_a(t) \tag{5.19}$$

and in the second-order matrix form as

$$\begin{bmatrix} J & 0 \\ 0 & 0 \end{bmatrix}\begin{Bmatrix} \ddot{\theta} \\ \ddot{i}_a \end{Bmatrix} + \begin{bmatrix} B_d & 0 \\ K_e & L_a \end{bmatrix}\begin{Bmatrix} \dot{\theta} \\ \dot{i}_a \end{Bmatrix} + \begin{bmatrix} 0 & -K_t \\ 0 & R_a \end{bmatrix}\begin{Bmatrix} \theta \\ i_a \end{Bmatrix} = \begin{Bmatrix} T_L \\ v_a \end{Bmatrix} \tag{5.20}$$

Notice that the top equation is chosen to be the mechanical θ-equation, but it can be the electrical i_a-equation if so desired.

Because this model does not contain the stiffness term, the equations are better expressed in terms of ω instead of θ. After introducing $\omega = \dot{\theta}$, the system equations become

$$J\dot{\omega} + B_d\omega - K_t i_a = T_L(t)$$

$$L_a\frac{di_a}{dt} + R_a i_a + K_e\omega = v_a(t) \tag{5.21}$$

and in the matrix form,

$$\begin{bmatrix} J & 0 \\ 0 & L_a \end{bmatrix}\begin{Bmatrix} \dot{\omega} \\ \dot{i}_a \end{Bmatrix} + \begin{bmatrix} B_d & -K_t \\ K_e & R_a \end{bmatrix}\begin{Bmatrix} \omega \\ i_a \end{Bmatrix} = \begin{Bmatrix} T_l \\ v_a \end{Bmatrix} \tag{5.22}$$

We now proceed to determine the transfer functions. Because the system is second-ordered, we will also obtain the natural frequency and the damping ratio. Taking the Laplace transform of the matrix equation with all the initial conditions set to zero, then the terms are collected as

$$\begin{bmatrix} Js + B_d & -K_t \\ K_e & L_a s + R_a \end{bmatrix}\begin{Bmatrix} \Omega(s) \\ I_a(s) \end{Bmatrix} = \begin{Bmatrix} T_L(s) \\ V_a(s) \end{Bmatrix} \tag{5.23}$$

Using Cramer's rule, the solutions are

$$\Omega(s) = \frac{1}{\Delta(s)} \begin{vmatrix} T_L(s) & -K_t \\ V_a(s) & L_a s + R_a \end{vmatrix} = \frac{L_a s + R_a}{\Delta(s)} T_L(s) + \frac{K_t}{\Delta(s)} V_a(s) \quad (5.24)$$

$$I_a(s) = \frac{1}{\Delta(s)} \begin{vmatrix} Js + B_d & T_L(s) \\ K_e & V_a(s) \end{vmatrix} = -\frac{K_e}{\Delta(s)} T_L(s) + \frac{Js + B_d}{\Delta(s)} V_a(s) \quad (5.25)$$

where

$$\Delta(s) = (Js + B_d)(L_a s + R_a) + K_e K_i \quad (5.26)$$

After some simple algebra,

$$\Delta(s) = JL_a s^2 + (B_d L_a + JR_a)s + B_d R_a + K_e K_t \quad (5.27)$$

Thus, the transfer functions concerning the output angular speed are

$$G_{11}(s) = \left.\frac{\Omega(s)}{T_L(s)}\right|_{V_a(s)=0} = \frac{L_a s + R_a}{\Delta(s)}, \quad G_{12}(s) = \left.\frac{\Omega(s)}{V_a(s)}\right|_{T_L(s)=0} = \frac{K_t}{\Delta(s)} \quad (5.28)$$

and the transfer functions for the output armature current are

$$G_{21}(s) = \left.\frac{I_a(s)}{T_L(s)}\right|_{V_a(s)=0} = -\frac{K_e}{\Delta(s)}, \quad G_{22}(s) = \left.\frac{I_a(s)}{V_a(s)}\right|_{T_L(s)=0} = \frac{Js + B_d}{\Delta(s)} \quad (5.29)$$

The corresponding transfer matrix is

$$\mathbf{G}(s) = \begin{bmatrix} G_{11}(s) & G_{12}(s) \\ G_{21}(s) & G_{22}(s) \end{bmatrix} \quad (5.30)$$

From the superposition principle (for linear systems), we have

$$\Omega(s) = \left.\frac{\Omega(s)}{T_L(s)}\right|_{V_a(s)=0} T_L(s) + \left.\frac{\Omega(s)}{V_a(s)}\right|_{T_L(s)=0} V_a(s) \quad (5.31)$$

$$I_a(s) = \left.\frac{I_a(s)}{T_L(s)}\right|_{V_a(s)=0} T_L(s) + \left.\frac{I_a(s)}{V_a(s)}\right|_{T_L(s)=0} V_a(s) \quad (5.32)$$

The *characteristic polynomial* $\Delta(s)$ is the denominator of any transfer function of the system. All transfer functions of the system must have the same characteristic polynomial. It is the characteristic of the system. The *characteristic equation* is simply $\Delta(s) = 0$. For a second-order system, the two important parameters are natural frequency ω_n and damping ratio ζ. For this system, the characteristic equation is

$$JL_a s^2 + (B_d L_a + JR_a)s + B_d R_a + K_e K_t = 0 \quad (5.33)$$

Because the characteristic equation of a second-order system in standard form is

$$s^2 + 2\zeta\omega_n s + \omega_n^2 = 0 \quad (5.34)$$

it is best to manipulate this characteristic equation so that its coefficient of the s^2-term is unity (one):

$$s^2 + \frac{B_d L_a + J R_a}{J L_a} s + \frac{B_d R_a + K_e K_t}{J L_a} = 0 \tag{5.35}$$

Identifying the two equations terms by terms, the results are

$$\omega_n^2 = \frac{B_d R_a + K_e K_t}{J L_a} \tag{5.36}$$

$$2\zeta\omega_n = \frac{B_d L_a + J R_a}{J L_a} \tag{5.37}$$

Thus the natural frequency ω_n and the damping ratio ζ are

$$\omega_n = \sqrt{\frac{B_d R_a + K_e K_t}{J L_a}} \tag{5.38}$$

$$\zeta = \frac{B_d L_a + J R_a}{2 J L_a} \sqrt{\frac{J L_a}{B_d R_a + K_e K_t}} = \frac{B_d L_a + J R_a}{2\sqrt{J L_a(B_d R_a + K_e K_t)}} \tag{5.39}$$

The transfer function relating the input voltage to the output angular displacement, $\Theta(s)/V_a(s)$, can also be obtained, as an example. Because $\omega = \dot\theta$,

$$\frac{\Omega(s)}{V_a(s)} = \frac{s\Theta(s)}{V_a(s)} = \frac{K_t}{J L_a s^2 + (B_d L_a + J R_a)s + B_d R_a + K_e K_t} \tag{5.40}$$

Thus,

$$\frac{\Theta(s)}{V_a(s)} = \frac{K_t/J L_a}{s\left[s^2 + \dfrac{B_d L_a + J R_a}{J L_a} s + \dfrac{B_d R_a + K_e K_t}{J L_a}\right]} \tag{5.41}$$

A more complex model: flexible and damped shaft

The electromechanical system is shown in Fig. 5.13 with the shaft now modeled as flexible and damped. The angular displacements of the rotor and the disk are θ_1 and θ_2, respectively. The system has one input, v_a, and three outputs, i_a, θ_1, and θ_2, as the independent variables. Thus, the system is governed by three independent differential equations.

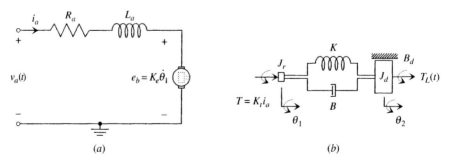

FIGURE 5.13 System with flexible and damped shaft: (*a*) electrical, (*b*) mechanical.

The electrical circuit remains essentially unchanged, except for the change in angular displacement; thus with $e_b = K_e \dot\theta_1$, we have

$$L_a \frac{di_a}{dt} + R_a i_a + K_e \dot\theta_1 = v_a \tag{5.42}$$

For the mechanical part, with $T(t) = K_t i_a$, the equations are given readily by using the skeleton approach as

$$\begin{bmatrix} J_r & 0 \\ 0 & J_d \end{bmatrix} \begin{Bmatrix} \ddot\theta_1 \\ \ddot\theta_2 \end{Bmatrix} + \begin{bmatrix} B & -B \\ -B & B+B_d \end{bmatrix} \begin{Bmatrix} \dot\theta_1 \\ \dot\theta_2 \end{Bmatrix} + \begin{bmatrix} K & -K \\ -K & K \end{bmatrix} \begin{Bmatrix} \theta_1 \\ \theta_2 \end{Bmatrix} = \begin{Bmatrix} K_t i_a \\ 0 \end{Bmatrix} \tag{5.43}$$

Arranging Eqs. (5.42) and (5.43) together in the second-order matrix form as

$$\begin{bmatrix} J_r & 0 & 0 \\ 0 & J_d & 0 \\ 0 & 0 & 0 \end{bmatrix} \begin{Bmatrix} \ddot\theta_1 \\ \ddot\theta_2 \\ \ddot i_a \end{Bmatrix} + \begin{bmatrix} B & -B & 0 \\ -B & B+B_d & 0 \\ K_e & 0 & L_a \end{bmatrix} \begin{Bmatrix} \dot\theta_1 \\ \dot\theta_2 \\ \dot i_a \end{Bmatrix}$$
$$+ \begin{bmatrix} K & -K & -K_t \\ -K & K & 0 \\ 0 & 0 & R_a \end{bmatrix} \begin{Bmatrix} \theta_1 \\ \theta_2 \\ i_a \end{Bmatrix} = \begin{Bmatrix} 0 \\ 0 \\ v_a \end{Bmatrix} \tag{5.44}$$

Taking Laplace transform of the above equation using all initial conditions set to zero, and then collecting terms,

$$\begin{bmatrix} J_r s^2 + Bs + K & -(Bs+K) & -K_t \\ -(Bs+K) & J_d s^2 + (B+B_d)s + K & 0 \\ K_e s & 0 & L_a s + R_a \end{bmatrix} \begin{Bmatrix} \Theta_1(s) \\ \Theta_2(s) \\ I_a(s) \end{Bmatrix} = \begin{Bmatrix} 0 \\ 0 \\ V_a(s) \end{Bmatrix} \tag{5.45}$$

Using Cramer's rule, we obtain

$$\Theta_2(s) = \frac{K_t(Bs+K)}{\Delta(s)} V_a(s) \tag{5.46}$$

where

$$\Delta(s) = (J_r s^2 + Bs + K)[J_d s^2 + (B+B_d)s + K](L_a s + R_a)$$
$$+ K_t K_e s[J_d s^2 + (B+B_d)s + K] - (Bs+K)^2(L_a s + R_a) \tag{5.47}$$

The transfer function is

$$\frac{\Theta_2(s)}{V_a(s)} = \frac{K_t(Bs+K)}{\Delta(s)} \tag{5.48}$$

Field-Controlled DC Motors

A simple model of a field-controlled DC motor is shown in Fig. 5.14. For the electrical part, the armature current, i_a, is constant and the voltage applied to field windings, v_f, is varied. For the mechanical part, the shaft is assumed massless, rigid, and undamped. Also, for simplicity, the disturbance input load torque, T_L, is ignored. For

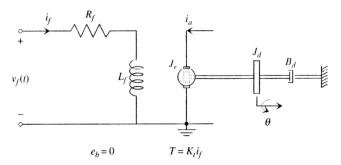

FIGURE 5.14 System with a field-controlled DC motor; the back emf is zero ($e_b = 0$).

this system,

B_d = coefficient of torsional viscous damping associated with the disk
i_a = armature current (constant)
i_f = field current
$J = J_d + J_r$ = combined mass moment of inertia of the rotor and the disk
J_d = mass moment of inertia of the disk
J_r = mass moment of inertia of the rotor (of the motor)
L_f = field inductance
R_f = field resistance
v_f = field voltage (input)

In a field-controlled DC motor, there is no back emf. In an armature-controlled DC motor, the rotor, which is a current-carrying conductor, moves in the magnetic field of the armature circuit; hence the back emf is created in its own *armature* circuit. On the other hand, in a field-controlled DC motor, the current-carrying conductor moves in the armature magnetic field; thus there is no back emf created in its *field* circuit.

In a field-controlled DC motor, the motor torque is still proportional to the current but it is the field current i_f,

$$T = K_t i_f \tag{5.49}$$

The system under consideration has one input, $v_f(t)$, and two outputs, $\omega(t)$ and $i_f(t)$, as the independent variables. For the electrical part, after applying KVL, we have

i_f-equation $\qquad\qquad R_f i_f + L_f \dfrac{di_f}{dt} = v_f \qquad\qquad$ (5.50)

Next, the mechanical part is analyzed. The free-body diagram is the same as in the case of armature-controlled DC motor. Similarly, applying the moment equation about the center of mass along the longitudinal axis,

$$+ \overrightarrow{\sum} M_c = I_c \alpha \overrightarrow{} +$$
$$T - B_d \dot{\theta} = J \ddot{\theta} \tag{5.51}$$

After introducing the motor torque relation $T = K_t i_f$ and rearranging the differential equation in the standard input-output form,

$$J\ddot{\theta} + B_d \dot{\theta} - K_t i_f = 0 \quad (\theta\text{-equation}) \tag{5.52}$$

or

$$J\dot{\omega} + B_d \omega - K_t i_f = 0 \quad (\omega\text{-equation}) \tag{5.53}$$

Finally, the system differential equations are given in the second-order matrix form as

$$\begin{bmatrix} J & 0 \\ 0 & L_f \end{bmatrix} \begin{Bmatrix} \dot{\omega} \\ \dot{i}_f \end{Bmatrix} + \begin{bmatrix} B_d & -K_t \\ 0 & R_f \end{bmatrix} \begin{Bmatrix} \omega \\ i_f \end{Bmatrix} = \begin{Bmatrix} 0 \\ v_f \end{Bmatrix} \tag{5.54}$$

Taking the Laplace transform of the matrix equation and letting all the initial conditions be zero, and then collecting terms,

$$\begin{bmatrix} Js + B_d & -K_t \\ 0 & L_f s + R_f \end{bmatrix} \begin{Bmatrix} \Omega(s) \\ I_f(s) \end{Bmatrix} = \begin{Bmatrix} 0 \\ V_f(s) \end{Bmatrix} \tag{5.55}$$

The solutions are

$$\Omega(s) = \frac{K_t V_f(s)}{\Delta(s)} \tag{5.56}$$

and

$$I_f(s) = \frac{(Js + B_d) V_f(s)}{\Delta(s)} \tag{5.57}$$

where

$$\Delta(s) = (Js + B_d)(L_f s + R_f)$$
$$\Delta(s) = J L_f s^2 + (B_d L_f + J R_f)s + B_d R_f \tag{5.58}$$

The transfer function relating the input voltage to the output angular speed is

$$G_1(s) = \frac{\Omega(s)}{V_f(s)} = \frac{K_t}{\Delta(s)} = \frac{K_t}{J L_f s^2 + (B_d L_f + J R_f)s + B_d R_f} \tag{5.59}$$

and the transfer function relating the input voltage to the output armature current is

$$G_2(s) = \frac{I_f(s)}{V_f(s)} = \frac{Js + B_d}{\Delta(s)} = \frac{Js + B_d}{J L_f s^2 + (B_d L_f + J R_f)s + B_d R_f} \tag{5.60}$$

The transfer matrix is

$$\mathbf{G}(s) = \begin{Bmatrix} G_1(s) \\ G_2(s) \end{Bmatrix} \quad (5.61)$$

For this system, the characteristic equation is

$$JL_f s^2 + (B_d L_f + JR_f)s + B_d R_f = 0 \quad (5.62)$$

Dividing both sides of the equation by JL_f so that its coefficient of the s^2-term is unity (one),

$$s^2 + \frac{B_d L_f + JR_f}{JL_f} s + \frac{B_d R_f}{JL_f} = 0 \quad (5.63)$$

Thus, the natural frequency and the damping ratio are

$$\omega_n = \sqrt{\frac{B_d R_f}{JL_f}} \quad (5.64)$$

$$\zeta = \frac{B_d L_f + JR_f}{2\omega_n JL_f} = \frac{B_d L_f + JR_f}{2\sqrt{JL_f B_d R_f}} \quad (5.65)$$

Because $\omega = \dot{\theta}$,

$$\frac{\Omega(s)}{V_f(s)} = \frac{s\Theta(s)}{V_f(s)} = \frac{K_t}{JL_f s^2 + (B_d L_f + JR_f)s + B_d R_f} \quad (5.66)$$

the transfer function relating the input voltage to the output angular displacement is

$$\frac{\Theta(s)}{V_f(s)} = \frac{K_t/JL_f}{s\left[s^2 + \dfrac{B_d L_f + JR_f}{JL_f} s + \dfrac{B_d R_f}{JL_f}\right]} \quad (5.67)$$

PROBLEM SET 5.4

5.7. The differential equations of an electromechanical system are given as

$$L_a \frac{di_a}{dt} + R_a i_a + K_e \dot{\theta} = v_a$$

$$J\ddot{\theta} + B\dot{\theta} - K_t i_a = 0$$

(a) Define $\dot{\theta} = \omega$, then show that the differential equations can be given in the standard second-order matrix form with the *electrical* equation on *top*, as

$$\begin{bmatrix} L_a & 0 \\ 0 & J \end{bmatrix} \begin{Bmatrix} \dot{i}_a \\ \dot{\omega} \end{Bmatrix} + \begin{bmatrix} R_a & K_e \\ -K_t & B \end{bmatrix} \begin{Bmatrix} i_a \\ \omega \end{Bmatrix} = \begin{Bmatrix} v_a \\ 0 \end{Bmatrix}$$

(b) Obtain the differential equations in the standard second-order matrix form with the *mechanical* equation on *top*.

5.8. Using the differential equations from Problem 5.7, determine the state-space representation, assuming that the output is the angular speed.

5.9. Consider the electromechanical system shown in Fig. P.5.9. Assume that the mass moment of inertia of the rotor is J_r and the shaft between the rotor and the disk J is *rigid*.
(a) Is this an armature-controlled DC motor or a field-controlled DC motor?
(b) Derive the differential equations.
(c) Obtain all the transfer functions relating the input voltage v_a to the outputs.

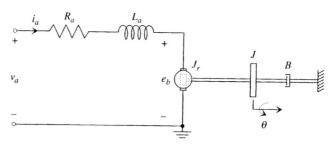

FIGURE P5.9

5.10. Consider the electromechanical system shown in Fig. P5.10. Assume that the mass moment of inertia of the rotor is J_r and the shaft between the rotor and the disk J is flexible and of stiffness K. The angular displacements of the rotor and the disk are θ_1 and θ_2, respectively.
(a) Derive the differential equations.
(b) Obtain all the transfer functions relating the input voltage v_a to the outputs.

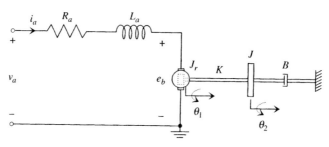

FIGURE P5.10

5.11. Consider the electromechanical system shown in Fig. P5.11. The mass moment of inertia of the rotor is J_r and that of the two identical disks is J. The shaft is flexible and the stiffness is K for the two segments, and the damping is modeled as a viscous damper. Derive the differential equations.

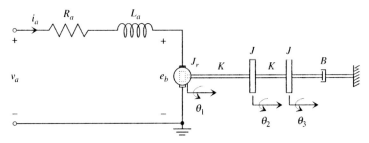

FIGURE P5.11

5.5 IMPEDANCE METHODS

The impedance method provides an alternative for obtaining the mathematical models of both electrical and mechanical systems. The impedance method is straightforward for electrical systems. For mechanical systems, the impedance method should be used in conjunction with the **linear graph**. However, many students are unfamiliar with the linear graph. Thus, for mechanical systems, definitions of impedances are discussed, but the impedance method for obtaining the mathematical models (transfer functions and differential equations) is not shown. Only the impedance method for electrical systems is presented.

Electrical Systems

For electrical circuits, the electrical impedance $Z(s)$ is defined as

$$Z(s) = \frac{V(s)}{I(s)} \tag{5.68}$$

where $V(s)$ and $I(s)$ are the Laplace transforms of the voltage $v(t)$ and current $i(t)$, respectively. All the Laplace transforms are complex in general.

The following are fundamental relations of electric elements:

Resistance: $v_R = Ri_R$

$$Z_R(s) = \frac{V_R(s)}{I_R(s)} = R \tag{5.69a}$$

Inductance: $v_L = L\frac{di_L}{dt}$

$$Z_L(s) = \frac{V_L(s)}{I_L(s)} = Ls \tag{5.69b}$$

Capacitance: $i_C = C\frac{dv_c}{dt}$

$$Z_C(s) = \frac{V_C(s)}{I_C(s)} = \frac{1}{Cs} \tag{5.69c}$$

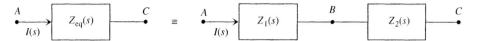

FIGURE 5.15 Equivalent impedance for two impedances in series.

Equivalence

The use of equivalence and impedance is particularly convenient in electrical circuits.

THEOREM 5.1. If there are n impedances in *series*, then the equivalent impedance Z_{eq} is equal to the sum of all the individual impedances Z_i:

$$Z_{eq}(s) = Z_1(s) + Z_2(s) + \cdots + Z_n(s) \tag{5.70}$$

As an example, consider a system of two impedances Z_1 and Z_2 in series as shown in Fig. 5.15. The system of two impedances is equivalent to a single impedance where

$$Z_{eq}(s) = Z_1(s) + Z_2(s) \tag{5.71}$$

Proof. From the figure, we have

$$V_{AC} = V_{AB} + V_{BC}$$

$$Z_{eq}I = Z_1 I + Z_2 I = (Z_1 + Z_2)I$$

$$Z_{eq} = Z_1 + Z_2$$

THEOREM 5.2. If there are n impedances in *parallel*, then the equivalent impedance Z_{eq} is equal to the sum of all the reciprocals of the individual impedances Z_i:

$$\frac{1}{Z_{eq}(s)} = \frac{1}{Z_1(s)} + \frac{1}{Z_2(s)} + \cdots + \frac{1}{Z_n(s)} \tag{5.72}$$

As an example, consider a system of two impedances Z_1 and Z_2 in parallel as shown in Fig. 5.16. The equivalent impedance of a system of two impedances in parallel is

$$\frac{1}{Z_{eq}} = \frac{1}{Z_1} + \frac{1}{Z_2} \tag{5.73a}$$

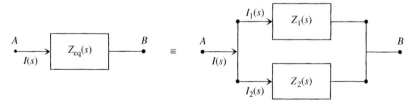

FIGURE 5.16 Equivalent impedance for two impedances in parallel.

or

$$Z_{eq} = \frac{Z_1 Z_2}{Z_1 + Z_2} \tag{5.73b}$$

Proof. From Fig. 5.16, we have

$$I = I_1 + I_2$$

Thus,

$$\frac{V_{AB}}{Z_{eq}} = \frac{V_{AB}}{Z_1} + \frac{V_{AB}}{Z_2} = \left(\frac{1}{Z_1} + \frac{1}{Z_2}\right) V_{AB}$$

$$\frac{1}{Z_{eq}} = \frac{1}{Z_1} + \frac{1}{Z_2} = \frac{Z_2 + Z_1}{Z_1 Z_2}$$

$$Z_{eq} = \frac{Z_1 Z_2}{Z_1 + Z_2}$$

Voltage Divider and Transfer Function

Consider a fundamental circuit with two impedances Z_1 and Z_2, as shown in Fig. 5.17. The current is

$$I(s) = \frac{V_a(s)}{Z_1(s) + Z_2(s)} \tag{5.74}$$

Thus the output voltage is

$$V_o(s) = Z_2(s) I(s) = \frac{Z_2(s)}{Z_1(s) + Z_2(s)} V_a(s) \tag{5.75}$$

The transfer function is

$$\frac{V_o(s)}{V_a(s)} = \frac{Z_2(s)}{Z_1(s) + Z_2(s)} \tag{5.76}$$

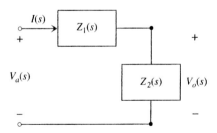

FIGURE 5.17 Fundamental circuit with two impedances.

Notes:

- In using the impedance method, all the variables must be functions of s.
- Although the impedance method is powerful, the algebra may get very involved for systems with many components.

EXAMPLE 5.5. Consider the electrical system shown in Fig. 5.18. The system consists of a resistor of resistance R, an inductor of inductance L, a capacitor of capacitance C, and the applied voltage is v_a. Using Kirchhoff's law and the impedance method, determine the transfer function and the differential equation.

Solution. The system has one input and one output: the input is the applied voltage v_a, and the output is the current i. Thus the system is governed by one differential equation. For this circuit, KVL is used,

$$\sum v_{\text{drop}} = 0$$

$$v_L + v_R + v_C - v_a = 0$$

In terms of impedance,

$$(Z_L + Z_R + Z_c)I(s) - V_a(s) = 0$$

Introducing the impedance relations term by term, then rearranging the result in the input-output form as

$$\left(Ls + R + \frac{1}{Cs}\right)I(s) = V_a(s)$$

The transfer function is

$$\frac{I(s)}{V_a(s)} = \frac{1}{R + Ls + 1/(Cs)} = \frac{Cs}{LCs^2 + RCs + 1}$$

Cross-multiplying,

$$(LCs^2 + RCs + 1)I(s) = CsV_a(s)$$

Now going back from the s domain to the time domain, the differential equation is

$$LC\frac{d^2i}{dt^2} + RC\frac{di}{dt} + i = C\frac{dv_a}{dt}$$

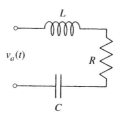

FIGURE 5.18 RLC circuit.

EXAMPLE 5.6. Consider the electrical system shown in Fig. 5.19, where the impedances are labeled. The applied voltage is v_a, and the components are resistor R and capacitors C_1 and C_2. Using the impedance method, determine the transfer function $V_o(s)/V_a(s)$.

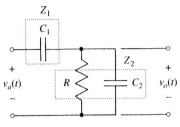

FIGURE 5.19 Circuit.

Solution. Using the impedance method, the transfer function is given as

$$\frac{V_o(s)}{V_a(s)} = \frac{Z_2(s)}{Z_1(s) + Z_2(s)} = \frac{\dfrac{R \cdot 1/(C_2 s)}{R + 1/(C_2 s)}}{\dfrac{1}{C_1 s} + \dfrac{R \cdot 1/(C_2 s)}{R + 1/(C_2 s)}}$$

$$\frac{V_o(s)}{V_a(s)} = \frac{RC_1 s}{R(C_1 + C_2)s + 1}$$

Mechanical Systems

Analogously to electrical impedance, mechanical impedance may be defined as

$$Z(s) = \frac{V(s)}{F(s)} \tag{5.77}$$

where $V(s)$ and $F(s)$ are the Laplace transforms of velocity $v(t)$ and force $f(t)$, respectively. This definition is the most logical although other definitions exist elsewhere. Because displacements are often used in mechanical systems, we thus have

$$Z(s) = \frac{V(s)}{F(s)} = \frac{sX(s)}{F(s)} \tag{5.78}$$

where $X(s)$ is the Laplace transform of the displacement $x(t)$.

Analogous to electrical elements, the following are fundamental relations of mechanical elements:

Viscous damping: $f_c = c\dot{x}_c = cv_c$

$$Z_c(s) = \frac{V_c(s)}{F_c(s)} = \frac{1}{c} \tag{5.79a}$$

Spring: $f_k = kx_k = k \int v_k \, dt$

$$Z_k(s) = \frac{V_k(s)}{F_k(s)} = \frac{s}{k} \qquad (5.79b)$$

Mass: $f_m = m\ddot{x}_m = m\dot{v}_m$

$$Z_m(s) = \frac{V_m(s)}{F_m(s)} = \frac{1}{ms} \qquad (5.79c)$$

Mechanical elements are not exactly equivalent to electrical elements. Comparing Eqs. (5.69) and (5.79), term by term, note that the corresponding electrical and mechanical elements are not equivalent, although they have similar physical effects. The electrical resistor and mechanical damper serve as an example. They are both damping elements, because they provide energy dissipation (they dissipate energy, or take energy out of the system). However, the mathematical expressions for their impedances are different, as shown in Eqs. (5.69a) and (5.79a). Therefore, trying to come up with an exact equivalence of electrical and mechanical systems is not fruitful.

PROBLEM SET 5.5

5.12. Reconsider the electrical system shown in Fig. P5.3 (Problem 5.3). Using the impedance method, determine the transfer function $V_{R1}(s)/V_a(s)$.

5.13. Reconsider the electrical system shown in Fig. P5.4 (Problem 5.4). Using the impedance method, determine the transfer function $V_{L1}(s)/V_a(s)$.

SUMMARY

The following are the elemental relations for passive electrical elements:

Resistor

$$v_R = Ri_R$$

Inductor

$$v_L = L\frac{di_L}{dt}$$

Capacitor

$$i_C = C\frac{dv_C}{dt}$$

For electrical, electronic, and electromechanical systems, Kirchhoff's laws are the fundamental laws for deriving the governing differential equations. Kirchhoff's voltage law (KVL) is used for the loop method, whereas Kirchhoff's current law (KCL) is for the node method.

Kirchhoff's current law (KCL), node method: *The algebraic sum of all currents entering (in) a circuit node is zero,*

$$\sum_k (i_k)_{in} = 0$$

Kirchhoff's voltage law (KVL), loop method: *The sum of all voltage drops around a circuit loop is zero,*

$$\sum v_{drop} = 0$$

Operational amplifiers (op-amps) are versatile electronic components. The op-amp equation is

$$v_o = K(v_+ - v_-) \implies v_+ \approx v_-$$

where the gain K is a large positive number, typically 10^5 to 10^6.

Electromechanical systems consist of electrical and mechanical components. For armature-controlled DC motors, which are commonly used, the back emf voltage e_b developed across the motor is given as

$$e_b = K_e \omega = K_e \dot{\theta}$$

and the torque T generated is

$$T = K_t i_a$$

In certain applications, the impedance method may be used conveniently to obtain the mathematical model of electrical systems.

PROBLEMS

5.14. Consider the electrical system shown in Fig. P5.14, where the applied voltage v_a is the input.
   (a) Determine the differential equation.
   (b) Obtain the transfer function $V_R(s)/V_a(s)$, which relates the input to the output voltage across the resistor.

CHAPTER 5: Electrical, Electronic, and Electromechanical Systems 297

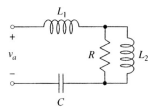

FIGURE P5.14

5.15. Consider the electrical system shown in Fig. P5.15 where the applied voltage v_a is the input.
   (a) Determine the differential equation.
   (b) Obtain the transfer function $V_C(s)/V_a(s)$, which relates the input to the output voltage across the capacitor.

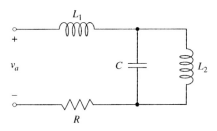

FIGURE P5.15

5.16. Consider the electrical system shown in Fig. P5.16, where the applied voltage v_a is the input.
   (a) Determine the differential equation.
   (b) Determine the transfer function $V_L(s)/V_a(s)$.

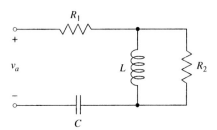

FIGURE P5.16

5.17. Consider the electrical system shown in Fig. P5.17, where the applied voltage v_a is the input.
   (a) Determine the differential equation.
   (b) Obtain the transfer function $V_L(s)/V_a(s)$.

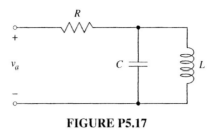

FIGURE P5.17

5.18. Consider the electrical system shown in Fig. P5.3 (Problem 5.3). Obtain the state-space representation assuming the output is the current.

5.19. Consider the electrical system shown in Fig. P5.19, with the applied voltage v_a as the input.
  (a) Determine the total number of independent differential equations in voltages.
  (b) Derive the differential equations.
  (c) Express the differential equations in the second-order matrix form.
  (d) Determine the transfer function $V_{L2}(s)/V_a(s)$.

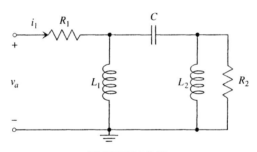

FIGURE P5.19

5.20. Consider the electrical system shown in Fig. P5.20, with the applied voltage v_a as the input.
  (a) Determine the total number of independent differential equations in voltages.
  (b) Derive the differential equations.
  (c) Express the differential equations in the second-order matrix form.
  (d) Determine the transfer function relating the input voltage to the output voltage across the resistor R_2.

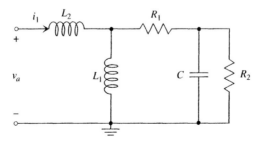

FIGURE P5.20

5.21. The system shown in Fig. P5.21, with the switch S closed, has reached the steady state condition for a while. Determine the values of the currents through R_1 and R_2.

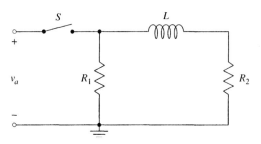

FIGURE P5.21

5.22. Consider the system shown in Fig. P5.22. At the time $t = 0$, the switch S is closed. Determine the values of the current (a) just *after* the switch is closed, $t = 0^+$, and (b) at steady state, $t \to \infty$.

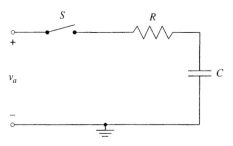

FIGURE P5.22

5.23. The capacitor C of the system shown in Fig. P5.23 has been charged to 12 V. Suddenly, the switch is closed at $t = 0$. Derive the differential equation and the initial condition for the capacitor voltage.

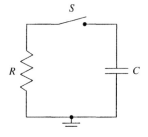

FIGURE P5.23

5.24. The capacitor C of the system shown in Fig. P5.24 has been charged to 12 V. Suddenly, the switch is closed. Derive the differential equation for the capacitor voltage, and determine the initial condition.

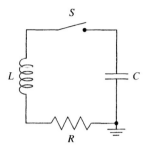

FIGURE P5.24

5.25. For the system shown in Fig. P5.25, i is the current source (input). Derive the differential equation.

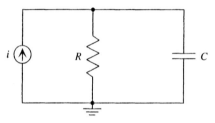

FIGURE P5.25

5.26. For the system shown in Fig. P5.26, i is the current source (input). Derive the differential equation.

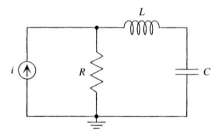

FIGURE P5.26

5.27. Consider the electrical system shown in Fig. P5.27. The voltage of the source (battery) is $v_a = 12$ V. The internal resistance of the battery is quite large, $R_i = 10^7$ Ω. The load consists of a capacitor $C = 10^3$ μF and two resistors $R = 10^3$ Ω. Derive the differential equation for the capacitor voltage when the switch is closed.

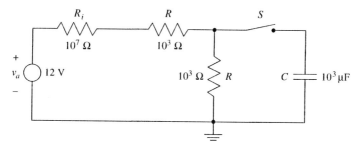

FIGURE P5.27

5.28. Consider the electrical system shown in Fig. P5.28. The voltage of the source (battery) is $v_a = 12$ V. The internal resistance of the battery is quite large, $R_i = 10^7\ \Omega$. The load consists of a capacitor $C = 10^2\ \mu\text{F}$, a resistor $R = 10^3\ \Omega$, and an inductor $L = 10^3\ \mu\text{H}$. Derive the differential equation for the capacitor voltage when the switch is closed.

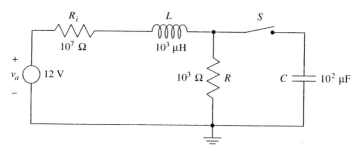

FIGURE P5.28

5.29. Consider the electrical system shown in Fig. P5.29, with the applied voltage v_a as the input. Determine the transfer function relating the input voltage v_a to the output voltage v_o and then the differential equation.

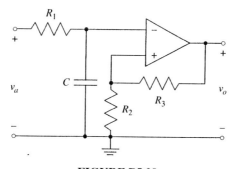

FIGURE P5.29

5.30. Consider the electrical system shown in Fig. P5.30, with the applied voltage v_a as the input.

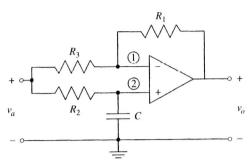

FIGURE P5.30

(a) Derive the differential equation.
(b) Determine all the transfer functions relating the input voltage v_a to the output voltages.

5.31. The differential equations of an electromechanical system are given as

$$\begin{bmatrix} L_a & 0 \\ 0 & J \end{bmatrix} \begin{Bmatrix} \dot{i}_a \\ \dot{\omega} \end{Bmatrix} + \begin{bmatrix} R_a & K_e \\ -K_t & B \end{bmatrix} \begin{Bmatrix} i_a \\ \omega \end{Bmatrix} = \begin{Bmatrix} v_a \\ 0 \end{Bmatrix}$$

Determine all the transfer functions.

5.32. The differential equations of an electromechanical system are given as

$$\begin{bmatrix} L_a & 0 \\ 0 & J \end{bmatrix} \begin{Bmatrix} \dot{i}_a \\ \dot{\omega} \end{Bmatrix} + \begin{bmatrix} R_a & K_e \\ -K_t & B \end{bmatrix} \begin{Bmatrix} i_a \\ \omega \end{Bmatrix} = \begin{Bmatrix} v_a \\ T_L \end{Bmatrix}$$

Determine the transfer function $\Omega(s)/T_L(s)$.

5.33. The differential equations of an electromechanical system are given in the standard second-order matrix form as

$$\begin{bmatrix} J & 0 \\ 0 & 0 \end{bmatrix} \begin{Bmatrix} \ddot{\theta} \\ \ddot{i}_a \end{Bmatrix} + \begin{bmatrix} B & 0 \\ K_e & L_a \end{bmatrix} \begin{Bmatrix} \dot{\theta} \\ \dot{i}_a \end{Bmatrix} + \begin{bmatrix} 0 & -K_t \\ 0 & R_a \end{bmatrix} \begin{Bmatrix} \theta \\ i_a \end{Bmatrix} = \begin{Bmatrix} T_L \\ v_a \end{Bmatrix}$$

Determine the transfer function $I_a(s)/V_a(s)$.

5.34. The differential equations of an electromechanical system are given in the standard second-order matrix form as

$$\begin{bmatrix} J & 0 \\ 0 & 0 \end{bmatrix} \begin{Bmatrix} \ddot{\theta} \\ \ddot{i}_a \end{Bmatrix} + \begin{bmatrix} B & 0 \\ K_e & L_a \end{bmatrix} \begin{Bmatrix} \dot{\theta} \\ \dot{i}_a \end{Bmatrix} + \begin{bmatrix} 0 & -K_t \\ 0 & R_a \end{bmatrix} \begin{Bmatrix} \theta \\ i_a \end{Bmatrix} = \begin{Bmatrix} 0 \\ v_a \end{Bmatrix}$$

Determine the transfer function $\Omega(s)/V_a(s)$.

5.35. Using the differential equations from Problem 5.34, determine the state-space representation, assuming that the output is the angular displacement.

5.36. Reconsider the electromechanical system shown in Fig. P5.10 (Problem 5.10). Determine the state-space representation, assuming that the outputs are the angular displacements.

5.37. The electromechanical system shown (Fig. P5.37) consists of a motor whose rotor is connected to a disk by a *flexible* and *damped* shaft. The polar moments of inertia of the rotor and disk are J_r and J, respectively. The shaft is modeled as a torsional spring of stiffness K and a torsional viscous damper of coefficient B in parallel. Obtain:
(a) the differential equations,
(b) the transfer function $\Theta_2(s)/V_a(s)$,
(c) the state-space representation where the outputs are θ_1 and $\dot{\theta}_1$.

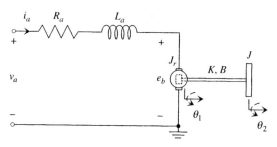

FIGURE P5.37

CHAPTER 6

Fluid and Thermal Systems

The modeling of fluid and thermal systems is presented in this chapter. The three major systems are pneumatic systems, liquid-level systems, and thermal (heat-transfer) systems. The fundamentals of thermodynamics, chemistry, fluid dynamics, and heat transfer are discussed in detail. Topics include ideal gases, real gases, mixtures of gases, conservation of energy, specific heats, isothermal processes, polytropic processes, conduction, and convection. The lumped-parameter model may be used for uniform-temperature approximation in solids if the Biot number, Bi, is less than 0.1.

6.1 INTRODUCTION

Fluids are general terms used to represent gases and liquids. All gases are *compressible*, but liquids can be considered *incompressible*. For a fixed volume and constant temperature, the density of a gas changes if the pressure is changed. Although real liquids are compressible, the changes in density are insignificant when the pressure is varied. Thus, liquids may be modeled as incompressible for constant-temperature processes. In other words, the density of a liquid is not a function of pressure. It is a function of temperature, but for most problems in hydraulics and liquid-level systems, the temperature is approximately constant. The study of hydraulics focuses on liquids flowing in conduits or pipes, whereas that of liquid-level systems has to do with the height or level of liquid in storage tanks.

In thermal systems, energy and heat are stored within the system and transferred in and out of the system. The mathematical modeling of a thermal system is often complicated because of the complex distribution of temperature throughout the system. Precise analysis requires a distributed-parameter model governed by partial differential equations. For simple analysis, however, only a simple model that represents the gross system behavior is required.

In this chapter, simple system models that represent the gross system behavior are discussed, whereas the details and complex concepts of thermal sciences are ignored. For both fluid and thermal systems, only fundamental facts in thermodynamics, fluid dynamics, and heat transfer are used.

6.2 THERMODYNAMICS

Simple principles of thermodynamics are explored in this section. Only fundamental facts are discussed; the details and complex concepts of thermodynamics are ignored.

Chemistry

A review of the fundamentals of chemistry is useful here. A typical atom is composed of protons, neutrons, and electrons. A molecule, in turn, is composed of two or more atoms. For example, an oxygen atom is O, whereas an oxygen molecule is O_2, which is a **diatomic molecule** (*di-* indicates two).

A pure substance is called an **element** if it cannot be broken down into simpler substances by chemical or physical changes. For example, oxygen is an element. A monoatomic oxygen element contains oxygen atoms, O. A **compound** is a pure substance that can be broken down into simpler elements. Water is an example of a compound, because it decomposes into hydrogen and oxygen under the proper physical or chemical conditions. A **mixture** is a simple combination of two or more elements or compounds. The "wet" air we normally breath is a mixture of "dry" air and water vapor (H_2O, a compound). Dry air itself is a mixture consisting mainly of nitrogen (N_2, a diatomic molecule), oxygen (O_2, a diatomic molecule), and other gases.

A **mole** of an element or a compound contains 6.023×10^{23} atoms or molecules. This number is called *Avogadro's number.* There are 6.022×10^{23} *atoms* in 12.01 g of carbon C and the same number of atoms in 16.00 g of oxygen O, and 6.023×10^{23} *molecules* in 32.00 g of oxygen O_2. Thus, the **molecular mass** of carbon C is $M_m = 12.01$ g/mol, and those of oxygen O and oxygen O_2 are $M_m = 16.00$ and 32.00 g/mol, respectively. Notice that the symbol M_m is used for molecular mass to avoid confusion with molarity M (the ratio of number of moles of solute to 1.0 liter of solution).

The unit **mole** or **mol** actually means **g-mol**. A mole, or 12.01 g of carbon, contains 6.022×10^{23} atoms, and 12.01 kg of carbon C certainly contains much *more* than 6.022×10^{23} atoms (1 kg-mol contains a thousand times more than 1 g-mol does). However, the molecular mass of carbon may be given as $M_m = 12.01$ g/mol or 12.01 g/g-mol or 12.01 kg/kg-mol or 12.01 lb_m/lb_m-mol.

Ideal Gases

An ideal gas is a hypothetical gas whose quantity pV/T is constant where p is the pressure, V is the (total) volume, and T is the absolute temperature (K or °R). At very low pressures and very high temperatures, all real gases behave as ideal gases. Thus, at low pressures and moderate temperatures, real gases may be approximated as ideal gases to simplify calculations. The relationship for an ideal gas is

$$pV = nR_u T \tag{6.1}$$

n is the number of moles and R_u is the *universal gas constant*. The numerical value of the universal gas constant is $R_u = 8314.3$ J/(kg · mol-K) or 1545.3 ft-lb$_f$/(lb$_m$ · mol-°R).

An alternative form of ideal gas is obtained by dividing the above equation by the mass $m = nM_m$

$$pv = R_s T \tag{6.2}$$

where M_m = molecular mass
$v = V/m$ = specific volume
$R_s = R_u/M_m$ = specific gas constant

Gaseous Mixtures

Mixtures of gases are important in many engineering applications. The following are elemental relations of mixtures of ideal gases.

Mole fraction and molecular mass

The total mass m of a mixture is the sum of the masses of each component:

$$m = m_1 + m_2 + \cdots + m_i = \sum_{i=1}^{k} m_i \tag{6.3}$$

The number of moles n is defined as

$$n = \frac{m}{M_m} \tag{6.4}$$

Thus, the molecular mass M_m is

$$M_m = \frac{m}{n} \tag{6.5}$$

For the components,

$$m_i = n_i M_{mi} \tag{6.6}$$

Thus,

$$m = \sum_{i=1}^{k} m_i = n_i M_{mi} \tag{6.7}$$

Dividing by n,

$$M_m = \frac{m}{n} = \sum_{i=1}^{k} \frac{n_i}{n} M_{mi} = \sum_{i=1}^{k} x_i M_{mi} \tag{6.8}$$

where the mole fraction of each component is defined as

$$x_i = \frac{n_i}{n} \tag{6.9}$$

Thus, the molecular mass of a mixture of gases is equal to the sum of each individual mole fraction multiplied by its molecular mass.

Mole fraction, partial pressure, and partial volume

In a gaseous mixture, each component is assumed to exist at its partial pressure p_i at temperature T and to take up the whole volume V,

$$p_i V = n_i R_u T \tag{6.10}$$

For the mixture as a whole,

$$pV = n R_u T \tag{6.11}$$

After dividing, the ratio of the partial pressure to the total pressure is equal to the mole fraction

$$\frac{p_i}{p} = \frac{n_i}{n} \tag{6.12}$$

Each component of the mixture is assumed to separate into its compartment of volume V_i but to exist at temperature T and total pressure p,

$$pV_i = n_i R_u T \tag{6.13}$$

Similarly, after dividing, the ratio of the partial volume to the total volume is equal to the mole fraction

$$\frac{V_i}{V} = \frac{n_i}{n} \tag{6.14}$$

In summary,

$$x_i = \frac{n_i}{n} = \frac{p_i}{p} = \frac{V_i}{V} \tag{6.15}$$

EXAMPLE 6.1. Dry air is an odorless and colorless gaseous mixture whose composition is given below. (The air we typically breath is not dry air because it contains moisture; water vapor behaves as if it were inert.) Determine (a) the molecular mass of dry air; (b) the specific gas constant for dry air in the English system.

Gas	Volumetric fraction, %	Mole fraction x_i	Molecular mass M_{mi}, g/g-mol
Nitrogen, N_2	78.03	0.7803	28.02
Oxygen, O_2	20.99	0.2099	32.00
Argon, Ar	0.94	0.0094	39.94
Carbon dioxide, CO_2	0.03	0.0003	44.00
Hydrogen, H_2	0.01	0.0001	2.01

Solution. The molecular mass of dry air is given as

$$M_m = \sum_{i=1}^{k} \frac{n_i}{n} M_{mi} = \sum_{i=1}^{k} x_i M_{mi}$$

$$M_m = 0.7803(28.02) + 0.2099(32.00) + 0.0094(39.94)$$
$$+ 0.0003(44.00) + 0.0001(2.01) = 28.967 \text{ g/g-mol}$$

or 28.967 lb/lb-mol; the units are different, but the numerical values are exactly the same. Thus, the specific gas constant for air is

$$R_s = R_{air} = \frac{R_u}{M_m} = \frac{1545.33 \text{ (ft-lb}_f)/(\text{lb}_m \cdot \text{mol-}°R)}{28.97 \text{ lb}_m/(\text{lb}_m \cdot \text{mol})} = 53.34 \text{ ft-lb}_f/(\text{lb}_m\text{-}°R)$$

Intensive and Extensive Properties

All properties of a system may be divided into two types: intensive and extensive. Those properties that are independent of the amount of material in the system are called *intensive properties*. They are not additive, e.g., pressure p, temperature T, specific volume v (volume per unit mass), and density ρ (mass per unit volume) where $\rho = 1/v$. On the other hand, those properties which are proportional to the mass of a system are called *extensive properties*. They are additive, e.g., volume $V = m/\rho$.

Conservation of Energy

The **first law of thermodynamics,** or the general law of conservation of energy, is given as

$$Q_{in} - W_{out} = \Delta E_{1 \to 2} = E_2 - E_1 \qquad (6.16)$$

where $\Delta E_{1 \to 2}$ is the change in the energy of the system from the reference state 1 to state 2. The sign convention and subscripts are used to avoid any ambiguity. In this expression, Q_{in} is positive if heat is added to the system and negative if heat is liberated by the system. The process in which heat is supplied to the system is called **endothermic** (*endo* means *in* and *thermic* means *heat*), and that in which heat is given off by the system is called **exothermic**. Without work, the energy of the system increases in endothermic reaction, $\Delta E_{1\to 2} > 0$, and decreases in exothermic reaction, $\Delta E_{1\to 2} < 0$. Note that the process is called **endergonic** if the **free energy** G—not just heat—increases and **exergonic** if G decreases. Also, W_{out} is positive if work is done *by* the system and negative if work is done *to* the system. Notice that the negative sign is included in front of the work term. The sign convention clearly shows that, without any heat, the energy of the system decreases when the system performs work.

For convenience in applications, the first law of thermodynamics can be modified as

$$\Delta E_{1\to 2} = (Q_{\text{in}} - Q_{\text{out}}) + (W_{\text{in}} - W_{\text{out}}) \tag{6.17}$$

The *energy*, or total energy, E is the sum of the *internal energy* (internal thermal energy) U, kinetic energy KE, potential energy PE, chemical energy CE, and others,

$$E = U + \text{KE} + \text{PE} + \text{CE} + \cdots \tag{6.18}$$

where the potential energy PE consists of the elastic potential energy V_e and gravitational energy V_g:

$$\text{PE} = V_e + V_g \tag{6.19}$$

In many closed thermodynamic systems, the only significant change in energy is the change in internal thermal energy U, and all other changes are negligible. The change of total energy reduces to that of internal energy,

$$\Delta E_{1\to 2} = \Delta U_{1\to 2} \tag{6.20}$$

Thus, the first law of thermodynamics reduces to

$$Q_{\text{in}} + W_{\text{out}} = \Delta U_{1\to 2} \tag{6.21}$$

or

$$\Delta U_{1\to 2} = (Q_{\text{in}} - Q_{\text{out}}) + (W_{\text{in}} - W_{\text{out}}) \tag{6.22}$$

Specific Heats

The two specific heats in thermodynamics are **constant-volume specific heat** and **constant-pressure specific heat**. By definition, the constant-volume specific heat c_v is given as

$$c_v = \left(\frac{\partial u}{\partial T}\right)_v \tag{6.23}$$

where u is the specific internal energy. The constant-pressure specific heat c_p is defined as

$$c_p = \left(\frac{\partial h}{\partial T}\right)_p \tag{6.24}$$

where h is the specific enthalpy,
$$h = u + pv \tag{6.25}$$

Real gases

In general, for a real gas or nonideal gas, the internal energy u and the enthalpy h are functions of two thermodynamic properties,
$$u = u(T, v) \qquad h = h(T, p) \tag{6.26}$$

From the definitions of specific heats, the differentials are
$$du_v = c_v\, dT_v \qquad dh_p = c_p\, dT_p \tag{6.27}$$

To obtain u, we must integrate the differential and must have the function relating c_v and T along the particular line of constant v. Similarly, for h we must have the function relating c_p and T along the particular line of constant p for the integration. The integration processes are rather complex for real gases.

Ideal gases

For an ideal gas, both the internal energy u and enthalpy h are functions of T only,
$$u = u(T) \qquad h = h(T) \tag{6.28}$$

The specific heats are also functions of T only,
$$c_v = c_v(T) \qquad c_p = c_p(T) \tag{6.29}$$

Thus, the subscripts on the differentials may be dropped:
$$du = c_v\, dT \qquad dh = c_p\, dT \tag{6.30}$$

Liquids and solids

A liquid can be approximated as an incompressible fluid. For the fluid that can be approximated as incompressible, the constant-pressure specific heat c_p can be shown as
$$c_p = \left(\frac{\partial u}{\partial T}\right)_v \tag{6.31}$$

The same holds true with solids. Recall that the constant-volume specific heat c_v is defined as
$$c_v = \left(\frac{\partial u}{\partial T}\right)_v$$

Thus, the numerical values of c_v and c_p are the same even though they are different by their formal definitions. Note again that c_p is defined as $c_p = (\partial h/\partial T)_p$. For solids and incompressible liquids, the subscripts are thus often dropped for simplicity:
$$c = c_v = c_p \tag{6.32}$$

6.3 FLUID MECHANICS

For the newtonian, isotropic fluid, the shear stress τ_{xy}, in a cartesian coordinate system, is given as

$$\tau_{xy} = 2\mu\varepsilon_{xy} = 2\mu\left(\frac{\partial u}{\partial y} + \frac{\partial v}{\partial x}\right) \quad (6.33)$$

where ε_{xy} = shear strain
μ = *dynamic* (or absolute) viscosity
u, v = velocity components in the x and y directions, respectively

The Navier-Stokes equation (also called the momentum equation) for fluid motion is given in vector form as

$$\rho\frac{D\mathbf{V}}{Dt} = -\nabla p_k + \mu\nabla^2\mathbf{V} + \frac{\mu}{3}\nabla(\nabla \cdot \mathbf{V}) \quad (6.34a)$$

or

$$\frac{D\mathbf{V}}{Dt} = -\frac{1}{\rho}\nabla p_k + \nu\nabla^2\mathbf{V} + \frac{\nu}{3}\nabla(\nabla \cdot \mathbf{V}) \quad (6.34b)$$

where $p_k = p + \rho g h$ = *kinetic* pressure
p = pressure
ρ = density
h = height
$\nu = \mu/\rho$ = *kinematic* viscosity
$\mathbf{V}$ = velocity vector
∇ = gradient operator (vector)
$\nabla^2 = \nabla \cdot \nabla$ = *Laplacian* operator (scalar)
$\nabla \cdot \mathbf{V}$ = divergence of vector $\mathbf{V}$ (scalar)

In a *cartesian coordinate system* (x, y, z), we may use u, v, w to represent the velocity components in the x, y, z directions. Thus, the Navier-Stokes equations are written in detail as

$$\rho\left(\frac{\partial}{\partial t} + u\frac{\partial}{\partial x} + v\frac{\partial}{\partial y} + w\frac{\partial}{\partial z}\right)u = -\frac{\partial p_k}{\partial x} + \mu\left(\frac{\partial^2}{\partial x^2} + \frac{\partial^2}{\partial x^2} + \frac{\partial^2}{\partial x^2}\right)u \quad (6.35)$$

$$\rho\left(\frac{\partial}{\partial t} + u\frac{\partial}{\partial x} + v\frac{\partial}{\partial y} + w\frac{\partial}{\partial z}\right)v = -\frac{\partial p_k}{\partial y} + \mu\left(\frac{\partial^2}{\partial x^2} + \frac{\partial^2}{\partial x^2} + \frac{\partial^2}{\partial x^2}\right)v \quad (6.36)$$

$$\rho\left(\frac{\partial}{\partial t} + u\frac{\partial}{\partial x} + v\frac{\partial}{\partial y} + w\frac{\partial}{\partial z}\right)w = -\frac{\partial p_k}{\partial w} + \mu\left(\frac{\partial^2}{\partial x^2} + \frac{\partial^2}{\partial x^2} + \frac{\partial^2}{\partial x^2}\right)w \quad (6.37)$$

The Navier-Stokes equations in a *cylindrical coordinate system* (r, θ, z) are relatively more complicated. If we use v_r, v_θ, v_z to denote the velocity components in

312   Dynamic Systems: Modeling and Analysis

the r, θ, z directions, then the Navier-Stokes equations are given as

$$\rho\left[\left(\frac{\partial}{\partial t} + v_r\frac{\partial}{\partial r} + \frac{v_\theta}{r}\frac{\partial}{\partial \theta} + v_z\frac{\partial}{\partial z}\right)v_r - \frac{v_\theta^2}{r}\right]$$
$$= -\frac{\partial p_k}{\partial r} + \mu\left\{\left[\frac{1}{r}\frac{\partial}{\partial r}\left(r\frac{\partial}{\partial r}\right) + \frac{1}{r^2}\frac{\partial^2}{\partial \theta^2} + \frac{\partial^2}{\partial z^2}\right]v_r - \frac{v_r}{r^2} - \frac{2}{r^2}\frac{\partial v_\theta}{\partial \theta}\right\} \quad (6.38)$$

$$\rho\left[\left(\frac{\partial}{\partial t} + v_r\frac{\partial}{\partial r} + \frac{v_\theta}{r}\frac{\partial}{\partial \theta} + v_z\frac{\partial}{\partial z}\right)v_\theta + \frac{v_r v_\theta}{r}\right]$$
$$= -\frac{1}{r}\frac{\partial p_k}{\partial \theta} + \mu\left\{\left[\frac{1}{r}\frac{\partial}{\partial r}\left(r\frac{\partial}{\partial r}\right) + \frac{1}{r^2}\frac{\partial^2}{\partial \theta^2} + \frac{\partial^2}{\partial z^2}\right]v_\theta + \frac{2}{r^2}\frac{\partial v_r}{\partial \theta} - \frac{v_\theta}{r^2}\right\} \quad (6.39)$$

$$\rho\left(\frac{\partial}{\partial t} + v_r\frac{\partial}{\partial r} + \frac{v_\theta}{r}\frac{\partial}{\partial \theta} + v_z\frac{\partial}{\partial z}\right)v_z$$
$$= -\frac{\partial p_k}{\partial z} + \mu\left[\frac{1}{r}\frac{\partial}{\partial r}\left(r\frac{\partial}{\partial r}\right) + \frac{1}{r^2}\frac{\partial^2}{\partial \theta^2} + \frac{\partial^2}{\partial z^2}\right]v_z \quad (6.40)$$

For flow in a pipe, however, we often use x to designate the longitudinal position along the pipe; thus x is identical (or equivalent) to z, or $x \equiv z$. Also, it is customary to denote u, v, w as the cylindrical velocity components in the x, r, θ directions. The Navier-Stokes equations, in this cylindrical coordinate system (x, r, θ), then become

$$\rho\left(\frac{\partial}{\partial t} + u\frac{\partial}{\partial x} + v\frac{\partial}{\partial r} + \frac{w}{r}\frac{\partial}{\partial \theta}\right)u$$
$$= -\frac{\partial p_k}{\partial x} + \mu\left[\frac{\partial^2}{\partial x^2} + \frac{1}{r}\frac{\partial}{\partial r}\left(r\frac{\partial}{\partial r}\right) + \frac{1}{r^2}\frac{\partial^2}{\partial \theta^2}\right]u \quad (6.41)$$

$$\rho\left[\left(\frac{\partial}{\partial t} + u\frac{\partial}{\partial x} + v\frac{\partial}{\partial r} + \frac{w}{r}\frac{\partial}{\partial \theta}\right)v - \frac{w^2}{r}\right]$$
$$= -\frac{\partial p_k}{\partial r} + \mu\left\{\left[\frac{\partial^2}{\partial x^2} + \frac{1}{r}\frac{\partial}{\partial r}\left(r\frac{\partial}{\partial r}\right) + \frac{1}{r^2}\frac{\partial^2}{\partial \theta^2}\right]v - \frac{v}{r^2} - \frac{2}{r^2}\frac{\partial w}{\partial \theta}\right\} \quad (6.42)$$

$$\rho\left[\left(\frac{\partial}{\partial t} + u\frac{\partial}{\partial x} + v\frac{\partial}{\partial r} + \frac{w}{r}\frac{\partial}{\partial \theta}\right)w - \frac{vw}{r}\right]$$
$$= -\frac{1}{r}\frac{\partial p_k}{\partial \theta} + \mu\left\{\left[\frac{\partial^2}{\partial x^2} + \frac{1}{r}\frac{\partial}{\partial r}\left(r\frac{\partial}{\partial r}\right) + \frac{1}{r^2}\frac{\partial^2}{\partial \theta^2}\right]w + \frac{2}{r^2}\frac{\partial v}{\partial \theta} - \frac{w}{r^2}\right\} \quad (6.43)$$

Laminar and Turbulent Flows

The fully developed flow in a circular cylindrical tube (or pipe), as shown in Fig. 6.1, is important. The internal flow in a pipe is **laminar** when the Reynolds number,

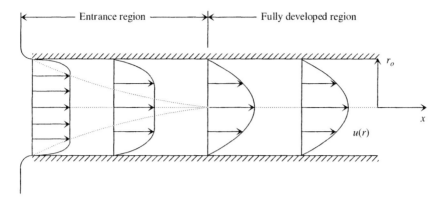

FIGURE 6.1 Laminar flow in a pipe.

Re = UD/ν, is less than approximately 2×10^3; U is the average velocity, and $D = 2r_o$ is the diameter of the pipe. Note that *turbulent* flow occurs when Re is greater than 1×10^4. The flow develops, with increasing values of the Reynolds number, as

$$\text{laminar flow} \rightarrow \text{critical zone} \rightarrow \text{intermediate zone} \rightarrow \text{turbulent flow}$$

where the zone in between laminar and turbulent regions is divided into two subregions: the critical zone and the intermediate zone.

Fluid Resistance

The fluid resistance associated with the fluid flow over a solid object is an important topic in engineering. This fluid resistance arises from the shear stresses. In this section, we will obtain the relation for fluid resistance, which involves volume flow rate and pressure drop for internal flow through a pipe (Fig. 6.1).

For fully developed laminar flow, all streamlines are parallel to the wall, thus $v = w = 0$; and for steady flow, $\partial(\cdot)/\partial t = 0$. Thus, Eq. (6.41) reduces to

$$0 = -\frac{\partial p_k}{\partial x} + \mu \frac{1}{r}\frac{d}{dr}\left(r\frac{du}{dr}\right) \tag{6.44}$$

since u is a function of r only, or $u = u(r)$.

The mathematical model of a fluid system should be simple; however, it must yield an accurate transient response as well as a steady-state response. To obtain a simple yet effective model, the unsteady terms in the Navier-Stokes equations can be ignored.

Note that Eq. (6.44) can be obtained by a different approach, where a free-body diagram and Newton's second law are utilized. Since the flow is axisymmetric, a differential annular ring can be chosen as the control volume. After applying Newton's second law to the control volume and setting the acceleration to zero for fully developed flow, we obtain

$$-\frac{\partial p_k}{\partial x} + \frac{\tau_{rx}}{r} + \frac{d\tau_{rx}}{dr} = 0 \tag{6.45}$$

$$\frac{\partial p_k}{\partial x} = \frac{1}{r}\frac{d(r\tau_{rx})}{dr} \tag{6.46}$$

Substituting

$$\tau_{rx} = \mu \frac{du}{dr} \tag{6.47}$$

into Eq. (6.46) yields Eq. (6.44).

The next step is to show that the pressure p_k is only a function of x. In general, $p_k = p_k(x, r, \theta; t)$; however, for steady and axisymmetric flow, all quantities are independent of θ, thus $p_k = p_k(x, r)$ at the most. Furthermore, $v = w = 0$, so Eq. (6.42) reduces to

$$0 = -\frac{\partial p_k}{\partial r} \tag{6.48}$$

In other words, p cannot be a function of r; therefore, p can only be a function of x, $p_k = p_k(x)$. Hence, Eq. (6.44) becomes

$$\mu \frac{1}{r}\frac{d}{dr}\left[r\frac{du(r)}{dr}\right] = \frac{dp_k(x)}{dx} \tag{6.49}$$

Since the left-hand side may be a function of r and the right-hand side may be a function of x, they both must be equal to a constant:

$$\mu \frac{1}{r}\frac{d}{dr}\left[r\frac{du(r)}{dr}\right] = \frac{dp_k(x)}{dx} = \text{constant} \tag{6.50}$$

We now proceed in integrating Eq. (6.50) to obtain

$$u(r) = \frac{r^2}{4\mu}\frac{dp_k(x)}{dx} + \frac{c_1}{\mu}\ln r + c_2 \tag{6.51}$$

After applying the boundary conditions, the velocity distribution for steady, fully developed, axisymmetric, laminar flow is given as

$$u(r) = -\frac{R^2}{4\mu}\frac{dp}{dx}\left[1 - \left(\frac{r}{r_o}\right)^2\right] \tag{6.52}$$

where the subscript k (of the pressure) has been dropped for simplicity. The relation between the pressure gradient dp/dx and the volume flow rate Q is obtained as

$$\frac{dp}{dx} = -\frac{8\mu Q}{\pi r_o^4} \tag{6.53}$$

The negative sign indicates that p is decreased with increasing x. The pressure difference, Δp (positive), may be defined as $\Delta p = p_1 - p_2$; thus

$$\frac{dp}{dx} = -\frac{p_1 - p_2}{L} = -\frac{\Delta p}{L} = \text{constant} \tag{6.54}$$

where points 1 and 2 are at upstream and downstream, respectively, and L is the length of the pipe. Thus,

$$Q = \frac{\pi r_o^4}{8\mu L}\Delta p = \frac{\pi D^4}{128\mu L}\Delta p \tag{6.55}$$

Finally, the fluid resistance R can be defined as

$$R = \frac{\Delta p}{\rho Q} \tag{6.56}$$

For steady, fully developed, axisymmetric laminar flow, the fluid resistance is given as

$$R = \frac{\Delta p}{\rho Q} = \frac{128\nu L}{\pi D^4} = \text{constant} \tag{6.57}$$

where $\nu = \mu/\rho$. Hence, the plot of the pressure difference Δp versus the mass flow rate ρQ is simply a straight line where the constant R is the slope.

The fluid resistance R for turbulent flow is, however, highly *nonlinear*. (The plot Δp versus ρQ is not a straight line.) The friction, or fluid resistance, is a strong function of the fluid-solid interface's roughness. The Moody diagram can be used to obtain the **friction factor,** f, for various values of **relative roughness,** e/D. Although the fluid resistance is nonlinear, it can be *linearized*, then given by Eq. (6.56), for a small pressure difference Δp in the neighborhood of an operating mass flow rate ρQ.

Note that the friction factor for laminar flow, which is also included in the Moody diagram, is independent of the relative roughness.

6.4 PNEUMATIC SYSTEMS

Fundamentals

Isothermal processes with ideal gases
The ideal gas relation is

$$pV = nR_u T$$

$$pv = \frac{p}{\rho} = R_s T \tag{6.58}$$

$$\rho = \frac{1}{R_s T}p \tag{6.59}$$

Thus, for an **isothermal** (constant temperature) process with an ideal gas, we have

$$\frac{d\rho}{dt} = \frac{d}{dt}\left(\frac{1}{R_s T}p\right) = \frac{1}{R_s T}\frac{dp}{dt} \tag{6.60}$$

Polytropic processes with ideal gases

The **polytropic process** is the most general thermodynamic process for general fluids. It is

$$pv^n = p/\rho^n = \text{constant} = K \tag{6.61}$$

where this power n is called the **polytropic exponent.** Note that isothermal processes with ideal gases are special cases of polytropic processes where $n = 1$:

$$pv = \frac{p}{\rho} = R_s T = \text{constant} = K \tag{6.62}$$

We have

$$\frac{d\rho}{dt} = \frac{d\rho}{dp}\frac{dp}{dt} \tag{6.63}$$

For a polytropic process,

$$\frac{dp}{d\rho} = nK\rho^{n-1} = \frac{nK\rho^n}{\rho} = \frac{np}{\rho} \tag{6.64}$$

Thus,

$$\frac{d\rho}{dt} = \frac{1}{n}\frac{\rho}{p}\frac{dp}{dt} \tag{6.65}$$

For an ideal gas,

$$pv = \frac{p}{\rho} = R_s T \tag{6.66}$$

Finally,

$$\frac{d\rho}{dt} = \frac{1}{n}\frac{1}{R_s T}\frac{dp}{dt} \tag{6.67}$$

and

$$\frac{\rho}{np} = \frac{d\rho}{dp} = \frac{1}{nR_s T} \tag{6.68}$$

The development here is general and systematic for polytropic processes and ideal gases. It provides a convenient check using special cases. When $n = 1$, the general result of a polytropic process reduces to that of an isothermal (constant temperature) process.

Pneumatic capacitance

The pneumatic capacitance C is defined as the ratio of the change in stored mass to that of pressure,

$$C = \frac{dm}{dp} \tag{6.69}$$

Thus, for a polytropic process,

$$C = \frac{d(\rho V)}{dp} = V\frac{d\rho}{dp} = V\frac{\rho}{np} \tag{6.70}$$

and an ideal gas,

$$C = V\frac{\rho}{np} = \frac{V}{nR_sT} \tag{6.71}$$

Mathematical Modeling

Precise modeling of pneumatic systems is rather difficult because these systems are often highly nonlinear. However, a simple model that represents the system behavior may be adequate.

Consider the tank shown in Fig. 6.2. The gas flows into the tank through the valve R by pressure gradient. Applying the law of conservation of mass,

$$\frac{dm}{dt} = \rho_i q_i \tag{6.72}$$

Introducing $m = \rho V$,

$$\frac{d}{dt}(\rho V) = \rho_i q_i \tag{6.73}$$

Although the valve resistance R is nonlinear, it may be linearized about an operating condition, then given as

$$\rho_i q_i = \frac{\Delta p}{R} = \frac{p_i - p}{R} \tag{6.74}$$

Thus, for constant volume V,

$$V\frac{d\rho}{dt} = \frac{p_i - p}{R} \tag{6.75}$$

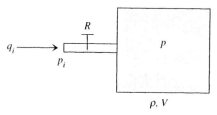

FIGURE 6.2 A pneumatic system: a gas flows into a rigid container.

318 Dynamic Systems: Modeling and Analysis

For an isothermal (constant temperature) process, it has been given previously as

$$\frac{d\rho}{dt} = \frac{1}{R_s T}\frac{dp}{dt}$$

Thus,

$$V\frac{d\rho}{dt} = \frac{V}{R_s T}\frac{dp}{dt} = \frac{p_i - p}{R} \tag{6.76}$$

or

$$\frac{V}{R_s T}\frac{dp}{dt} + \frac{p}{R} = \frac{p_i}{R} \tag{6.77}$$

The differential equation can be rearranged as

$$\frac{RV}{R_s T}\dot{p} + p = p_i \tag{6.78}$$

or

$$RCp + p = p_i \tag{6.79}$$

where the time constant is $\tau = RC = RV/(R_s T)$. The transfer function is given as

$$\frac{P(s)}{P_i(s)} = \frac{1}{RC_s + 1} \tag{6.80}$$

We now reconsider the pneumatic tank shown in Fig. 6.1 but with polytropic processes.

$$\frac{d\rho}{dt} = \frac{1}{n}\frac{1}{R_s T}\frac{dp}{dt} \tag{6.81}$$

Thus,

$$\frac{dm}{dt} = V\frac{d\rho}{dt} = V\frac{1}{nR_s T}\frac{dp}{dt} \tag{6.82}$$

Therefore, the differential equation of the system consisting of a polytropic process with an ideal gas is

$$\frac{V}{nR_s T}\frac{dp}{dt} + \frac{p}{R} = \frac{p_i}{R} \tag{6.83}$$

The differential equation can be rearranged as

$$\frac{RV}{nR_s T}\dot{p} + p = p_i \tag{6.84}$$

or

$$RC\dot{p} + P = p_i \tag{6.85}$$

where the time constant is $\tau = RC = RV/(nR_s T)$.

CHAPTER 6: Fluid and Thermal Systems   319

EXAMPLE 6.2. A PNEUMATIC SYSTEM. Dry air passes through a valve into a rigid 1,000 ft³ container (Fig. 6.3) at a constant temperature of 77°F (25°C). The constant pressure at the inlet of the valve is p_i, which is greater than p. The valve resistance is approximately linear, $R = 2$ lb$_f$-hr/(ft²-lb$_m$). Assume ideal gas, isothermal process, and low pressures. Determine (a) the mathematical model of the pressure p and (b) the time constant.

Solution. From conservation of mass,

$$\frac{dm}{dt} = \frac{d}{dt}(\rho V) = \rho_i q_i$$

Assuming a linear relationship, the valve resistance R is defined as

$$\rho_i q_i = \frac{p_i - p}{R}$$

Thus,

$$V\frac{d\rho}{dt} = \frac{p_i - p}{R}$$

Air at standard pressure and temperature can be approximated as an ideal gas,

$$pv = \frac{p}{\rho} = R_{air}T \qquad \rho = \frac{1}{R_{air}T}p$$

For the constant temperature process,

$$\frac{d\rho}{dt} = \frac{1}{R_{air}T}\frac{dp}{dt}$$

Finally, the differential equation of the system is given as

$$\frac{V}{R_{air}T}\frac{dp}{dt} + \frac{p}{R} = \frac{p_i}{R}$$

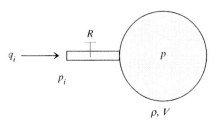

FIGURE 6.3 A pneumatic system.

The differential equation can be rearranged as

$$\frac{RV}{R_{air}T}\dot{p} + p = p_i$$

Thus, the time constant is

$$\tau = \frac{RV}{R_{air}T} = \frac{2\frac{\text{lb}_f\text{-hr}}{\text{ft}^2\text{-lb}_m}(1{,}000 \text{ ft}^3)}{53.35\frac{\text{ft-lb}_f}{\text{lb}_m\text{-}°\text{R}}(537 \text{ °R})} = 0.0698 \text{ hr or } 4.19 \text{ min.}$$

6.5 LIQUID-LEVEL SYSTEMS

Fundamentals

We will only obtain simple system models that represent the gross system behavior, ignoring the details of fluid flow and complex concepts of fluid dynamics. Although dynamic models are to be obtained, the use of *hydrostatic pressure* as opposed to *dynamic pressure* is justifiable here.

Mathematical Modeling

Consider the liquid-level system shown in Fig. 6.4. This system is used to illustrate the basic principles for obtaining the mathematical model of liquid-level systems. The system is a storage tank of cross-sectional area A whose liquid level, or height, is h. The liquid enters the tank from the top and leaves the tank at the bottom through the valve, whose fluid resistance is R. The volume flow rate in and volume flow rate

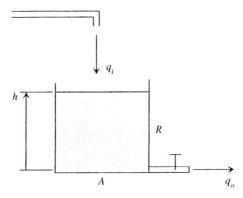

FIGURE 6.4 A liquid-level system.

out are q_i and q_o, respectively. The fluid density ρ is constant. The system has one input, q_i, and one output, h. Thus the system is governed by a single differential equation in h. The basic physical law can be given in mathematical form as

$$\frac{d}{dt}(\rho A h) = \rho q_i - \rho q_o \tag{6.86}$$

This is the **principle of conservation of mass:** *the time rate of change of the fluid mass inside the tank is equal to the mass flow rate in minus the mass flow rate out.*

Next, the variable q_o is expressed in terms of the height h. For simplicity, assuming a linear relationship, the fluid valve resistance R is defined as

$$\rho q_o = \frac{\Delta p}{R} \tag{6.87}$$

where Δp is the pressure difference across the valve and ρq_o is the mass flow rate out of the valve. In this case,

$$\rho q_o = \frac{\Delta p}{R} = \frac{p_1 - p_2}{R} = \frac{(p_a + \rho g h) - p_a}{R} = \frac{\rho g h}{R} \tag{6.88}$$

where point 1 is in the upstream side of the valve where liquid enters, and point 2 is in the downstream side where liquid leaves. We estimate that p_1 is equal to the hydrostatic pressure at point 1 and that p_2 is equal to atmospheric pressure, p_a, since the tank is open to the atmosphere.

$$p_1 = p_a + \rho g h, \qquad p_2 = p_a \tag{6.89}$$

Assuming ρA is constant, the equation becomes

$$\rho A \frac{dh}{dt} = \rho q_i - \frac{\rho g h}{R} \tag{6.90}$$

The model is, finally, obtained by canceling ρ and rearranging the equation in the standard input-output differential equation form as

$$A\dot{h} + \frac{g}{R}h = q_i \tag{6.91}$$

or

$$\frac{RA}{g}\dot{h} + h = \frac{R}{g}q_i \tag{6.92}$$

where the time constant can be identified readily as $\tau = RA/g$.

EXAMPLE 6.3. A TWO-TANK LIQUID-LEVEL SYSTEM. Consider the liquid-level system shown in Fig. 6.5. The heights of tanks 1 and 2 are h_1 and h_2, respectively. The flow resistances at the valves are R_1 and R_2. Assume a linear relationship for resistance. Determine the differential equations.

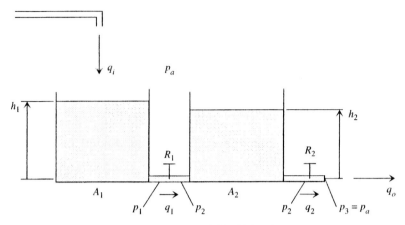

FIGURE 6.5 Two-tank liquid-level system.

Solution. The law of conservation of mass is applied to tank 1:

$$\frac{d}{dt}(\rho A_1 h_1) = \rho q_i - \rho q_1$$

$$= \rho q_i - \frac{p_1 - p_2}{R_1} = \rho q_i - \frac{(p_a + \rho g h_1) - (p_a + \rho g h_2)}{R_1}$$

Thus,

$$A_1 \dot{h}_1 + \frac{g h_1}{R_1} - \frac{g h_2}{R_1} = q_i \qquad (a)$$

Conservation of mass is applied to tank 2:

$$\frac{d}{dt}(\rho A_2 h_2) = \rho q_1 - \rho q_2$$

$$= \frac{p_1 - p_2}{R_1} - \frac{p_2 - p_3}{R_2}$$

$$= \frac{(p_a + \rho g h_1) - (p_a + \rho g h_2)}{R_1} - \frac{(p_a + \rho g h_2) - p_a}{R_2}$$

Thus,

$$A_2 \dot{h}_2 - \frac{g h_1}{R_1} + \left(\frac{1}{R_1} + \frac{1}{R_2}\right) g h_2 = 0 \qquad (b)$$

The system differential equations are Eqs. (a) and (b).

For a dynamic system with a set of simultaneous differential equations, the equations are often put in matrix form so that the correctness can be checked easily. For the system of this example, the coefficient matrices must be symmetric, and all the parameters in the same matrix must have the same dimensions. The second-order matrix form is used for this purpose. After rearranging, the model of the dynamic system is obtained as

$$\begin{bmatrix} A_1 & 0 \\ 0 & A_2 \end{bmatrix} \begin{Bmatrix} \dot{h}_1 \\ \dot{h}_2 \end{Bmatrix} + \begin{bmatrix} g/R_1 & -g/R_1 \\ -g/R_1 & g/R_1 + g/R_2 \end{bmatrix} \begin{Bmatrix} h_1 \\ h_2 \end{Bmatrix} = \begin{Bmatrix} q_i \\ 0 \end{Bmatrix}$$

The system is governed by two dependent differential equations, and each equation is of first order. Thus the order of the system is second.

Note that unlike first-order systems, the time constants of a system of higher order (greater than one) cannot be identified readily. The system of differential equations must first be solved simultaneously, and then the time constants can be determined.

6.6 THERMAL SYSTEMS: HEAT TRANSFER

Fundamentals

Heat transfer is a branch of the thermal sciences. It can be argued that there are only two basic and distinct modes in heat transfer: **conduction** and **radiation** (Rohsenow and Choi, 1961). Conduction is the transfer of heat by molecular motion. In both solids and fluids (gases and liquids), heat is conducted by elastic collision of molecules, an energy diffusion process. The word **convection** implies fluid motion. On the other hand, radiation, or more precisely thermal radiation, belongs to the general category of electromagnetic radiation. In thermal radiation, the energy is transmitted mainly in infrared waves (infrared region). The mechanisms of heat transfer anywhere in the fluid, even within solid bodies, are only conduction and radiation. The processes of conduction and radiation occur simultaneously. In many engineering problems, however, either conduction *or* radiation is dominant.

It is common practice, however, that the modes of heat transfer are conveniently put into three groups: conduction (diffusion through a substance), convection (fluid transport), and thermal radiation (electromagnetic radiation in the infrared region).

Conduction

In general, heat transfer, in an arbitrary x-direction, is governed by the **Fourier law of conduction** or **Fourier's equation** as

$$\frac{q}{A_s} = -k \frac{\partial T}{\partial x} \tag{6.93}$$

where the subscript s denotes surface, and the minus sign indicates that the heat flow is in the direction of decreasing temperature, i.e., negative $\partial T/\partial x$ for positive q/A_s. The equation was named after Fourier, although both Biot and Fourier had suggested it. In this equation,

k = thermal conductivity of the solid or the fluid, Btu/(hr-ft-°F)
T = temperature, °F
x = coordinate for an arbitrary x-direction, ft
q/A_s = heat transfer rate per unit area, Btu/(hr-ft^2)

Although Fourier's equation is valid for both solids and fluids (gases or liquids), in practice we often use it only for solids where k is of the solid.

Convection

Convection usually refers to heat transfer in fluids (gas and liquid). In theory, heat transfer for fluids at the solid-fluid interface is still governed by Fourier's equation as

$$\left(\frac{q}{A}\right)_s = -k_f \left(\frac{\partial T_f}{\partial x}\right)_s \qquad (6.94)$$

where the subscript f indicates fluid, and s, again, stands for the surface of the solid-fluid interface. It is often difficult to evaluate the fluid temperature distribution, T_f, or the partial derivative $\partial T_f/\partial x$ at the surface. It is, however, more practical to determine the heat transfer for fluid by using **Newton's equation,** which was first suggested by Newton as

$$\left(\frac{q}{A}\right)_s = h(T_s - T_f) \qquad (6.95)$$

In this equation,

h = surface coefficient of heat transfer, Btu/(hr-ft^2-°F)
T_s = surface temperature, °F
T_f = fluid temperature, °F
$(q/A)_s$ = rate of heat transfer per unit area, evaluated at the surface, Btu/(hr-ft^2)

The surface coefficient of heat transfer h, for many engineering problems, can be found in tabulated tables in the literature.

A few words must be said about the fluid temperature. If the fluid is infinite, T_f is the fluid temperature at a distance far away from the surface: $T_f = T_\infty$. If the fluid is flowing in a conduit such as a pipe, T_f is the mixed-mean temperature $T_f = T_m$, which is defined as

$$T_m = T_s - \frac{\int_0^R (T_s - T) v 2\pi r \, dr}{\int_0^R v 2\pi r \, dr} \qquad (6.96)$$

where $v = v(r)$ and $T = T(r)$ are the fluid velocity and temperature, respectively; both are functions of radius r; R is the pipe radius.

Thermal radiation

Radiation, or more precisely thermal radiation, is a subset of electromagnetic radiation. In thermal radiation, most of the energy transmitted is mainly in infrared waves or in the infrared region of the electromagnetic spectrum.

Modern theories explain radiant energy on the basis of quantum theory and electromagnetic wave theory. Thermal radiation is based on the **Stefan-Boltzmann law,** which expresses a proportionality between the energy flux, or heat-transfer rate per unit area, emitted by a body radiator and the fourth power of its absolute temperature as

$$\frac{q}{A_s} = e\sigma T^4 \qquad (6.97)$$

where σ is a universal physical constant called the **Stefan-Boltzmann constant**,

$$\sigma = 5.6696 \times 10^{-8} \frac{\text{W}}{\text{m}^2\text{K}^4} \qquad (6.98)$$

The emissivity of the surface e lies between zero and unity, depending on the nature of the surface. The value of e of copper, for example, is 0.3. In general, the emissivity is larger for rough and smaller for smooth surfaces. It is also a weak function of temperature.

When a small body of emissivity e_1 at a temperature T_1 is completely enclosed by walls of emissivity e_2 at a temperature T_2, the net rate of loss (or gain) of radiant energy per unit area is

$$\left.\frac{q}{A_s}\right|_{\text{net}} = e_1\sigma T_1^4 - e_2\sigma T_2^4 = \sigma\left(e_1 T_1^4 - e_2 T_2^4\right) \qquad (6.99)$$

For other complex cases, the law is modified by several empirical factors.

Mathematical Modeling

The mathematical modeling of a thermal system is often complicated because of the complex distribution of temperature. Precise analysis requires a distributed-parameter model that is governed by a partial differential equation. In general, the solutions of partial differential equations are difficult to obtain.

If the gross system behavior is of interest, we only need to obtain a simple model that adequately represents the system. In many applications, we can correctly assume that the temperature of the body involved is uniform. Thus the model is represented by a lumped-parameter model that is governed by an ordinary differential equation.

Biot number

When a solid with an initially uniform temperature is submerged in a hot bath at fluid temperature T_f, the solid temperature distribution varies with time. If the solid is "thin or small," the temperature gradients within the body are negligible and a single value of temperature T can be used to describe the temperature at any point of the body. To provide a criterion for a solid being "thin or small," we introduce the **Biot number**

$$Bi = \frac{hL_c}{k} \qquad (6.100)$$

where k is of the solid. The **characteristic length** L_c for some simple shapes is

Spheres: $\qquad L_c = \dfrac{V}{A_s} = \dfrac{4/3\pi r^3}{4\pi r^2} = \dfrac{1}{3}r \qquad (6.101)$

Solid cylinders: $$L_c = \frac{V}{A_s} = \frac{\pi r^2 L}{\pi(2r)L + 2\pi r^2} = \frac{rL}{2(r+L)} \quad (6.102)$$

Cubes: $$L_c = \frac{V}{A_s} = \frac{L^3}{6L^2} = \frac{L}{6} \quad (6.103)$$

where A_s is the surface of the solid-fluid interface (area of the solid exposed to the fluid); r is the radius; L is the (true) length.

An accepted criterion for a solid with approximate uniform temperature (negligible temperature gradients) is

$$Bi = \frac{hL_c}{k} < 0.1 \quad (6.104)$$

A value smaller than 0.1 for the Biot number can be used if a more conservative result is needed.

Differential equations

For thermal systems with pure heat transfer and no work involved,

$$W_{in} = W_{out} = 0 \quad (6.105)$$

thus the principle of conservation of energy is given as

$$\Delta U_{1 \to 2} = Q_{in} - Q_{out} \quad (6.106)$$

Taking the time derivative on both sides of the equation to obtain the time rates of change,

$$\frac{dU}{dt} = \dot{Q}_{in} - \dot{Q}_{out} \quad (6.107)$$

For simplicity, the dots are dropped and the symbols are redefined; thus

$$\frac{dU}{dt} = q_{in} - q_{out} \quad (6.108)$$

where the internal energy is given as

$$U = mu = \rho V u = \rho V c T \quad (6.109)$$

Thus, the conservation of energy for heat-transfer systems is

$$\frac{d}{dt}(\rho V c T) = q_{in} - q_{out} \quad (6.110)$$

EXAMPLE 6.4. MODELING OF A HEAT-TRANSFER PROCESS. A cold sphere of copper of diameter $d = 0.06$ m is suddenly placed in a large hot liquid bath of temperature T_∞. Since the sphere is heated, the temperature of the sphere, T, rises as a function of time. The following properties of pure copper are given (at 20°C): density $\rho = 8954$ kg/m³, specific heat $c = 0.3831$ kJ/(kg-°C), and thermal conductivity $k = 385$ W/(m-°C). The surface coefficient of heat transfer is given as

$h = 25$ W/(m²-°C). The properties are weak functions of temperature; i.e., they do not change much as temperature varies.

(a) Is it valid to use a lumped-parameter model?
(b) Obtain a mathematical model, i.e., the differential equation in $T(t)$.
(c) Determine the time constant.

Solution

(a) Check the Biot number to see whether the lumped-parameter model is appropriate, i.e., whether we can assume the temperature within the sphere can be approximated as uniform. The characteristic length is

$$L_c = \frac{V}{A_s} = (4/3)\pi r^3 / 4\pi r^2 = \frac{1}{3}r = \frac{1}{3}\left(\frac{0.06}{2}\right) = 0.01 \text{ m}$$

and the Biot number is given as

$$Bi = \frac{hL_c}{k} = \frac{25(0.01)}{385} = 6.49 \times 10^{-4} < 0.1$$

Thus, the temperature within the sphere can be assumed uniform.

(b) Applying conservation of energy,

$$\frac{d}{dt}(\rho V c T) = q_{in}$$

Using the concept of convection heat transfer,

$$\rho V c \frac{dT}{dt} = hA_s(T_\infty - T)$$

After rearranging, the differential equation is

$$\rho V c \dot{T} + hA_s T = hA_s T_\infty$$

or

$$\frac{\rho V c}{hA_s}\dot{T} + T = T_\infty$$

(c) The time constant is

$$\tau = \frac{\rho V c}{hA_s} = \frac{\rho c L_c}{h} = \frac{8954(0.3831)(0.01)}{25}$$

$$= 1.372 \text{ kJ/W} = 1.372 \times 10^3 \text{ s or } 22.9 \text{ min.}$$

(The temperature will reach 98 percent of the steady-state value at 4τ, or 1.5 hr.)

6.7 ANALOGOUS SYSTEMS

Although the physical systems may be completely different, their mathematical models are similar. These systems are **analogous;** their differential equations or transfer functions have the same form. This concept of analogous systems is useful because the understanding of a mechanical system may lead to a deeper appreciation for the analogous electrical system or fluid system, and vice versa.

A simple liquid-level system and a mass-damper system are examples of analogous fluid and mechanical systems. The governing differential equation of the liquid-level system has been given as

$$A\dot{h} + \frac{g}{R}h = q_i$$

where h is the height of the liquid in a tank of constant area A, R is the fluid resistance of the valve, and q_i is the volume flow rate applied to the tank. The mechanical system consists of a mass m, which is subject to an applied force f, which is connected to a viscous damper of damping coefficient b. The mathematical model of the mechanical system is given as

$$m\dot{v} + bv = f$$

We can see that the mass m is analogous to the tank area A; the damping coefficient b is analogous to the reciprocal of the fluid resistance R; the force f is analogous to the volume flow rate q_i.

Although the systems may be analogous, the reader should be aware that the corresponding elements of two different physical systems are not really *equivalent* mathematically. With the above example, the damping coefficient b is not exactly equivalent to the fluid resistance R. As discussed in Chapter 5, mechanical elements are not exactly equivalent to electrical elements either. For example, although both the mechanical damper and electrical resistor dissipate energy, their mathematical expressions for their impedances are quite different. As another example, although the mass (mechanical) and capacitor (electrical) are analogous, the former has only one terminal while the latter is a two-terminal element.

In the past, the study of analogous systems was particularly helpful, because building and testing an analogous electrical system is much easier than working with the real mechanical or fluid system. Obtaining and assembling electrical components together with the instrumentation is simpler. However, with the advent of computer and powerful simulation software packages, studying analogous systems may be no longer necessary.

SUMMARY

The modeling of fluid and thermal systems was presented in this chapter. The three major systems discussed were pneumatic systems, liquid-level systems, and thermal (heat-transfer) systems. The fundamentals of thermodynamics, chemistry, fluid dynamics, and heat transfer were discussed in detail. Topics included ideal gases, real gases, mixtures of gases, conservation of energy, specific heats, isothermal processes, polytropic processes, conduction, and convection.

For an ideal gas of mass m, the relationship among pressure p, total volume V, and temperature T is

$$pV = nR_uT$$

where n is the number of moles, $n = m/M_m$, in which M_m is the molecular mass, and R_u is the universal gas constant. An alternate form is given as

$$pv = R_s T$$

where the specific gas constant is $R_s = R_u/M_m$, and the specific volume is $v = V/m$.

The principle of conservation of energy, or the first law of thermodynamics is given as

$$Q_{in} - W_{out} = \Delta E_{1 \to 2} = E_2 - E_1$$

where $\Delta E_{1 \to 2}$ is the change in the total energy of the system from the reference state 1 to state 2. As given, the heat Q_{in} is positive if it is added *to* the system, whereas the work W_{out} is positive if it is done *by* the system. This principle can also be given as

$$(Q_{in} - Q_{out}) + (W_{in} - W_{out}) = \Delta E_{1 \to 2}$$

In many applications of interest, the change of total energy E reduces to that of internal thermal energy U because only the change in the latter is significant,

$$\Delta E_{1 \to 2} = \Delta U_{1 \to 2}$$

Thus,

$$(Q_{in} - Q_{out}) + (W_{in} - W_{out}) = \Delta U_{1 \to 2}$$

For real gases, the specific heat constants, c_v and c_p, are functions of two thermodynamic properties, e.g., T and v, or T and p. However, for ideal gases, these constants are functions of temperature only, $c_v = c_v(T)$ and $c_p = c_p(T)$. For solids and liquids, a single letter without subscript, c, is often used because these two constants are equal in numerical values, $c_v = c_p = c$.

The relationship between the time rate of change of density ρ and that of pressure p for a polytropic process is given as

$$\frac{d\rho}{dt} = \frac{1}{n} \frac{1}{R_s T} \frac{dp}{dt}$$

which reduces to

$$\frac{d\rho}{dt} = \frac{1}{R_s T} \frac{dp}{dt}$$

for a constant-temperature (isothermal) process because the polytropic exponent is unity, $n = 1$.

The viscosity of a fluid can be represented by the dynamic viscosity μ or the kinematic viscosity $v = \mu/\rho$. This viscosity gives rise to the fluid resistance, which can be defined as

$$R = \frac{\Delta p}{\rho Q}$$

For a steady, fully developed, axisymmetric laminar flow, the fluid resistance is given as

$$R = \frac{\Delta p}{\rho Q} = \frac{128 \nu L}{\pi D^4} = \text{constant}$$

whereas the resistance for a turbulent flow is not constant. For a small deviation from the operating point, a constant fluid resistance, however, can be obtained through linearization.

For a pneumatic system, with a single container of volume V, the principle of conservation of mass is given as

$$\frac{d}{dt}(\rho V) = \rho q_i - \rho q_o$$

where ρq_i and ρq_o are the mass flow rate in and the mass flow rate out of the gas, respectively.

For a liquid-level tank of height h and cross sectional area A, the principle of conservation of mass is expressed similarly as

$$\frac{d}{dt}(\rho A h) = \rho q_i - \rho q_o$$

where ρq_i and ρq_o are the mass flow rate in and the mass flow rate out of the liquid, respectively.

Heat transfer can be categorized into three groups, and the relations are given as follows.

Conduction: $\quad \dfrac{q}{A_s} = -k \dfrac{\partial T}{\partial x}$

Convection: $\quad \left(\dfrac{q}{A}\right)_s = h(T_s - T_f)$

Radiation: $\quad \dfrac{q}{A_s} = e \sigma T^4$

In these expressions, the subscripts s and f denote the surface of the solid and the property of the fluid, respectively.

Similar to other dynamic systems, a simple mathematical model of a heat-transfer system can be obtained by using the lumped parameter method. In this method, the temperature of a solid is approximated uniform, and the criterion is that the Biot number, Bi, must be less than 0.1,

$$Bi = \frac{h L_c}{k} < 0.1$$

This approximation is valid if the body is relatively small or thin.

The characteristic length L_c for some simple shapes is

Spheres: $\quad L_c = \tfrac{1}{3} r$

Nonhollow cylinders: $$L_c = \frac{rL}{2(r+L)}$$

Rectangular cubes: $$L_c = \frac{L}{6}$$

where r is the radius and L is the real physical length.

For a solid or liquid, with pure heat transfer and no work involved, the principle of conservation of energy is given as

$$\frac{d}{dt}(\rho V c T) = q_{\text{in}} - q_{\text{out}}$$

PROBLEMS

6.1. Determine the density of dry air at atmospheric pressure and 76°F.

6.2. Determine the density of dry air at a standard condition: 1 atm and 25°C.

6.3. Imagine a gaseous mixture whose volumetric composition is 78% nitrogen, 21% oxygen, and 1% carbon dioxide. Determine (a) the mole fraction of each component, and (b) the molecular mass of the mixture.

6.4. Dry air passes through a valve into a rigid cubic container of 1 m on each side (Fig. P6.4) at a constant temperature of 20°C (68°F). The constant pressure at the inlet of the valve is p_i, which is greater than p. The valve resistance is approximately linear, $R = 1000$ Pa-s/kg. Assume ideal gas, isothermal process, and low pressures. Determine (a) a mathematical model of the pressure p and (b) the time constant.

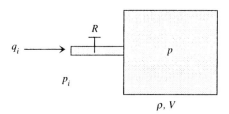

FIGURE P6.4

6.5. Consider the one-tank liquid-level system shown in Fig. P6.5. The system is a storage tank of cross-sectional area A, whose liquid level or height is h. The liquid enters the tank from the top and leaves the tank at the bottom through two valves whose linear flow resistances are R_1 and R_2. The volume-flow-rate in is q_i. The fluid density ρ is constant. Determine a mathematical model or the differential equation in h and the time constant.

6.6. Consider the liquid-level system shown in Fig. P6.6. Tanks 1 and 2 are of heights h_1 and h_2, respectively, and constant areas A_1 and A_2, respectively. The two valves have linear flow resistances of R_1 and R_2. A volume flow rate in, q_i, is applied to tank 2.

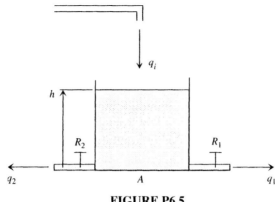

FIGURE P6.5

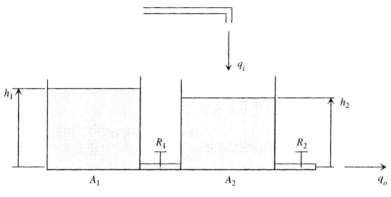

FIGURE P6.6

(a) Derive the differential equations.
(b) Express the equations in the second-order matrix form.
(c) Obtain a state-space representation assuming the output is h_1.

6.7. Consider the liquid-level system shown in Fig. P6.7. Tanks 1 and 2 are of heights h_1 and h_2, respectively, and constant areas A_1 and A_2, respectively. There are three valves whose linear flow resistances are R_1, R_2, and R_3. A volume flow rate in, q_i, is applied to tank 1. (a) Derive the differential equations, then put them in (b) the second-order matrix form and (c) the state-space representation where the output is h_2.

6.8. Consider a metal cube of equal sides $L = 10$ in. and $k = 212$ Btu/(hr-ft-°F) at low temperature T that is immersed in a large liquid bath at high temperature T_∞. The surface coefficient of heat transfer is given as $h = 5$ Btu/(hr-ft²-°F). Obtain the Biot number, and then determine whether a lumped-parameter model is appropriate.

6.9. Initially in a hot oven ($T_i = 200°C$), a pure aluminum sphere of diameter $d = 0.08$ m is removed and placed in air at room temperature ($T_\infty = 20°C$ or 68°F). Since it is cooled, the temperature of the sphere, T, drops as a function of time. The following properties

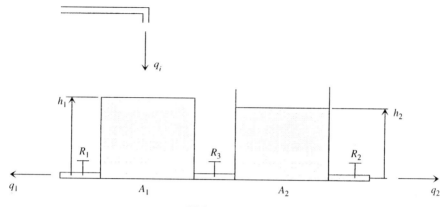

FIGURE P6.7

of pure aluminum are given (at 20°C): $\rho = 2707$ kg/m³, $c = 0.896$ kJ/(kg-°C), $k = 204$ W/(m-°C). The surface coefficient of heat transfer is given as $h = 3.5$ W/(m²-°C). The properties are weak functions of temperature; i.e. they do not change much as temperature varies.
(a) Is it valid to use a lumped-parameter model?
(b) Obtain a mathematical model, i.e., the differential equation in $T(t)$.
(c) Determine the time constant.

6.10. Initially in an air environment ($T_i = 20°C$), a pure-copper cylinder of diameter $d = 0.02$ m and length $L = 0.10$ m is removed and placed in a hot oven ($T_\infty = 300°C$). Since it is heated, the temperature of the cylinder, T, rises as a function of time. The following properties of pure copper are given (at 20°C): $\rho = 8954$ kg/m³, $c = 0.3831$ kJ/(kg-°C), $k = 385$ W/(m-°C). The surface coefficient of heat transfer is given as $h = 3.5$ W/(m²-°C). The properties are weak functions of temperature; i.e., they do not change much as temperature varies.
(a) Is it valid to use a lumped-parameter model?
(b) Obtain a mathematical model, i.e., the differential equation in $T(t)$.
(c) Determine the time constant.

6.11. Consider a cold metal sphere of radius $r = 0.5$ in. with $k = 212$ Btu/(hr-ft-°F) at temperature T that is immersed in a large hot liquid bath at temperature T_∞. For the metal, the density is $\rho = 513.2$ lb$_m$/ft³ and the specific heat is $c = 2.93$ Btu/(lb$_m$-°F). The surface coefficient of heat transfer is given as $h = 5$ Btu/(hr-ft²-°F). Assume the heat loss from the hot bath to the environment is negligible. Determine the differential equation and the time constant.

CHAPTER 7

System Response

This chapter is concerned with the response of dynamic systems corresponding to specified inputs. The types of input signals that are frequently used include impulse, step, ramp, and sinusoidal functions, all of which are defined and discussed in detail in Chapter 1. The nature of the selected input signal depends on the physical characteristics of the disturbance to which the system is subjected. For instance, if a certain system is subjected to an input of large magnitude that is applied during a very short period of time, then an impulse function may be selected to approximately model the input.

Two types of system response are discussed in this chapter: *transient response* and *frequency response*. Analytical determination of the response is possible through application of the Laplace transformation, the method of undetermined coefficients, and what is known as the **convolution integral.** Whereas the first two methods are thoroughly covered in Chapter 1, the third is discussed in a later section (Section 7.6). Among these three techniques, the Laplace transform is the most commonly used. This is partly because of the fact that the method of undetermined coefficients is limited to special types of forcing functions, as listed in Table 1.1.

7.1 TYPES OF RESPONSE

Consider an nth-order dynamic system that is modeled by its input-output equation in the form

$$y^{(n)} + a_1 y^{(n-1)} + \cdots + a_{n-1}\dot{y} + a_n y = f(t) \qquad (7.1)$$

and that is subjected to prescribed initial conditions, $y^{(n-1)}(0^-), \ldots, \dot{y}(0^-), y(0^-)$. The function $f(t)$ is called the **applied forcing function,** and the solution to Eq. (7.1) is known as the **forced response** of the system. In the event that the system is subjected to zero input, i.e., $f(t) = 0$, then Eq. (7.1) reduces to a homogeneous differential equation in the form

$$y^{(n)} + a_1 y^{(n-1)} + \cdots + a_{n-1}\dot{y} + a_n y = 0 \qquad (7.2)$$

the solution of which is referred to as the **free response**. The portion of the response that approaches zero after a sufficiently long time (i.e., as $t \to \infty$) is referred to as the **transient response**. The behavior of the system's response as $t \to \infty$ is known as the **steady-state response**.

EXAMPLE 7.1. Suppose that the response of a dynamic system is given by

$$x(t) = e^{-t} - 2e^{-3t} + \sin 2t$$

Then the portion of the response that approaches zero as $t \to \infty$ is $e^{-t} - 2e^{-3t}$, which represents the transient response. On the other hand, as $t \to \infty$, the response behaves as $\sin 2t$, which is the steady-state response.

7.2 TRANSIENT RESPONSE OF FIRST-ORDER SYSTEMS

Consider a linear, first-order dynamic system for which the governing equation is given by

$$\tau \dot{y} + y = f(t), \qquad \tau > 0 \qquad (7.3)$$

and subjected to initial condition $y(0^-) = y_0$. Recall that τ is known as the **time constant** and provides a measure of how fast the system responds to an input. Although Eq. (7.3) does not exactly agree with the standard form, Eq. (1.21), it contains the same information about the system. As mentioned earlier, the system's response corresponding to a specific input may then be obtained via Laplace transformation. To that end, two types of input signals, step and ramp functions, will be considered here.

Free Response

We speak of free response when the system represented by Eq. (7.3) is not subjected to a forcing function and is only driven by its initial condition, $y(0^-) = y_0$. In this event, Laplace transformation of Eq. (7.3) with $f = 0$ yields

$$\tau[sY(s) - y_0] + Y(s) = 0 \implies Y(s) = \frac{\tau y_0}{\tau s + 1} = \frac{y_0}{s + 1/\tau}$$

the inverse Laplace transform of which is given by

$$y(t) = y_0 e^{-(t/\tau)} \qquad \text{Free response of a first-order system}$$

Because $\tau > 0$, the free response will start at y_0 and decay exponentially to zero as t increases, as Fig. 7.1 illustrates.

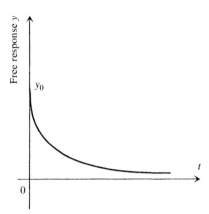

FIGURE 7.1 Free response of a first-order system.

Forced Response

Let us now extend the idea to when the system is not only driven by its initial condition but also an applied forcing function. In particular, in what follows, the response of a first-order system is sought when it is subjected to a step input and a ramp input. The terms *unit-step response* and *unit-impulse response* are used in the event that the system is subjected to a unit-step or a unit-impulse input, respectively, and zero initial conditions. Otherwise, in general, they are referred to as *step response* and *impulse response*, respectively.

Step Response

Assume that the input is $f(t) = Au(t)$ where $u(t)$ denotes the unit-step function. Analytically speaking,

$$f(t) = \begin{cases} A & \text{if } t > 0 \\ 0 & \text{if } t < 0 \end{cases}$$

Taking the Laplace transform of Eq. (7.3), and taking into account the prescribed initial condition, we have

$$\tau[sY(s) - y_0] + Y(s) = \frac{A}{s} \implies Y(s) = \frac{\tau y_0 s + A}{s(\tau s + 1)} \qquad (7.4)$$

Using the partial fraction method, one obtains

$$Y(s) = \frac{\tau y_0 s + A}{s(\tau s + 1)} = \frac{y_0 s + A/\tau}{s(s + 1/\tau)} = \frac{c_1}{s} + \frac{c_2}{s + 1/\tau}$$

where $\quad c_1 = [sY(s)]_{s=0} = \left[\dfrac{\tau y_0 s + A}{\tau s + 1}\right]_{s=0} = A$

and $\quad c_2 = \left[\left(s + \dfrac{1}{\tau}\right)Y(s)\right]_{s=-1/\tau} = \left[\dfrac{y_0 s + A/\tau}{s}\right]_{s=-1/\tau} = y_0 - A$

Back substitution into the expression of $Y(s)$ results in

$$Y(s) = \frac{A}{s} + (y_0 - A)\frac{1}{s + 1/\tau}$$

Term-by-term inverse Laplace transformation yields the **step response** as follows:

$$y_{\text{step}}(t) = A + (y_0 - A)e^{-(t/\tau)}, \qquad t \geq 0 \tag{7.5}$$

Because $\tau > 0$, the second term in Eq. (7.5) involves a negative exponent and, hence, approaches zero as $t \to \infty$. The second term is what we referred to as the transient response, while the first term, A, represents the steady-state response, so that $(y_{\text{step}})_{ss} = A$. Consequently, the system response tracks the step input after a sufficient length of time. Define the **steady-state error,** e_{ss}, as the difference between the input, $f(t)$, and the steady-state response of the system. In relation to the current problem, this error is then expressed as

$$e_{ss} = f(t) - (y_{\text{step}})_{ss} = A - A = 0$$

which only confirms the earlier remark that the response eventually tracks the step input. To illustrate the results, assume that $y_0 = 0$, which causes Eq. (7.5) to reduce to

$$y_{\text{step}}(t) = A(1 - e^{-(t/\tau)}), \qquad t \geq 0 \tag{7.6}$$

The system response, as well as the associated steady-state error, are shown in Fig. 7.2.

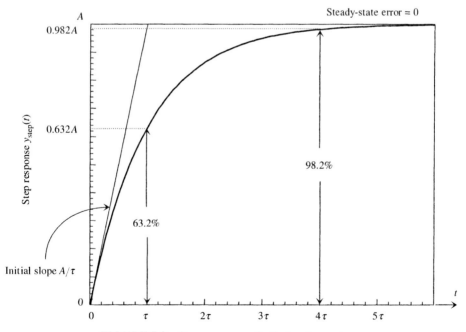

FIGURE 7.2 Step response of a first-order system.

Observations

The significance of the time constant in the step response of a first-order system may easily be realized from the following information. Referring to Fig. 7.2, as well as Eq. (7.6), after one time constant (i.e., $t = \tau$) observe that

$$y(\tau) = A(1 - e^{-1}) = 0.632A$$

which implies that the response reaches 63.2 percent of its final value. Similar evaluations at a few subsequent times result in

$$y(2\tau) = 0.865A, \quad y(3\tau) = 0.95A, \quad y(4\tau) = 0.982A, \quad y(5\tau) = 0.995A$$

By inspection, it is seen that beyond $t = 4\tau$, the step response stays within 0.02 (2 percent) of its final value, A. From the practical standpoint, the time required for the response to reach steady-state can be considered as four time constants.

Ramp Response

Consider the system in Eq. (7.3) and assume that the input is

$$f(t) = \begin{cases} At & \text{if } t > 0 \\ 0 & \text{if } t < 0 \end{cases}$$

which represents a ramp function. Laplace transformation of Eq. (7.3) and solving for $Y(s)$ yields

$$\tau[sY(s) - y_0] + Y(s) = \frac{A}{s^2} \implies Y(s) = \frac{A/\tau + y_0 s^2}{s^2(s + 1/\tau)} = \frac{B_2}{s^2} + \frac{B_1}{s} + \frac{C}{s + 1/\tau} \tag{7.7}$$

where the residues, B_1, B_2, and C, are calculated as

$$B_2 = A, \quad B_1 = -\tau A, \quad C = y_0 + \tau A$$

Substituting into Eq. (7.7) a term-by-term inverse Laplace transformation, and assuming $y_0 = 0$, results in

$$y_{\text{ramp}}(t) = At - A\tau + A\tau e^{-(t/\tau)}$$

Thus, the steady-state response is given by

$$(y_{\text{ramp}})_{ss}(t) = At - A\tau$$

and the corresponding steady-state error is determined as

$$e_{ss} = f(t) - (y_{\text{ramp}})_{ss}(t)$$
$$= At - (At - A\tau) = A\tau$$

which is clearly dependent on the slope A of the ramp function, as well as the time constant. System response, together with the ramp input are shown in Fig. 7.3. Observe that, because e_{ss} is constant, the asymptotic behavior of the response, characterized by $(y_{\text{ramp}})_{ss} = At - A\tau$, and the ramp input At are parallel lines.

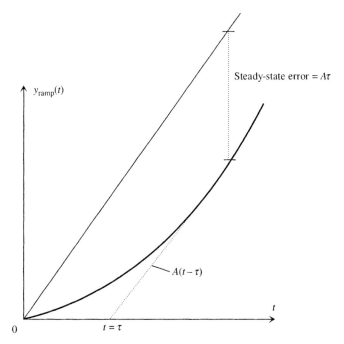

FIGURE 7.3 Ramp response of a first-order system.

7.3 TRANSIENT RESPONSE OF SECOND-ORDER SYSTEMS

Consider a linear, second-order dynamic system whose governing equation in standard normalized form, as in Eq. (1.22), is given by

$$\ddot{x} + 2\zeta\omega_n\dot{x} + \omega_n^2 x = \tilde{f}(t) \tag{7.8}$$

Assume that the system is subjected to prescribed initial conditions $x(0)$ and $\dot{x}(0)$ and an applied forcing function $f(t)$. Recall from Section 1.9 that ζ and ω_n are known as the *damping ratio* and the *undamped natural frequency* of the system, respectively. The damping ratio is defined as the ratio of the system's actual and critical damping, to be discussed later. Equation (7.8) represents mathematical models of many physical systems, although quite different in nature, with a single degree of freedom. These may range from a spring-mass-damper mechanical system to an RLC electrical circuit. The damping ratio and the undamped natural frequency for an RLC circuit were identified in Example 1.12. A similar analysis may be conducted for a single-degree-of-freedom mechanical system, as shown in Fig. 7.4. The system is subjected to given initial conditions and an applied external force. The equation of motion, as well as the prescribed initial conditions, are given as

$$m\ddot{x} + b\dot{x} + kx = f(t), \qquad x(0) = x_0, \qquad \dot{x}(0) = v_0 \tag{7.9}$$

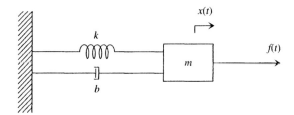

FIGURE 7.4 Single-degree-of-freedom mechanical system.

Dividing by m and comparing with the standard form, Eq. (7.8), yields

$$\ddot{x} + \frac{b}{m}\dot{x} + \frac{k}{m}x = \frac{1}{m}f(t) \implies \begin{cases} \dfrac{k}{m} = \omega_n^2 \\ \dfrac{b}{m} = 2\zeta\omega_n \end{cases}$$

solution of which gives

$$\omega_n = \sqrt{\frac{k}{m}}, \quad \zeta = \frac{b}{2\sqrt{mk}} \tag{7.10}$$

Free Response

In the event that $f = 0$ in Eq. (7.8), the response of the second-order system being considered is referred to as the **free response**. Laplace transformation of Eq. (7.8), taking into account the given initial conditions, yields

$$\left[s^2 X(s) - sx(0^-) - \dot{x}(0^-)\right] + 2\zeta\omega_n\left[sX(s) - x(0^-)\right] + \omega_n^2 X(s) = 0$$

Using the notations of x_0 and v_0 for the initial conditions and solving the above equation for $X(s)$, one obtains

$$X(s) = \frac{(s + 2\zeta\omega_n)x_0 + v_0}{s^2 + 2\zeta\omega_n s + \omega_n^2} \tag{7.11}$$

The objective here is to find the response $x(t)$ via inverse Laplace transformation of $X(s)$. However, the manner in which this problem is approached depends entirely on whether $X(s)$ possesses real or complex poles. To that end, consider the **characteristic equation** that generates these poles:

$$s^2 + 2\zeta\omega_n s + \omega_n^2 = 0 \tag{7.12}$$

Solving Eq. (7.12) leads to

$$s_{1,2} = -\zeta\omega_n \pm \sqrt{(\zeta\omega_n)^2 - \omega_n^2} = -\zeta\omega_n \pm \omega_n\sqrt{\zeta^2 - 1} \tag{7.13}$$

In what follows, we conduct a thorough investigation of the free response associated with specific values, or range of values, of the damping ratio. The system is said to be **overdamped** if $\zeta > 1$, **critically damped** if $\zeta = 1$, **underdamped** if $0 < \zeta < 1$, and **undamped** when $\zeta = 0$.

Case 1. Overdamped, $\zeta > 1$. In this case, the poles of $X(s)$, defined by Eq. (7.13), are real and distinct. Consequently, the expression of $X(s)$, Eq. (7.11), may be rewritten as

$$X(s) = \frac{(s + 2\zeta\omega_n)x_0 + v_0}{s^2 + 2\zeta\omega_n s + \omega_n^2} = \frac{(s + 2\zeta\omega_n)x_0 + v_0}{(s - s_1)(s - s_2)}$$

where $s_1 = -\zeta\omega_n + \omega_n\sqrt{\zeta^2 - 1}$ and $s_2 = -\zeta\omega_n - \omega_n\sqrt{\zeta^2 - 1}$. Applying the partial fraction method to $X(s)$ results in

$$X(s) = \frac{(s + 2\zeta\omega_n)x_0 + v_0}{(s - s_1)(s - s_2)} = \frac{A_1}{s - s_1} + \frac{A_2}{s - s_2} \quad (7.14)$$

where

$$A_1 = [(s - s_1)X(s)]_{s=s_1} = \left.\frac{(s + 2\zeta\omega_n)x_0 + v_0}{s - s_2}\right|_{s=s_1} = \frac{\omega_n(\zeta + \sqrt{\zeta^2 - 1})x_0 + v_0}{2\omega_n\sqrt{\zeta^2 - 1}}$$

and

$$A_2 = [(s - s_2)X(s)]_{s=s_2} = \left.\frac{(s + 2\zeta\omega_n)x_0 + v_0}{s - s_1}\right|_{s=s_2} = -\frac{\omega_n(\zeta - \sqrt{\zeta^2 - 1})x_0 + v_0}{2\omega_n\sqrt{\zeta^2 - 1}}$$

Insert these expressions for A_1 and A_2 into Eq. (7.14) and take the term-by-term inverse Laplace transform to obtain

$$x(t) = \left[\frac{\omega_n(\zeta + \sqrt{\zeta^2 - 1})x_0 + v_0}{2\omega_n\sqrt{\zeta^2 - 1}}\right]e^{-\omega_n(\zeta - \sqrt{\zeta^2-1})t}$$
$$- \left[\frac{\omega_n(\zeta - \sqrt{\zeta^2 - 1})x_0 + v_0}{2\omega_n\sqrt{\zeta^2 - 1}}\right]e^{-\omega_n(\zeta + \sqrt{\zeta^2-1})t} \quad \text{Free response of an overdamped system}$$

(7.15)

Note that because $\zeta > \sqrt{\zeta^2 - 1}$, the exponents involved in both terms of Eq. (7.15) are negative. This implies that the response will decay toward zero after a sufficiently long time.

Case 2. Critically damped, $\zeta = 1$. Treatment of this case is quite different from that of the previous case, mainly because for $\zeta = 1$, Eq. (7.13) generates a double root: $s_{1,2} = -\omega_n$. In this event, directly substitute $\zeta = 1$ into Eq. (7.11) to obtain

$$X(s) = \frac{(s + 2\omega_n)x_0 + v_0}{s^2 + 2\omega_n s + \omega_n^2} = \frac{(s + \omega_n)x_0 + \omega_n x_0 + v_0}{(s + \omega_n)^2}$$

which may be rewritten as

$$X(s) = \frac{(s + \omega_n)x_0}{(s + \omega_n)^2} + \frac{\omega_n x_0 + v_0}{(s + \omega_n)^2} = \frac{x_0}{s + \omega_n} + \frac{\omega_n x_0 + v_0}{(s + \omega_n)^2}$$

Term-by-term inverse Laplace transformation yields

$$x(t) = x_0 e^{-\omega_n t} + (\omega_n x_0 + v_0)t e^{-\omega_n t} \quad \text{Free response of a critically damped system} \quad (7.16)$$

As $t \to \infty$, the first term approaches zero. This is also true for the second term, because the rate at which $e^{-\omega_n t} \to 0$ is considerably faster than that at which $t \to \infty$. As a result, after a sufficient amount of time, the response will converge to zero.

Case 3. Underdamped, $0 < \zeta < 1$. In the event that $0 < \zeta < 1$, the solutions to the characteristic equation, Eq. (7.12), form a pair of complex conjugates. Then recall from Chapter 1 that $s^2 + 2\zeta\omega_n s + \omega_n^2$ is called irreducible, and completion of the square is advised prior to inverse Laplace transformation. In this case, as previously discussed in Section 1.9, it is customary to define

$$\sigma = \zeta\omega_n \quad \text{and} \quad \omega_d = \omega_n\sqrt{1 - \zeta^2}$$

where ω_d is known as the **damped natural frequency**. Rewrite $s^2 + 2\zeta\omega_n s + \omega_n^2$ as follows:

$$s^2 + 2\zeta\omega_n s + \omega_n^2 = (s + \zeta\omega_n)^2 - \zeta^2\omega_n^2 + \omega_n^2$$
$$= (s + \sigma)^2 + \omega_d^2$$

With this information, the expression of $X(s)$ can be rewritten in the form

$$X(s) = \frac{(s + 2\zeta\omega_n)x_0 + v_0}{s^2 + 2\zeta\omega_n s + \omega_n^2} = \frac{(s + 2\sigma)x_0 + v_0}{(s + \sigma)^2 + \omega_d^2}$$

Algebraic manipulation of this expression yields

$$X(s) = \frac{(s + \sigma)x_0}{(s + \sigma)^2 + \omega_d^2} + \frac{\sigma x_0 + v_0}{(s + \sigma)^2 + \omega_d^2}$$

$$= \frac{(s + \sigma)x_0}{(s + \sigma)^2 + \omega_d^2} + \left(\frac{\sigma x_0 + v_0}{\omega_d}\right)\frac{\omega_d}{(s + \sigma)^2 + \omega_d^2}$$

Consequently, term-by-term inversion and using Table 1.2, we have

$$x(t) = x_0 e^{-\sigma t}\cos\omega_d t + \left(\frac{\sigma x_0 + v_0}{\omega_d}\right)e^{-\sigma t}\sin\omega_d t$$

$$= e^{-\sigma t}\left[x_0 \cos\omega_d t + \left(\frac{\sigma x_0 + v_0}{\omega_d}\right)\sin\omega_d t\right] \quad \text{Free response of an underdamped system} \quad (7.17)$$

As in the two previous cases, the response approaches zero as $t \to \infty$. This is because the quantity in brackets is finite (bounded) and $e^{-\sigma t} \to 0$ eventually.

Case 4. Undamped, $\zeta = 0$. When the system has no inherent damping (i.e., $\zeta = 0$), the expression of $X(s)$, Eq. (7.11), reduces to

$$X(s) = \frac{sx_0 + v_0}{s^2 + \omega_n^2} = x_0\left(\frac{s}{s^2 + \omega_n^2}\right) + \frac{v_0}{\omega_n}\left(\frac{\omega_n}{s^2 + \omega_n^2}\right)$$

the inverse Laplace transform of which is given by

$$x(t) = x_0 \cos \omega_n t + \frac{v_0}{\omega_n} \sin \omega_n t \qquad \text{Free response of an undamped system} \qquad (7.18)$$

Notice that because there is no inherent damping in the system, unlike the first three damping cases, the response will not decay to zero. In fact, it exhibits oscillatory behavior with time.

In order to gain better insight into the general behavior of the responses in all four possible cases, we present them in graphical form in Fig. 7.5. For the purpose of illustration of results, the following parameter values are used:

$$\omega_n = 1, \qquad x_0 = 0.5, \qquad v_0 = 0$$

The four cases discussed above may now be interpreted in relation to the mechanical system of Fig. 7.4. The characteristic equation associated with the system's equation of motion, Eq. (7.9), is given by

$$m\lambda^2 + b\lambda + k = 0 \tag{7.19}$$

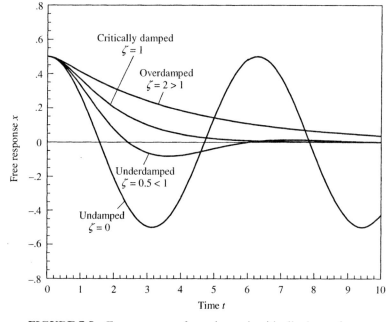

FIGURE 7.5 Free response of overdamped, critically damped, underdamped, and undamped second-order systems; $x_0 \neq 0$, $v_0 = 0$.

solutions of which are the characteristic values, that is,

$$\lambda_{1,2} = \frac{-b \pm \sqrt{b^2 - 4mk}}{2m} \quad (7.20)$$

In reference to case 2, a critically damped system, we learned that there exists a pole of multiplicity two. This means that the quantity under the radical must vanish. In relation to Eq. (7.20), this implies that the discriminant is zero:

$$b^2 - 4mk = 0 \implies b_{cr} = 2\sqrt{mk} \quad (7.21)$$

This specific value of b, denoted by b_{cr}, is known as the **critical damping**. At this time, recall from an earlier remark that the damping ratio is defined as the ratio of actual damping and critical damping. The actual damping of the mechanical system under consideration is characterized by the damping coefficient b, and critical damping is given by Eq. (7.21). Consequently,

$$\zeta = \frac{\text{value of actual damping} = b}{\text{value of critical damping} = 2\sqrt{mk}} = \frac{b}{2\sqrt{mk}}$$

which agrees with the earlier result, Eq. (7.10). The system is said to be overdamped if $b > 2\sqrt{mk}$, critically damped if $b = 2\sqrt{mk}$, underdamped if $b < 2\sqrt{mk}$, and undamped if $b = 0$.

EXAMPLE 7.2. Consider the single-degree-of-freedom mechanical system shown in Fig. 7.4 with $f = 0$. Assume the following parameter values: $m = 1$ kg, $b = 2$ N·s/m, $k = 2$ N/m, and initial conditions $x_0 = 1$ m, $v_0 = 0$. Determine the system's free response.

Solution. A simple inspection of the damping ratio reveals that

$$\zeta = \frac{b}{2\sqrt{mk}} = \frac{1}{\sqrt{2}} = 0.707 < 1$$

indicating that the system is underdamped, and the results of case 3 will apply. Furthermore, the undamped natural frequency is calculated as

$$\omega_n = \sqrt{\frac{k}{m}} = \sqrt{2} \text{ rad/s}$$

using which the damped natural frequency may be obtained as

$$\omega_d = \omega_n \sqrt{1 - \zeta^2} = \sqrt{2}\sqrt{1 - (1/2)} = 1 \text{ rad/s}$$

Subsequently, substitution of the given parameter values, as well as ζ and ω_d, into Eq. (7.17) yields the system's free response. This is represented analytically as

$$x(t) = e^{-t}\cos t + e^{-t}\sin t = e^{-t}(\cos t + \sin t) \text{ m}$$

and graphically in Fig. 7.6.

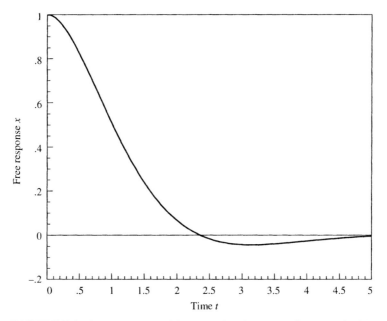

FIGURE 7.6 Free response of the second-order system in Example 7.2.

Forced Response

We speak of forced response when the dynamic system is driven by an applied forcing function, and possibly nonzero initial conditions. In what follows, the response of second-order systems to two types of inputs will be investigated: unit-impulse and unit-step.

Response to a Unit Impulse

Consider a linear, second-order system governed by Eq. (7.8) and subjected to an impulsive forcing function. Then the governing equation is represented by

$$\ddot{x} + 2\zeta\omega_n\dot{x} + \omega_n^2 x = f(t) = \delta(t) \qquad (7.22)$$

subject to initial conditions $x(0^-) = x_0$ and $\dot{x}(0^-) = v_0$. Laplace transformation of Eq. (7.22), taking into consideration the specified initial conditions, and rearranging terms, yields

$$[s^2 X(s) - sx(0^-) - \dot{x}(0^-)] + 2\zeta\omega_n[sX(s) - x(0^-)] + \omega_n^2 X(s) = F(s)$$

$$\Rightarrow X(s) = \left[\frac{1}{s^2 + 2\zeta\omega_n s + \omega_n^2}\right] F(s) + \frac{(s + 2\zeta\omega_n)x_0 + v_0}{s^2 + 2\zeta\omega_n s + \omega_n^2}$$

Thus, in the general case, the system's response to a unit impulse and general initial conditions is determined via inverse Laplace transformation of the previous expression, using $F(s) = \Delta(s) = 1$, as

$$x(t) = L^{-1}\left\{\frac{1}{s^2 + 2\zeta\omega_n s + \omega_n^2} + \frac{(s + 2\zeta\omega_n)x_0 + v_0}{s^2 + 2\zeta\omega_n s + \omega_n^2}\right\}$$

Response of a second-order system to a unit impulse and general initial conditions

(7.23)

In the event that initial conditions are zero, the system's transfer function is obtained directly through Laplace transformation of Eq. (7.22), as

$$G(s) = \frac{X(s)}{F(s)} = \frac{1}{s^2 + 2\zeta\omega_n s + \omega_n^2} \qquad (7.24)$$

which is precisely the first quantity appearing within the brackets in Eq. (7.23). Note that the right-hand side of Eq. (7.23) consists of two terms. The first, involving the transfer function, characterizes the response to a nonzero input when initial conditions are zero. The second, involving x_0 and v_0, signifies the contribution of initial conditions when the system is subjected to zero input. Thus, it characterizes the system's response due to initial conditions and zero input. With the assumption of zero initial conditions, and noting that $F(s) = 1$, the response is obtained directly from Eq. (7.24) as

$$x(t) = L^{-1}\{G(s)\} = L^{-1}\left\{\frac{1}{s^2 + 2\zeta\omega_n s + \omega_n^2}\right\}$$

Unit-impulse response of a second-order system—zero initial conditions

(7.25)

and is known as the **unit-impulse response**. The unit-impulse response of a dynamic system is the inverse Laplace transform of the system's transfer function, provided that the initial conditions are zero. A thorough discussion of the impulse response of second-order systems for all four possible cases of damping follows.

Case 1. Overdamped, $\zeta > 1$. Following an analysis similar to that leading to Eq. (7.17), the expression of $x(t)$ given by Eq. (7.23) takes the specific form

$$x(t) = \frac{1}{2\omega_n\sqrt{\zeta^2 - 1}}e^{-\omega_n(\zeta - \sqrt{\zeta^2-1})t} - \frac{1}{2\omega_n\sqrt{\zeta^2 - 1}}e^{-\omega_n(\zeta + \sqrt{\zeta^2-1})t}$$

Response to unit impulse

$$+ \left[\frac{\omega_n\left(\zeta + \sqrt{\zeta^2 - 1}\right)x_0 + v_0}{2\omega_n\sqrt{\zeta^2 - 1}}\right]e^{-\omega_n(\zeta - \sqrt{\zeta^2-1})t} \qquad (7.26)$$

$$- \left[\frac{\omega_n\left(\zeta - \sqrt{\zeta^2 - 1}\right)x_0 + v_0}{2\omega_n\sqrt{\zeta^2 - 1}}\right]e^{-\omega_n(\zeta + \sqrt{\zeta^2-1})t}$$

Response to initial conditions

CHAPTER 7: System Response   347

Case 2. Critically damped, $\zeta = 1$. The inverse Laplace transform in Eq. (7.23) results in

$$x(t) = \underbrace{te^{-\omega_n t}}_{\text{Response to unit impulse}} + \underbrace{x_0 e^{-\omega_n t} + (\omega_n x_0 + v_0)te^{-\omega_n t}}_{\text{Response to initial conditions}} \quad (7.27)$$

Case 3. Underdamped, $0 < \zeta < 1$.

$$x(t) = \underbrace{\frac{1}{\omega_d} e^{-\sigma t} \sin \omega_d t}_{\text{Response to unit impulse}} + \underbrace{e^{-\sigma t}\left[x_0 \cos \omega_d t + \left(\frac{\sigma x_0 + v_0}{\omega_d}\right)\sin \omega_d t\right]}_{\text{Response to initial conditions}} \quad (7.28)$$

Case 4. Undamped, $\zeta = 0$.

$$x(t) = \underbrace{\frac{1}{\omega_n} \sin \omega_n t}_{\text{Response to unit impulse}} + \underbrace{x_0 \cos \omega_n t + \frac{v_0}{\omega_n} \sin \omega_n t}_{\text{Response to initial conditions}} \quad (7.29)$$

Graphical interpretations of these four responses are presented in Fig. 7.7. Once again, for the purpose of illustration of results, the following parameter values have been employed:

$$\omega_n = 1, \quad x_0 = 0.5, \quad v_0 = 0$$

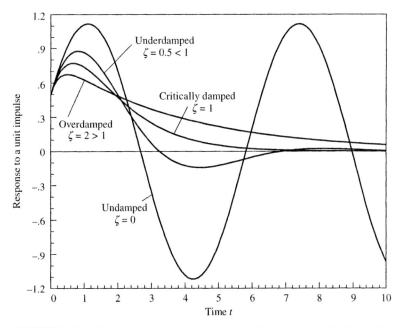

FIGURE 7.7 Response of overdamped, critically damped, underdamped, and undamped second-order systems to a unit impulse; $x_0 \neq 0$, $v_0 = 0$.

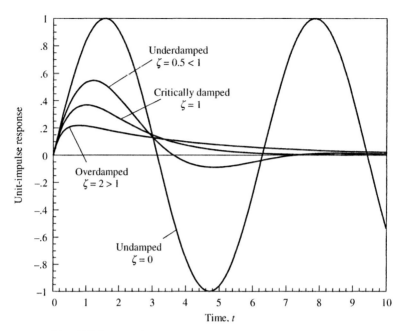

FIGURE 7.8 Unit-impulse response of overdamped, critically damped, underdamped, and undamped second-order systems; zero initial conditions.

Unit-Impulse Response

As mentioned earlier, the *unit-impulse response* refers to the system's response to a unit impulse and is subjected to zero initial conditions. The analytical expressions of the response for all four cases of damping may readily be determined by setting the initial conditions to zero in Eqs. (7.26)–(7.29). The graphical representation of this type of response, for all damping cases, is depicted in Fig. 7.8.

As observed in the free-response analysis, the unit-impulse response of a second-order system approaches zero for damping (cases 1–3) and exhibits oscillatory behavior in the absence of damping (case 4).

Note: It can be shown that the free response of a second-order mechanical system subjected to initial conditions $x_0 = 0$, $v_0 = 1$ is the same as its response when subjected to a unit-impulse forcing function and zero initial conditions (see Problem Set 7.1–7.3). This is because an impulse applied to the mass of the system causes an instantaneous change of momentum, that is, an initial velocity.

EXAMPLE 7.3. Consider the torsional mechanical system shown in Fig. 4.39. The system is subjected to an impulsive applied torque of magnitude 1 and initial conditions $\theta(0) = 0.5$ rad and $\dot{\theta}(0) = 0$. Assume the following parameter values: $J =$

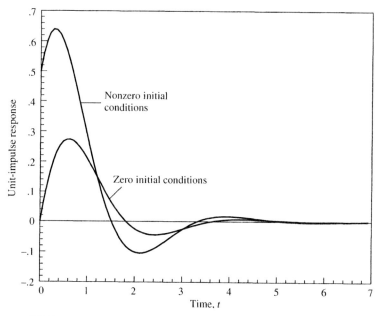

FIGURE 7.9 Unit-impulse response of the torsional system in Example 7.3.

$1 \text{ kg} \cdot \text{m}^2$, $B = 2 \text{ N} \cdot \text{m} \cdot \text{s}$, and $K = 4 \text{ N} \cdot \text{m}$. Determine the system's response. Also, find the response for the case of zero initial conditions.

Solution. Substitution of the given numerical values in the equation of motion yields

$$J\ddot{\theta} + B\dot{\theta} + K\theta = T(t) \implies \ddot{\theta} + 2\dot{\theta} + 4\theta = \delta(t)$$

Comparison with the standard form results in

$$\omega_n^2 = 4, \quad 2\zeta\omega_n = 2 \implies \omega_n = 2 \text{ rad/s}, \quad \zeta = 0.5 < 1$$

indicating that the system is underdamped (case 3), and the response is described by Eq. (7.28). Insert the prescribed initial conditions into this equation, use θ in place of x, and $\sigma = \zeta\omega_n = 1$, $\omega_d = \omega_n\sqrt{1-\zeta^2} = \sqrt{3}$, to obtain

$$\theta(t) = \underbrace{\frac{1}{\sqrt{3}}e^{-t}\sin\sqrt{3}t}_{\text{Response to unit impulse}} + \underbrace{e^{-t}\left[0.5\cos\sqrt{3}t + \left(\frac{0.5}{\sqrt{3}}\right)\sin\sqrt{3}t\right]}_{\text{Response to initial conditions}} \quad \text{rad}$$

Furthermore, assuming zero initial conditions, the corresponding unit-impulse response is described by the first term in the previous equation. Figure 7.9 illustrates time variations of the two responses.

Response to a Unit Step

Consider a linear, second-order system with its governing equation in the form of Eq. (7.8). Suppose the system is subjected to a unit-step input, so that

$$\ddot{x} + 2\zeta\omega_n\dot{x} + \omega_n^2 x = f(t) = u(t) \tag{7.30}$$

and initial conditions $x(0^-) = x_0$ and $\dot{x}(0^-) = v_0$. Taking the Laplace transform of Eq. (7.30) and using the given initial conditions, we have

$$x(t) = L^{-1}\left\{\frac{1}{s(s^2 + 2\zeta\omega_n s + \omega_n^2)} + \frac{(s + 2\zeta\omega_n)x_0 + v_0}{s^2 + 2\zeta\omega_n s + \omega_n^2}\right\} \quad \text{Response of a second-order system to a unit step; general initial conditions}$$

(7.31)

In the event that initial conditions are zero, the response is known as the **unit-step response**. In this case, the transfer function is determined from Eq. (7.30) as

$$G(s) = \frac{X(s)}{F(s)} = \frac{1}{s^2 + 2\zeta\omega_n s + \omega_n^2}$$

and the unit-step response is given by

$$x(t) = L^{-1}\{F(s)G(s)\} = L^{-1}\left\{\frac{1}{s} \cdot \frac{1}{s^2 + 2\zeta\omega_n s + \omega_n^2}\right\} \quad \text{Unit-step response of a second-order system; zero initial conditions}$$

(7.32)

Note that this is indeed the first expression on the right-hand side of Eq. (7.31), as expected. The second term in Eq. (7.31) provides the response for when the system is driven by its initial conditions and zero input. As was done previously, four possible cases of damping are discussed below.

Case 1. Overdamped, $\zeta > 1$

$$x(t) = \frac{1}{\omega_n^2}\left\{1 + \frac{1}{2\sqrt{\zeta^2 - 1}}\left[\frac{1}{-\zeta + \sqrt{\zeta^2 - 1}}e^{-\omega_n(\zeta - \sqrt{\zeta^2 - 1})t}\right.\right.$$

$$\left.\left. + \frac{1}{\zeta + \sqrt{\zeta^2 - 1}}e^{-\omega_n(\zeta + \sqrt{\zeta^2 - 1})t}\right]\right\}$$

Response to unit step

$$+ \left[\frac{\omega_n\left(\zeta + \sqrt{\zeta^2 - 1}\right)x_0 + v_0}{2\omega_n\sqrt{\zeta^2 - 1}}\right]e^{-\omega_n(\zeta - \sqrt{\zeta^2 - 1})t}$$

$$- \left[\frac{\omega_n\left(\zeta - \sqrt{\zeta^2 - 1}\right)x_0 + v_0}{2\omega_n\sqrt{\zeta^2 - 1}}\right]e^{-\omega_n(\zeta + \sqrt{\zeta^2 - 1})t}$$

(7.33)

Response to initial conditions

CHAPTER 7: System Response   351

Case 2. Critically damped, $\zeta = 1$

$$x(t) = \underbrace{\frac{1}{\omega_n^2}[1 - e^{-\omega_n t}(1 - \omega_n t)]}_{\text{Response to unit step}} + \underbrace{x_0 e^{-\omega_n t} + (\omega_n x_0 + v_0)t e^{-\omega_n t}}_{\text{Response to initial conditions}} \quad (7.34)$$

Case 3. Underdamped, $0 < \zeta < 1$

$$x(t) = \underbrace{\frac{1}{\omega_n^2}\left\{1 - e^{-\sigma t}\left[\cos \omega_d t + \frac{\zeta}{\sqrt{1-\zeta^2}}\sin \omega_d t\right]\right\}}_{\text{Response to unit step}}$$

$$+ \underbrace{e^{-\sigma t}\left[x_0 \cos \omega_d t + \left(\frac{\sigma x_0 + v_0}{\omega_d}\right)\sin \omega_d t\right]}_{\text{Response to initial conditions}} \quad (7.35)$$

An alternative, and perhaps more convenient, form of Eq. (7.35) may be obtained as

$$x(t) = \underbrace{\frac{1}{\omega_n^2}\left\{1 - e^{-\sigma t}\frac{1}{\sqrt{1-\zeta^2}}\sin(\omega_d t + \phi)\right\}}_{\text{Response to unit step}}$$

$$+ \underbrace{e^{-\sigma t}\frac{\sqrt{(\sigma x_0 + v_0)^2 + \omega_d^2 x_0^2}}{\omega_d}\sin(\omega_d t + \psi)}_{\text{Response to initial conditions}} \quad (7.36)$$

where

$$\phi = \tan^{-1}\frac{\sqrt{1-\zeta^2}}{\zeta} \quad \text{and} \quad \psi = \tan^{-1}\frac{x_0 \omega_d}{\sigma x_0 + v_0}$$

Case 4. Undamped, $\zeta = 0$

$$x(t) = \underbrace{\frac{1}{\omega_n^2}(1 - \cos \omega_n t)}_{\text{Response to unit step}} + \underbrace{x_0 \cos \omega_n t + \frac{v_0}{\omega_n}\sin \omega_n t}_{\text{Response to initial conditions}} \quad (7.37)$$

The step responses defined by Eqs. (7.33)–(7.37) are plotted versus time and illustrated in Fig. 7.10. Parameter values used are as before; i.e., $\omega_n = 1$, $x_0 = 0.5$, and $v_0 = 0$. These plots suggest that as long as the system has inherent damping (cases 1–3), its step response converges to a fixed constant, the steady-state value (or final value). Based on Eqs. (7.33)–(7.36), the final value is $1/\omega_n^2$ because $x(t) \to 1/\omega_n^2$ as $t \to \infty$. With the numerical values used here, specifically $\omega_n = 1$, this steady-state value is 1. In the absence of damping, the response oscillates indefinitely about 1.

352 Dynamic Systems: Modeling and Analysis

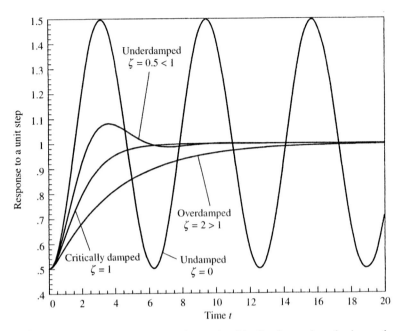

FIGURE 7.10 Response of overdamped, critically damped, underdamped, and undamped second-order systems to a unit step; $x_0 \neq 0$, $v_0 = 0$.

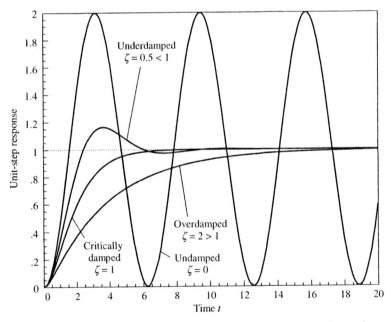

FIGURE 7.11 Unit-step response of overdamped, critically damped, underdamped, and undamped second-order systems; zero initial conditions.

Unit-Step Response

The **unit-step response** refers to the system's response to a unit step and is subjected to zero initial conditions. In this event, the analytical expressions of the response for all four cases of damping are determined by setting the initial conditions equal to zero in Eqs. (7.33)–(7.37). The unit-step response, in all cases of damping, is depicted in Fig. 7.11.

EXAMPLE 7.4. Consider the RLC circuit shown in Fig. 7.12. Assume that the charge and the current are initially zero, i.e., $q(0^-) = 0 = i(0^-)$, and that the applied voltage is constant E. Determine the system's response, $q(t)$.

Solution. The governing differential equation in terms of q for the circuit is

$$L\ddot{q} + R\dot{q} + \frac{1}{C}q = E \quad \xRightarrow{\text{divide by } L} \quad \ddot{q} + \frac{R}{L}\dot{q} + \frac{1}{LC}q = \frac{E}{L}$$

which is in the standard form of Eq. (7.8) with

$$\omega_n = \sqrt{\frac{1}{LC}} \quad \text{and} \quad \zeta = \frac{R}{2}\sqrt{\frac{C}{L}}$$

as discussed in Section 1.4. Because the applied forcing function is E/L, it is regarded as a step function, not a unit-step function. Laplace transformation of the governing equation, using zero initial conditions, yields

$$Q(s) = \frac{E}{L} \cdot \frac{1}{s[s^2 + (R/L)s + 1/(LC)]}$$

which in turn implies

$$q(t) = \frac{E}{L} L^{-1}\left\{\frac{1}{s[s^2 + (R/L)s + 1/(LC)]}\right\}$$

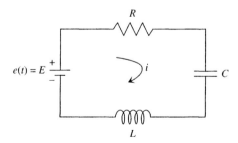

FIGURE 7.12 RLC circuit.

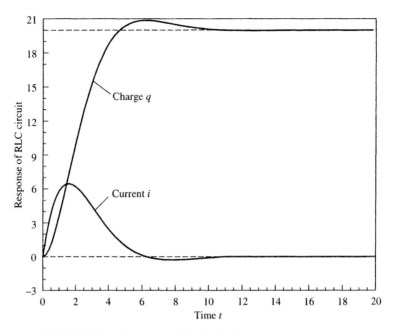

FIGURE 7.13 Response of the RLC circuit to a step change in voltage.

which is simply a constant multiple of the expression in Eq. (7.32), a unit-step response. Now, suppose for a given set of parameter values we have $R^2C < 4L$; i.e., $\zeta < 1$. Then, the inverse Laplace transform is the first term in Eq. (7.35). Note that

$$\omega_d = \omega_n \sqrt{1 - \zeta^2} = \sqrt{\frac{1}{LC} - \left(\frac{R}{2L}\right)^2} \quad \text{and} \quad \sigma = \zeta \omega_n = \frac{R}{2L}$$

As a result, the response is described analytically as

$$q(t) = EC\left\{1 - e^{-(R/2L)t}\left[\cos\sqrt{\frac{1}{LC} - \left(\frac{R}{2L}\right)^2}\,t + \frac{R\sqrt{C}}{\sqrt{4L - R^2C}}\sin\sqrt{\frac{1}{LC} - \left(\frac{R}{2L}\right)^2}\,t\right]\right\}$$

For instance, given parameter values $R = 1\,\Omega$, $C = 2$ F, $L = 1$ H, and $E = 10$ V, the system is seen to be underdamped and its response takes the specific form

$$q(t) = 20\{1 - e^{-0.5t}[\cos 0.5t + \sin 0.5t]\} \quad \text{A}\cdot\text{s}$$

time variations of which are presented in Fig. 7.13. Furthermore, recalling that $i = dq/dt$, the circuit current is also plotted in the figure. It is then observed that while $q(t) \to \omega_n^2 E = 20$ A·s as $t \to \infty$, $i(t) \to 0$.

PROBLEM SET 7.1–7.3

7.1. Consider an underdamped second-order system:
$$\ddot{x} + 2\zeta\omega_n \dot{x} + \omega_n^2 x = 0, \qquad 0 < \zeta < 1$$
subject to initial conditions $x(0^-) = x_0$ and $\dot{x}(0^-) = 0$. Show that the free response is given by
$$x(t) = \frac{x_0}{\sqrt{1-\zeta^2}} e^{-\zeta\omega_n t} \cos(\omega_d t - \phi)$$
$$\text{with } \phi = \tan^{-1}\frac{\zeta}{\sqrt{1-\zeta^2}}$$

7.2. Consider the following second-order system:
$$\ddot{x} + 4\dot{x} + 3x = 0$$
subject to initial conditions $x(0^-) = 0$ and $\dot{x}(0^-) = 1$. Determine whether it is underdamped, overdamped, or critically damped, and determine its free response.

7.3. The governing equation and the initial condition of a dynamic system are defined by
$$2\dot{h} + h = q_i(t), \qquad h(0^-) = 1$$
Assuming that the input $q_i(t)$ is a unit-ramp function, determine the system's response.

7.4. Repeat Problem 7.3 for $q_i(t) = 2u_s(t)$.

7.5. Find the response of a system governed by
$$\ddot{\theta} + 2\dot{\theta} + 4\theta = \delta(t)$$
and subjected to initial conditions $\theta(0^-) = 0$ and $\dot{\theta}(0^-) = 1$.

7.6. Consider an RL circuit subjected to zero initial current and a unit-step input voltage. Determine the analytical expression for the current at any time.

7.7. The equations of motion for a translational mechanical system are given by
$$\begin{cases} \ddot{x}_1 + 2\dot{x}_1 + (x_1 - x_2) = f(t) \\ 2\dot{x}_2 + (x_2 - x_1) = 0 \end{cases}$$
(a) Assuming zero initial conditions, find the transfer function $X_1(s)/F(s)$.
(b) Assuming $f(t) = u_s(t)$, determine the response $x_1(t)$. What is the steady-state value of $x_1(t)$?

7.8. Show that the free response of a second-order mechanical system subjected to initial conditions $x_0 = 0$, $v_0 = 1$ m/s is the same as its response when subjected to a unit-impulse forcing function and zero initial conditions. Verify the result for all four cases of damping.

7.9. Consider the following two scenarios concerning a second-order mechanical system:
(a) The system is subjected to $-u(t)$ and zero initial conditions, where $u(t)$ denotes the unit-step forcing function.
(b) The system is subjected to no applied external force, and initial conditions are $x_0 = 1/\omega_n^2$ m, $v_0 = 0$. Show that the corresponding responses simply differ by a constant.

7.4 TRANSIENT RESPONSE OF HIGHER-ORDER SYSTEMS

In this section we turn our attention to those linear dynamic systems whose models are governed by more than a single differential equation and may be subjected to one or more applied forcing functions. In general, if there are n independent variables in the model, then determination of a total of n responses to a given set of forcing functions may be desirable. However, the governing differential equations are generally *coupled* and cannot be treated independently of one another. Consequently, using the Laplace transform becomes essential. To that end, suppose $x_1, x_2, \ldots, x_n$ denote the independent variables in the model, and $f_1, f_2, \ldots, f_m$ represent the forcing functions (inputs). The procedure is then as follows: (1) Take the Laplace transform of each differential equation, taking into account the initial conditions, to obtain a system of algebraic equations in terms of $X_1(s), \ldots, X_n(s)$ and $F_1(s), \ldots, F_m(s)$, the transform functions of the variables and inputs, respectively. (2) Proceed as in Section 3.4, utilizing Cramer's rule, to determine each $X_i(s)$, $i = 1, 2, \ldots, n$, in terms of $F_1(s), \ldots, F_m(s)$. (3) Determine the responses (i.e., $x_i(t)$, $i = 1, 2, \ldots, n$) via inverse Laplace transformation of their transform functions (i.e., $L^{-1}\{X_i(s)\}$). The following example illustrates the specific application of these steps.

EXAMPLE 7.5. The governing equations for an electromechanical system are defined as

$$\begin{cases} L\dfrac{di}{dt} + Ri + K_1\omega = v(t) \\ J\dfrac{d\omega}{dt} + B\omega - K_2 i = 0 \end{cases}$$

Suppose the applied voltage $v(t)$ is a unit-step function, and initial conditions are zero. Determine responses $i(t)$ and $\omega(t)$. Assume the following parameter values: $L = 1\text{ H}, J = 1\text{ kg} \cdot \text{m}^2, B = 2\text{ N} \cdot \text{m} \cdot \text{s}, R = 1\,\Omega, K_1 = 1\text{ V} \cdot \text{s}, K_2 = 1\text{ N} \cdot \text{m/A}$.

Solution. Laplace transformation of the governing equations, using zero initial conditions, and application of Cramer's rule, yields

$$I(s) = \left[\frac{Js + B}{LJs^2 + (LB + RJ)s + RB + K_1 K_2}\right] V(s),$$

$$\Omega(s) = \left[\frac{K_2}{LJs^2 + (LB + RJ)s + RB + K_1 K_2}\right] V(s)$$

Substitute for the given parameter values and take into account $V(s) = 1/s$ to obtain

$$I(s) = \frac{s + 2}{s(s^2 + 3s + 3)}, \qquad \Omega(s) = \frac{1}{s(s^2 + 3s + 3)}$$

Consequently, the responses $i(t)$ and $\omega(t)$ will be determined via inverse Laplace transformation of the above expressions. First, note that $s^2 + 3s + 3$ has complex roots of $-3/2 \pm j(\sqrt{3}/2)$ and hence is *irreducible*. Forming the proper partial fractions for the two transform functions, we have

$$I(s) = \frac{s+2}{s(s^2+3s+3)} = \frac{A_1}{s} + \frac{B_1 s + C_1}{s^2+3s+3}$$

$$\Omega(s) = \frac{1}{s(s^2+3s+3)} = \frac{A_2}{s} + \frac{B_2 s + C_2}{s^2+3s+3}$$

where the constants are determined as $A_1 = \frac{2}{3}, B_1 = -\frac{2}{3}, C_1 = -1, A_2 = \frac{1}{3}, B_2 = -\frac{1}{3}$, and $C_2 = -1$. Substitute and rewrite the expressions in familiar forms, as

$$I(s) = \frac{2}{3}\left(\frac{1}{s}\right) - \frac{(2/3)s+1}{s^2+3s+3} = \frac{2}{3}\left(\frac{1}{s}\right) - \frac{2}{3}\frac{s+3/2}{(s+3/2)^2+(\sqrt{3}/2)^2}$$

$$\Omega(s) = \frac{1}{3}\left(\frac{1}{s}\right) - \frac{(1/3)s+1}{s^2+3s+3}$$

$$= \frac{1}{3}\left(\frac{1}{s}\right) - \frac{1}{3}\left[\frac{s+3/2}{(s+3/2)^2+(\sqrt{3}/2)^2} + \sqrt{3}\frac{\sqrt{3}/2}{(s+3/2)^2+(\sqrt{3}/2)^2}\right]$$

Inverse Laplace transformation of these will result in the analytical expressions for the responses, as

$$i(t) = \frac{2}{3}\left[1 - e^{-(3/2)t}\cos\frac{\sqrt{3}}{2}t\right] \quad \text{A}$$

$$\omega(t) = \frac{1}{3}\left\{1 - e^{-(3/2)t}\left[\cos\frac{\sqrt{3}}{2}t + \sqrt{3}\sin\frac{\sqrt{3}}{2}t\right]\right\} \quad \text{rad/s}$$

Figure 7.14 shows the time variations of $i(t)$ and $\omega(t)$, indicating steady-state values of 0.67 A and 0.33 rad/s, respectively. These, of course, are consistent with the analytical results obtained previously.

EXAMPLE 7.6. Consider the two-degree-of-freedom mechanical system shown in Fig. P3.9, subjected to two applied forces, $f_1(t)$ and $f_2(t)$, and zero initial conditions. Determine system responses $x_1(t)$ and $x_2(t)$ when

(a) $f_1(t) = 0$, $f_2(t) = u_s(t)$.
(b) $f_1(t) = u_s(t)$, $f_2(t) = u_s(t)$.

Use the following parameter values: $m_1 = 1$ kg $= m_2$, $b_1 = 2$ N·s/m, $b_2 = 1$ N·s/m, $k_1 = 1$ N/m $= k_2$.

Solution. From previous work, the equations of motion are given as

$$\begin{cases} m_1\ddot{x}_1 + b_1\dot{x}_1 + k_1 x_1 - k_2(x_2 - x_1) - b_2(\dot{x}_2 - \dot{x}_1) = f_1(t) \\ m_2\ddot{x}_2 + b_2(\dot{x}_2 - \dot{x}_1) + k_2(x_2 - x_1) = f_2(t) \end{cases}$$

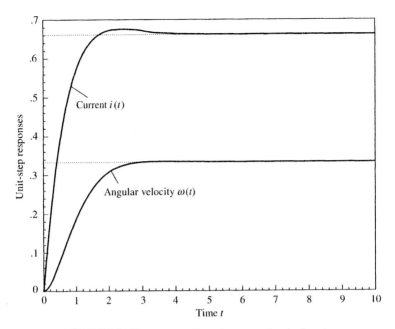

FIGURE 7.14 Responses of an electromechanical system.

Take the Laplace transform, use initial conditions of zero, and solve for $X_1(s)$ and $X_2(s)$ to obtain

$$X_1(s) = \underbrace{\left(\frac{m_2 s^2 + b_2 s + k_2}{\Delta(s)}\right)}_{G_{11}(s)} F_1(s) + \underbrace{\left(\frac{b_2 s + k_2}{\Delta(s)}\right)}_{G_{12}(s)} F_2(s) \quad (7.38)$$

$$X_2(s) = \underbrace{\left(\frac{m_1 s^2 + (b_1 + b_2)s + k_1 + k_2}{\Delta(s)}\right)}_{G_{22}(s)} F_2(s) + \underbrace{\left(\frac{b_2 s + k_2}{\Delta(s)}\right)}_{G_{21}(s)} F_1(s) \quad (7.39)$$

where

$$\Delta(s) = \begin{vmatrix} m_1 s^2 + (b_1 + b_2)s + k_1 + k_2 & -(b_2 s + k_2) \\ -(b_2 s + k_2) & m_2 s^2 + b_2 s + k_2 \end{vmatrix}$$

(a) Because $f_1(t) = 0$, then $x_1(t)$ and $x_2(t)$ represent responses to $f_2(t)$ only. For the purpose of clarity, introduce notations $x_{12}(t)$ and $x_{22}(t)$, where the second subscripts denote the applied forcing function being considered. Knowing that $F_2(s) = 1/s$, the previous expressions reduce to

$$X_{12}(s) = G_{12}(s) \cdot \frac{1}{s} = \frac{b_2 s + k_2}{\Delta(s)} \cdot \frac{1}{s}$$

(and) $\quad X_{22}(s) = G_{22}(s) \cdot \dfrac{1}{s} = \dfrac{m_1 s^2 + (b_1 + b_2)s + k_1 + k_2}{\Delta(s)} \cdot \dfrac{1}{s}$

Inserting the parameter values results in $\Delta(s) = s^4 + 4s^3 + 5s^2 + 3s + 1$. Given that the roots of $\Delta(s)$ are calculated as $-1, -2.3247, -0.3376 \pm j0.5623$, the partial fractions are constructed as

$$X_{12}(s) = \frac{s+1}{s(s^4 + 4s^3 + 5s^2 + 3s + 1)}$$

$$= \frac{A_1}{s} + \frac{B_1}{s+1} + \frac{C_1}{s+2.3247} + \frac{D_1 s + E_1}{(s+0.3376)^2 + (0.5623)^2}$$

and

$$X_{22}(s) = \frac{s^2 + 3s + 2}{s(s^4 + 4s^3 + 5s^2 + 3s + 1)}$$

$$= \frac{A_2}{s} + \frac{B_2}{s+1} + \frac{C_2}{s+2.3247} + \frac{D_2 s + E_2}{(s+0.3376)^2 + (0.5623)^2}$$

where the constants are evaluated as $A_1 = 1$, $B_1 = 0$, $C_1 = -0.1$, $D_1 = -0.9$, $E_1 = -0.84$, $A_2 = 2$, $B_2 = 0$, $C_2 = 0.0328$, $D_2 = -2.0328$, and $E_2 = -1.2966$.

Substitute back into the partial fractions and take the inverse Laplace transform to obtain the responses

$$x_{12}(t) = 1 - 0.1e^{-2.3247t}$$
$$- 0.9e^{-0.3376t}\left[\cos 0.5623t + 1.06 \sin 0.5623t\right] \quad \text{m}$$

and

$$x_{22}(t) = 2 + 0.0328e^{-2.3247t}$$
$$- 2.0328e^{-0.3376t}\left[\cos 0.5623t + 0.5339 \sin 0.5623t\right] \quad \text{m}$$

whose plots versus time are shown in Fig. 7.15. Furthermore, observe that the steady-state values are 1 and 2 meters, respectively.

(b) In part (a), only $f_2(t)$ was applied to the system, and the two responses were determined via transfer functions $G_{12}(s)$ and $G_{22}(s)$. When both $f_1(t)$ and $f_2(t)$ are applied, all four transfer functions will contribute to the determination of the responses. The idea is to find x_{11} and x_{12} corresponding to f_1 and f_2 and use the **principle of superposition** to determine $x_1(t) = x_{11}(t) + x_{12}(t)$. Similarly, $x_2(t) = x_{21}(t) + x_{22}(t)$. Because x_{12} and x_{22} have been found in part (a), it suffices to determine x_{11} and x_{21}. To that end, we proceed as follows:

$$X_{11}(s) = G_{11}(s) \cdot \frac{1}{s} = \frac{m_2 s^2 + b_2 s + k_2}{\Delta(s)} \cdot \frac{1}{s}$$

and

$$X_{21}(s) = G_{21}(s) \cdot \frac{1}{s} = \frac{b_2 s + k_2}{\Delta(s)} \cdot \frac{1}{s}$$

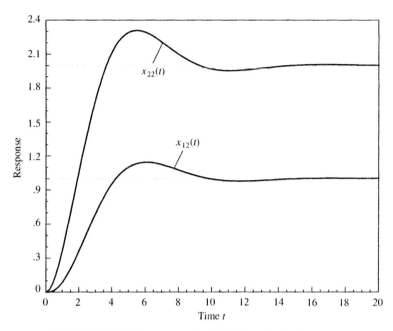

FIGURE 7.15 Responses of a TDOF mechanical system.

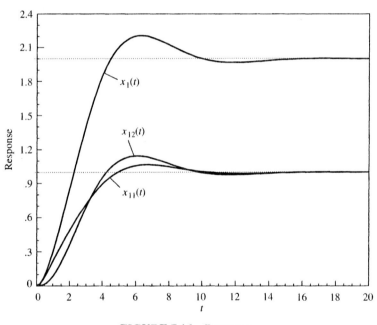

FIGURE 7.16 Response.

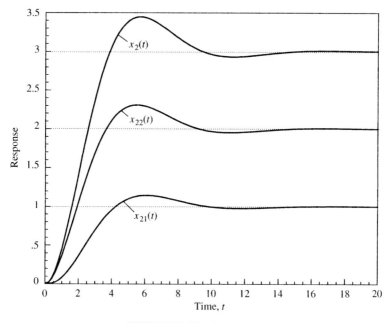

FIGURE 7.17 Response.

Note that $X_{21}(s)$ is identical to $X_{12}(s)$, so that $x_{21}(t) = x_{12}(t)$. However, upon insertion of the parameter values, x_{11} is obtained as

$$x_{11}(t) = L^{-1}\left\{\frac{s^2 + s + 1}{s^4 + 4s^3 + 5s^2 + 3s + 1} \cdot \frac{1}{s}\right\}$$
$$= 1 - e^{-t} + 0.3106 e^{-2.3247t}$$
$$- 0.3106 e^{-0.3376t}\left[\cos 0.5623t + 2.1917 \sin 0.5623t\right] \quad \text{m}$$

Figure 7.16 shows responses x_{11} and x_{12}, as well as their superposition, x_1. Similarly, Fig. 7.17 contains the plots of x_{21} and x_{22}, and superposition, x_2.

PROBLEM SET 7.4

7.10. Consider the translational mechanical system shown in Fig. P7.10.
(a) Show that the equations of motion are given by

$$\begin{bmatrix} m & 0 \\ 0 & 0 \end{bmatrix}\begin{Bmatrix} \ddot{x}_1 \\ \ddot{x}_2 \end{Bmatrix} + \begin{bmatrix} b_1 & 0 \\ 0 & b_2 \end{bmatrix}\begin{Bmatrix} \dot{x}_1 \\ \dot{x}_2 \end{Bmatrix} + \begin{bmatrix} k_1 & -k_1 \\ -k_1 & k_1 + k_2 \end{bmatrix}\begin{Bmatrix} x_1 \\ x_2 \end{Bmatrix} = \begin{Bmatrix} f \\ 0 \end{Bmatrix}$$

(b) Assuming zero initial conditions, find the transfer functions $X_1(s)/F(s)$ and $X_2(s)/F(s)$.
(c) Assume the following parameter values: $m = 1$ kg, $k_1 = 1$ N/m, $k_2 = 2$ N/m, $b_1 = 2$ N·s/m, and $b_2 = 1$ N·s/m. Assuming $f(t) = u_s(t)$, determine the

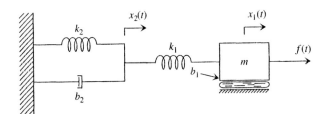

FIGURE P7.10 Mechanical system.

responses $x_1(t)$ and $x_2(t)$. What are their steady-state values? Use the fact that the roots of $s^3 + 5s^2 + 7s + 2$ are -0.3820, -2, and -2.618.

7.11. Repeat Problem 7.10 for $f(t) = \delta(t)$.

7.12. Consider the mechanical system of Problem 7.10. Assume that the system is not subjected to an applied force and is driven only by its initial conditions $x_1(0^-) = 0$, $\dot{x}_1(0^-) = 1$ m/s, $x_2(0^-) = 0$, and $\dot{x}_2(0^-) = 0$. Using the parameter values of Problem 7.10, determine the corresponding free responses $x_1(t)$ and $x_2(t)$.

7.5 STEADY-STATE RESPONSE TO SINUSOIDAL INPUT

The *steady-state response of a linear dynamic system subject to a sinusoidal input* may be best obtained by a technique known as the **FRF method,** which is based on the **frequency response function** (FRF). Although the FRF may be obtained via Fourier transform, it is more convenient to use Laplace transformation instead. To obtain the FRF, simply substitute $s = j\omega$ in the system's transfer function. Because the input is sinusoidal, it has one of the following forms:

$$f(t) = F_0 \cos \omega t, \qquad f(t) = F_0 \sin \omega t, \qquad f(t) = F_0 e^{j\omega t}$$

where F_0 and ω are real and denote the amplitude and the *forcing frequency,* respectively. The steady-state response of a system to a sinusoidal input is also known as the system's **frequency response.** Assuming that the input is $f(t) = F_0 \sin \omega t$, the steady-state response $x_{ss}(t)$ is obtained as follows: Suppose a linear system is represented by its transfer function $G(s) = X(s)/F(s)$. From prior discussions, we assume that $G(s)$ is expressed in the general form

$$G(s) = K\frac{(s + z_1)\cdots(s + z_m)}{(s + p_1)\cdots(s + p_n)}$$

in which K is a constant and the real part of each of the p_i is positive. Then, using partial fractions and noting that $F(s) = L\{F_0 \sin \omega t\} = F_0\omega/(s^2 + \omega^2)$, we have

$$X(s) = G(s)F(s) = G(s)\frac{F_0\omega}{s^2 + \omega^2}$$

$$= \frac{c_1}{s + p_1} + \cdots + \frac{c_n}{s + p_n} + \frac{d_1}{s + j\omega} + \frac{d_2}{s - j\omega}$$

where it can readily be shown that constants d_1 and d_2 form a pair of complex conjugates; i.e., $d_2 = \overline{d_1}$. Subsequently, through inverse Laplace transformation, one obtains

$$x(t) = c_1 e^{-p_1 t} + \cdots + c_n e^{-p_n t} + d_1 e^{-j\omega t} + \overline{d_1} e^{j\omega t} \quad (7.40)$$

As mentioned below, it is the steady-state response that is of interest to us. Because the real parts of the p_i are assumed to be positive, all terms in Eq. (7.40), except for the last two terms, are exponential decays. Thus, as $t \to \infty$, we have

Steady-state response $\quad x_{ss}(t) = d_1 e^{-j\omega t} + \overline{d_1} e^{j\omega t} \quad (7.41)$

where the *residues* are calculated as

$$d_1 = \{(s + j\omega)X(s)\}_{s=-j\omega} = -\frac{F_0}{2j} G(-j\omega), \quad \overline{d_1} = \frac{F_0}{2j} G(j\omega)$$

Recall from Section 1.3 that the polar form of $G(j\omega)$ is given by

$$G(j\omega) = |G(j\omega)| \angle \phi = |G(j\omega)| e^{j\phi(\omega)}$$

where

$$\phi(\omega) = \tan^{-1}\left\{ \frac{\text{imaginary part of } G(j\omega)}{\text{real part of } G(j\omega)} \right\}$$

Substituting this information in Eq. (7.41) yields

$$\begin{aligned} x_{ss}(t) &= -\frac{F_0}{2j} G(-j\omega) e^{-j\omega t} + \frac{F_0}{2j} G(j\omega) e^{j\omega t} \\ &= -\frac{F_0}{2j} |G(j\omega)| e^{-j\phi} e^{-j\omega t} + \frac{F_0}{2j} |G(j\omega)| e^{j\phi} e^{j\omega t} \quad (7.42) \\ &= F_0 |G(j\omega)| \left(\frac{e^{(\omega t + \phi)j} - e^{-(\omega t + \phi)j}}{2j} \right) = F_0 |G(j\omega)| \sin(\omega t + \phi) \end{aligned}$$

so that the steady-state response is

$$x_{ss}(t; \omega) = F_0 |G(j\omega)| \sin[\omega t + \phi(\omega)] \quad (7.43)$$

For simplicity, we often write the steady-state response as

$$x_{ss}(t) = F_0 |G(j\omega)| \sin(\omega t + \phi)$$

In summary, for a *linear system,* if the input function is

$$f(t) = F_0 \sin \omega t \quad (7.44)$$

the steady-state response is

$$x_{ss}(t) = F_0 |G(j\omega)| \sin(\omega t + \phi)$$

Following similar reasoning, if the input function is

$$r(t) = A \cos(\omega t + \theta)$$

then the steady-state response is

$$x_{ss}(t) = A|G(j\omega)|\cos[(\omega t + \theta) + \phi]$$

Note that the steady-state response is the same as the particular solution if damping is present, whereas the inverse Laplace transform gives the complete solution, not the particular. The terms *complete solution* and *total solution* are used interchangeably. If the steady-state solution is desired, it would be a fatal mistake to take the inverse Laplace transform of $X(s)$, the transform of the output $x(t)$. This inverse Laplace transform gives the complete solution, a combination of homogeneous and particular solutions. Note that for a damped system, after a sufficiently long time, the transient homogeneous solution will damp out, and the particular solution characterizes the steady-state solution.

EXAMPLE 7.7. STEADY-STATE RESPONSE OF A DYNAMIC SYSTEM. The differential equation of a dynamic system is given as

$$\ddot{x} + 2\zeta\omega_n\dot{x} + \omega_n^2 x = f(t) = A\sin\omega t$$

Determine the steady-state solution, $x_{ss}(t)$, using the numerical values $\omega_n = 1$, $\zeta = 0.5$, $A = 10$, and $\omega = 2$.

Solution. The transfer function (TF) is

$$G(s) = \frac{X(s)}{F(s)} = \frac{1}{s^2 + 2\zeta\omega_n s + \omega_n^2} = \frac{1}{s^2 + s + 1}$$

The frequency response function (FRF) is then obtained as

$$G(j\omega)\Big|_{\omega=2} = \frac{1}{1 - \omega^2 + j\omega}\Big|_{\omega=2} = \frac{1}{-3 + j2}$$

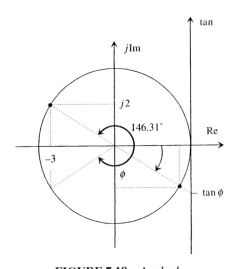

FIGURE 7.18 Angle ϕ.

Magnitude $\quad |G(j2)| = \dfrac{1}{\sqrt{2^2 + 3^2}} = 0.277$

Phase angle $\quad \phi = \angle G(j2) = -\tan^{-1}\left(\dfrac{2}{-3}\right)$

$\qquad\qquad\qquad = -(-33.69° + 180°) = -146.31°$

Note that there are two possible values for $\tan^{-1}(2/3)$, as shown in Fig. 7.18, namely, $-33.69°$ and $146.31°$. However, one must select the proper angle based on the values of the real and imaginary parts of $G(j\omega)$. Rewrite $G(j\omega)$ as

$$G(j\omega)\big|_{\omega=2} = \dfrac{1}{-3 + j2} \cdot \dfrac{-3 - j2}{-3 - j2} = \dfrac{-3 - j2}{13}$$

which indicates that the real and imaginary parts are both negative. Thus, the corresponding angle is selected in the third quadrant, i.e., $\phi = -146.31°$. Therefore, the steady-state response is

$$x_{ss}(t) = A\,|G(j\omega)|\sin(\omega t + \phi) = 2.77 \sin\!\left(2t - 146.31°\right)$$

Bode Plot

The frequency response is often presented as a *Bode plot* where the magnitude plot is on log-log scale (or dB-log scale) and the phase plot is on linear-log scale (see Fig. 7.19). The unit dB (decibel) is defined as

$$x \text{ (in dB)} = 20 \log x$$

Recall that *standard (common) logarithm*, $\log x$, is the log of x to base 10:

$$\log x \equiv \log_{10} x$$

and *natural logarithm*, $\ln x$, is the log of x to base e ($e = 2.71828$):

$$\ln x \equiv \log_e x$$

The relationship between the standard and natural logarithms is given by

$$\ln x = 2.30 \log x$$

A *Bode plot* consists of two separate plots:

(a) The **Bode magnitude plot**, which is a plot of the magnitude (on log or dB scale) versus the frequency ω (on log scale).
(b) The **Bode phase plot**, which is a plot of the phase (on linear scale) versus the frequency ω (on log scale).

We now present the Bode plot of the frequency response function for the two fundamental systems, first-order and second-order. The transfer function and the FRF are general in nature; thus they can represent a mechanical or electrical system. More applications of frequency response in mechanical vibrations are treated in Chapter 8.

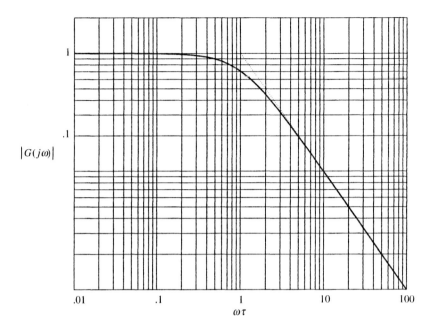

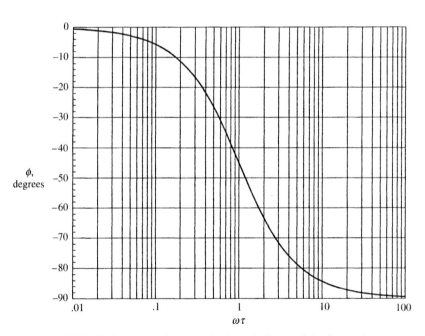

FIGURE 7.19 Bode plot (magnitude and phase) of the first-order system, $G(j\omega) = 1/(1 + j\omega\tau)$.

First-Order System

Suppose the differential equation of a first-order dynamic system is given as

$$\tau \dot{v} + v = f(t) \tag{7.45}$$

This equation is general because it is given in terms of the time constant. It is applicable to any system, mechanical or electrical. The right-hand side of the equation is adjusted for convenience. (For a mass-spring-damper system, it is equivalent to multiplying the applied forcing function by a constant b.) The TF is

$$G(s) = \frac{V(s)}{F(s)} = \frac{1}{\tau s + 1}$$

and the FRF is given by

$$G(j\omega) = \frac{1}{1 + j\omega\tau}$$

Suppose the applied forcing function is a sinusoidal function

$$f(t) = F_0 \cos \omega t$$

Then the steady-state response is

$$v_{ss}(t) = \underbrace{F_0 |G(j\omega)|}_{|v_{ss}|} \cos(\omega t + \phi)$$

where

$$|G(j\omega)| = \frac{1}{\sqrt{1 + (\omega\tau)^2}} \quad \text{and} \quad \phi = -\tan^{-1} \omega\tau$$

The Bode plot for $G(j\omega)$ of the first-order system is shown in Fig. 7.19.

Second-Order System

Suppose the differential equation of a second-order dynamic system is given as

$$\ddot{x} + 2\zeta\omega_n \dot{x} + \omega_n^2 x = \omega_n^2 f(t) \tag{7.46}$$

This equation is general because it is given in terms of the natural frequency and the damping ratio. Thus, it is applicable to any system, mechanical or electrical. The right-hand side of the equation is adjusted for convenience. (For a mass-spring-damper system, it is equivalent to multiplying the applied forcing function by a constant k.) The TF is

$$G(s) = \frac{X(s)}{F(s)} = \frac{\omega_n^2}{s^2 + 2\zeta\omega_n s + \omega_n^2} \tag{7.47}$$

so that the FRF is

$$G(j\omega) = \frac{1}{1 - (\omega/\omega_n)^2 + j2\zeta(\omega/\omega_n)}$$

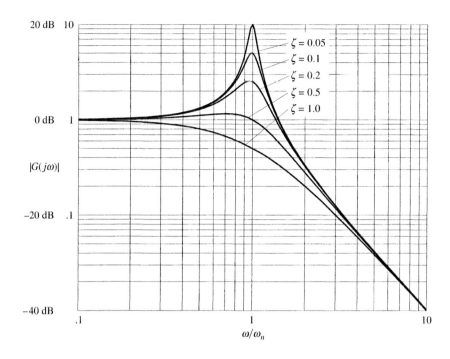

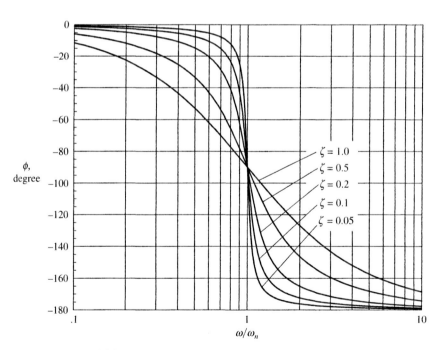

FIGURE 7.20 Bode plot (magnitude and phase) of the second-order system $G(j\omega) = 1/[1 - (\omega/\omega_n)^2 + j2\zeta(\omega/\omega_n)]$.

where ω/ω_n is referred to as the **normalized frequency.** Suppose that the applied forcing function is in the form

$$f(t) = F_0 \cos \omega t$$

Then the steady-state response is

$$x_{ss} = \underbrace{F_0 |G(j\omega)|}_{|x_{ss}|} \cos(\omega t + \phi)$$

where

$$|G(j\omega)| = \frac{1}{\sqrt{[1 - (\omega/\omega_n)^2]^2 + [2\zeta(\omega/\omega_n)]^2}} \quad \text{and} \quad \phi = -\tan^{-1} \frac{2\zeta(\omega/\omega_n)}{1 - (\omega/\omega_n)^2}$$

The Bode plot for $G(j\omega)$ is shown in Fig. 7.20. Inspection of magnitude curves for systems with damping ratios less than the *critical damping ratio* (i.e., $\zeta < \zeta_{cr} = 0.707$) reveals that each reaches its *maximum peak* at some normalized frequency. This frequency, as well as the value of the peak itself, may be determined through differentiation of the expression of $|G(j\omega)|$ with respect to ω/ω_n and setting it equal to zero. First, let $v = \omega/\omega_n$ so that

$$|G(j\omega)| = \frac{1}{\sqrt{(1 - v^2)^2 + (2\zeta v)^2}}$$

Differentiate with respect to v and set equal to zero to obtain

$$\frac{1}{2\sqrt{(1 - v^2)^2 + (2\zeta v)^2}} \left[-4v(1 - v^2) + 8\zeta^2 v \right] = 0$$

which implies

$$-4v(1 - v^2) + 8\zeta^2 v = 0 \quad \Longrightarrow \quad v = 0, \, v = \sqrt{1 - 2\zeta^2}$$

Application of the second derivative test yields that the critical point associated with $v = \sqrt{1 - 2\zeta^2}$ provides an absolute maximum. (It is also clear from the magnitude curves of Fig. 7.20 that $v = 0$ corresponds to a relative minimum.) The frequency at which the maximum peak occurs is then obtained as

$$v = \sqrt{1 - 2\zeta^2} \quad \Longrightarrow \quad \frac{\omega}{\omega_n} = \sqrt{1 - 2\zeta^2} \quad \Longrightarrow \quad \omega_r = \omega_n \sqrt{1 - 2\zeta^2}$$

The last expression provides the exact value of the **resonant frequency** ω_r for any damping ratio $\zeta < \zeta_{cr}$. Substitution of $v = \sqrt{1 - 2\zeta^2}$ into $|G(j\omega)|$ then gives the **resonant peak**

$$M_r = \frac{1}{2\zeta \sqrt{1 - \zeta^2}} \tag{7.48}$$

EXAMPLE 7.8. Consider a second-order system with a transfer function $G(s) = 1/(s^2 + 0.6s + 1)$. Comparison with Eq. (7.47) yields $\omega_n = 1$ rad/sec, $\zeta = 0.3$.

Following the results obtained previously, $|G(j\omega)|$ attains a maximum at

$$\frac{\omega}{\omega_n} = \sqrt{1 - 2\zeta^2} = 0.905$$

Because $\omega_n = 1$ rad/sec, the resonant frequency is simply $\omega_r = 0.905$ rad/sec. Furthermore, from Eq. (7.48), the resonant peak is calculated as $M_r = 1.747$.

7.6 RESPONSE TO AN ARBITRARY INPUT—CONVOLUTION INTEGRAL

This section is concerned with determination of the response of a dynamic system to an arbitrary input. In particular, we focus on second-order systems and draw the desired conclusions. The approach introduced and implemented here involves the concepts of convolution (see Section 1.7) and unit-impulse response. To this end, consider a linear, underdamped second-order system represented in the standard form, Eq. (7.8), as

$$\ddot{x} + 2\zeta\omega_n\dot{x} + \omega_n^2 x = f(t), \qquad 0 < \zeta < 1 \qquad (7.8)$$

where $f(t)$ is some arbitrary forcing function. Although an underdamped system is considered here, the same approach is still applicable to other types of systems. In order to find the response $x(t)$, we first consider this system when subjected to a unit-impulse forcing function applied at $t = \tilde{t}$:

$$\ddot{x} + 2\zeta\omega_n\dot{x} + \omega_n^2 x = \delta(t - \tilde{t}) \qquad (7.49)$$

and zero initial conditions, $x(0^-) = 0 = \dot{x}(0^-)$. Note that because the impulsive force does not generally occur at $t = 0$, the results of Section 7.3 do not apply; however, those obtained in Section 1.9 do. In particular, following the formulation in Eq. (1.73), the unit-impulse response is expressed in the form

$$x(t) = \begin{cases} 0 & \text{if } 0 \leq t < \tilde{t} \\ (1/\omega_d)e^{-\sigma(t-\tilde{t})} \sin\omega_d(t - \tilde{t}) & \text{if } t > \tilde{t} \end{cases} \qquad (7.50)$$

with $\sigma = \zeta\omega_n$ and $\omega_d = \omega_n\sqrt{1 - \zeta^2}$. A convenient notation for the second portion of $x(t)$ in Eq. (7.50) is

$$h(t - \tilde{t}) = \frac{1}{\omega_d} e^{-\sigma(t-\tilde{t})} \sin\omega_d(t - \tilde{t}), \qquad t > \tilde{t} \qquad (7.51)$$

The objective is then to determine the system's response to $f(t)$ using the unit-impulse response defined by Eq. (7.51). To that end, consider Fig. 7.21, the graphical representation of an arbitrary function f.

Divide the curve representing $f(t)$ into small segments so that the vertical projections of their endpoints onto the t-axis form the mesh points dividing the time interval $[0, t]$. Two typical such points are then labeled as $\tilde{t}$ and $\tilde{t} + d\tilde{t}$. Now consider the region below the curve bounded between $\tilde{t}$ and $\tilde{t} + d\tilde{t}$. Its area, measuring the impulse of $f(t)$, is approximately $f(\tilde{t})d\tilde{t}$ because the mesh points are dense. If this

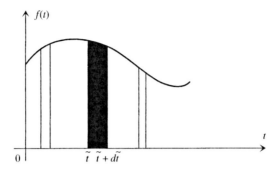

FIGURE 7.21 Arbitrary function.

impulse were to be the forcing function in Eq. (7.49), then the corresponding forced response would be

$$x(t) = \frac{\text{impulse of } f(t) \text{ between } \tilde{t} \text{ and } \tilde{t} + d\tilde{t}}{\text{impulse of } \delta(t - \tilde{t})} \cdot h(t - \tilde{t})$$

which reduces to

$$x(t) = f(\tilde{t}) d\tilde{t} \cdot h(t - \tilde{t}) \qquad (7.52)$$

because $\delta(t - \tilde{t})$ is a *unit* impulse. Extending this result to the entire group of regions that are generated by consecutive mesh points, the response to the forcing function only, labeled $x_{\text{forced}}(t)$, is expressed in the form of an integral as

$$x_{\text{forced}}(t) = \int_0^t f(\tilde{t}) \cdot h(t - \tilde{t}) d\tilde{t} \qquad (7.53)$$

Consequently, the *total response* corresponding to Eq. (7.8) is $x_{\text{total}}(t) = x_{\text{forced}}(t) + x_{\text{free}}(t)$. However, by assumption we have $x(0^-) = 0 = \dot{x}(0^-)$, so the free response defined by Eq. (7.17) reduces to zero. As a result, the total response is given by

$$x(t) = \int_0^t f(\tilde{t}) \cdot h(t - \tilde{t}) d\tilde{t} \qquad \text{Total response; zero initial conditions} \qquad (7.54)$$

Note that this integral is indeed the convolution of f and h, as defined in Section 1.7, and hence

$$x(t) = (f * h)(t)$$

The integral in Eq. (7.54) is also referred to as the **convolution integral**.

EXAMPLE 7.9. Find the response of the following undamped second-order system subjected to the unit-step forcing function and initial conditions $x(0^-) = x_0$, $\dot{x}(0^-) = v_0$:

$$\ddot{x} + \omega_n^2 x = u_s(t)$$

Solution. Note that since $\zeta = 0$, we have $\omega_d = \omega_n$ and $\sigma = 0$ so that the unit-impulse response is

$$h(t - \tilde{t}) = \frac{1}{\omega_n} \sin \omega_n(t - \tilde{t})$$

As a result, the system's response to the forcing function is given by Eq. (7.53) with $f = 1$, as

$$x_{\text{forced}}(t) = \frac{1}{\omega_n} \int_0^t \sin \omega_n(t - \tilde{t}) \, d\tilde{t} = \frac{1}{\omega_n^2}[1 - \cos \omega_n t]$$

The free response, corresponding to initial conditions, is described by Eq. (7.18) as

$$x_{\text{free}}(t) = x_0 \cos \omega_n t + \frac{v_0}{\omega_n} \sin \omega_n t$$

The total response of the system is then expressed as

$$x(t) = \underbrace{x_0 \cos \omega_n t + \frac{v_0}{\omega_n} \sin \omega_n t}_{x_{\text{free}}} + \underbrace{\frac{1}{\omega_n^2}[1 - \cos \omega_n t]}_{x_{\text{forced}}}$$

$$= \frac{1}{\omega_n^2} + \left(x_0 - \frac{1}{\omega_n^2}\right) \cos \omega_n t + \frac{v_0}{\omega_n} \sin \omega_n t$$

It should be mentioned that the problem under consideration is precisely the unit-step response of an undamped, second-order system, as studied in Section 7.3. Using the Laplace transformation technique, the response was obtained and presented as Eq. (7.37). Observe then that utilization of the convolution integral results in the exact same expression for the response, as expected.

EXAMPLE 7.10. Consider the underdamped second-order system below, subjected to zero initial conditions,

$$\ddot{x} + 2\zeta \omega_n \dot{x} + \omega_n^2 x = f(t), \qquad x(0^-) = 0, \qquad \dot{x}(0^-) = 0$$

where

$$f(t) = \begin{cases} 1 & \text{if } 0 \leq t \leq 1 \\ 0 & \text{if } t > 1 \end{cases}$$

Find the total response.

Solution. Because the forcing function exhibits two types of behavior in two time intervals, the response must be determined in two different time intervals: $0 \leq t \leq 1$ and $t > 1$.

1. $0 \le t \le 1$. The initial conditions are zero by assumption, hence the total response in this time interval is given by Eq. (7.54) with $f = 1$ as

$$x(t) = \int_0^t \frac{1}{\omega_d} e^{-\sigma(t-\tilde{t})} \sin\omega_d(t-\tilde{t})\,d\tilde{t}$$

$$\stackrel{\text{integration by parts}}{=} \frac{1}{\omega_d} \cdot \frac{1}{\sigma^2 + \omega_d^2}[\omega_d - e^{-\sigma t}(\omega_d \cos\omega_d t + \sigma \sin\omega_d t)]$$

Noting that $\sigma^2 + \omega_d^2 = \omega_n^2$, we have

$$x(t) = \frac{1}{\omega_n^2 \omega_d} \cdot [\omega_d - e^{-\sigma t}(\omega_d \cos\omega_d t + \sigma \sin\omega_d t)] \quad \text{for} \quad 0 \le t \le 1 \tag{7.55}$$

2. $t > 1$. Beyond $t = 1$ the system is subjected to no forcing function and hence experiences free motion. However, the information on $x(t)$ from Eq. (7.55) provides the initial conditions for $t > 1$. Therefore, the response is defined by Eq. (7.17) as

$$x(t) = e^{-\sigma(t-1)} \left[x_0 \cos\omega_d(t-1) + \frac{\sigma x_0 + v_0}{\omega_d} \sin\omega_d(t-1) \right] \quad \text{for} \quad t > 1 \tag{7.56}$$

in which $x_0 = x(1)$ and $v_0 = \dot{x}(1)$, where x is given by Eq. (7.55). Direct calculation yields

$$x_0 = \frac{1}{\omega_n^2 \omega_d} \cdot [\omega_d - e^{-\sigma}(\omega_d \cos\omega_d + \sigma \sin\omega_d)]$$

and

$$v_0 = \frac{1}{\omega_n^2 \omega_d} \cdot e^{-\sigma}[\omega_d(\sigma+1)\cos\omega_d + (\sigma + \omega_d^2)\sin\omega_d]$$

Substitution of these quantities into Eq. (7.56) then yields the response for $t > 1$.

EXAMPLE 7.11. Determine the total response of an undamped second-order system described by

$$\ddot{x} + \omega_n^2 x = \sin\omega_n t, \qquad x(0^-) = 0, \qquad \dot{x}(0^-) = v_0$$

Solution. First note that because $\zeta = 0$, we have $\omega_d = \omega_n$. The unit-impulse response is given by

$$h(t-\tilde{t}) = \frac{1}{\omega_n} \sin\omega_n(t-\tilde{t})$$

so that the system response to the forcing function is

$$x_{\text{forced}}(t) = \frac{1}{\omega_n} \int_0^t \sin\omega_n \tilde{t} \sin\omega_n(t-\tilde{t})\,d\tilde{t} \tag{7.57}$$

Using the trigonometric identity

$$\sin\alpha \sin\beta = \tfrac{1}{2}[\cos(\alpha-\beta) - \cos(\alpha+\beta)]$$

we have

$$\sin\omega_n \tilde{t}\sin\omega_n(t-\tilde{t}) = \tfrac{1}{2}[\cos\omega_n(2\tilde{t}-t) - \cos\omega_n t]$$

and Eq. (7.57) becomes

$$x_{\text{forced}}(t) = \frac{1}{2\omega_n}\int_0^t [\cos\omega_n(2\tilde{t}-t) - \cos\omega_n t]\,d\tilde{t}$$

$$= \frac{1}{2\omega_n}\left[\frac{1}{\omega_n}\sin\omega_n t - t\cos\omega_n t\right]$$

The free response, $x_{\text{free}}(t)$, due to initial conditions is given by Eq. (7.18), with $x_0 = 0$, as

$$x_{\text{free}}(t) = \frac{v_0}{\omega_n}\sin\omega_n t$$

Therefore, the total response of the system is obtained as

$$x(t) = \frac{v_0}{\omega_n}\sin\omega_n t + \frac{1}{2\omega_n}\left[\frac{1}{\omega_n}\sin\omega_n t - t\cos\omega_n t\right]$$

$$= \left(\frac{v_0}{\omega_n} + \frac{1}{2\omega_n^2}\right)\sin\omega_n t - \frac{1}{2\omega_n}t\cos\omega_n t$$

PROBLEM SET 7.5–7.6

7.13. The governing equation of a dynamic system is defined as

$$2\dot{x} + x = f(t)$$

where the applied forcing function is sinusoidal in the form $f(t) = F_0 \sin\omega t$.
(a) Assuming zero initial conditions, find the transfer function $X(s)/F(s)$.
(b) Find the frequency response function (FRF).
(c) Find the steady-state response of the system.

7.14. Reconsider the system of Problem 7.13. Assuming $F_0 = 10$ and $\omega = 3$, find the steady-state response.

7.15. Using the convolution integral, determine the response for the following system:

$$\ddot{x} + \omega_n^2 x = f(t)$$

subjected to zero initial conditions where

$$f(t) = \begin{cases} 1 & \text{if } 0 \le t \le 1 \\ 0 & \text{if } t > 1 \end{cases}$$

7.16. Consider the following underdamped, second-order system

$$\ddot{x} + 2\zeta\omega_n \dot{x} + \omega_n^2 x = u_s(t), \qquad 0 < \zeta < 1$$

subjected to zero initial conditions. Find the total response via the convolution integral.

7.17. Repeat Problem 7.16 for initial conditions $x(0^-) = x_0$ and $\dot{x}(0^-) = 0$.

7.7 SOLVING THE STATE EQUATION

We now turn our attention to solving the linear state equation, of which the state matrix has constant entries. This class of state equations is also known as **time-invariant**. The solution method is inspired by, and a generalization of, that employed to solve a single, first-order differential equation. To begin with, we will seek the solution to a homogeneous state equation, and later extend the idea to nonhomogeneous equations. It will be evident that the background material on matrix analysis (Chapter 2 and especially Section 2.4) plays a significant role in the analysis to follow.

Homogeneous State Equation

Consider a linear, homogeneous state equation in the form

$$\dot{\mathbf{x}} = \mathbf{A}\mathbf{x} \qquad (7.58)$$

subjected to an initial state vector $\mathbf{x}(0) = \mathbf{x}_0$. The state vector $\mathbf{x}$ and state matrix $\mathbf{A}$ are $n \times 1$ and $n \times n$, respectively. Recall that the solution to the scalar homogeneous differential equation

$$\dot{x} = ax(t), \qquad x(0) = x_0 \qquad (7.59)$$

is determined via the characteristic equation $\lambda - a = 0$. Then, the solution is in the general form

$$x(t) = ke^{at}$$

where constant k is evaluated using the initial condition, as $k = x(0) = x_0$. Consequently, the specific solution to Eq. (7.59), satisfying the initial condition, is expressed as

$$x(t) = x_0 e^{at} \qquad (7.60)$$

With this in mind, the solution to the homogeneous state equation, Eq. (7.58), is

$$\mathbf{x}(t) = e^{\mathbf{A}t}\mathbf{x}_0 \qquad (7.61)$$

where the order of multiplication of $e^{\mathbf{A}t}$ and $\mathbf{x}_0$ should be noted; that is, reversing the order makes the matrix product undefined. The matrix exponential $e^{\mathbf{A}t}$ was thoroughly studied in Section 2.4 and is defined by Eq. (2.20). As previously discussed, if the state matrix $\mathbf{A}$ is diagonal, then the exponential is easy to calculate. In a more general sense, if $\mathbf{A}$ has a complete set of n linearly independent eigenvectors, then $\mathbf{A}$

is diagonalizable and its exponential matrix may be determined analytically via the modal matrix, Eq. (2.25). The most important feature of Eq. (7.61) is that it enables one to obtain the state vector **x** in closed form.

Laplace Transform Approach

The homogeneous solution, Eq. (7.61), can also be obtained via Laplace transformation of the state equation. Generalizing the results on scalar differential equations, the Laplace transform of Eq. (7.58) can be written as

$$s\mathbf{X}(s) - \mathbf{x}(0) = \mathbf{A}\mathbf{X}(s)$$

where $\mathbf{X}(s)$ denotes the vector of Laplace transforms of components of $\mathbf{x}(t)$. Rearranging this last equation, we have

$$(s\mathbf{I} - \mathbf{A})\mathbf{X}(s) = \mathbf{x}_0$$

Based on the assumption that $\mathbf{A}$ has a set of n linearly independent eigenvectors, $\mathbf{A}$ is nonsingular, i.e., $|s\mathbf{I} - \mathbf{A}| \neq 0$. Premultiplication of the equation above by $(s\mathbf{I} - \mathbf{A})^{-1}$ yields

$$\mathbf{X}(s) = (s\mathbf{I} - \mathbf{A})^{-1}\mathbf{x}_0$$

inverse Laplace transformation of which results in the expression for the state vector, as

$$\mathbf{x}(t) = L^{-1}\{(s\mathbf{I} - \mathbf{A})^{-1}\}\mathbf{x}_0 \tag{7.62}$$

In order to establish a relation between the exponential of the state matrix and $(s\mathbf{I} - \mathbf{A})^{-1}$, compare Eqs. (7.61) and (7.62) to conclude that

$$e^{\mathbf{A}t} = L^{-1}\{(s\mathbf{I} - \mathbf{A})^{-1}\} \tag{7.63}$$

Note that $(s\mathbf{I} - \mathbf{A})^{-1}$ is an $n \times n$ matrix whose entries are generally functions of the Laplace variable s. Then, inverse Laplace transformation of all of these entries forms the $n \times n$ matrix $L^{-1}\{(s\mathbf{I} - \mathbf{A})^{-1}\}$.

State-Transition Matrix

The state vector $\mathbf{x}(t)$, the solution to Eq. (7.58), is sometimes expressed in the form

$$\mathbf{x}(t) = \mathbf{\Phi}(t)\mathbf{x}_0 \tag{7.64}$$

where $\mathbf{\Phi}(t)$ is $n \times n$ and referred to as the **state-transition matrix.** This matrix is usually defined as the solution to the first-order system of differential equations,

$$\dot{\mathbf{\Phi}} = \mathbf{A}\mathbf{\Phi}, \qquad \mathbf{\Phi}(0) = \mathbf{I} \tag{7.65}$$

which may readily be verified as follows: From Eq. (7.64) we have

$$\dot{\mathbf{x}} = \dot{\mathbf{\Phi}}(t)\mathbf{x}_0$$

In this last equation, substitute for $\dot{\Phi}(t)$ using Eq. (7.65) to obtain

$$\dot{x} = A\underbrace{\Phi(t)x_0}_{x(t)} = Ax$$

which indeed proves the earlier claim. Comparison of Eqs. (7.64) and (7.61) yields

$$\Phi(t) = e^{At} \qquad (7.66)$$

From Eq. (7.63) we further conclude that

$$\Phi(t) = e^{At} = L^{-1}\{(sI - A)^{-1}\}$$

EXAMPLE 7.12. Find the solution of the homogeneous state equation

$$\dot{x} = Ax, \qquad x = \begin{Bmatrix} x_1 \\ x_2 \end{Bmatrix}, \qquad A = \begin{bmatrix} 0 & 1 \\ -4 & -5 \end{bmatrix}$$

subjected to the initial state

$$x_0 = \begin{Bmatrix} 0 \\ -1 \end{Bmatrix}$$

using

(a) diagonalization of A.
(b) $L^{-1}\{(sI - A)^{-1}\}$.

Solution

(a) The eigenvalues and the corresponding eigenvectors of A are easily calculated as

$$\lambda_{1,2} = -4, -1, \qquad v_1 = \begin{bmatrix} -1 \\ 4 \end{bmatrix}, \qquad v_2 = \begin{bmatrix} 1 \\ -1 \end{bmatrix}$$

so that the modal matrix is formed as

$$P = \begin{bmatrix} -1 & 1 \\ 4 & -1 \end{bmatrix}$$

with the property that

$$P^{-1}AP = \Lambda = \begin{bmatrix} -4 & 0 \\ 0 & -1 \end{bmatrix}$$

Then, following the result of Eq. (2.25), we have

$$e^{At} = Pe^{\Lambda t}P^{-1} = \begin{bmatrix} -1 & 1 \\ 4 & -1 \end{bmatrix}\begin{bmatrix} e^{-4t} & 0 \\ 0 & e^{-t} \end{bmatrix}\frac{1}{3}\begin{bmatrix} 1 & 1 \\ 4 & 1 \end{bmatrix}$$

$$= \frac{1}{3}\begin{bmatrix} -e^{-4t} + 4e^{-t} & -e^{-4t} + e^{-t} \\ 4(e^{-4t} - e^{-t}) & 4e^{-4t} - e^{-t} \end{bmatrix}$$

Subsequently, the state vector is obtained via Eq. (7.61), as

$$\mathbf{x}(t) = e^{\mathbf{A}t}\mathbf{x}_0 = \frac{1}{3}\begin{Bmatrix} e^{-4t} - e^{-t} \\ -4e^{-4t} + e^{-t} \end{Bmatrix}$$

(b) We first note that

$$s\mathbf{I} - \mathbf{A} = \begin{bmatrix} s & -1 \\ 4 & s+5 \end{bmatrix} \implies (s\mathbf{I} - \mathbf{A})^{-1} = \frac{1}{(s+1)(s+4)}\begin{bmatrix} s+5 & 1 \\ -4 & s \end{bmatrix}$$

$$= \begin{bmatrix} \dfrac{s+5}{(s+1)(s+4)} & \dfrac{1}{(s+1)(s+4)} \\ \dfrac{-4}{(s+1)(s+4)} & \dfrac{s}{(s+1)(s+4)} \end{bmatrix}$$

As mentioned earlier, $L^{-1}\{(s\mathbf{I} - \mathbf{A})^{-1}\}$ is determined through inverse Laplace transformation of the four entries of this last matrix. Applying the partial fraction method to each entry, followed by Laplace inversion, one obtains

$$L^{-1}\{(s\mathbf{I} - \mathbf{A})^{-1}\} = L^{-1}\left\{\begin{bmatrix} \dfrac{4/3}{s+1} - \dfrac{1/3}{s+4} & \dfrac{1/3}{s+1} - \dfrac{1/3}{s+4} \\ \dfrac{-4/3}{s+1} + \dfrac{4/3}{s+4} & \dfrac{-1/3}{s+1} + \dfrac{4/3}{s+4} \end{bmatrix}\right\}$$

$$= \frac{1}{3}\begin{bmatrix} -e^{-4t} + 4e^{-t} & -e^{-4t} + e^{-t} \\ 4(e^{-4t} - e^{-t}) & 4e^{-4t} - e^{-t} \end{bmatrix}$$

which is identical to $e^{\mathbf{A}t}$ of part (a), as expected. Thus, the resulting state vector is the same as in part (a).

Nonhomogeneous State Equation

Consider the nonhomogeneous state equation

$$\dot{\mathbf{x}} = \mathbf{A}\mathbf{x} + \mathbf{B}\mathbf{u} \tag{7.67}$$

subjected to an initial state vector $\mathbf{x}(0) = \mathbf{x}_0$. Here, $\mathbf{x}$, $\mathbf{A}$, $\mathbf{B}$, and $\mathbf{u}$ are $n \times 1$, $n \times n$, $n \times m$, and $m \times 1$, respectively. To obtain the solution to this equation, let us recall the treatment of its scalar counterpart,

$$\dot{x} = ax(t) + bu(t), \qquad x(0) = x_0 \tag{7.68}$$

Because $u(t)$ is not necessarily of any specific nature, the method of undetermined coefficients does not generally apply. However, we first rewrite Eq. (7.68) as

$$\dot{x} - ax(t) = bu(t), \qquad x(0) = x_0 \tag{7.69}$$

In general, given a linear first-order differential equation in the form

$$\dot{x} + g(t)x = f(t)$$

its solution is given by

$$x(t) = e^{-h(t)}\left[\int_0^t e^{h(\tau)} f(\tau)\, d\tau + c\right] \quad \text{where} \quad h(t) = \int g(t)\, dt \quad (7.70)$$

Applying Eq. (7.70) to Eq. (7.69) with $g = -a$ and $f = bu$, we have $h(t) = \int g(t)\, dt = -at$ and

$$x(t) = e^{at}\left[\int_0^t e^{-a\tau} bu(\tau)\, d\tau + c\right]$$

Finally, apply the initial condition to get $x(0) = c = x_0$, and insert into the equation above to obtain

$$x(t) = e^{at}\left[\int_0^t e^{-a\tau} bu(\tau)\, d\tau + x_0\right] = e^{at} x_0 + e^{at}\int_0^t e^{-a\tau} bu(\tau)\, d\tau$$

$$= e^{at} x_0 + \int_0^t e^{a(t-\tau)} bu(\tau)\, d\tau \quad (7.71)$$

Extending the above methodology to a nonhomogeneous system, we write the solution to Eq. (7.67) as

$$\mathbf{x}(t) = e^{\mathbf{A}t}\mathbf{x}_0 + \int_0^t e^{\mathbf{A}(t-\tau)}\mathbf{B}\mathbf{u}(\tau)\, d\tau \quad (7.72)$$

which may also be represented in terms of the state-transition matrix $\mathbf{\Phi}(t) = e^{\mathbf{A}t}$ as

$$\mathbf{x}(t) = \mathbf{\Phi}(t)\mathbf{x}_0 + \int_0^t \mathbf{\Phi}(t-\tau)\mathbf{B}\mathbf{u}(\tau)\, d\tau \quad (7.73)$$

Analogous to Section 7.6, note that $\mathbf{\Phi}(t)\mathbf{x}_0$ and the integral term signify the contributions of the initial conditions and the input to the system response, respectively.

Laplace Transform Approach

Taking the Laplace transform of Eq. (7.67), taking into account the initial state vector, we have

$$s\mathbf{X}(s) - \mathbf{x}(0) = \mathbf{A}\mathbf{X}(s) + \mathbf{B}\mathbf{U}(s)$$

Rearranging this last equation and assuming $\mathbf{A}$ is nonsingular, one obtains

$$(s\mathbf{I} - \mathbf{A})\mathbf{X}(s) = \mathbf{x}_0 + \mathbf{B}\mathbf{U}(s)$$

$$\Longrightarrow \quad \mathbf{X}(s) = (s\mathbf{I} - \mathbf{A})^{-1}\mathbf{x}_0 + (s\mathbf{I} - \mathbf{A})^{-1}\mathbf{B}\mathbf{U}(s)$$

Inverse Laplace transformation yields the expression for the state vector

$$\mathbf{x}(t) = L^{-1}\{(s\mathbf{I} - \mathbf{A})^{-1}\}\mathbf{x}_0 + L^{-1}\{(s\mathbf{I} - \mathbf{A})^{-1}\mathbf{B}\mathbf{U}(s)\} \tag{7.74}$$

Although the inverse transform in the first term is simply $e^{\mathbf{A}t}$, evaluation of the second one requires additional work. Recalling the convolution theorem (see Chapter 1), the second term is

$$L^{-1}\{(s\mathbf{I} - \mathbf{A})^{-1}\mathbf{B}\mathbf{U}(s)\} = \int_0^t e^{\mathbf{A}(t-\tau)}\mathbf{B}\mathbf{u}(\tau)\,d\tau$$

Substitution into Eq. (7.74) yields

$$\mathbf{x}(t) = e^{\mathbf{A}t}\mathbf{x}_0 + \int_0^t e^{\mathbf{A}(t-\tau)}\mathbf{B}\mathbf{u}(\tau)\,d\tau \tag{7.75}$$

which completely agrees with Eqs. (7.72) and (7.73).

EXAMPLE 7.13. Solve the nonhomogeneous state equation described by

$$\dot{\mathbf{x}} = \mathbf{A}\mathbf{x} + \mathbf{B}u, \qquad \mathbf{x} = \begin{Bmatrix} x_1 \\ x_2 \end{Bmatrix}, \qquad \mathbf{A} = \begin{bmatrix} 0 & 1 \\ -4 & -5 \end{bmatrix},$$

$$\mathbf{B} = \begin{bmatrix} 0 \\ 1 \end{bmatrix}, \qquad u = \text{unit-step function}$$

subjected to the initial state

$$\mathbf{x}_0 = \begin{Bmatrix} 0 \\ -1 \end{Bmatrix}$$

Solution. Because the state matrix is the same as in the previous example, the modal matrix $\mathbf{P}$ and the exponential matrix $e^{\mathbf{A}t}$ remain unchanged. Also note that based on $e^{\mathbf{A}t}$ of the last example,

$$e^{\mathbf{A}(t-\tau)} = \frac{1}{3}\begin{bmatrix} -e^{-4(t-\tau)} + 4e^{-(t-\tau)} & -e^{-4(t-\tau)} + e^{-(t-\tau)} \\ 4[e^{-4(t-\tau)} - e^{-(t-\tau)}] & 4e^{-4(t-\tau)} - e^{-(t-\tau)} \end{bmatrix}$$

Via insertion of the available information into Eq. (7.72), the state vector is obtained as

$$\mathbf{x}(t) = \frac{1}{3}\begin{bmatrix} e^{-4t} - e^{-t} \\ -4e^{-4t} + e^{-t} \end{bmatrix} + \int_0^t \begin{bmatrix} -e^{-4(t-\tau)} + e^{-(t-\tau)} \\ 4e^{-4(t-\tau)} - e^{-(t-\tau)} \end{bmatrix} d\tau$$

Term-by-term integration of each component of the integrand vector gives

$$\mathbf{x}(t) = \frac{1}{3}\begin{bmatrix} e^{-4t} - e^{-t} \\ -4e^{-4t} + e^{-t} \end{bmatrix} + \begin{bmatrix} (1/4)e^{-4t} - e^{-t} + 3/4 \\ -e^{-4t} + e^{-t} \end{bmatrix}$$

$$= \begin{bmatrix} (7/12)e^{-4t} - (4/3)e^{-t} + 3/4 \\ (-7/3)e^{-4t} + (4/3)e^{-t} \end{bmatrix}$$

PROBLEM SET 7.7

7.18. Consider the following homogeneous system of equations

$$\dot{\mathbf{x}} = \begin{bmatrix} 1 & 2 \\ 0 & 3 \end{bmatrix} \mathbf{x}$$

subject to an initial vector

$$\mathbf{x}_0 = \begin{Bmatrix} 2 \\ 1 \end{Bmatrix}$$

(a) Determine $\mathbf{x}$ via the exponential of the coefficient matrix.
(b) Determine $\mathbf{x}$ via the Laplace transform approach.

7.19. The state equation for a certain dynamic system is given by

$$\dot{\mathbf{x}} = \mathbf{A}\mathbf{x} \quad \text{where} \quad \mathbf{A} = \begin{bmatrix} 0 & 1 \\ -2 & -3 \end{bmatrix}$$

with the initial state vector as

$$\mathbf{x}_0 = \begin{Bmatrix} 1 \\ 0 \end{Bmatrix}$$

Find the state vector $\mathbf{x}$ using
(a) The exponential of matrix $\mathbf{A}$
(b) The Laplace transform approach

7.20. Consider the nonhomogeneous state equation

$$\dot{\mathbf{x}} = \mathbf{A}\mathbf{x} + \mathbf{B}u, \quad \mathbf{B} = \begin{bmatrix} 0 \\ 1 \end{bmatrix}, \quad u = \text{unit-step function}$$

where the state matrix $\mathbf{A}$ is as in Problem 7.19. Assume that the initial state vector is $\mathbf{x}_0 = \mathbf{0}$. Find the state vector $\mathbf{x}(t)$ using
(a) The exponential of the state matrix
(b) The Laplace transform approach

7.8 MODAL DECOMPOSITION

The concept of modal decomposition plays a significant role in the analysis of a physical system whose equations of motion are represented in the standard second-order matrix form, as defined by Eq. (3.7). In particular, we are concerned with systems with no inherent damping, so that Eq. (3.7) takes the form

$$\mathbf{m}\ddot{\mathbf{x}} + \mathbf{k}\mathbf{x} = \mathbf{f} \tag{7.76}$$

Because the mass matrix is nonsingular, premultiplication of Eq. (7.76) by $\mathbf{m}^{-1}$ yields

$$\ddot{\mathbf{x}} + \underbrace{\mathbf{m}^{-1}\mathbf{k}}_{\bar{\mathbf{k}}}\mathbf{x} = \underbrace{\mathbf{m}^{-1}\mathbf{f}}_{\bar{\mathbf{f}}} \quad \Longrightarrow \quad \ddot{\mathbf{x}} + \bar{\mathbf{k}}\mathbf{x} = \bar{\mathbf{f}} \tag{7.77}$$

Often, these equations are *coupled* through the elements of matrix $\bar{\mathbf{k}}$. Because of this coupling phenomenon, solutions to these differential equations may not be easily determined and often require tremendous analytical effort. To that end, based on certain assumptions and through coordinate transformation, the coupled system is transformed into a *decoupled* one in which the differential equations are independent of one another and may be solved separately. For simplicity, we consider a homogeneous system of n second-order ODEs expressed as

$$\begin{Bmatrix} \ddot{x}_1 \\ \ddot{x}_2 \\ \vdots \\ \ddot{x}_n \end{Bmatrix} = \begin{bmatrix} \mathbf{A} \\ {}_{n \times n} \end{bmatrix} \begin{Bmatrix} x_1 \\ x_2 \\ \vdots \\ x_n \end{Bmatrix} \quad \text{or simply} \quad \ddot{\mathbf{x}} = \mathbf{A}\mathbf{x} \qquad (7.78)$$

subject to initial conditions

$$\mathbf{x}_0 = \begin{bmatrix} x_1(0) \\ x_2(0) \\ \vdots \\ x_n(0) \end{bmatrix}, \quad \dot{\mathbf{x}}_0 = \begin{bmatrix} \dot{x}_1(0) \\ \dot{x}_2(0) \\ \vdots \\ \dot{x}_n(0) \end{bmatrix}$$

Note that Eq. (7.78) is indeed in the form of Eq. (7.77) with $\mathbf{A} = -\bar{\mathbf{k}}$ and $\bar{\mathbf{f}} = \mathbf{0}$. Assuming that matrix $\mathbf{A}$ has n linearly independent eigenvectors, $\{\mathbf{v}_1, \ldots, \mathbf{v}_n\}$, then by Theorem 2.5, the *modal matrix* $\mathbf{P} = [\mathbf{v}_1 \quad \cdots \quad \mathbf{v}_n]$ diagonalizes $\mathbf{A}$. To that end, introduce the matrix transformation

$$\mathbf{x} = \mathbf{P}\mathbf{y} \qquad (7.79)$$

where $\mathbf{y}$ is an $n \times 1$ vector of the transformed coordinates. Differentiate Eq. (7.79) twice with respect to t and substitute into Eq. (7.78) to obtain

$$\mathbf{P}\ddot{\mathbf{y}} = \mathbf{A}\mathbf{P}\mathbf{y} \qquad (7.80)$$

Premultiplication of Eq. (7.80) by $\mathbf{P}^{-1}$ results in

$$\mathbf{P}^{-1}\mathbf{P}\ddot{\mathbf{y}} = \underbrace{\mathbf{P}^{-1}\mathbf{A}\mathbf{P}}_{\Lambda}\mathbf{y} \quad \Longrightarrow \quad \ddot{\mathbf{y}} = \Lambda\mathbf{y} \qquad (7.81)$$

Next, we identify the initial conditions associated with Eq. (7.81) by rewriting Eq. (7.79) as

$$\mathbf{y} = \mathbf{P}^{-1}\mathbf{x} \quad \Longrightarrow \quad \mathbf{y}(0) = \mathbf{P}^{-1}\mathbf{x}(0) = \mathbf{P}^{-1}\mathbf{x}_0$$
$$\dot{\mathbf{y}}(0) = \mathbf{P}^{-1}\dot{\mathbf{x}}(0) = \mathbf{P}^{-1}\dot{\mathbf{x}}_0 \qquad (7.82)$$

and

The newly constructed system given by Eq. (7.81) has the unique property that its coefficient matrix is diagonal, and hence its governing equations are independent. To further understand the structure of Eq. (7.81), let

$$\mathbf{y} = \begin{Bmatrix} y_1 \\ y_2 \\ \vdots \\ y_n \end{Bmatrix}, \quad \Lambda = \begin{bmatrix} \lambda_1 & & & \\ & \lambda_2 & & \\ & & \ddots & \\ & & & \lambda_n \end{bmatrix}, \quad \mathbf{y}(0) = \begin{bmatrix} y_{0,1} \\ y_{0,2} \\ \vdots \\ y_{0,n} \end{bmatrix}, \quad \dot{\mathbf{y}}(0) = \begin{bmatrix} \dot{y}_{0,1} \\ \dot{y}_{0,2} \\ \vdots \\ \dot{y}_{0,n} \end{bmatrix}$$

where $y_{0,i}$ and $\dot{y}_{0,i}$ ($i = 1, 2, \ldots, n$) denote components of $\mathbf{y}(0)$ and $\dot{\mathbf{y}}(0)$, respectively. Consequently, the ith row of Eq. (7.81) reads

$$\ddot{y}_i = \lambda_i y_i, \qquad \underbrace{y_i(0) = y_{0,i}, \; \dot{y}_i(0) = \dot{y}_{0,i}}_{\text{initial conditions}}, \qquad i = 1, 2, \ldots, n \qquad (7.83)$$

which is a homogeneous, linear, second-order ODE and can easily be solved via standard techniques such as Laplace transformation or undetermined coefficients. Once vector $\mathbf{y}$ is available, the transformation defined by Eq. (7.79) is used to determine the original vector $\mathbf{x}$.

EXAMPLE 7.14. The equations of motion of a mechanical system are given as

$$\begin{bmatrix} \ddot{x}_1 \\ \ddot{x}_2 \end{bmatrix} = \begin{bmatrix} -5 & 2 \\ 2 & -2 \end{bmatrix} \begin{bmatrix} x_1 \\ x_2 \end{bmatrix} \qquad (7.84)$$

subject to initial conditions

$$x_1(0^-) = 0 = x_2(0^-), \qquad \dot{x}_1(0^-) = 1 = \dot{x}_2(0^-)$$

Determine the response at any time $t \geq 0$.

Solution. The eigenvalues and corresponding eigenvectors of the coefficient matrix are calculated as

$$\lambda_{1,2} = -1, -6 \quad \text{and} \quad \mathbf{v}_1 = \begin{bmatrix} 1 \\ 2 \end{bmatrix}, \quad \mathbf{v}_2 = \begin{bmatrix} 2 \\ -1 \end{bmatrix}$$

Because the eigenvalues are distinct, the resulting eigenvectors are linearly independent. Form the modal matrix and use the diagonalization property to obtain

$$\mathbf{P} = [\mathbf{v}_1 \quad \mathbf{v}_2] = \begin{bmatrix} 1 & 2 \\ 2 & -1 \end{bmatrix} \implies \mathbf{P}^{-1}\mathbf{A}\mathbf{P} = \mathbf{\Lambda} = \begin{bmatrix} -1 & 0 \\ 0 & -6 \end{bmatrix}$$

Therefore, Eq. (7.84) is reduced to

$$\begin{Bmatrix} \ddot{y}_1 \\ \ddot{y}_2 \end{Bmatrix} = \begin{bmatrix} -1 & 0 \\ 0 & -6 \end{bmatrix} \begin{Bmatrix} y_1 \\ y_2 \end{Bmatrix} \qquad (7.85)$$

The initial conditions for the original system are

$$\mathbf{x}_0 = \begin{bmatrix} 0 \\ 0 \end{bmatrix} \quad \text{and} \quad \dot{\mathbf{x}}_0 = \begin{bmatrix} 1 \\ 1 \end{bmatrix}$$

Therefore, the initial conditions corresponding to Eq. (7.85) are obtained using Eq. (7.82), as

$$\mathbf{y}(0) = \mathbf{P}^{-1}\mathbf{x}_0 = \begin{bmatrix} 0 \\ 0 \end{bmatrix} \quad \text{and} \quad \dot{\mathbf{y}}(0) = \mathbf{P}^{-1}\dot{\mathbf{x}}_0 = \begin{bmatrix} 0.6 \\ 0.2 \end{bmatrix} \qquad (7.86)$$

Componentwise interpretation of Eq. (7.85) subjected to initial conditions of Eq. (7.86) yields

$$\begin{cases} \ddot{y}_1 = -y_1, & y_1(0) = 0, \quad \dot{y}_1(0) = 0.6 \\ \ddot{y}_2 = -6y_2, & y_2(0) = 0, \quad \dot{y}_2(0) = 0.2 \end{cases}$$

Solving the two equations independently, we have

$$\begin{cases} y_1(t) = 0.6 \sin t \\ y_2(t) = (0.2/\sqrt{6}) \sin(\sqrt{6}t) \end{cases} \implies \mathbf{y} = \begin{bmatrix} 0.6 \sin t \\ (0.2/\sqrt{6}) \sin(\sqrt{6}t) \end{bmatrix}$$

Ultimately, system response **x** is computed via the transformation

$$\mathbf{x} = \mathbf{Py} = \begin{bmatrix} 1 & 2 \\ 2 & -1 \end{bmatrix} \begin{bmatrix} 0.6 \sin t \\ (0.2/\sqrt{6}) \sin(\sqrt{6}t) \end{bmatrix} = \begin{bmatrix} 0.6 \sin t + (0.4/\sqrt{6}) \sin(\sqrt{6}t) \\ 1.2 \sin t - (0.2/\sqrt{6}) \sin(\sqrt{6}t) \end{bmatrix}$$

This approach may be easily extended and applied to the case of a nonhomogeneous system in the form

$$\ddot{\mathbf{x}} = \mathbf{A}\mathbf{x} + \mathbf{f} \tag{7.87}$$

which agrees with Eq. (7.77) with $\mathbf{A} = -\bar{\mathbf{k}}$ and $\mathbf{f} = \bar{\mathbf{f}}$. Proceeding as in the case of a homogeneous system, using the transformation $\mathbf{x} = \mathbf{Py}$, the decoupled system takes the form

$$\ddot{\mathbf{y}} = \Lambda \mathbf{y} + \mathbf{P}^{-1}\mathbf{f} \tag{7.88}$$

subjected to initial conditions, Eq. (7.82). This time, each row in Eq. (7.88) is a nonhomogeneous second-order differential equation that may be solved explicitly.

EXAMPLE 7.15. Given that the governing equations of a dynamic system are

$$\ddot{\mathbf{x}} = \begin{bmatrix} -5 & 2 \\ 2 & -2 \end{bmatrix} \mathbf{x} + \begin{bmatrix} 1 \\ 0 \end{bmatrix}$$

subjected to initial conditions $x_1(0^-) = 0 = x_2(0^-)$ and $\dot{x}_1(0^-) = 1 = \dot{x}_2(0^-)$, determine $\mathbf{x}(t)$.

Solution. Note that the coefficient matrix is identical to the one in the previous example, and hence the modal matrix remains unchanged. Following the formulation of Eq. (7.88), one obtains

$$\ddot{\mathbf{y}} = \begin{bmatrix} -1 & 0 \\ 0 & -6 \end{bmatrix} \mathbf{y} + \begin{bmatrix} 0.2 \\ 0.4 \end{bmatrix} \quad \text{subject to} \quad \mathbf{y}_0 = \begin{bmatrix} 0 \\ 0 \end{bmatrix} \quad \text{and} \quad \dot{\mathbf{y}}_0 = \begin{bmatrix} 0.6 \\ 0.2 \end{bmatrix}$$
$$\tag{7.89}$$

The two independent ODEs contained in Eq. (7.89) are then

$$\begin{cases} \ddot{y}_1 = -y_1 + 0.2, & y_1(0) = 0, & \dot{y}_1(0) = 0.6 \\ \ddot{y}_2 = -6y_2 + 0.4, & y_2(0) = 0, & \dot{y}_2(0) = 0.2 \end{cases}$$

Solving these initial-value problems, one obtains

$$y_1(t) = -0.2 \cos t + 0.6 \sin t + 0.2$$

and

$$y_2(t) = -\frac{0.2}{3} \cos \sqrt{6}t + \frac{0.2}{\sqrt{6}} \sin \sqrt{6}t + \frac{0.2}{3}$$

Transforming back to the $x_1 x_2$-coordinates, we have

$$\begin{cases} x_1(t) = -0.2\cos t + 0.6\sin t - \dfrac{0.4}{3}\cos\sqrt{6}t + \dfrac{0.4}{\sqrt{6}}\sin\sqrt{6}t + \dfrac{1}{3} \\ x_2(t) = -0.4\cos t + 1.2\sin t + \dfrac{0.2}{3}\cos\sqrt{6}t - \dfrac{0.2}{\sqrt{6}}\sin\sqrt{6}t + \dfrac{1}{3} \end{cases}$$

7.9 FOURIER ANALYSIS

It has frequently been mentioned that many characteristics of a dynamic system may be identified through inspection of its response to a given input. Often, these input signals are *periodic;* that is, they exhibit the same behavior after every certain amount of time over long periods of time. It is then desired to approximate such functions by a number of sinusoidal components, known as **harmonics**, of various frequencies. These components then constitute a series, referred to as a **trigonometric series** expansion, which approximates the function. In the event that the input signal is not periodic, and only occurs in a given finite time interval [0, b], it is then extended to a periodic function. Subsequently, the trigonometric series of the periodic extension agrees with the input signal in the interval [0, b].

Fourier Series

Consider a periodic function $f(t)$ with an arbitrary period p. Then, assuming that $f(t)$ has a trigonometric series representation, it is in the following form

$$f(t) = \frac{1}{2}a_0 + \sum_{n=1}^{\infty}\left(a_n \cos\frac{2n\pi t}{p} + b_n \sin\frac{2n\pi t}{p}\right) \qquad (7.90)$$

where the constants, known as **Euler constants**, are generated by the **Euler-Fourier formulas**, as

Constant term $\qquad a_0 = \dfrac{2}{p}\displaystyle\int_{-p/2}^{p/2} f(t)\,dt \qquad (7.91)$

$$a_n = \frac{2}{p}\int_{-p/2}^{p/2} f(t)\cos\frac{2n\pi t}{p}\,dt \qquad \text{for} \quad n = 1, 2, 3, \ldots \qquad (7.92)$$

and

$$b_n = \frac{2}{p}\int_{-p/2}^{p/2} f(t)\sin\frac{2n\pi t}{p}\,dt \qquad \text{for} \quad n = 1, 2, 3, \ldots \qquad (7.93)$$

Subsequently, Eq. (7.90), with constants generated through Eqs. (7.91)–(7.93), is known as the **Fourier series** of the function $f(t)$. A significant feature of the Fourier series is the fact that, unlike the Taylor series, it can be used to represent functions with discontinuities.

EXAMPLE 7.16. Determine the Fourier series expansion of the following periodic function whose definition in one period is given graphically in Fig. 7.22.

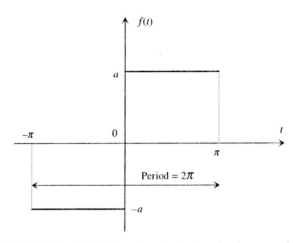

FIGURE 7.22 Definition of a periodic function in one period.

Solution. From Fig. 7.22, the analytical description of the function is

$$f(t) = \begin{cases} -a & \text{if } -\pi < t < 0 \\ a & \text{if } 0 < t < \pi \end{cases}$$

The Euler constants are generated by Eqs. (7.91)–(7.93), with $p = 2\pi$, as follows:

$$a_0 = \frac{1}{\pi}\int_{-\pi}^{\pi} f(t)\,dt = \frac{1}{\pi}\left[\int_{-\pi}^{0} -a\,dt + \int_{0}^{\pi} a\,dt\right] = 0$$

For $n = 1, 2, \ldots$

$$a_n = \frac{1}{\pi}\int_{-\pi}^{\pi} f(t)\cos nt\,dt = \frac{1}{\pi}\left[\int_{-\pi}^{0} -a\cos nt\,dt + \int_{0}^{\pi} a\cos nt\,dt\right]$$

$$= \frac{a}{n\pi}\left[(-\sin nt)_{-\pi}^{0} + (\sin nt)_{0}^{\pi}\right] = 0$$

For $n = 1, 2, \ldots$

$$b_n = \frac{1}{\pi}\int_{-\pi}^{\pi} f(t)\sin nt\,dt = \frac{1}{\pi}\left[\int_{-\pi}^{0} -a\sin nt\,dt + \int_{0}^{\pi} a\sin nt\,dt\right]$$

$$= \frac{a}{n\pi}\left[(\cos nt)_{-\pi}^{0} - (\cos nt)_{0}^{\pi}\right]$$

$$= \frac{2a}{n\pi}[1 - \cos n\pi] = \frac{2a}{n\pi}[1 - (-1)^n] = \begin{cases} 0 & \text{if } n = \text{even} \\ (4a)/(n\pi) & \text{if } n = \text{odd} \end{cases}$$

Subsequently, the Fourier series in Eq. (7.90) is reduced to

$$f(t) = \sum_{n=1}^{\infty} b_n \sin nt = \frac{4a}{\pi} \sum_{\substack{n=1 \\ n=\text{odd}}}^{\infty} \left(\frac{1}{n} \sin nt\right) = \frac{4a}{\pi} \left(\sin t + \frac{1}{3} \sin 3t + \cdots \right) \quad (7.94)$$

Naturally, the more terms in Eq. (7.94) that are taken into account, the more acceptable the accuracy of the approximation will be. Figure 7.23 shows the third and fifth partial sums in Eq. (7.94) as compared with the original function.

Fourier Cosine and Sine Series

In the event that the input signal is either an even or an odd periodic function, either b_n or a_n will vanish in the Fourier series expansion, respectively. The fact that the product of two even functions is even, of two odd functions is odd, and of an even and an odd function is odd leads to the desired results. In what follows, more details are presented.

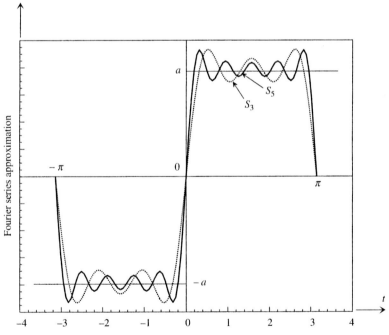

FIGURE 7.23 Third and fifth partial sums in Eq. (7.94) approximating $f(t)$.

Even periodic functions

Suppose $f(t)$ is even and periodic with an arbitrary period p. Then, the Euler constants are obtained as

$$a_n = \frac{2}{p}\underbrace{\int_{-p/2}^{p/2} f(t)\cos\frac{2n\pi t}{p}\,dt}_{\text{even}} = \frac{4}{p}\int_0^{p/2} f(t)\cos\frac{2n\pi t}{p}\,dt \quad \text{for} \quad n = 0, 1, 2, \ldots \tag{7.95}$$

and

$$b_n = \frac{2}{p}\underbrace{\int_{-p/2}^{p/2} f(t)\sin\frac{2n\pi t}{p}\,dt}_{\text{odd}} = 0 \quad \text{for} \quad n = 1, 2, 3, \ldots$$

Consequently, in the Fourier series of an even function, all coefficients of sine terms vanish. The resulting series, known as the **Fourier cosine series**, is then defined as

$$f(t) = \frac{1}{2}a_0 + \sum_{n=1}^{\infty} a_n \cos\frac{2n\pi t}{p} \tag{7.96}$$

where a_n $(n = 0, 1, \ldots)$ are given by Eq. (7.95).

Odd periodic functions

Suppose $f(t)$ is odd and periodic with an arbitrary period p. Then, the Euler coefficients are given by

$$a_n = \frac{2}{p}\underbrace{\int_{-p/2}^{p/2} f(t)\cos\frac{2n\pi t}{p}\,dt}_{\text{odd}} = 0 \quad \text{for} \quad n = 0, 1, 2, \ldots$$

and

$$b_n = \frac{2}{p}\underbrace{\int_{-p/2}^{p/2} f(t)\sin\frac{2n\pi t}{p}\,dt}_{\text{even}} = \frac{4}{p}\int_0^{p/2} f(t)\sin\frac{2n\pi t}{p}\,dt \quad \text{for} \quad n = 1, 2, 3, \ldots \tag{7.97}$$

Consequently, in the Fourier series of an odd function, all coefficients of cosine terms, as well as the constant term, vanish. The resulting series, known as the **Fourier sine series**, is then defined as

$$f(t) = \sum_{n=1}^{\infty} b_n \sin\frac{2n\pi t}{p} \tag{7.98}$$

where the b_n $(n = 1, 2, \ldots)$ are given by Eq. (7.97).

EXAMPLE 7.17. The input signal of Example 7.16 is clearly an odd periodic function and, hence, it is expected that it would be approximated by a Fourier sine series. This, of course, agrees with the results of the previous example.

Convergence of Fourier Series

Many periodic functions experience what is called a **jump** (or *simple discontinuity*) at a finite number of points in one period, as shown in Fig. 7.24.

We define the **left-hand limit** of a function $f(t)$ at the point $t = t_0$, denoted by $f(t_0^-)$, as

$$f(t_0^-) = \lim_{\Delta t_0 \to 0} f(t_0 - \Delta t_0), \qquad \Delta t_0 > 0 \tag{7.99}$$

and, similarly, the **right-hand limit** as

$$f(t_0^+) = \lim_{\Delta t_0 \to 0} f(t_0 + \Delta t_0), \qquad \Delta t_0 > 0 \tag{7.100}$$

Note that although the left- and right-hand limits indeed exist, they are not equal (see Fig. 7.24). We now state the fundamental theorem on convergence of the Fourier series of a periodic function.

THEOREM 7.1. DIRICHLET'S THEOREM. Let $f(t)$ be periodic with period 2π. Suppose $f(t)$ is defined in the interval $[-\pi, \pi)$, has finitely many points of discontinuity, maxima and minima, and is bounded. Then, at all t where $f(t)$ is continuous, its Fourier series converges to $f(t)$. At every point of discontinuity t_0 the Fourier series converges to the average of left- and right-hand limits,

$$\tfrac{1}{2}\left[f(t_0^-) + f(t_0^+)\right] \tag{7.101}$$

It is then readily seen that at the points of continuity, $f(t_0^-) = f(t_0^+) = f(t_0)$, and hence, by Eq. (7.101), the Fourier series converges to the function itself.

Interval Extension

Often, in physical applications the applied forcing function $f(t)$ only occurs, and is defined, on a finite interval $[0, b]$. Then, for the purpose of determining the system's

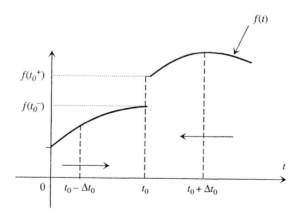

FIGURE 7.24 Jump (simple discontinuity).

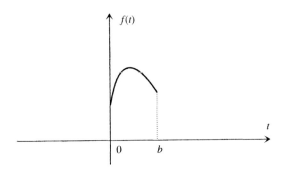

FIGURE 7.25 Function defined on a finite interval.

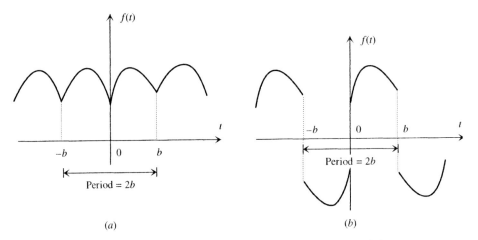

FIGURE 7.26 (a) Even and (b) odd periodic extensions of $f(t)$.

response, it is usually desired to represent this function by a Fourier series expansion. However, as previously mentioned, this Fourier series only agrees with $f(t)$ on $[0, b]$ and not outside the interval. Therefore, it would be ideal to make the original function into a periodic function and, as a result, use the Fourier series representation in Eq. (7.90). Of course, now this series agrees with $f(t)$ at all points in the domain of f. An even better alternative is to extend $f(t)$ into an even or odd periodic function with period $2b$ so that the corresponding Fourier series becomes a cosine or sine series, respectively, both of which are simpler in nature than the full Fourier series. To that end, consider the forcing function in Fig. 7.25. **Even** and **odd periodic extensions** of $f(t)$, each with a period of $2b$ and in agreement with $f(t)$ on $[0, b]$, are shown in Fig. 7.26.

EXAMPLE 7.18. Find the Fourier sine series expansion for the function $f(t) = t$ defined on the interval $(0, 1)$, shown in Fig. 7.27.

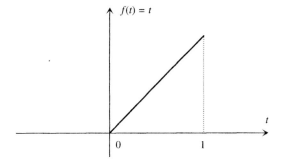

FIGURE 7.27 Function considered in Example 7.18.

Solution. Obtaining a Fourier sine series requires an odd periodic extension of $f(t)$ with a period $p = 2$, as outlined previously. This extension is illustrated in Fig. 7.28. Following prior results of this section, only the Euler constants b_n are to be calculated. These are given by Eq. (7.97) as

$$b_n = \frac{4}{p}\int_0^{p/2} f(t)\sin\frac{2n\pi t}{p}\,dt = 2\int_0^1 t\sin n\pi t\,dt$$

$$= 2\left\{-\frac{1}{n\pi}[t\cos n\pi t]_0^1 + \frac{1}{n\pi}\int_0^1 \cos n\pi t\,dt\right\} \qquad (7.102)$$

$$= -\frac{2}{n\pi}(-1)^n = \begin{cases} -2/(n\pi) & n\text{ even} \\ 2/(n\pi) & n\text{ odd} \end{cases}$$

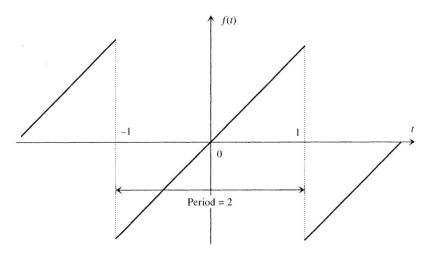

FIGURE 7.28 Odd periodic extension of the function in Fig. 7.27.

Using $p = 2$ and b_n from Eq. (7.102), the corresponding Fourier sine series is then determined via Eq. (7.98):

$$f(t) = \sum_{n=1}^{\infty} b_n \sin \frac{2n\pi t}{p} = \sum_{n=1}^{\infty} b_n \sin n\pi t$$

$$= \frac{2}{\pi}(\sin \pi t - \frac{1}{2} \sin 2\pi t + \frac{1}{3} \sin 3\pi t - \cdots)$$

$$= \frac{2}{\pi} \sum_{n=1}^{\infty} \frac{(-1)^{n+1}}{n} \sin n\pi t$$

System Response via Fourier Series

In the study of linear dynamic systems, periodic inputs with arbitrary periods are often encountered. Following the preceding discussions, we know that representing such functions with Fourier series leads to more than one single (sine or cosine) term. The system's response to each term in the series may then be determined via the FRF method discussed in Section 7.5. Ultimately, the *principle of superposition* is used to determine the total response of the system. To that end, we first introduce the following result.

THEOREM 7.2. Suppose the Fourier coefficients of $f_1(t)$ are denoted by $(a_n)_1$ and $(b_n)_1$, and those of $f_2(t)$ by $(a_n)_2$ and $(b_n)_2$. Then, the Fourier coefficients of $\alpha f_1(t) + \beta f_2(t)$ are given by

$$\alpha(a_n)_1 + \beta(b_n)_1 \quad \text{and} \quad \alpha(a_n)_2 + \beta(b_n)_2$$

EXAMPLE 7.19. Find the steady-state response of the second-order system

$$\ddot{x} + 2\dot{x} + 4x = f(t)$$

subjected to a periodic applied force as shown in Fig. 7.29.

Solution. The nature of the forcing function in Fig. 7.29 clearly suggests that an odd periodic extension must be obtained, which will then be represented by a Fourier sine series. In that event, the Euler constants are obtained via Eq. (7.97) with $p = 2\pi$, as

$$b_n = \frac{4}{p} \int_0^{p/2} f(t) \sin \frac{2n\pi t}{p} dt = \frac{2}{\pi} \int_0^{\pi} \sin nt \, dt$$

$$= \frac{2}{n\pi}[1 - (-1)^n] = \begin{cases} 0 & n \text{ even} \\ 4/(n\pi) & n \text{ odd} \end{cases}$$

and the Fourier sine series is determine subsequently via Eq. (7.98), as

$$f(t) = \sum_{\substack{n=1 \\ n=\text{odd}}}^{\infty} 4/(n\pi) \sin nt = \frac{4}{\pi}\left(\sin t + \frac{1}{3} \sin 3t + \cdots\right)$$

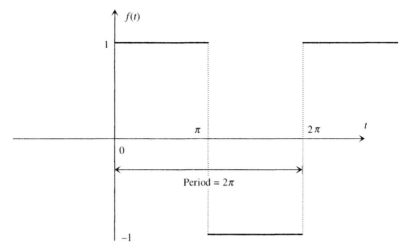

FIGURE 7.29 Applied forcing function in Example 7.19.

Next, consider the system when subjected to a typical term in this series. The governing equation in that case is in the following form:

$$\ddot{x} + 2\dot{x} + 4x = \frac{4}{n\pi}\sin nt = \underbrace{4}_{\omega_n^2}\underbrace{\left(\frac{1}{n\pi}\sin nt\right)}_{f_n(t)}, \quad n = 1, 3, 5, \ldots \quad (7.103)$$

which resembles Eq. (7.46), the normalized standard form of a second-order system. Direct comparison yields $\omega_n = 2$ rad/s, $\zeta = 0.5$, $\omega^{(n)} = n$, $F_{0n} = 1/(n\pi)$. Note that $\omega^{(n)}$ is chosen to denote the forcing frequency associated with $f_n(t)$ to avoid confusion with ω_n, the natural frequency. Consequently, from previous work in Section 7.5, the nth steady-state response $(x_{ss})_n$ is determined via Eq. (7.44), as

$$(x_{ss})_n = F_{0n}|G(j\omega^{(n)})|\sin[\omega^{(n)}t + \phi_n] \quad (7.104)$$

where the FRF is

$$G(j\omega^{(n)}) = \left.\frac{1}{s^2 + 2s + 4}\right|_{s=j\omega^{(n)}} = \frac{1}{1 - (n/2)^2 + j(n/2)}$$

Following the results of Section 7.5, Eq. (7.104) reduces to

$$(x_{ss})_n = \frac{1}{n\pi} \cdot \frac{1}{\sqrt{[1 - (n/2)^2]^2 + (n/2)^2}} \cdot \sin(nt + \phi_n)$$

where

$$\phi_n = \tan^{-1}\left\{\frac{n/2}{1 - (n/2)^2}\right\}$$

Ultimately, the total steady-state response is given by

$$x_{ss}(t) = \sum_{n=\text{odd}} (x_{ss})_n$$

PROBLEM SET 7.8–7.9

7.21. A mechanical system has the following equations of motion

$$\ddot{x} = Ax, \quad x = \begin{Bmatrix} x_1 \\ x_2 \end{Bmatrix}, \quad A = \begin{bmatrix} -1 & 2 \\ 2 & 2 \end{bmatrix}$$

subject to

$$x(0) = \begin{Bmatrix} 0 \\ 0 \end{Bmatrix} \quad \text{and} \quad \dot{x}(0) = \begin{Bmatrix} 0 \\ 1 \end{Bmatrix}$$

Determine x via modal decomposition.

7.22. Determine the Fourier series expansion of a periodic function, which, in one period, is defined as

$$f(t) = \begin{cases} 0 & \text{if } -\pi \le t < 0 \\ \sin t & \text{if } 0 < t \le \pi \end{cases}$$

7.23. Consider the signal described as in Fig. P7.23. Find an even periodic extension and write the corresponding Fourier cosine series.

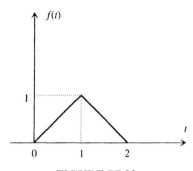

FIGURE P7.23

SUMMARY

The portion of the time response that approaches zero after a sufficiently long time is the *transient response*. The behavior of the response as $t \to \infty$ is the *steady-state response*. The time constant is a measure of the rate at which the system responds to an input. Referring to the step response of a first-order system, 98 percent of the final value is recovered after four time constants, which in practice is considered the time for the response to reach the steady state.

The *characteristic equation* for a second-order system is in the form

$$s^2 + 2\zeta\omega_n s + \omega_n^2 = 0$$

whose roots are the *characteristic values* $s_{1,2} = -\zeta\omega_n \pm \omega_n \sqrt{\zeta^2 - 1}$. The system is overdamped if $\zeta > 1$, critically damped if $\zeta = 1$, underdamped if $\zeta < 1$, and undamped if $\zeta = 0$. For an underdamped system, the characteristic values are $s_{1,2} = -\zeta\omega_n \pm j\omega_d$, where $\omega_d = \omega_n\sqrt{1-\zeta^2}$ is the *damped natural frequency*.

The *frequency response function* (FRF) is the system's transfer function evaluated at $s = j\omega$. The steady-state response of a dynamic system to a sinusoidal input is the system's *frequency response*. Given that a linear system's FRF is $G(j\omega)$ and the input is $f(t) = F_0 \sin \omega t$, the system's frequency response is

$$x_{ss}(t) = F_0 |G(j\omega)| \sin(\omega t + \phi)$$

where ϕ is the phase of the FRF. A combination of two plots, magnitude and phase, is called a *Bode plot*. The magnitude plot is a plot of the magnitude (log or dB scale) versus frequency ω (log scale). The phase plot is a plot of the phase (linear scale) versus the frequency ω.

Suppose the governing equation of a second-order system has been normalized to

$$\ddot{x} + 2\zeta\omega_n \dot{x} + \omega_n^2 x = \omega_n^2 f(t)$$

so that the transfer function and the FRF are

$$G(s) = \frac{\omega_n^2}{s^2 + 2\zeta\omega_n s + \omega_n^2}, \quad G(j\omega) = \frac{1}{1 - (\omega/\omega_n)^2 + j2\zeta(\omega/\omega_n)}$$

where ω/ω_n is the *normalized frequency*. Because the system is linear, its frequency response is determined via the magnitude and phase of the FRF. As long as the damping ratio of a system is less than the *critical damping*, i.e., $\zeta < \zeta_{cr} = 0.707$, the magnitude curve reaches its maxmum peak at a normalized frequency given as

$$\frac{\omega}{\omega_n} = \sqrt{1 - 2\zeta^2}$$

where $\omega_n\sqrt{1-2\zeta^2}$ is the *resonant frequency* ω_r. The resonant peak is $M_r = 1/2\zeta\sqrt{1-\zeta^2}$.

Consider a linear, underdamped, second-order system governed by

$$\ddot{x} + 2\zeta\omega_n \dot{x} + \omega_n^2 x = f(t), \quad 0 < \zeta < 1$$

and subject to zero initial conditions. Then, its *response to an arbitrary forcing function* $f(t)$ is given in terms of the *convolution integral*, as

$$x(t) = \int_0^t f(\tau) h(t - \tau) \, d\tau$$

where $h(t - \tau) = (1/\omega_d) e^{-\sigma(t-\tau)} \sin \omega_d(t - \tau)$, $t > \tau$, is the *unit-impulse response* of the system.

Consider the nonhomogeneous state equation

$$\dot{x} = Ax + Bu$$

subject to an initial state $\mathbf{x}_0$. The state vector $\mathbf{x}$ can be represented as

$$\mathbf{x}(t) = e^{\mathbf{A}t}\mathbf{x}_0 + \int_0^t e^{\mathbf{A}(t-\tau)}\mathbf{B}\mathbf{u}(\tau)\,d\tau$$

$$= \mathbf{\Phi}(t)\mathbf{x}_0 + \int_0^t \mathbf{\Phi}(t-\tau)\mathbf{B}\mathbf{u}(\tau)\,d\tau$$

where $\mathbf{\Phi}(t)$ is the *state-transition matrix*. The matrix exponential of the state matrix $\mathbf{A}$ can also be defined in relation to the inverse Laplace transform and the state-transition matrix, as

$$e^{\mathbf{A}t} = L^{-1}\{(s\mathbf{I} - \mathbf{A})^{-1}\} = \mathbf{\Phi}(t)$$

A coupled system of differential equations in the form

$$\mathbf{m\ddot{x}} + \mathbf{kx} = \mathbf{f}$$

can be decoupled via *modal decomposition*. The matrix transformation used is $\mathbf{x} = \mathbf{Py}$, where $\mathbf{P}$ is the matrix whose columns are the linearly independent eigenvectors of the matrix $-\mathbf{m}^{-1}\mathbf{k}$.

A periodic function $f(t)$ with an arbitrary period p can be represented by a trigonometric series, known as the *Fourier series*, as

$$f(t) = \frac{1}{2}a_0 + \sum_{n=1}^{\infty}\left[a_n \cos\frac{2n\pi t}{p} + b_n \sin\frac{2n\pi t}{p}\right]$$

where the constants, called the *Euler constants*, are defined as

$$a_n = \frac{2}{p}\int_{-p/2}^{p/2} f(t)\cos\frac{2n\pi t}{p}\,dt, \quad n = 0, 1, \ldots$$

$$b_n = \frac{2}{p}\int_{-p/2}^{p/2} f(t)\sin\frac{2n\pi t}{p}\,dt, \quad n = 1, 2, \ldots$$

For the special cases when $f(t)$ is even or odd, either all b_n's or all a_n's are zero, respectively. The resulting series is then called the *Fourier cosine series* or *Fourier sine series*, respectively. If a function experiences a finite number of simple discontinuities (jumps), its Fourier series converges to the average value of the left-hand and right-hand limits of the function at those points.

PROBLEMS

7.24. Consider the electrical system shown in Fig. P7.24. Assume that the capacitors are initially uncharged (i.e., zero initial conditions). Voltages e_i and e_o denote the input and output, respectively.
  (a) Write the governing equations and obtain the expression for the system transfer function.

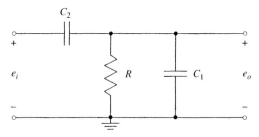

FIGURE P7.24

(b) Given that the input voltage is described by

$$e_i(t) = \begin{cases} E & \text{if } t > 0 \\ 0 & \text{otherwise} \end{cases}$$

find the output voltage $e_o(t)$. E is constant.

7.25. The single-degree-of-freedom (SDOF) mechanical system shown in Fig. P7.25 is subjected to initial conditions $x(0^-) = 0$ and $\dot{x}(0^-) = 1$ m/s.
   (a) Obtain the equation of motion.
   (b) Calculate the (undamped) natural frequency and the damping ratio.
   (c) Determine whether the system is underdamped, critically damped, or overdamped. Find the (free) response accordingly.

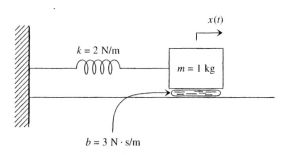

FIGURE P7.25

7.26. The motion of an undamped torsional system is described by

$$J\ddot{\theta} + K\theta = 0, \qquad J = 1 \text{ kg} \cdot \text{m}^2, \qquad K = 4 \text{ N} \cdot \text{m}$$

subject to initial conditions $\theta(0^-) = 0$, $\dot{\theta}(0^-) = 0.5$ rad/s.
   (a) Determine the natural frequency.
   (b) Find the system response.

7.27. The equation of motion of a torsional system is given as

$$J\ddot{\theta} + B\dot{\theta} = u_s(t), \qquad J = 1 \text{ kg} \cdot \text{m}^2, \qquad B = 4 \text{ N} \cdot \text{m} \cdot \text{s}$$

where $u_s(t)$ denotes the unit-step input to the system. Assume zero initial conditions.
(a) Rewrite the equation of motion as a first-order equation in terms of angular velocity.
(b) Find the system *time constant*.
(c) What is the final value of the response? What percentage of this final value will the response curve reach after exactly one time constant?
(d) Approximately how much time is required for the response to come within 2 percent of its final value?

7.28. Consider the mechanical system shown in Fig. P7.28 subjected to zero initial conditions. The system is subject to a sinusoidal input $f(t)$ with amplitude $F = 5$ N and frequency $\omega = 1.5$ rad/s.
(a) Find the equation of motion.
(b) Obtain the (sinusoidal) transfer function.
(c) Determine the steady-state response of the system.

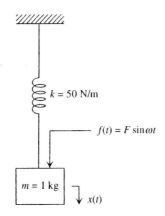

FIGURE P7.28

7.29. Find the response of the mechanical system of Problem 7.28 when subjected to a *unit-ramp* input.

7.30. The method of *modal decomposition*, discussed in Section 7.8, may also be employed to determine the response of a system described by its state equation (first-order system). Through appropriate modification of the procedure outlined in that section, find the response of a system with the following state equation:

$$\begin{cases} \dot{\mathbf{x}} = \mathbf{A}\mathbf{x} + \mathbf{B}u \\ \mathbf{x}(0) = \mathbf{x}_0 \end{cases}$$

where

$$\mathbf{A} = \begin{bmatrix} 0 & 1 \\ 2 & -1 \end{bmatrix}, \qquad \mathbf{B} = \begin{bmatrix} 0 \\ 1 \end{bmatrix}, \qquad \mathbf{x}_0 = \begin{bmatrix} 1 \\ -1 \end{bmatrix}, \qquad u = \text{unit step}$$

7.31. Find the Fourier series representation of a periodic function ($p = 2$) with the geometric description shown in Fig. P7.31. The function repeats itself outside this interval.

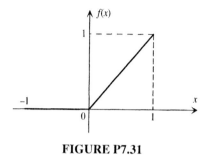

FIGURE P7.31

7.32. Find the Fourier series of the function $f(x) = |x|$ in $(-\pi, \pi)$, of period 2π, as shown in Fig. P7.32.

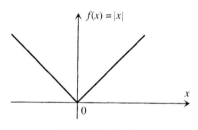

FIGURE P7.32

7.33. Consider an RLC circuit, as shown in Fig. P7.33a, where i, R, C, and L denote current, resistance, capacitance, and inductance, respectively. Assume that initial conditions are zero and that all elements are *linear*. Determine the steady-state response of the system if $e(t)$ represents a *periodic* applied voltage with the graphical description given in Fig. P7.33b.

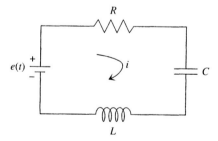

FIGURE P7.33(a)

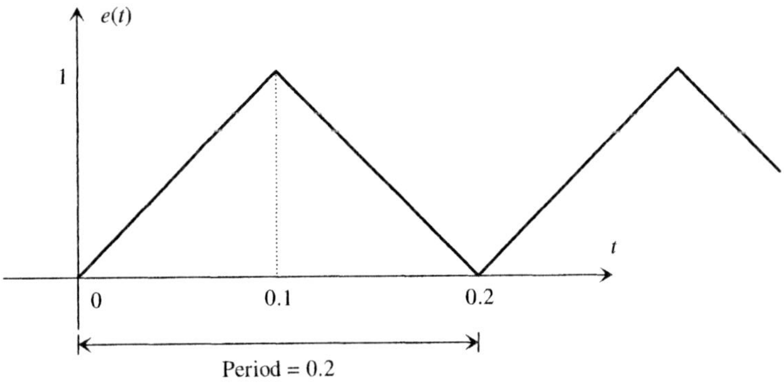

FIGURE P7.33(b)

(Triangular wave $e(t)$ with peak 1 at $t=0.1$, zero at $t=0$ and $t=0.2$; Period = 0.2)

7.34. A dynamic system is represented by a transfer function, as

$$G(s) = \frac{X(s)}{U(s)} = \frac{2}{s^2 + 3s + 2}$$

where x and u denote the system's output and input, respectively.
(a) Determine the state-space form.
(b) Solve the state equation via the exponential of the state matrix.

7.35. Consider the following system of second-order differential equations subjected to the indicated vectors of initial conditions

$$\ddot{\mathbf{x}} = \mathbf{A}\mathbf{x}, \quad \mathbf{A} = \begin{bmatrix} -3 & 2 \\ 2 & 0 \end{bmatrix}, \quad \mathbf{x} = \begin{Bmatrix} x_1 \\ x_2 \end{Bmatrix}, \quad \mathbf{x}(0) = \begin{Bmatrix} 0 \\ 1 \end{Bmatrix}, \quad \dot{\mathbf{x}}(0) = \begin{Bmatrix} 0 \\ 0 \end{Bmatrix}$$

Determine **x** via modal decomposition.

7.36. Consider the following nonhomogeneous system of second-order differential equations and the vectors of initial conditions

$$\ddot{\mathbf{x}} = \mathbf{A}\mathbf{x} + \mathbf{f}, \quad \mathbf{A} = \begin{bmatrix} -3 & 2 \\ 2 & 0 \end{bmatrix}, \quad \mathbf{x} = \begin{Bmatrix} x_1 \\ x_2 \end{Bmatrix},$$

$$\mathbf{f} = \begin{bmatrix} 2 \\ 0 \end{bmatrix}, \quad \mathbf{x}(0) = \begin{Bmatrix} 0 \\ 1 \end{Bmatrix}, \quad \dot{\mathbf{x}}(0) = \begin{Bmatrix} 0 \\ 0 \end{Bmatrix}$$

Find **x** using modal decomposition.

CHAPTER 8

Introduction to Vibrations

This chapter provides an introduction to mechanical vibrations, a vital subject with many applications in engineering. The chapter includes various topics such as natural frequencies (related to eigenvalues) and mode shapes (eigenvectors), modal matrices, orthogonality, orthonormality, the logarithmic decrement method, the beat phenomenon, frequency response and the Bode plot, the vibration isolation system, systems with support motion, rotating unbalanced mass, damped vibration absorbers, and modal analysis. The advantages and disadvantages of modal analysis and the Laplace transform are also discussed.

8.1 INTRODUCTION

The building that we work in or live in would shake noticeably if a relatively strong earthquake occurred. The car that we drive would vibrate when we traveled on a bumpy road. When the motor mounts of the car's engine deteriorated, or if the engine were out of tune, or if the wheels were unbalanced or out of alignment, the steering wheel would shake unpleasantly, most noticeably when the car stopped at a traffic light. In medical applications, the classical and traditional way of measuring blood pressure is by the *auscultatory* method with a *stethoscope* and a *sphygmomanometer*. The vibration at the brachial artery is caused by the turbulent flow of the blood going through the constricted blood vessel. As we look around, we can see many more important systems and applications that involve vibrations. To understand and to appreciate how these systems work, or to learn how to mitigate vibration problems, one needs to learn the theory and applications of vibrations. This chapter will introduce the reader to the world of vibrations by presenting the essentials of mechanical vibrations.

It is assumed that the reader is familiar with all the fundamental aspects of single-degree-of-freedom (SDOF) systems. These aspects include natural frequency ω_n, damping ratio ζ, free responses, and forced responses. A free response is the response caused by the initial conditions, whereas a forced response is that driven by the applied forcing function. In forced response, the system may have zero or nonzero initial conditions. Both the undamped and damped cases are important in free and forced vibrations. These cases are conveniently classified as undamped ($\zeta = 0$), underdamped ($0 < \zeta < 1$), critically damped ($\zeta = 1$), and overdamped ($\zeta > 1$).

We begin with a discussion of natural frequencies and mode shapes of multiple-degree-of-freedom (MDOF) systems. The two-degree-of-freedom (TDOF) system is an excellent representative of multiple-degree-of-freedom (MDOF) systems; thus many of our examples will be TDOF. The solutions to an *eigenvalue problem* are called *eigensolutions*. The natural frequencies are related to the *eigenvalues,* and the mode shapes are the same as *eigenvectors* (for discrete systems). To obtain the natural frequencies, the analyst would solve for the eigenvalues first. The natural frequencies are listed in increasing order, with the smallest frequency as the first natural frequency. The first natural frequency for the vibrational mode must be nonzero, because the zero frequency corresponds to the rigid-body mode. Many fundamental systems are used as illustrative examples. Among real-world examples, models of an automobile suspension system are included.

Following is a discussion of a *modal matrix,* which is composed of mode shapes. There is more than one way to normalize the mode shapes or the modal matrix. The normalization for sketching the mode shapes is different than that for obtaining the system response.

A modal matrix possesses orthogonality properties that have many important applications. One such application involves checking for the correctness of natural frequencies and mode shapes using the modal matrix. Another critical application is to obtain a system response using the method called *modal analysis,* which uses modal expansion together with orthonormality. When the modal matrix is normalized for modal analysis, the orthogonality becomes orthonormality. The important topic of *damping* is also discussed in conjunction with the modal matrix.

Various important topics are treated next: the logarithmic decrement method, beat phenomenon, vibration isolation system, system with support motion, rotating unbalanced mass, and damped vibration absorber. Transient response and/or frequency response are considered. For frequency response, the Bode plot is very useful.

For completeness, modal analysis is presented at the end of the chapter. Modal analysis is a powerful method of obtaining closed-form solutions for vibration problems. Modal analysis is conveniently broken into three steps:

1. Obtaining the natural frequencies and mode shapes
2. Obtaining the orthonormality property
3. Obtaining the response by using modal expansion

8.2 NATURAL FREQUENCIES AND MODE SHAPES

We begin with a TDOF system in the most general form and then give specific examples. The differential equations of a general TDOF system, which is free and undamped, are given as

$$\begin{bmatrix} m_{11} & m_{12} \\ m_{21} & m_{22} \end{bmatrix} \begin{Bmatrix} \ddot{x}_1 \\ \ddot{x}_2 \end{Bmatrix} + \begin{bmatrix} k_{11} & k_{12} \\ k_{21} & k_{22} \end{bmatrix} \begin{Bmatrix} x_1 \\ x_2 \end{Bmatrix} = \begin{Bmatrix} 0 \\ 0 \end{Bmatrix} \quad (8.1)$$

This section shows how to obtain the natural frequencies and the corresponding mode shapes.

When the system vibrates in a natural mode, the solution is in the form

$$\mathbf{x} = \mathbf{X} e^{j\omega t} \quad (8.2)$$

Thus

$$\begin{Bmatrix} x_1 \\ x_2 \end{Bmatrix} = \begin{Bmatrix} X_1 \\ X_2 \end{Bmatrix} e^{j\omega t} \quad (8.3a)$$

$$\begin{Bmatrix} \ddot{x}_1 \\ \ddot{x}_2 \end{Bmatrix} = -\omega^2 \begin{Bmatrix} X_1 \\ X_2 \end{Bmatrix} e^{j\omega t} \quad (8.3b)$$

where ω and $\mathbf{X}$ are the natural frequency and the corresponding mode shape, respectively. Note that this is a direct result from the method of separation of variables for vibration systems.

Introducing Eqs. (8.3) into Eq. (8.1), we obtain

$$\begin{bmatrix} -m_{11}\omega^2 + k_{11} & -m_{12}\omega^2 + k_{12} \\ -m_{21}\omega^2 + k_{21} & -m_{22}\omega^2 + k_{22} \end{bmatrix} \begin{Bmatrix} X_1 \\ X_2 \end{Bmatrix} e^{j\omega t} = \begin{Bmatrix} 0 \\ 0 \end{Bmatrix} \quad (8.4)$$

Thus

$$\begin{bmatrix} k_{11} - m_{11}\omega^2 & k_{12} - m_{12}\omega^2 \\ k_{21} - m_{21}\omega^2 & k_{22} - m_{22}\omega^2 \end{bmatrix} \underbrace{\begin{Bmatrix} X_1 \\ X_2 \end{Bmatrix}}_{\neq 0} \underbrace{e^{j\omega t}}_{\neq 0} = \begin{Bmatrix} 0 \\ 0 \end{Bmatrix} \quad (8.5)$$

At this point, the natural frequencies are given as the solutions, or roots, of the following equation:

$$\begin{vmatrix} k_{11} - m_{11}\omega^2 & k_{12} - m_{12}\omega^2 \\ k_{21} - m_{21}\omega^2 & k_{22} - m_{22}\omega^2 \end{vmatrix} = 0 \quad (8.6)$$

$$(k_{11} - m_{11}\omega^2)(k_{22} - m_{22}\omega^2) - (k_{21} - m_{21}\omega^2)(k_{12} - m_{12}\omega^2) = 0 \quad (8.7)$$

This equation is called the *frequency equation*. The equation is expanded and all the terms are lined up as

$$\begin{aligned} m_{11}m_{22}\omega^4 & -(k_{11}m_{22} + k_{22}m_{11})\omega^2 & +k_{11}k_{22} \\ -m_{12}m_{21}\omega^4 & +(k_{12}m_{21} + k_{21}m_{12})\omega^2 & -k_{12}k_{21} & = 0 \end{aligned} \quad (8.8)$$

For vibration problems, the mass and stiffness matrices are symmetric; thus $m_{12} = m_{21}$ and $k_{12} = k_{21}$. Therefore, after adding the terms, term by term, we obtain

$$(m_{11}m_{22} - m_{12}^2)\omega^4 - (k_{11}m_{22} + k_{22}m_{11} - 2k_{12}m_{12})\omega^2 + k_{11}k_{22} - k_{12}^2 = 0 \quad (8.9)$$

or

$$\omega^4 - \frac{(k_{11}m_{22} + k_{22}m_{11} - 2k_{12}m_{12})}{m_{11}m_{22} - m_{12}^2}\omega^2 + \frac{k_{11}k_{22} - k_{12}^2}{m_{11}m_{22} - m_{12}^2} = 0 \quad (8.10)$$

Using the quadratic formula, we obtain

$$\begin{matrix}\omega_1^2 \\ \omega_2^2\end{matrix} = \frac{k_{11}m_{22} + k_{22}m_{11} - 2k_{12}m_{12}}{2(m_{11}m_{22} - m_{12}^2)}$$

$$\mp \sqrt{\left[\frac{k_{11}m_{22} + k_{22}m_{11} - 2k_{12}m_{12}}{2(m_{11}m_{22} - m_{12}^2)}\right]^2 - \frac{k_{11}k_{22} - k_{12}^2}{m_{11}m_{22} - m_{12}^2}} \quad (8.11)$$

where the natural frequencies are the square roots of the expressions.
The mode shapes will be obtained next. From Eq. (8.5), we have

$$\begin{bmatrix} k_{11} - m_{11}\omega^2 & k_{12} - m_{12}\omega^2 \\ k_{21} - m_{21}\omega^2 & k_{22} - m_{22}\omega^2 \end{bmatrix} \begin{Bmatrix} X_1 \\ X_2 \end{Bmatrix} = \begin{Bmatrix} 0 \\ 0 \end{Bmatrix} \quad (8.12)$$

Thus,

$$(k_{11} - m_{11}\omega^2)X_1 - (k_{12} - m_{12}\omega^2)X_2 = 0 \quad (8.13a)$$

$$(k_{21} - m_{21}\omega^2)X_1 - (k_{22} - m_{22}\omega^2)X_2 = 0 \quad (8.13b)$$

Since these two equations are dependent, either one can be used. Using the top equation, Eq. (8.13a),

$$\frac{X_2}{X_1} = \frac{k_{11} - m_{11}\omega^2}{k_{12} - m_{12}\omega^2} \quad (8.14)$$

Thus for the first mode,

$$\left(\frac{X_2}{X_1}\right)_1 = \frac{k_{11} - m_{11}\omega_1^2}{k_{12} - m_{12}\omega_1^2} \quad (8.15a)$$

and for the second mode,

$$\left(\frac{X_2}{X_1}\right)_2 = \frac{k_{11} - m_{11}\omega_2^2}{k_{12} - m_{12}\omega_2^2} \quad (8.15b)$$

The mode shapes are normalized as

$$\begin{Bmatrix} X_1 \\ X_2 \end{Bmatrix} = \begin{Bmatrix} X_1/X_1 \\ X_2/X_1 \end{Bmatrix} = \begin{Bmatrix} 1 \\ X_2/X_1 \end{Bmatrix} \tag{8.16}$$

Finally, the first mode shape is

$$\begin{Bmatrix} X_1 \\ X_2 \end{Bmatrix}_1 = \begin{Bmatrix} 1 \\ X_2/X_1 \end{Bmatrix}_1 \tag{8.17a}$$

and the second mode shape is

$$\begin{Bmatrix} X_1 \\ X_2 \end{Bmatrix}_2 = \begin{Bmatrix} 1 \\ X_2/X_1 \end{Bmatrix}_2 \tag{8.17b}$$

where the ratios of X_2 to X_1 are given by Eqs. (8.15).

The general solution has been given for a general TDOF system to show the general procedure for MDOF systems. However, for a specific TDOF (or MDOF) system, instead of deducing from the general solution it is better to start from scratch, working directly with the system under consideration.

Unsymmetrical Systems

The following are examples of fundamental unsymmetrical systems. For unconstrained systems, the rigid-body mode corresponds to the zero natural frequency.

EXAMPLE 8.1. Consider the unconstrained and unsymmetrical system shown in Fig. 8.1.

(a) Obtain the natural frequencies.
(b) Obtain and sketch the mode shapes.

Solution

(a) The differential equations are given in matrix form as

$$\begin{bmatrix} 2m & 0 \\ 0 & m \end{bmatrix} \begin{Bmatrix} \ddot{x}_1 \\ \ddot{x}_2 \end{Bmatrix} + \begin{bmatrix} k & -k \\ -k & k \end{bmatrix} \begin{Bmatrix} x_1 \\ x_2 \end{Bmatrix} = \begin{Bmatrix} 0 \\ 0 \end{Bmatrix}$$

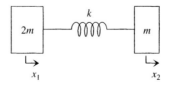

FIGURE 8.1 TDOF system.

The solution is in the form of Eq. (8.3a), thus

$$\begin{bmatrix} k - 2m\omega^2 & -k \\ -k & k - m\omega^2 \end{bmatrix} \underbrace{\begin{Bmatrix} X_1 \\ X_2 \end{Bmatrix}}_{\neq 0} \underbrace{e^{j\omega t}}_{\neq 0} = \begin{Bmatrix} 0 \\ 0 \end{Bmatrix}$$

The frequency equation is

$$\begin{vmatrix} k - 2m\omega^2 & -k \\ -k & k - m\omega^2 \end{vmatrix} = 0$$

$$(k - 2m\omega^2)(k - m\omega^2) - k^2 = 0$$

The equation is expanded and factored as

$$2m^2\omega^4 - 3km\omega^2 = m\omega^2(2m\omega^2 - 3k) = 0$$

Thus, the natural frequencies are given as

$$m\omega^2 = 0 \implies \omega_1 = 0$$

$$2m\omega^2 = 3k \implies \omega_2 = \sqrt{\frac{3k}{2m}}$$

(b) We have

$$\begin{bmatrix} k - 2m\omega^2 & -k \\ -k & k - m\omega^2 \end{bmatrix} \begin{Bmatrix} X_1 \\ X_2 \end{Bmatrix} = \begin{Bmatrix} 0 \\ 0 \end{Bmatrix}$$

Since these two equations are dependent, we can use either one. Using the top equation,

$$(k - 2m\omega^2)X_1 - kX_2 = 0$$

Thus

$$\frac{X_2}{X_1} = \frac{k - 2m\omega^2}{k}$$

First mode: $m\omega_1^2 = 0$

$$\left(\frac{X_2}{X_1}\right)_1 = \frac{k - 2m\omega_1^2}{k} = \frac{k - 0}{k} = 1$$

The first mode shape is normalized as

$$\begin{Bmatrix} X_1 \\ X_2 \end{Bmatrix}_1 = \begin{Bmatrix} 1 \\ X_2/X_1 \end{Bmatrix}_1 = \begin{Bmatrix} 1 \\ 1 \end{Bmatrix}$$

Second mode: $2m\omega_2^2 = 3k$

$$\left(\frac{X_2}{X_1}\right)_2 = \frac{k - 2m\omega_2^2}{k} = \frac{k - 3k}{k} = -2$$

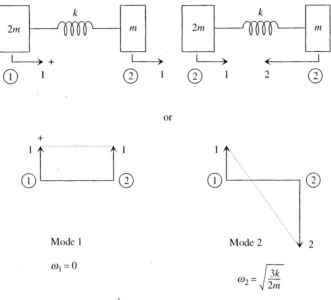

FIGURE 8.2 Mode shapes.

The second normalized mode shape is

$$\begin{Bmatrix} X_1 \\ X_2 \end{Bmatrix}_2 = \begin{Bmatrix} 1 \\ X_2/X_1 \end{Bmatrix}_2 = \begin{Bmatrix} 1 \\ -2 \end{Bmatrix}$$

The mode shapes are shown in Fig. 8.2.

EXAMPLE 8.2. Consider the constrained and unsymmetrical system shown in Fig. 8.3.

(a) Obtain the natural frequencies.
(b) Obtain and sketch the mode shapes.

Solution

(a) The differential equations are given as

$$\begin{bmatrix} m & 0 \\ 0 & m \end{bmatrix} \begin{Bmatrix} \ddot{x}_1 \\ \ddot{x}_2 \end{Bmatrix} + \begin{bmatrix} 2k & -k \\ -k & k \end{bmatrix} \begin{Bmatrix} x_1 \\ x_2 \end{Bmatrix} = \begin{Bmatrix} 0 \\ 0 \end{Bmatrix}$$

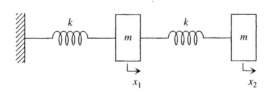

FIGURE 8.3 TDOF system.

The solution is in the form of Eq. (8.3a), thus

$$\begin{bmatrix} 2k - m\omega^2 & -k \\ -k & k - m\omega^2 \end{bmatrix} \underbrace{\begin{Bmatrix} X_1 \\ X_2 \end{Bmatrix}}_{\neq 0} \underbrace{e^{j\omega t}}_{\neq 0} = \begin{Bmatrix} 0 \\ 0 \end{Bmatrix}$$

The frequency equation is given as

$$\begin{vmatrix} 2k - m\omega^2 & -k \\ -k & k - m\omega^2 \end{vmatrix} = 0 \implies m^2\omega^4 - 3km\omega^2 + k^2 = 0$$

$$\xrightarrow{\text{Divide by } m^2} \omega^4 - \frac{3k}{m}\omega^2 + \frac{k^2}{m^2} = 0$$

Using the quadratic formula, we obtain

$$\begin{matrix} \omega_1^2 \\ \omega_2^2 \end{matrix} = \frac{3k}{2m} \mp \sqrt{\left(\frac{3k}{2m}\right)^2 - \frac{k^2}{m^2}} = \frac{3 \mp \sqrt{5}}{2}\frac{k}{m}$$

Thus, the natural frequencies are given as

$$m\omega_1^2 = \frac{3 - \sqrt{5}}{2}k = 0.38197k \implies \omega_1 = 0.61803\sqrt{\frac{k}{m}}$$

$$m\omega_2^2 = \frac{3 + \sqrt{5}}{2}k = 2.61803k \implies \omega_2 = 1.61803\sqrt{\frac{k}{m}}$$

(b) We have

$$\begin{bmatrix} 2k - m\omega^2 & -k \\ -k & k - m\omega^2 \end{bmatrix} \begin{Bmatrix} X_1 \\ X_2 \end{Bmatrix} = \begin{Bmatrix} 0 \\ 0 \end{Bmatrix}$$

Using the top equation,

$$(2k - m\omega^2)X_1 - kX_2 = 0$$

Thus

$$\frac{X_2}{X_1} = \frac{2k - m\omega^2}{k}$$

First mode: $m\omega_1^2 = 0.38197k$

$$\left(\frac{X_2}{X_1}\right)_1 = \frac{2k - m\omega_1^2}{k} = \frac{2k - 0.38197k}{k} = 1.61803$$

The first mode shape is normalized as

$$\begin{Bmatrix} X_1 \\ X_2 \end{Bmatrix}_1 = \begin{Bmatrix} 1 \\ X_2/X_1 \end{Bmatrix} = \begin{Bmatrix} 1 \\ 1.61803 \end{Bmatrix}$$

Second mode: $m\omega_2^2 = 2.61803k$

$$\left(\frac{X_2}{X_1}\right)_2 = \frac{2k - m\omega_2^2}{k} = \frac{2k - 2.61803k}{k} = -0.61803$$

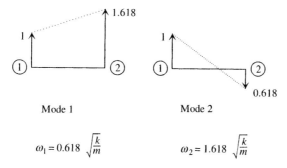

FIGURE 8.4 Mode shapes.

The second normalized mode shape is

$$\begin{Bmatrix} X_1 \\ X_2 \end{Bmatrix}_2 = \begin{Bmatrix} 1 \\ X_2/X_1 \end{Bmatrix}_2 = \begin{Bmatrix} 1 \\ -0.61803 \end{Bmatrix}$$

The mode shapes are shown in Fig. 8.4.

EXAMPLE 8.3. Consider the mechanical system shown in Fig. 8.5.

(a) Obtain the natural frequencies.
(b) Obtain and sketch the mode shapes.

Solution

(a) The differential equations are given as

$$\begin{bmatrix} 2m & 0 \\ 0 & m \end{bmatrix} \begin{Bmatrix} \ddot{x}_1 \\ \ddot{x}_2 \end{Bmatrix} + \begin{bmatrix} 3k & -k \\ -k & k \end{bmatrix} \begin{Bmatrix} x_1 \\ x_2 \end{Bmatrix} = \begin{Bmatrix} 0 \\ 0 \end{Bmatrix}$$

The solution is in the form of Eq. (8.3a), thus

$$\begin{bmatrix} 3k - 2m\omega^2 & -k \\ -k & k - m\omega^2 \end{bmatrix} \underbrace{\begin{Bmatrix} X_1 \\ X_2 \end{Bmatrix}}_{\neq 0} \underbrace{e^{j\omega t}}_{\neq 0} = \begin{Bmatrix} 0 \\ 0 \end{Bmatrix}$$

The frequency equation is given as

$$(3k - 2m\omega^2)(k - m\omega^2) - k^2 = 0$$

$$\omega^4 - \frac{2.5k}{m}\omega^2 + \frac{k^2}{m^2} = 0$$

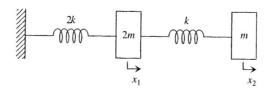

FIGURE 8.5 TDOF system.

Using the quadratic formula, we obtain

$$\begin{matrix}\omega_1^2\\ \omega_2^2\end{matrix} = \frac{1.25k}{m} \mp \sqrt{\left(\frac{1.25k}{m}\right)^2 - \frac{k^2}{m^2}} = 1.25 \mp 0.75 \frac{k}{m}$$

Thus the natural frequencies are given as

$$m\omega_1^2 = \frac{k}{2} \implies \omega_1 = \sqrt{\frac{k}{2m}}$$

$$m\omega_2^2 = 2k \implies \omega_2 = \sqrt{\frac{2k}{m}}$$

(b) We have

$$\begin{bmatrix} 3k - 2m\omega^2 & -k \\ -k & k - m\omega^2 \end{bmatrix} \begin{Bmatrix} X_1 \\ X_2 \end{Bmatrix} = \begin{Bmatrix} 0 \\ 0 \end{Bmatrix}$$

Using the top equation,

$$(3k - 2m\omega^2)X_1 - kX_2 = 0$$

Thus

$$\frac{X_2}{X_1} = \frac{3k - 2m\omega^2}{k}$$

First mode: $m\omega_1^2 = k/2$

$$\left(\frac{X_2}{X_1}\right)_1 = \frac{3k - 2m\omega_1^2}{k} = \frac{3k - k}{k} = 2.0$$

The first mode shape is normalized as

$$\begin{Bmatrix} X_1 \\ X_2 \end{Bmatrix}_1 = \begin{Bmatrix} 1 \\ X_2/X_1 \end{Bmatrix}_1 = \begin{Bmatrix} 1 \\ 2.0 \end{Bmatrix}$$

Second mode: $m\omega_2^2 = 2k$

$$\left(\frac{X_2}{X_1}\right)_2 = \frac{3k - 2m\omega_2^2}{k} = \frac{3k - 4k}{k} = -1.0$$

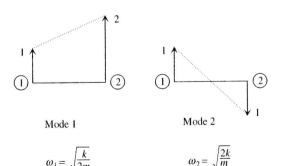

FIGURE 8.6 Mode shapes.

The second normalized mode shape is

$$\left\{\begin{array}{c}X_1\\X_2\end{array}\right\}_2 = \left\{\begin{array}{c}1\\X_2/X_1\end{array}\right\}_2 = \left\{\begin{array}{c}1\\-1.0\end{array}\right\}$$

The mode shapes are shown in Fig. 8.6.

Symmetrical Systems

The following examples show symmetrical systems. The concept of symmetry provides physical insight; thus, it can be used to obtain readily certain natural frequencies and mode shapes.

EXAMPLE 8.4. Consider the unconstrained and symmetrical system shown in Fig. 8.7.

(a) Obtain the natural frequencies.
(b) Obtain and sketch the mode shapes.

Solution

(a) The differential equations are given as

$$\begin{bmatrix}m & 0\\0 & m\end{bmatrix}\left\{\begin{array}{c}\ddot{x}_1\\\ddot{x}_2\end{array}\right\} + \begin{bmatrix}k & -k\\-k & k\end{bmatrix}\left\{\begin{array}{c}x_1\\x_2\end{array}\right\} = \left\{\begin{array}{c}0\\0\end{array}\right\}$$

The solution is in the form of Eq. (8.3a), thus

$$\begin{bmatrix}k-m\omega^2 & -k\\-k & k-m\omega^2\end{bmatrix}\underbrace{\left\{\begin{array}{c}X_1\\X_2\end{array}\right\}}_{\neq 0}\underbrace{e^{j\omega t}}_{\neq 0} = \left\{\begin{array}{c}0\\0\end{array}\right\}$$

The frequency equation is given as

$$(k-m\omega^2)^2 - k^2 = 0$$

Three different methods will be shown for obtaining the natural frequencies of this problem.
Method 1:

$$(k-m\omega^2)^2 = k^2$$
$$k-m\omega^2 = \pm k$$

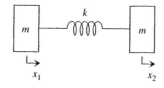

FIGURE 8.7 TDOF system.

Case "+": $k - m\omega^2 = k$

$$m\omega^2 = 0$$

Thus the first natural frequency is

$$\omega_1 = 0$$

which is the rigid-body mode. Since the frequency is zero, the system simply translates in this first mode. It simply does not oscillate.

Case "−": $k - m\omega^2 = -k$

$$m\omega^2 = 2k$$

Thus the second natural frequency is

$$\omega_2 = \sqrt{\frac{2k}{m}}$$

Method 2: The equation is expanded as

$$(k - m\omega^2)^2 - k^2 = 0$$
$$k^2 - 2km\omega^2 + m^2\omega^4 - k^2 = 0$$
$$m^2\omega^4 - 2km\omega^2 = m\omega^2(m\omega^2 - 2k) = 0$$

Thus,

$$m\omega^2 = 0 \implies \omega_1 = 0$$
$$m\omega^2 = 2k \implies \omega_2 = \sqrt{\frac{2k}{m}}$$

Method 3: Use the concept of unconstraint and symmetry. Since the system is unconstrained, the first mode is a rigid-body mode; thus the first natural frequency is zero, $\omega_1 = 0$. Since the system is symmetrical, the second mode behaves as if there were two identical subsystems. Each is a SDOF system with half of the spring (Fig. 8.8). The stiffness of this spring is $2k$ because the length is cut in half. Thus, the second natural frequency is $\omega_2 = \sqrt{2k/m}$.

(b) We have

$$\begin{bmatrix} k - m\omega^2 & -k \\ -k & k - m\omega^2 \end{bmatrix} \begin{Bmatrix} X_1 \\ X_2 \end{Bmatrix} = \begin{Bmatrix} 0 \\ 0 \end{Bmatrix}$$

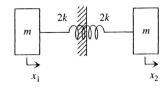

FIGURE 8.8 TDOF system.

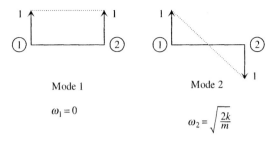

FIGURE 8.9 Mode shapes.

Using the top equation,

$$(k - m\omega^2)X_1 - kX_2 = 0$$

Thus

$$\frac{X_2}{X_1} = \frac{k - m\omega^2}{k}$$

First mode: $m\omega_1^2 = 0$

$$\left(\frac{X_2}{X_1}\right)_1 = \frac{k - m\omega^2}{k} = \frac{k}{k} = 1$$

The first mode shape is normalized as

$$\begin{Bmatrix} X_1 \\ X_2 \end{Bmatrix}_1 = \begin{Bmatrix} 1 \\ X_2/X_1 \end{Bmatrix}_1 = \begin{Bmatrix} 1 \\ 1 \end{Bmatrix}$$

Second mode: $m\omega_2^2 = 2k$

$$\left(\frac{X_2}{X_1}\right)_2 = \frac{k - m\omega^2}{k} = \frac{k - 2k}{k} = -1$$

The second normalized mode shape is

$$\begin{Bmatrix} X_1 \\ X_2 \end{Bmatrix}_2 = \begin{Bmatrix} 1 \\ X_2/X_1 \end{Bmatrix}_2 = \begin{Bmatrix} 1 \\ -1 \end{Bmatrix}$$

The mode shapes are shown in Fig. 8.9.

EXAMPLE 8.5. Consider the constrained and symmetrical system shown in Fig. 8.10.

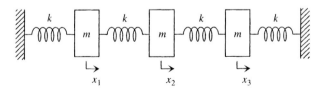

FIGURE 8.10 Three-DOF system.

(a) Obtain the natural frequencies.
(b) Obtain and sketch the mode shapes.

Solution

(a) The differential equations of the free, undamped system are given, in the second-order matrix form, as

$$\begin{bmatrix} m & 0 & 0 \\ 0 & m & 0 \\ 0 & 0 & m \end{bmatrix} \begin{Bmatrix} \ddot{x}_1 \\ \ddot{x}_2 \\ \ddot{x}_3 \end{Bmatrix} + \begin{bmatrix} 2k & -k & 0 \\ -k & 2k & -k \\ 0 & -k & 2k \end{bmatrix} \begin{Bmatrix} x_1 \\ x_2 \\ x_3 \end{Bmatrix} = \begin{Bmatrix} 0 \\ 0 \\ 0 \end{Bmatrix}$$

The solution is in the form

$$\begin{Bmatrix} x_1 \\ x_2 \\ x_3 \end{Bmatrix} = \begin{Bmatrix} X_1 \\ X_2 \\ X_3 \end{Bmatrix} e^{j\omega t}$$

thus

$$\begin{Bmatrix} \ddot{x}_1 \\ \ddot{x}_2 \\ \ddot{x}_3 \end{Bmatrix} = -\omega^2 \begin{Bmatrix} X_1 \\ X_2 \\ X_3 \end{Bmatrix} e^{j\omega t}$$

where ω and **X** are the natural frequency and the corresponding mode shape, respectively. Therefore,

$$\underbrace{\begin{bmatrix} 2k - m\omega^2 & -k & 0 \\ -k & 2k - m\omega^2 & -k \\ 0 & -k & 2k - m\omega^2 \end{bmatrix}}_{\det[\cdots]=0} \begin{Bmatrix} X_1 \\ X_2 \\ X_3 \end{Bmatrix} \underbrace{e^{j\omega t}}_{\neq 0} = \begin{Bmatrix} 0 \\ 0 \\ 0 \end{Bmatrix}$$

$\neq 0$

The frequency equation is given as

$$\begin{vmatrix} 2k - m\omega^2 & -k & 0 \\ -k & 2k - m\omega^2 & -k \\ 0 & -k & 2k - m\omega^2 \end{vmatrix} = 0$$

$$(2k - m\omega^2)^3 - 2k^2(2k - m\omega^2) = 0$$

After factoring, we have

$$(2k - m\omega^2)\left[(2k - m\omega^2)^2 - 2k^2\right] = 0$$

$$2k - m\omega^2 = 0 \implies m\omega^2 = 2k$$

Other frequencies are given by

$$2k - m\omega^2 = \pm k\sqrt{2} \implies m\omega^2 = (2 \mp \sqrt{2})k$$

Thus, the three natural frequencies of the system are given as

$$\omega_1 = \sqrt{(2-\sqrt{2})\frac{k}{m}}, \quad \omega_2 = \sqrt{\frac{2k}{m}}, \quad \omega_3 = \sqrt{(2+\sqrt{2})\frac{k}{m}}$$

Note that from symmetry, we would expect one of the modes to behave as if there were two identical SDOF subsystems, separated by the mass m in the middle (Fig. 8.11). Thus, the corresponding natural frequency is $\sqrt{2k/m}$.

(b) We have

$$\begin{bmatrix} 2k - m\omega^2 & -k & 0 \\ -k & 2k - m\omega^2 & -k \\ 0 & -k & 2k - m\omega^2 \end{bmatrix} \begin{Bmatrix} X_1 \\ X_2 \\ X_3 \end{Bmatrix} = \begin{Bmatrix} 0 \\ 0 \\ 0 \end{Bmatrix}$$

The three dependent equations are

$$(2k - m\omega^2)X_1 - kX_2 = 0$$

$$-kX_1 + (2k - m\omega^2)X_2 - kX_3 = 0$$

$$-kX_2 + (2k - m\omega^2)X_3 = 0$$

Thus,

$$\frac{X_2}{X_1} = \frac{2k - m\omega^2}{k}$$

$$\frac{X_3}{X_2} = \frac{k}{2k - m\omega^2} \quad \text{if } 2k - m\omega^2 \neq 0$$

$$\frac{X_3}{X_1} = \frac{X_3}{X_2}\frac{X_2}{X_1} = \frac{k}{2k - m\omega^2}\frac{2k - m\omega^2}{k} = 1 \quad \text{if } 2k - m\omega^2 \neq 0$$

First mode: $m\omega_1^2 = 2k - k\sqrt{2}$

$$\left(\frac{X_2}{X_1}\right)_1 = \frac{2k - m\omega^2}{k} = \frac{2k - 2k + k\sqrt{2}}{k} = \sqrt{2}$$

$$\left(\frac{X_3}{X_1}\right)_1 = 1$$

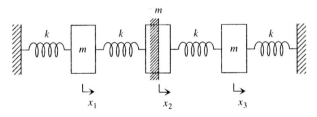

FIGURE 8.11 Symmetry.

Therefore, the first normalized mode shape is

$$\begin{Bmatrix} X_1 \\ X_2 \\ X_3 \end{Bmatrix}_1 = \begin{Bmatrix} 1 \\ X_2/X_1 \\ X_3/X_1 \end{Bmatrix}_1 = \begin{Bmatrix} 1 \\ \sqrt{2} \\ 1 \end{Bmatrix}$$

Second mode: $m\omega_2^2 = 2k$

$$-kX_2 = 0$$
$$-kX_1 - kX_3 = 0$$
$$-kX_2 = 0$$

Thus,

$$\left(\frac{X_2}{X_1}\right)_2 = 0$$

$$\left(\frac{X_3}{X_1}\right)_2 = -1$$

Therefore, the second normalized mode shape is

$$\begin{Bmatrix} X_1 \\ X_2 \\ X_3 \end{Bmatrix}_2 = \begin{Bmatrix} 1 \\ X_2/X_1 \\ X_3/X_1 \end{Bmatrix}_2 = \begin{Bmatrix} 1 \\ 0 \\ -1 \end{Bmatrix}$$

Third mode: $m\omega_3^2 = 2k + k\sqrt{2}$

$$\left(\frac{X_2}{X_1}\right)_3 = \frac{2k - m\omega^2}{k} = \frac{2k - 2k - k\sqrt{2}}{k} = -\sqrt{2}$$

$$\left(\frac{X_3}{X_1}\right)_3 = 1$$

Therefore, the third normalized mode shape is

$$\begin{Bmatrix} X_1 \\ X_2 \\ X_3 \end{Bmatrix}_3 = \begin{Bmatrix} 1 \\ X_2/X_1 \\ X_3/X_1 \end{Bmatrix}_3 = \begin{Bmatrix} 1 \\ -\sqrt{2} \\ 1 \end{Bmatrix}$$

The mode shapes are shown in Fig. 8.12.

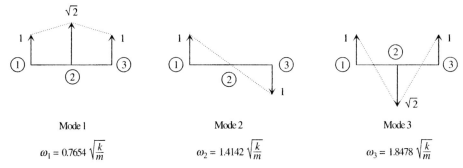

Mode 1 $\omega_1 = 0.7654 \sqrt{\frac{k}{m}}$

Mode 2 $\omega_2 = 1.4142 \sqrt{\frac{k}{m}}$

Mode 3 $\omega_3 = 1.8478 \sqrt{\frac{k}{m}}$

FIGURE 8.12 Mode shapes.

EXAMPLE 8.6. The double pendulum system shown in Fig. 8.13 consists of two uniform rods and a spring connecting the two. The rods are of length L and are assumed to be massless. A concentrated mass m is attached at the tip of each rod. The spring k is connected at distance a from the pivot point. Assume the displacement angles of the pendulums are small, so that the spring is always horizontal.

(a) Draw the system free-body diagram, derive the governing differential equations, then express the equations in the second-order matrix form.
(b) Obtain the natural frequencies.
(c) Obtain and sketch the mode shapes.

Solution

(a) The system free-body diagram is shown in Fig. 8.14, where it is assumed that the spring remains horizontal and the displacement angles are small,

$$\sin\theta_1 \approx \theta_1 \qquad \sin\theta_2 \approx \theta_2 \qquad \cos\theta_1 \approx 1 \qquad \cos\theta_2 \approx 1$$

and $\theta_2 > \theta_1$ (the spring is in tension). Applying the moment equation about the fixed points O_1 and O_2,

$$+\circlearrowleft \sum M_O = I_O \alpha \circlearrowleft +$$

$$-L\theta_1 \cdot mg + a \cdot ka(\theta_2 - \theta_1) = mL^2\ddot{\theta}_1$$

$$-L\theta_2 \cdot mg - a \cdot ka(\theta_2 - \theta_1) = mL^2\ddot{\theta}_2$$

The equations are rearranged in the standard input-output form as

$$mL^2\ddot{\theta}_1 + \left(mgL + ka^2\right)\theta_1 - ka^2\theta_2 = 0$$

$$mL^2\ddot{\theta}_2 - ka^2\theta_1 + \left(mgL + ka^2\right)\theta_2 = 0$$

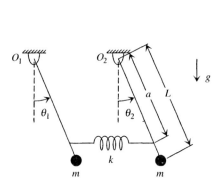

FIGURE 8.13 Double pendulum system.

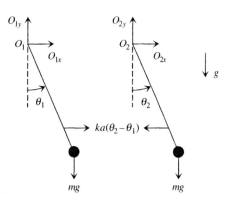

FIGURE 8.14 System free-body diagram.

or in the second-order matrix form as

$$\begin{bmatrix} mL^2 & 0 \\ 0 & mL^2 \end{bmatrix} \begin{Bmatrix} \ddot{\theta}_1 \\ \ddot{\theta}_2 \end{Bmatrix} + \begin{bmatrix} mgL + ka^2 & -ka^2 \\ -ka^2 & mgL + ka^2 \end{bmatrix} \begin{Bmatrix} \theta_1 \\ \theta_2 \end{Bmatrix} = \begin{Bmatrix} 0 \\ 0 \end{Bmatrix}$$

Dividing all the coefficients by mL^2, we obtain

$$\begin{bmatrix} 1 & 0 \\ 0 & 1 \end{bmatrix} \begin{Bmatrix} \ddot{\theta}_1 \\ \ddot{\theta}_2 \end{Bmatrix} + \begin{bmatrix} g/L + ka^2/(mL^2) & -ka^2/(mL^2) \\ -ka^2/(mL^2) & g/L + ka^2/(mL^2) \end{bmatrix} \begin{Bmatrix} \theta_1 \\ \theta_2 \end{Bmatrix} = \begin{Bmatrix} 0 \\ 0 \end{Bmatrix}$$

Notice that all the elements of the stiffness matrix have dimensions of the square of the frequency, (rad/sec)^2.

(b) The solution is in the form

$$\begin{Bmatrix} \theta_1 \\ \theta_2 \end{Bmatrix} = \begin{Bmatrix} \Theta_1 \\ \Theta_2 \end{Bmatrix} e^{j\omega t}$$

Thus,

$$\begin{Bmatrix} \ddot{\theta}_1 \\ \ddot{\theta}_2 \end{Bmatrix} = -\omega^2 \begin{Bmatrix} \Theta_1 \\ \Theta_2 \end{Bmatrix} e^{j\omega t}$$

where ω and Θ are the natural frequency and the corresponding mode shape, respectively.

$$\underbrace{\begin{bmatrix} g/L + ka^2/(mL^2) - \omega^2 & -ka^2/(mL^2) \\ -ka^2/(mL^2) & g/L + ka^2/(mL^2) - \omega^2 \end{bmatrix}}_{\det[\cdots]=0} \underbrace{\begin{Bmatrix} \Theta_1 \\ \Theta_2 \end{Bmatrix}}_{\neq 0} \underbrace{e^{j\omega t}}_{\neq 0} = \begin{Bmatrix} 0 \\ 0 \end{Bmatrix}$$

The frequency equation is given as

$$\left(\frac{g}{L} + \frac{ka^2}{mL^2} - \omega^2 \right)^2 - \left(\frac{ka^2}{mL^2} \right)^2 = 0$$

$$\frac{g}{L} + \frac{ka^2}{mL^2} - \omega^2 = \pm \frac{ka^2}{mL^2}$$

Case "+": $g/L + ka^2/(mL^2) - \omega^2 = +ka^2/(mL^2)$

$$\omega^2 = \frac{g}{L}$$

Thus, the first natural frequency is

$$\omega_1 = \sqrt{\frac{g}{L}}$$

Case "−": $g/L + ka^2/mL^2 - \omega^2 = -ka^2/(mL^2)$

$$\omega^2 = \frac{g}{L} + \frac{2ka^2}{mL^2}$$

Thus, the second natural frequency is

$$\omega_2 = \sqrt{\frac{g}{L} + \frac{2ka^2}{mL^2}}$$

(c) We have

$$\begin{bmatrix} g/L + ka^2/(mL^2) - \omega^2 & -ka^2/(mL^2) \\ -ka^2/(mL^2) & g/L + ka^2/(mL^2) - \omega^2 \end{bmatrix} \begin{Bmatrix} \Theta_1 \\ \Theta_2 \end{Bmatrix} = \begin{Bmatrix} 0 \\ 0 \end{Bmatrix}$$

Using the top equation,

$$\left(\frac{g}{L} + \frac{ka^2}{mL^2} - \omega^2 \right) \Theta_1 - \frac{ka^2}{mL^2} \Theta_2 = 0$$

Thus,

$$\frac{\Theta_2}{\Theta_1} = \frac{g/L + ka^2/(mL^2) - \omega^2}{ka^2/(mL^2)}$$

First mode: $\omega_1^2 = g/L$

$$\left(\frac{\Theta_2}{\Theta_1} \right)_1 = \frac{g/L + ka^2/(mL^2) - \omega_1^2}{ka^2/(mL^2)} = \frac{g/L + ka^2/(mL^2) - g/L}{ka^2/(mL^2)} = 1$$

The first mode shape is normalized as

$$\begin{Bmatrix} \Theta_1 \\ \Theta_2 \end{Bmatrix}_1 = \begin{Bmatrix} 1 \\ \Theta_2/\Theta_1 \end{Bmatrix}_1 = \begin{Bmatrix} 1 \\ 1 \end{Bmatrix}$$

Second mode: $\omega_2^2 = g/L + 2ka^2/(mL^2)$

$$\left(\frac{\Theta_2}{\Theta_1} \right)_2 = \frac{g/L + ka^2/(mL^2) - \omega^2}{ka^2/(mL^2)}$$

$$= \frac{g/L + ka^2/(mL^2) - g/L - 2ka^2/(mL^2)}{ka^2/(mL^2)} = -1$$

The first mode shape is normalized as

$$\begin{Bmatrix} \Theta_1 \\ \Theta_2 \end{Bmatrix}_2 = \begin{Bmatrix} 1 \\ \Theta_2/\Theta_1 \end{Bmatrix}_2 = \begin{Bmatrix} 1 \\ -1 \end{Bmatrix}$$

The mode shapes are shown in Fig. 8.15.

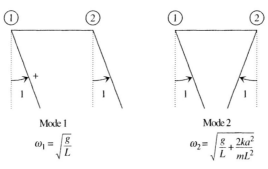

FIGURE 8.15 Mode shapes.

Automobile Suspension

As an important application, the automobile suspension system is considered. A simple model is considered first, then a more complex model.

Simple model

A simple two-degree-of-freedom model is used for an automobile suspension system. Figure 8.16 shows a schematic diagram of the system and its free-body diagram. Assuming the two rear wheels are identical, $k_1 = k$ represents the stiffness of the combined rear suspension. Similarly, $k_2 = k$ represents the stiffness of the combined front suspension. For simplicity, the center of mass of the automobile is assumed at the center. Thus, the chassis and the body of the automobile are represented by a uniform rod of mass m and length L. This simple model can be used for a motocycle suspension system. The natural frequencies and mode shapes of the system are to be investigated.

As usual, the displacements are measured from the static equilibrium position. Thus, the gravity g does *not* enter into the problem. The notation is given as

$x = $ (translational) displacement of the center of mass, c
$x_1 = $ (translational) displacement of the rear end
$x_2 = $ (translational) displacement of the front end
$\theta = $ angular displacement of the rod

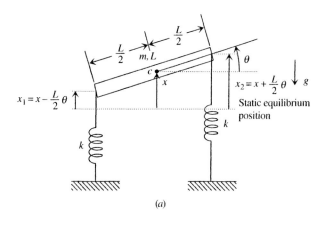

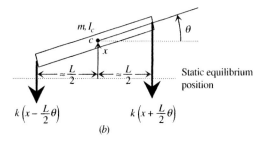

FIGURE 8.16 TDOF system: (*a*) system diagram, (*b*) free-body diagram.

CHAPTER 8: Introduction to Vibrations   421

We assume that the motion is in the plane of the paper and the rod is rigid; thus there are only two degrees of freedom. The set of two generalized coordinates can be chosen as (x, θ) or (x_1, x_2), where the former, (x, θ), is selected for convenience.

Assuming small motions for linearity, we can see that the springs deform in a nearly vertical direction. The approximations for the angle are

$$\theta \ll 1 \text{ rad}, \qquad \sin\theta \approx \theta, \qquad \cos\theta \approx 1$$

Applying Newton's second law,

$$+\uparrow \sum F_x = m a_{cx} \uparrow +$$

$$-k\left(x - \frac{L}{2}\theta\right) - k\left(x + \frac{L}{2}\theta\right) = m\ddot{x} \tag{8.18}$$

and the moment equation about the center of mass,

$$+\circlearrowleft \sum M_c = I_c \alpha \circlearrowleft +$$

$$\frac{L}{2} \cdot k\left(x - \frac{L}{2}\theta\right) - \frac{L}{2} \cdot k\left(x + \frac{L}{2}\theta\right) = I_c \ddot{\theta} \tag{8.19}$$

where

$$I_c = \frac{mL^2}{12} \tag{8.20}$$

Thus,

$$m\ddot{x} + 2kx = 0 \tag{8.21a}$$

$$I_c \ddot{\theta} + \frac{kL^2}{2}\theta = 0 \tag{8.21b}$$

or

$$\begin{bmatrix} m & 0 \\ 0 & I_c \end{bmatrix} \begin{Bmatrix} \ddot{x} \\ \ddot{\theta} \end{Bmatrix} + \begin{bmatrix} 2k & 0 \\ 0 & kL^2/2 \end{bmatrix} \begin{Bmatrix} x \\ \theta \end{Bmatrix} = \begin{Bmatrix} 0 \\ 0 \end{Bmatrix} \tag{8.22}$$

The two equations, in x and θ, are independent (not coupled); thus, they can be solved separately (not simultaneously). It is evident as given below. The solution is in the form

$$\begin{Bmatrix} x \\ \theta \end{Bmatrix} = \begin{Bmatrix} X \\ \Theta \end{Bmatrix} e^{j\omega t} \tag{8.23a}$$

Thus,

$$\begin{Bmatrix} \ddot{x} \\ \ddot{\theta} \end{Bmatrix} = -\omega^2 \begin{Bmatrix} X \\ \Theta \end{Bmatrix} e^{j\omega t} \tag{8.23b}$$

where ω and $\{X \quad \Theta\}$ are the natural frequency and the corresponding mode shape, respectively. Introducing Eq. (8.23) into (8.22), we obtain

$$\underbrace{\begin{bmatrix} 2k - m\omega^2 & 0 \\ 0 & kL^2/2 - I_c\omega^2 \end{bmatrix}}_{\det[\cdots]=0} \underbrace{\begin{Bmatrix} X \\ \Theta \end{Bmatrix}}_{\neq 0} \underbrace{e^{j\omega t}}_{\neq 0} = \begin{Bmatrix} 0 \\ 0 \end{Bmatrix} \qquad (8.24)$$

The frequency equation is given as

$$\underbrace{(2k - m\omega^2)}_{0}\underbrace{\left(\frac{kL^2}{2} - I_c\omega^2\right)}_{0} = 0 \qquad (8.25)$$

Thus,

$$m\omega^2 = 2k \implies \omega_1 = \sqrt{\frac{2k}{m}} = 1.41421\sqrt{\frac{k}{m}} \qquad (8.26a)$$

$$I_c\omega^2 = \frac{kL^2}{2} \implies \omega_2 = \sqrt{\frac{kL^2}{2I_c}} = \sqrt{\frac{6k}{m}} = 2.44949\sqrt{\frac{k}{m}} \qquad (8.26b)$$

Note that the mode shapes are not relevant for this problem because the translational displacement of the mass center, x, and the angular displacement of the rigid body, θ, are uncoupled. As it is shown by the top equation of Eq. (8.24),

$$(2k - m\omega^2)X - 0\Theta = 0 \qquad \text{or} \qquad \frac{\Theta}{X} = \frac{2k - m\omega^2}{0} \qquad (8.27)$$

More complex model

A more complex model is shown in Fig. 8.17. This model takes into account that the stiffness of the rear is different than that of the front, $k_1 \neq k_2$, and the center of mass is not at the middle of the rod. Since the engine is usually heavy and located in the front, the figure shows $L_2 < L_1$. Again, it is desired to obtain the natural frequencies and the corresponding mode shapes. Assuming small motions, springs deform in a nearly vertical direction, and

$$\theta \ll 1 \text{ rad}, \qquad \sin\theta \approx \theta$$

Applying Newton's second law,

$$+\uparrow \sum F_x = ma_{cx} \uparrow +$$
$$-k_1(x - L_1\theta) - k_2(x + L_2\theta) = m\ddot{x} \qquad (8.28)$$

and the moment equation about the center of mass,

$$+\circlearrowleft \sum M_c = I_c\alpha \circlearrowleft +$$
$$L_1 \cdot k_1(x - L_1\theta) - L_2 \cdot k_2(x + L_2\theta) = I_c\ddot{\theta} \qquad (8.29)$$

Rearranging the equations, we obtain

$$m\ddot{x} + (k_1 + k_2)x + (k_2L_2 - k_1L_1)\theta = 0 \qquad (8.30a)$$
$$I_c\ddot{\theta} + (k_2L_2 - k_1L_1)x + (k_2L_2^2 + k_1L_1^2)\theta = 0 \qquad (8.30b)$$

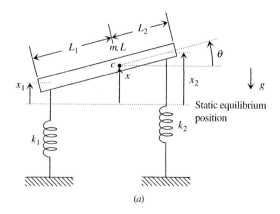

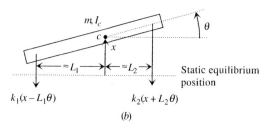

FIGURE 8.17 More complex model: (*a*) system diagram, (*b*) free-body diagram.

In the second-order matrix form, the equations become

$$\begin{bmatrix} m & 0 \\ 0 & I_c \end{bmatrix} \begin{Bmatrix} \ddot{x} \\ \ddot{\theta} \end{Bmatrix} + \begin{bmatrix} k_1 + k_2 & k_2L_2 - k_1L_1 \\ k_2L_2 - k_1L_1 & k_2L_2^2 + k_1L_1^2 \end{bmatrix} \begin{Bmatrix} x \\ \theta \end{Bmatrix} = \begin{Bmatrix} 0 \\ 0 \end{Bmatrix} \quad (8.31)$$

(b) The solution is in the form of Eq. (8.23), thus

$$\underbrace{\begin{bmatrix} k_1 + k_2 - m\omega^2 & k_2L_2 - k_1L_1 \\ k_2L_2 - k_1L_1 & k_2L_2^2 + k_1L_1^2 - I_c\omega^2 \end{bmatrix}}_{\det[\cdots]=0} \underbrace{\begin{Bmatrix} X \\ \Theta \end{Bmatrix}}_{\neq 0} \underbrace{e^{j\omega t}}_{\neq 0} = \begin{Bmatrix} 0 \\ 0 \end{Bmatrix} \quad (8.32)$$

The frequency equation is given as

$$(k_1 + k_2 - m\omega^2)(k_2L_2^2 + k_1L_1^2 - I_c\omega^2) - (k_2L_2 - k_1L_1)^2 = 0 \quad (8.33)$$

$$mI_c\omega^4 - \left[(k_1 + k_2)I_c + (k_1L_1^2 + k_2L_2^2)m\right]\omega^2 - k_1k_2(L_1 + L_2)^2 = 0 \quad (8.34)$$

$$\omega^4 - \left(\frac{k_1 + k_2}{m} + \frac{k_1L_1^2 + k_2L_2^2}{I_c}\right)\omega^2 + \frac{k_1k_2(L_1 + L_2)^2}{mI_c} = 0 \quad (8.35)$$

The natural frequencies are given by

$$\begin{matrix} \omega_1^2 \\ \omega_2^2 \end{matrix} = \frac{k_1 + k_2}{2m} + \frac{k_1L_1^2 + k_2L_2^2}{2I_c} \mp \sqrt{\left(\frac{k_1 + k_2}{2m} + \frac{k_1L_1^2 + k_2L_2^2}{2I_c}\right)^2 - \frac{k_1k_2(L_1 + L_2)^2}{mI_c}} \quad (8.36)$$

424 Dynamic Systems: Modeling and Analysis

The mode shapes are obtained next. We have

$$\begin{bmatrix} k_1 + k_2 - m\omega^2 & k_2 L_2 - k_1 L_1 \\ k_2 L_2 - k_1 L_1 & k_2 L_2^2 + k_1 L_1^2 - I_c \omega^2 \end{bmatrix} \begin{Bmatrix} X \\ \Theta \end{Bmatrix} = \begin{Bmatrix} 0 \\ 0 \end{Bmatrix} \quad (8.37)$$

Using the top equation,

$$(k_1 + k_2 - m\omega^2)X + (k_2 L_2 - k_1 L_1)\Theta = 0 \quad (8.38)$$

Thus,

$$\frac{\Theta}{X} = \frac{k_1 + k_2 - m\omega^2}{k_1 L_1 - k_2 L_2} \quad (8.39)$$

Thus for the first mode,

$$\left(\frac{\Theta}{X}\right)_1 = \frac{k_1 + k_2 - m\omega_1^2}{k_1 L_1 - k_2 L_2} \quad (8.40a)$$

and for the second mode,

$$\left(\frac{\Theta}{X}\right)_2 = \frac{k_1 + k_2 - m\omega_2^2}{k_1 L_1 - k_2 L_2} \quad (8.40b)$$

The mode shapes are normalized as

$$\begin{Bmatrix} X \\ \Theta \end{Bmatrix} = \begin{Bmatrix} X/X \\ \Theta/X \end{Bmatrix} = \begin{Bmatrix} 1 \\ \Theta/X \end{Bmatrix} \quad (8.41)$$

Finally, the first mode shape is

$$\begin{Bmatrix} X \\ \Theta \end{Bmatrix}_1 = \begin{Bmatrix} 1 \\ \Theta/X \end{Bmatrix}_1 \quad (8.42a)$$

and the second mode shape is

$$\begin{Bmatrix} X \\ \Theta \end{Bmatrix}_2 = \begin{Bmatrix} 1 \\ \Theta/X \end{Bmatrix}_2 \quad (8.42b)$$

EXAMPLE 8.7. As an illustrative example, consider the system shown in Fig. 8.17 with the following numerical values: $m = 1$ kg, $L_1 = 2$ m, $L_2 = 1$ m, $I_c = 2$ kg-m^2, $k_1 = k_2 = 1$ N/m.

(a) Obtain the natural frequencies.
(b) Obtain and sketch the mode shapes.

Solution

(a) The frequency equation is

$$\omega^4 - \left(\frac{k_1 + k_2}{m} + \frac{k_1 L_1^2 + k_2 L_2^2}{I_c}\right)\omega^2 + \frac{k_1 k_2 (L_1 + L_2)^2}{m I_c} = 0$$

After substituting the numerical values, we obtain

$$\omega^4 - 4.5\omega^2 + 4.5 = 0$$

$$\begin{matrix}\omega_1^2\\\omega_2^2\end{matrix} = \frac{4.5}{2} \mp \sqrt{\left(\frac{4.5}{2}\right)^2 - 4.5} = 2.25 \mp 0.75$$

$$\omega_1^2 = 1.50 \implies \omega_1 = 1.22474 \text{ rad/s}$$

$$\omega_2^2 = 3 \implies \omega_2 = 1.73205 \text{ rad/s}$$

(b) For the mode shapes, we have

$$\frac{\Theta}{X} = \frac{k_1 + k_2 - m\omega^2}{k_1 L_1 - k_2 L_2}$$

After substituting the numerical values, we obtain

$$\frac{\Theta}{X} = \frac{2 - \omega^2}{2 - 1} = 2 - \omega^2$$

First mode: $\omega_1^2 = 1.5 \text{ (rad/s)}^2$

$$\left(\frac{\Theta}{X}\right)_1 = 2 - \omega_1^2 = 2 - 1.5 = 0.5$$

The first mode shape is normalized as

$$\left\{\begin{matrix}X\\\Theta\end{matrix}\right\}_1 = \left\{\begin{matrix}1\\\Theta/X\end{matrix}\right\}_1 = \left\{\begin{matrix}1\\0.5\end{matrix}\right\}$$

Second mode: $\omega_2^2 = 3.0 \text{ (rad/s)}^2$

$$\left(\frac{\Theta}{X}\right)_2 = 2 - \omega_2^2 = 2 - 3.0 = -1$$

The second normalized mode shape is

$$\left\{\begin{matrix}X\\\Theta\end{matrix}\right\}_2 = \left\{\begin{matrix}1\\\Theta/X\end{matrix}\right\}_2 = \left\{\begin{matrix}1\\-1\end{matrix}\right\}$$

The mode shapes are shown in Fig. 8.18.

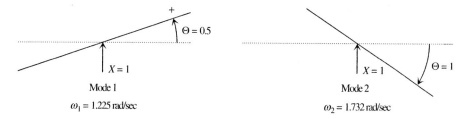

FIGURE 8.18 Mode shapes.

PROBLEM SET 8.1–8.2

8.1. Consider the mechanical system shown in Fig. P8.1
   (a) Obtain the natural frequencies.
   (b) Obtain and sketch the mode shapes.

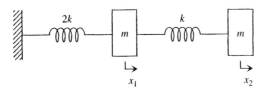

FIGURE P8.1 TDOF system.

8.2. Consider the mechanical system shown in Fig. P8.2
   (a) Obtain the natural frequencies.
   (b) Obtain and sketch the mode shapes.

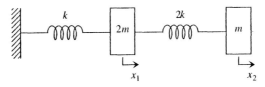

FIGURE P8.2 TDOF system.

8.3. The double pendulum system shown in Fig. P8.3 consists of two uniform rods and a spring connecting the two. The rods are of length L and are assumed to be massless. A concentrated mass m is attached at the tip of each rod. The spring k is connected at $a = (3/4)L$. Assume that the displacement angles of the pendulums are small and that the spring is always horizontal. The differential equations are given, in the second-order matrix form, as

$$\begin{bmatrix} 1 & 0 \\ 0 & 1 \end{bmatrix} \begin{Bmatrix} \ddot{\theta}_1 \\ \ddot{\theta}_2 \end{Bmatrix} + \begin{bmatrix} g/L + 9k/(16m) & -9k/(16m) \\ -9k/(16m) & g/L + 9k/(16m) \end{bmatrix} \begin{Bmatrix} \theta_1 \\ \theta_2 \end{Bmatrix} = \begin{Bmatrix} 0 \\ 0 \end{Bmatrix}$$

   (a) Obtain the natural frequencies.
   (b) Obtain and sketch the mode shapes.

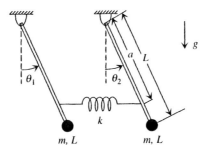

FIGURE P8.3 Pendulum system.

8.4. Consider the pendulum system shown Fig. P8.4. The system consists of two rods and a spring. The two rods are identical, and each is of mass m and length L. The spring is of stiffness k connecting the two rods at a distance a from the pivot point. A concentrated mass m is attached at the tip of each rod. Assume that the displacement angles of the pendulums are small and that the spring is always horizontal.
  (a) Draw the system free-body diagram, derive the governing differential equations, then express the equations in the second-order matrix form.
  (b) Obtain the natural frequencies.
  (c) Obtain and sketch the mode shapes.

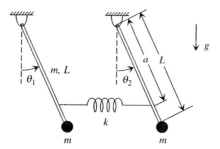

FIGURE P8.4 Pendulum system.

8.3 MODAL MATRIX

A modal matrix possesses orthogonality properties that have many important applications. One of the useful applications is to check for the correctness of the eigensolutions using a modal matrix. Another critical application is to obtain the system response using the method called modal analysis, which uses modal expansion together with orthonormality. When the modal matrix is normalized for modal analysis, the orthogonality becomes orthonormality. The important topic of damping is also discussed in conjunction with the modal matrix.

Orthogonality

For the undamped free vibration (eigenvalue problem), the differential equations in matrix form are

$$\mathbf{m}\ddot{\mathbf{x}} + \mathbf{k}\mathbf{x} = \mathbf{0} \tag{8.43}$$

The system vibrates in one of its natural modes, say the rth mode; thus

$$\mathbf{x}_r = \mathbf{X}_r e^{j\omega_r t} \tag{8.44a}$$

$$\ddot{\mathbf{x}}_r = -\omega_r^2 \mathbf{X}_r e^{j\omega_r t} = -\omega_r^2 \mathbf{x}_r \tag{8.44b}$$

where ω_r and $\mathbf{X}_r$ are the natural frequency and mode shape of the rth mode, respectively.

For orthogonality, in contrast to solving for the eigensolutions, we write

$$-\omega_r^2 \mathbf{m} \mathbf{x}_r + \mathbf{k} \mathbf{x}_r = 0 \tag{8.45}$$

Thus,

$$\mathbf{k} \mathbf{x}_r = \omega_r^2 \mathbf{m} \mathbf{x}_r \tag{8.46}$$

or

$$\mathbf{k} \underbrace{\mathbf{X}_r e^{j\omega_r t}}_{\neq 0} = \omega_r^2 \mathbf{m} \underbrace{\mathbf{X}_r e^{j\omega_r t}}_{\neq 0} \tag{8.47}$$

Thus,

$$\mathbf{k} \mathbf{X}_r = \omega_r^2 \mathbf{m} \mathbf{X}_r \tag{8.48}$$

This is an important result. Also considering the sth mode, we have

$$\mathbf{X}_s^T \left(\mathbf{k} \mathbf{X}_r = \omega_r^2 \mathbf{m} \mathbf{X}_r \right) \tag{8.49a}$$

$$\mathbf{X}_r^T \left(\mathbf{k} \mathbf{X}_s = \omega_s^2 \mathbf{m} \mathbf{X}_s \right) \tag{8.49b}$$

Thus,

$$\mathbf{X}_s^T \mathbf{k} \mathbf{X}_r = \omega_r^2 \mathbf{X}_s^T \mathbf{m} \mathbf{X}_r \tag{8.50a}$$

$$\mathbf{X}_r^T \mathbf{k} \mathbf{X}_s = \omega_s^2 \mathbf{X}_r^T \mathbf{m} \mathbf{X}_s \tag{8.50b}$$

Taking the transpose of Eq. (8.50b), we obtain

$$\mathbf{X}_s^T \underbrace{\mathbf{k}^T}_{\mathbf{k}} \mathbf{X}_r = \omega_s^2 \mathbf{X}_s^T \underbrace{\mathbf{m}^T}_{\mathbf{m}} \mathbf{X}_r \tag{8.51}$$

For vibration problems, the mass and stiffness matrices are symmetric:

$$\mathbf{m}^T = \mathbf{m} \quad \mathbf{k}^T = \mathbf{k} \tag{8.52}$$

For example,

$$\mathbf{m} = \begin{bmatrix} m & 0 \\ 0 & m \end{bmatrix} = \mathbf{m}^T \quad \mathbf{k} = \begin{bmatrix} 2k & -k \\ -k & k \end{bmatrix} = \mathbf{k}^T$$

Thus, Eq. (8.51) becomes

$$\mathbf{X}_s^T \mathbf{k} \mathbf{X}_r = \omega_s^2 \mathbf{X}_s^T \mathbf{m} \mathbf{X}_r \tag{8.53}$$

Subtracting Eq. (8.53) from (8.50a), we have

$$0 = \left(\omega_r^2 - \omega_s^2 \right) \underbrace{\mathbf{X}_s^T \mathbf{m} \mathbf{X}_r}_{0 \text{ if } r = s} \tag{8.54}$$

Assuming there is no degeneracy, meaning two different modes must correspond to two different natural frequencies, we have

$$\omega_r \neq \omega_s \quad \text{if } r \neq s \tag{8.55}$$

Thus,

$$\omega_r^2 - \omega_s^2 = \begin{cases} \text{nonzero} & \text{if } r \neq s \\ 0 & \text{if } r = s \end{cases} \quad (8.56)$$

The orthogonality with respect to the mass matrix is given as

$$\mathbf{X}_s^T \mathbf{m} \mathbf{X}_r = \begin{cases} 0 & \text{if } r \neq s \\ \text{nonzero} & \text{if } r = s \end{cases} \quad (8.57)$$

For the orthogonality with respect to the stiffness matrix, a similar (but distinct) step is used. We write

$$\mathbf{X}_s^T \mathbf{m} \mathbf{X}_r = \frac{1}{\omega_r^2} \mathbf{X}_s^T \mathbf{k} \mathbf{X}_r \quad (8.58a)$$

$$\mathbf{X}_r^T \mathbf{m} \mathbf{X}_s = \frac{1}{\omega_s^2} \mathbf{X}_r^T \mathbf{k} \mathbf{X}_s \quad (8.58b)$$

Taking the transpose of Eq. (8.58b), we obtain

$$\mathbf{X}_s^T \underbrace{\mathbf{m}^T}_{\mathbf{m}} \mathbf{X}_r = \frac{1}{\omega_s^2} \mathbf{X}_s^T \underbrace{\mathbf{k}^T}_{\mathbf{k}} \mathbf{X}_r \quad (8.59)$$

thus Eq. (8.59) becomes

$$\mathbf{x}_s^T \mathbf{m} \mathbf{X}_r = \frac{1}{\omega_s^2} \mathbf{X}_s^T \mathbf{k} \mathbf{X}_r \quad (8.60)$$

Subtracting Eq. (8.60) from (8.58a), we have

$$0 = \underbrace{\left(\frac{1}{\omega_r^2} - \frac{1}{\omega_s^2} \right)}_{0 \text{ if } r=s} \mathbf{X}_s^T \mathbf{k} \mathbf{X}_r \quad (8.61)$$

Again, with no degeneracy and a nontrivial solution, either term of the product must be zero, thus

$$\frac{1}{\omega_r^2} - \frac{1}{\omega_s^2} = \begin{cases} \text{nonzero} & \text{if } r \neq s \\ 0 & \text{if } r = s \end{cases} \quad (8.62)$$

Therefore, the orthogonality with respect to the stiffness matrix $\mathbf{k}$ is given as

$$\mathbf{X}_s^T \mathbf{k} \mathbf{X}_r = \begin{cases} 0 & \text{if } r \neq s \\ \text{nonzero} & \text{if } r = s \end{cases} \quad (8.63)$$

For generality, the symbol for the modal vector is changed from $\mathbf{X}_r$ to $\mathbf{u}_r$.

$$\mathbf{X}_r \longrightarrow \mathbf{u}_r \ (r\text{th mode shape}) \quad (8.64)$$

where the arrow $\longrightarrow$ is used to the denote the change. Thus, the orthogonality with respect to the mass and stiffness matrices becomes

$$\mathbf{u}_s^T \mathbf{m} \mathbf{u}_r = \begin{cases} 0 & \text{if } r \neq s \\ \text{nonzero} & \text{if } r = s \end{cases} \quad (8.65a)$$

$$\mathbf{u}_s^T \mathbf{k} \mathbf{u}_r = \begin{cases} 0 & \text{if } r \neq s \\ \text{nonzero} & \text{if } r = s \end{cases} \quad (8.65b)$$

For MDOF systems with n modes, the modal matrix $\mathbf{U}$ is defined as

$$\mathbf{U} = \begin{bmatrix} \mathbf{u}_1 & \mathbf{u}_2 & \cdots & \mathbf{u}_r & \cdots & \mathbf{u}_n \end{bmatrix} \quad (8.66)$$

As an example, the mode shapes of a TDOF system are

$$\mathbf{u}_1 = \mathbf{X}_1 = \begin{Bmatrix} X_1 \\ X_2 \end{Bmatrix}_1 = \begin{Bmatrix} 1 \\ X_2/X_1 \end{Bmatrix}_1 \quad (8.67)$$

$$\mathbf{u}_2 = \mathbf{X}_2 = \begin{Bmatrix} X_1 \\ X_2 \end{Bmatrix}_2 = \begin{Bmatrix} 1 \\ X_2/X_1 \end{Bmatrix}_2 \quad (8.68)$$

Thus the modal matrix is

$$\mathbf{U} = \begin{bmatrix} \mathbf{u}_1 & \mathbf{u}_2 \end{bmatrix} \quad (8.69)$$

The modal matrix must be orthogonal with respect to the mass matrix $\mathbf{m}$ and the stiffness matrix $\mathbf{k}$,

$$\mathbf{U}^T \mathbf{m} \mathbf{U} = [M_r] = \text{diagonal matrix} \quad (8.70)$$

$$\mathbf{U}^T \mathbf{k} \mathbf{U} = [K_r] = \text{diagonal matrix} \quad (8.71)$$

where M_r = *modal mass* of the rth mode
K_r = *modal stiffness* of the rth mode
$\omega_r = \sqrt{K_r/M_r}$ = (modal) *natural frequency* of the rth mode

In general, the matrix $\mathbf{U}$ is *not* orthogonal with respect to itself if the mass and stiffness matrices are not proportional to an identity matrix:

$$\mathbf{U}^T \mathbf{U} \neq \text{diagonal matrix} \quad \text{if } \mathbf{k} \neq \alpha \mathbf{I} \text{ or } \mathbf{m} \neq \beta \mathbf{I} \quad (8.72)$$

where α and β are proportional constants.

Orthonormality

As mentioned previously, the way the mode shapes are normalized for modal analysis is different than that for sketching the mode shapes. The *modal vector* $\mathbf{u}_r$ is changed to $\boldsymbol{\phi}_r$:

$$\mathbf{u}_r \to \boldsymbol{\phi}_r \quad (r\text{th mode shape}) \quad (8.73)$$

Thus,

$$\boldsymbol{\phi}_s^T \mathbf{m} \boldsymbol{\phi}_r = \delta_{rs} = \begin{cases} 0 & \text{if } r \neq s \\ 1 & \text{if } r = s \end{cases} \quad (8.74a)$$

$$\boldsymbol{\phi}_s^T \mathbf{k} \boldsymbol{\phi}_r = \omega_r^2 \delta_{rs} = \begin{cases} 0 & \text{if } r \neq s \\ \omega_r^2 & \text{if } r = s \end{cases} \quad (8.74b)$$

where δ_{rs} is the *Kronecker delta*. The modal matrix is changed from $\mathbf{U}$ to $\boldsymbol{\Phi}$,

$$\mathbf{U} \to \boldsymbol{\Phi} \quad (8.75)$$

The *orthonormality* for the modal matrix is thus given as

$$\Phi^T m \Phi = I = [1] \qquad (8.76a)$$

$$\Phi^T k \Phi = [\omega_r^2] \qquad (8.76b)$$

where I is the identity matrix. For example,

$$I = \begin{bmatrix} 1 & 0 \\ 0 & 1 \end{bmatrix}$$

Let us consider the modal matrix of a TDOF system

$$U = \begin{bmatrix} u_1 & u_2 \end{bmatrix}$$

which becomes

$$\Phi = \begin{bmatrix} \alpha_1 & \alpha_2 \end{bmatrix} \qquad (8.77)$$

where the two constants α_1 and α_2 are to be determined using the orthonormality.

Damping

The modal matrix is orthogonal or orthonormal with respect to the damping matrix c *only* if the damping matrix is proportional to either the mass matrix m or the stiffness matrix k, or to both the matrices,

$$U^T c U = [C_r] = \text{diagonal matrix} \qquad (8.78)$$

if $c = \alpha m$ or $c = \beta m$ or $c = \alpha m + \beta m$

or

$$\Phi^T c \Phi = [2\zeta_r \omega_r] \qquad (8.79)$$

where, again, α and β are proportional constants and

C_r = modal damping coefficient of the rth mode
ζ_r = modal damping ratio of the rth mode

It must be emphasized that, in general, the damping matrix c is neither proportional to the mass matrix m nor the stiffness matrix k; thus the modal matrix is *not* orthogonal or orthonormal with respect to c,

$$U^T c U \neq \text{diagonal matrix} \qquad \text{if } c \neq \alpha m \text{ or } c \neq \beta m \qquad (8.80)$$

or

$$\Phi^T c \Phi \neq [2\zeta_r \omega_r] \qquad (8.81)$$

EXAMPLE 8.8. Consider the mechanical system shown in Fig. 8.19. From the eigensolutions worked out previously, the mode shapes are given as

$$u_1 = \begin{Bmatrix} 1 \\ 1.61803 \end{Bmatrix} \qquad u_2 = \begin{Bmatrix} 1 \\ -0.61803 \end{Bmatrix}$$

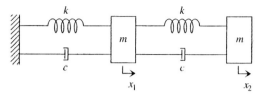

FIGURE 8.19 TDOF system.

(a) Show that the following matrices are diagonal:

$$\mathbf{U}^T\mathbf{U} \quad \mathbf{U}^T\mathbf{m}\mathbf{U} = [M_r] \quad \mathbf{U}^T\mathbf{k}\mathbf{U} = [K_r] \quad \mathbf{U}^T\mathbf{c}\mathbf{U}$$

(b) Determine the normalized modal matrix $\mathbf{\Phi}$.

Solution

(a) The mass, stiffness, and damping matrices are

$$\mathbf{m} = \begin{bmatrix} m & 0 \\ 0 & m \end{bmatrix} = m\begin{bmatrix} 1 & 0 \\ 0 & 1 \end{bmatrix} = m\mathbf{I} \quad \mathbf{k} = \begin{bmatrix} 2k & -k \\ -k & k \end{bmatrix} = k\begin{bmatrix} 2 & -1 \\ -1 & 1 \end{bmatrix}$$

$$\mathbf{c} = \begin{bmatrix} 2c & -c \\ -c & c \end{bmatrix} = c\begin{bmatrix} 2 & -1 \\ -1 & 1 \end{bmatrix}$$

and the modal matrix is

$$\mathbf{U} = [\mathbf{u}_1 \quad \mathbf{u}_2] = \begin{bmatrix} 1 & 1 \\ 1.61803 & -0.61803 \end{bmatrix}$$

Since the mass matrix $\mathbf{m}$ is proportional to an identity matrix, the modal matrix $\mathbf{U}$ is orthogonal with itself,

$$\mathbf{U}^T\mathbf{U} = \begin{bmatrix} 1 & 1.61803 \\ 1 & -0.61803 \end{bmatrix}\begin{bmatrix} 1 & 1 \\ 1.61803 & -0.61803 \end{bmatrix}$$

$$= \begin{bmatrix} 3.6180 & 0 \\ 0 & 1.3820 \end{bmatrix} = \text{diagonal matrix}$$

The modal matrix must be orthogonal with respect to the mass matrix,

$$\mathbf{U}^T\mathbf{m}\mathbf{U} = \begin{bmatrix} 1 & 1.61803 \\ 1 & -0.61803 \end{bmatrix}\underbrace{\begin{bmatrix} m & 0 \\ 0 & m \end{bmatrix}\begin{bmatrix} m & 0 \\ 0 & m \end{bmatrix}}_{m\mathbf{I}}\begin{bmatrix} 1 & 1 \\ 1.61803 & -0.61803 \end{bmatrix}$$

$$\mathbf{U}^T\mathbf{m}\mathbf{U} = m\begin{bmatrix} 1 & 1.61803 \\ 1 & -0.61803 \end{bmatrix}\begin{bmatrix} 1 & 1 \\ 1.61803 & -0.61803 \end{bmatrix}$$

$$= \begin{bmatrix} 3.6180m & 0 \\ 0 & 0.3820m \end{bmatrix} = \text{diagonal matrix}$$

and to the stiffness matrix

$$\mathbf{U}^T\mathbf{k}\mathbf{U} = \begin{bmatrix} 1 & 1.61803 \\ 1 & -0.61803 \end{bmatrix}\begin{bmatrix} 2k & -k \\ -k & k \end{bmatrix}\begin{bmatrix} 1 & 1 \\ 1.61803 & -0.61803 \end{bmatrix}$$

$$\mathbf{U}^T\mathbf{k}\mathbf{U} = \begin{bmatrix} 1 & 1.61803 \\ 1 & -0.61803 \end{bmatrix}\begin{bmatrix} 0.38197k & 2.61803k \\ 1.61803k & -1.61803k \end{bmatrix}$$

$$= \begin{bmatrix} 1.3820k & 0 \\ 0 & 3.61803k \end{bmatrix} = \text{diagonal matrix}$$

The damping matrix **c** is proportional to *both* the stiffness matrix **k** and the mass matrix **m**, thus the modal matrix is orthogonal with respect to the damping matrix,

$$\mathbf{U}^T\mathbf{c}\mathbf{U} = \begin{bmatrix} 1 & 1.61803 \\ 1 & -0.61803 \end{bmatrix}\begin{bmatrix} 2c & -c \\ -c & c \end{bmatrix}\begin{bmatrix} 1 & 1 \\ 1.61803 & -0.61803 \end{bmatrix}$$

$$\mathbf{U}^T\mathbf{c}\mathbf{U} = \begin{bmatrix} 1 & 1.61803 \\ 1 & -0.61803 \end{bmatrix}\begin{bmatrix} 0.38197c & 2.61803c \\ 1.61803c & -1.61803c \end{bmatrix}$$

$$= \begin{bmatrix} 1.3820c & 0 \\ 0 & 3.6180c \end{bmatrix} = \text{diagonal matrix}$$

(b) We have

$$\boldsymbol{\phi}_1 = \alpha_1 \begin{Bmatrix} 1 \\ 1.61803 \end{Bmatrix} \qquad \boldsymbol{\phi}_2 = \alpha_2 \begin{Bmatrix} 1 \\ -0.61803 \end{Bmatrix}$$

$$\boldsymbol{\Phi} = [\boldsymbol{\phi}_1 \quad \boldsymbol{\phi}_2] = \begin{bmatrix} \alpha_1 & \alpha_2 \\ 1.61803\alpha_1 & -0.61803\alpha_2 \end{bmatrix}$$

$$\boldsymbol{\Phi}^T\mathbf{m}\boldsymbol{\Phi} = \mathbf{I}$$

$$\begin{bmatrix} \alpha_1 & 1.61803\alpha_1 \\ \alpha_2 & -0.61803\alpha_2 \end{bmatrix}\begin{bmatrix} m & 0 \\ 0 & m \end{bmatrix}\begin{bmatrix} \alpha_1 & \alpha_2 \\ 1.61803\alpha_1 & -0.61803\alpha_2 \end{bmatrix} = \begin{bmatrix} 1 & 0 \\ 0 & 1 \end{bmatrix}$$

$$m\begin{bmatrix} \alpha_1 & 1.61803\alpha_1 \\ \alpha_2 & -0.61803\alpha_2 \end{bmatrix}\begin{bmatrix} \alpha_1 & \alpha_2 \\ 1.61803\alpha_1 & -0.61803\alpha_2 \end{bmatrix} = \begin{bmatrix} 1 & 0 \\ 0 & 1 \end{bmatrix}$$

$$m\begin{bmatrix} 3.61802\alpha_1^2 & \alpha_1\alpha_2 - 1.61803(0.61803)\alpha_1\alpha_2 \\ \alpha_1\alpha_2 - 1.61803(0.61803)\alpha_1\alpha_2 & 1.38196\alpha_2^2 \end{bmatrix} = \begin{bmatrix} 1 & 0 \\ 0 & 1 \end{bmatrix}$$

Thus,

$$3.61802 m\alpha_1^2 = 1 \implies \alpha_1 = \frac{0.52573}{\sqrt{m}}$$

$$1.38196 m\alpha_2^2 = 1 \implies \alpha_2 = \frac{0.85065}{\sqrt{m}}$$

Notice that the off-diagonal terms of the matrix on the left must be zero if the mode shapes are correct. The reason is that the damping matrix **c** is proportional to the stiffness matrix **k**, thus the modal matrix Φ must be orthonormal with respect to **c**. This is how we check for the correctness of the eigensolutions. The modal matrix is given as

$$\Phi = \frac{1}{\sqrt{m}} \begin{bmatrix} 0.52573 & 0.85065 \\ 0.85065 & -0.52573 \end{bmatrix}$$

As a final check, we have

$$\Phi^T m \Phi = \begin{bmatrix} 0.52573 & 0.85065 \\ 0.85065 & -0.52573 \end{bmatrix} \begin{bmatrix} 0.52573 & 0.85065 \\ 0.85065 & -0.52573 \end{bmatrix} = \begin{bmatrix} 1 & 0 \\ 0 & 1 \end{bmatrix}$$

EXAMPLE 8.9. Consider the system shown in Fig. 8.20, where the applied force is an impulse function, $f(t) = F_o \delta(t)$. From the eigensolutions worked out previously, the modal matrix is given as

$$U = \begin{bmatrix} 1 & 1 \\ 2.0 & -1.0 \end{bmatrix}$$

(a) Show that

$$U^T U \neq \text{diagonal matrix}$$
$$U^T c U \neq \text{diagonal matrix}$$

(b) Determine the normalized modal matrix Φ.
(c) Given that $\mathbf{f}(t)$ is the applied force matrix, determine

$$N(t) = \Phi^T \mathbf{f}(t)$$

Solution. For this problem, the mass, stiffness, and damping matrices are

$$\mathbf{m} = \begin{bmatrix} 2m & 0 \\ 0 & m \end{bmatrix} = m \begin{bmatrix} 2 & 0 \\ 0 & 1 \end{bmatrix} \qquad \mathbf{k} = \begin{bmatrix} 3k & -k \\ -k & k \end{bmatrix} = k \begin{bmatrix} 3 & -1 \\ -1 & 1 \end{bmatrix}$$

$$\mathbf{c} = \begin{bmatrix} 0 & -c \\ -c & c \end{bmatrix} = c \begin{bmatrix} 0 & -1 \\ -1 & 1 \end{bmatrix}$$

and the applied force matrix is

$$\mathbf{f}(t) = \begin{Bmatrix} F_o \delta(t) \\ 0 \end{Bmatrix}$$

(a) Since the mass matrix is *not* proportional to an identity matrix, the modal matrix is *not* orthogonal with itself,

$$U^T U = \begin{bmatrix} 1 & 2.0 \\ 1 & -1.0 \end{bmatrix} \begin{bmatrix} 1 & 1 \\ 2.0 & -1.0 \end{bmatrix} = \begin{bmatrix} 5 & -1 \\ -1 & 2 \end{bmatrix} \neq \text{diagonal matrix}$$

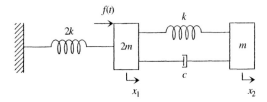

FIGURE 8.20 TDOF system.

The damping matrix **c** is proportional neither to the stiffness matrix **k** nor to the mass matrix **m**, thus the modal matrix is not orthogonal with respect to the damping matrix,

$$\mathbf{U}^T\mathbf{c}\mathbf{U} = \begin{bmatrix} 1 & 2.0 \\ 1 & -1.0 \end{bmatrix}\begin{bmatrix} 0 & -c \\ -c & c \end{bmatrix}\begin{bmatrix} 1 & 1 \\ 2.0 & -1.0 \end{bmatrix} = \begin{bmatrix} 1 & 2.0 \\ 1 & -1.0 \end{bmatrix}\begin{bmatrix} -2c & c \\ c & -2c \end{bmatrix}$$

$$\mathbf{U}^T\mathbf{c}\mathbf{U} = \begin{bmatrix} 0 & -3c \\ -3c & 3c \end{bmatrix} \neq \text{diagonal matrix}$$

(b) We have

$$\boldsymbol{\phi}_1 = \alpha_1 \begin{Bmatrix} 1 \\ 2.0 \end{Bmatrix} \qquad \boldsymbol{\phi}_2 = \alpha_2 \begin{Bmatrix} 1 \\ -1.0 \end{Bmatrix}$$

$$\boldsymbol{\Phi} = [\boldsymbol{\phi}_1 \quad \boldsymbol{\phi}_2] = \begin{bmatrix} \alpha_1 & \alpha_2 \\ 2\alpha_1 & -\alpha_2 \end{bmatrix}$$

$$\boldsymbol{\Phi}^T\mathbf{m}\boldsymbol{\Phi} = \mathbf{I}$$

$$\begin{bmatrix} \alpha_1 & 2\alpha_1 \\ \alpha_2 & -\alpha_2 \end{bmatrix}\begin{bmatrix} 2m & 0 \\ 0 & m \end{bmatrix}\begin{bmatrix} \alpha_1 & \alpha_2 \\ 2\alpha_1 & -\alpha_2 \end{bmatrix} = \begin{bmatrix} 1 & 0 \\ 0 & 1 \end{bmatrix}$$

$$\begin{bmatrix} \alpha_1 & 2\alpha_1 \\ \alpha_2 & -\alpha_2 \end{bmatrix}\begin{bmatrix} 2m\alpha_1 & 2m\alpha_2 \\ 2m\alpha_1 & -m\alpha_2 \end{bmatrix} = \begin{bmatrix} 1 & 0 \\ 0 & 1 \end{bmatrix}$$

$$\begin{bmatrix} 6m\alpha_1^2 & 0 \\ 0 & 3m\alpha_2^2 \end{bmatrix} = \begin{bmatrix} 1 & 0 \\ 0 & 1 \end{bmatrix}$$

Thus,

$$6m\alpha_1^2 = 1 \implies \alpha_1 = \frac{1}{\sqrt{6m}}$$

$$3m\alpha_2^2 = 1 \implies \alpha_2 = \frac{1}{\sqrt{3m}}$$

$$\boldsymbol{\Phi} = \frac{1}{\sqrt{m}}\begin{bmatrix} 1/\sqrt{6} & 1/\sqrt{3} \\ 2/\sqrt{6} & -1/\sqrt{3} \end{bmatrix}$$

(c)

$$\mathbf{N}(t) = \boldsymbol{\Phi}^T\mathbf{f}(t)\frac{1}{\sqrt{m}}\begin{bmatrix} 1/\sqrt{6} & 1/\sqrt{3} \\ 2/\sqrt{6} & -1/\sqrt{3} \end{bmatrix}\begin{Bmatrix} F_o\delta(t) \\ 0 \end{Bmatrix} = \frac{F_o}{\sqrt{m}}\begin{Bmatrix} 1/\sqrt{6} \\ 1/\sqrt{3} \end{Bmatrix}\delta(t)$$

EXAMPLE 8.10. Consider the system shown in Fig. 8.21. From the eigensolutions worked out previously, the normalized modal matrix is given as

$$\boldsymbol{\Phi} = \frac{1}{\sqrt{m}}\begin{bmatrix} 1/\sqrt{6} & 1/\sqrt{3} \\ 2/\sqrt{6} & -1/\sqrt{3} \end{bmatrix}$$

Determine the modal damping ratios, ζ_1 and ζ_2.

Solution. For this problem, the damping matrix is

$$\mathbf{c} = \begin{bmatrix} 3c & -c \\ -c & c \end{bmatrix}$$

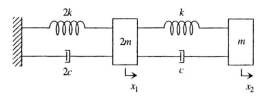

FIGURE 8.21 TDOF system.

We have

$$\mathbf{\Phi}^T \mathbf{c} \mathbf{\Phi} = [2\zeta_r \omega_r] = \text{diagonal matrix}$$

$$\frac{1}{m}\begin{bmatrix} 1/\sqrt{6} & 2/\sqrt{6} \\ 1/\sqrt{3} & -1/\sqrt{3} \end{bmatrix}\begin{bmatrix} 3c & -c \\ -c & c \end{bmatrix}\begin{bmatrix} 1/\sqrt{6} & 1/\sqrt{3} \\ 2/\sqrt{6} & -1/\sqrt{3} \end{bmatrix} = \begin{bmatrix} 2\zeta_1 \omega_1 & 0 \\ 0 & 2\zeta_2 \omega_2 \end{bmatrix}$$

$$\frac{c}{m}\begin{bmatrix} 1/\sqrt{6} & 2/\sqrt{6} \\ 1/\sqrt{3} & -1/\sqrt{3} \end{bmatrix}\begin{bmatrix} 1/\sqrt{6} & 4/\sqrt{3} \\ 1/\sqrt{6} & -2/\sqrt{3} \end{bmatrix} = \begin{bmatrix} 2\zeta_1 \omega_1 & 0 \\ 0 & 2\zeta_2 \omega_2 \end{bmatrix}$$

$$\begin{bmatrix} c/(2m) & 0 \\ 0 & 2c/m \end{bmatrix} = \begin{bmatrix} 2\zeta_1 \omega_1 & 0 \\ 0 & 2\zeta_2 \omega_2 \end{bmatrix}$$

Thus,

$$2\zeta_1 \omega_1 = \frac{c}{2m} \implies \zeta_1 = \frac{c}{4m\omega_1}$$

$$2\zeta_2 \omega_2 = \frac{2c}{m} \implies \zeta_2 = \frac{c}{m\omega_2}$$

PROBLEM SET 8.3

8.5. Consider the mechanical system shown in Fig. P8.5. From the eigensolutions worked out previously, the modal matrix is given as

$$\mathbf{U} = \begin{bmatrix} 1 & 1 \\ 1.61803 & -0.61803 \end{bmatrix}$$

Show that

$$\mathbf{U}^T \mathbf{U} = \text{diagonal matrix}$$

$$\mathbf{U}^T \mathbf{c} \mathbf{U} \neq \text{diagonal matrix}$$

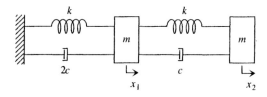

FIGURE P8.5 TDOF system.

8.6. Consider the mechanical system shown in Fig. P8.6. From the eigensolutions worked out previously, the modal matrix is given as

$$\mathbf{U} = \begin{bmatrix} 1 & 1 \\ 2.0 & -1.0 \end{bmatrix}$$

Show that

$$\mathbf{U}^T \mathbf{U} \neq \text{diagonal matrix}$$

$$\mathbf{U}^T \mathbf{c} \mathbf{U} = \text{diagonal matrix}$$

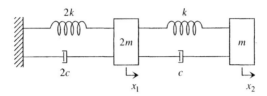

FIGURE P8.6

8.7. Show that

$$\zeta_r = \frac{C_r}{2M_r \omega_r} = \frac{C_r}{2\sqrt{K_r M_r}}$$

where M_r = *modal mass* of the *r*th mode
K_r = *modal stiffness* of the *r*th mode
C_r = *modal damping coefficient* of the *r*th mode
ζ_r = *modal damping ratio* of the *r*th mode
ω_r = $\sqrt{K_r/M_r}$ = (modal) *natural frequency* of the *r*th mode

8.4 LOGARITHMIC DECREMENT

The amount of damping of a vibration system, or of a dynamic system, can be determined experimentally. We will focus our attention on an underdamped second-order system with viscous damping. With this damping, the response of the single-degree-of-freedom system is sinusoidal with an exponential decayed envelope. This envelope is in the form of $Ae^{-\zeta \omega_n t}$ where A is a constant. This section shows how the **log decrement method** is used to obtain the damping ratio ζ from the system response.

The free vibration response of such a system is shown in Fig. 8.22. The two consecutive displacements x_1 and x_2 occur at their corresponding times t_1 and t_2, which are separated by one period T. Although these displacements need not be at the peak, they are chosen for convenience. Thus, the ratio of these two displacements are given as

$$\frac{x_1}{x_2} = \frac{e^{-\sigma t_1} A \cos(\omega_d t_1 + \phi)}{e^{-\sigma t_2} A \cos(\omega_d t_2 + \phi)}$$

$$= \frac{e^{-\sigma t_1} \cos(\omega_d t_1 + \phi)}{e^{-\sigma(t_1+T)} \cos[\omega_d (t_1 + T) + \phi]} = e^{\sigma T} = e^{2\pi\zeta/\sqrt{1-\zeta^2}} \quad (8.82)$$

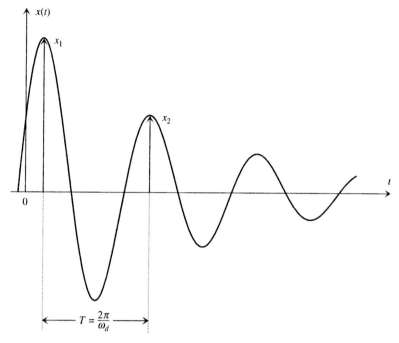

FIGURE 8.22 Free vibration response of an underdamped, second-order system.

since

$$t_2 = t_1 + T$$

$$T = \frac{2\pi}{\omega_d} \implies \omega_d T = 2\pi$$

$$\cos[\omega_d(t_1 + T) + \phi] = \cos(\omega_d t_1 + \phi)$$

$$\sigma T = \zeta \omega_n \frac{2\pi}{\omega_n \sqrt{1-\zeta^2}} = \frac{2\pi\zeta}{\sqrt{1-\zeta^2}} \quad (8.83)$$

Thus,

$$\delta = \ln \frac{x_1}{x_2} = \frac{2\pi\zeta}{\sqrt{1-\zeta^2}} \quad (8.84)$$

where δ is the **log decrement.** After some algebra, we obtain the damping ratio ζ as a function of the log decrement δ as

$$\zeta = \frac{\delta}{\sqrt{4\pi^2 + \delta^2}} \quad (8.85)$$

If the damping is small, the log decrement is also small; therefore, we have the following approximation:

$$\zeta \approx \frac{\delta}{2\pi} \tag{8.86}$$

In certain applications, it may be desirable to obtain the damping ratio by measuring the values of nonsuccessive displacements. Since

$$\frac{x_1}{x_2} = \frac{x_2}{x_3} = \cdots = \frac{x_n}{x_{n+1}} = e^{2\pi\zeta/\sqrt{1-\zeta^2}} \tag{8.87}$$

$$\frac{x_1}{x_{n+1}} = \frac{x_1}{x_2}\frac{x_2}{x_3}\cdots\frac{x_n}{x_{n+1}} = \left(e^{2\pi\zeta/\sqrt{1-\zeta^2}}\right)^n = e^{2\pi\zeta n/\sqrt{1-\zeta^2}} \tag{8.88}$$

$$\delta = \ln\frac{x_1}{x_2} = \ln\frac{x_2}{x_3} = \cdots = \ln\frac{x_n}{x_{n+1}} = \frac{2\pi\zeta}{\sqrt{1-\zeta^2}} \tag{8.89}$$

Thus,

$$\ln\frac{x_1}{x_{n+1}} = n\delta = \frac{2\pi\zeta n}{\sqrt{1-\zeta^2}} \tag{8.90}$$

Therefore,

$$\delta = \frac{1}{n}\ln\frac{x_1}{x_{n+1}} \tag{8.91}$$

EXAMPLE 8.11. Consider the free vibration response of an SDOF system (Fig. 8.23). Of the original value, the amplitude drops to 50 percent and 25 percent after one and two periods, respectively.

(a) Determine the log decrement.
(b) Determine the exact value of the damping ratio (using the exact formula).
(c) Determine the approximate value of the damping ratio (using the approximate formula), then compute the error in percent.

Solution

(a) The log decrement is given as

$$\delta = \ln\frac{x_1}{x_2} = \ln\frac{x_1}{0.5x_1} = 0.69315$$

or

$$\delta = \frac{1}{2}\ln\frac{x_1}{x_{2+1}} = \frac{1}{2}\ln\frac{x_1}{x_3} = \frac{1}{2}\ln\frac{x_1}{0.25x_1} = 0.69315$$

(b) The exact value of damping ratio is

$$\zeta = \frac{\delta}{\sqrt{4\pi^2 + \delta^2}} = \frac{0.69315}{\sqrt{4\pi^2 + 0.69315^2}} = 0.10965$$

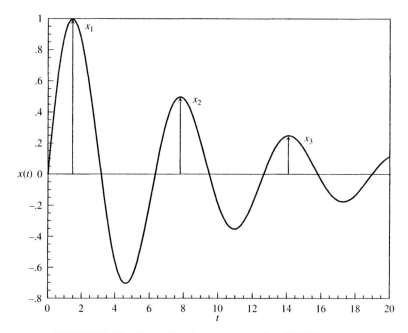

FIGURE 8.23 Free vibration response of an SDOF system.

(c) Since the damping is relatively small, the approximation may be used,

$$\zeta \approx \frac{\delta}{2\pi} = \frac{0.69315}{2\pi} = 0.11032$$

and the error is +0.6 percent.

PROBLEM SET 8.4

8.8. The free vibration response of a structure is in the form

$$x(t) = Ae^{-\sigma t}\sin(\omega_d t + \theta)$$

Show that the damping ratio is given as

$$\zeta = \frac{\delta}{\sqrt{4\pi^2 + \delta^2}}$$

where δ is the log decrement.

8.9. Consider the free vibration response of an SDOF system Fig. P8.9. Of the original value, the amplitude drops to 45 percent and 20.25 percent after one and two periods, respectively. Determine: (a) the log decrement, (b) the exact value of the damping ratio (using the exact formula), and (c) the approximate value of the damping ratio (using the approximate formula), and then compute the error in percent.

CHAPTER 8: Introduction to Vibrations   441

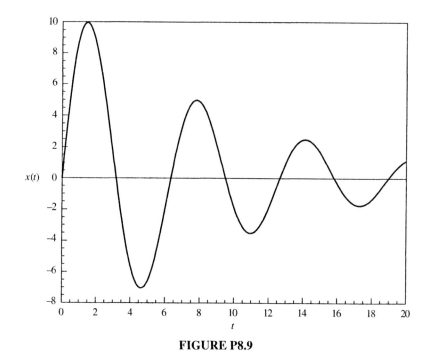

FIGURE P8.9

8.5 BEAT PHENOMENON

When the values of the two natural frequencies of a vibration system are nearly equal, the **beat phenomenon** occurs. The resulting free response is an oscillating signal of a certain frequency, whose amplitude itself is another sinusoidal signal of a smaller frequency (larger period). As an illustrative example, consider the double pendulum system shown in Fig. 8.24. This section shows the beat phenomenon associated with the system when the two modes are weakly coupled. The differential equations were derived as

$$\begin{bmatrix} mL^2 & 0 \\ 0 & mL^2 \end{bmatrix} \begin{Bmatrix} \ddot{\phi}_1 \\ \ddot{\phi}_2 \end{Bmatrix} + \begin{bmatrix} mgL + kl^2 & -kl^2 \\ -kl^2 & mgL + kl^2 \end{bmatrix} \begin{Bmatrix} \phi_1 \\ \phi_2 \end{Bmatrix} = \begin{Bmatrix} 0 \\ 0 \end{Bmatrix} \quad (8.92)$$

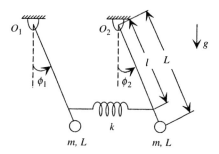

FIGURE 8.24 Pendulum system.

The natural frequencies were obtained as

$$\omega_1 = \sqrt{\frac{g}{L}} \qquad \omega_2 = \sqrt{\frac{g}{L} + \frac{2kl^2}{mL^2}} \qquad (8.93)$$

Suppose the initial conditions are

$$\phi_1(0^-) = \phi_o \qquad \phi_2(0^-) = 0 \qquad \dot{\phi}_1(0^-) = \dot{\phi}_2(0^-) = 0 \qquad (8.94)$$

the solution can be shown as

$$\phi_1(t) = \underbrace{\phi_o \cos\frac{\omega_2 - \omega_1}{2}t}_{\text{amplitude}} \cos\frac{\omega_2 + \omega_1}{2}t = A_1(t)\cos\frac{\omega_2 + \omega_1}{2}t \qquad (8.95a)$$

$$\theta_2(t) = \underbrace{\phi_o \sin\frac{\omega_2 - \omega_1}{2}t}_{\text{amplitude}} \sin\frac{\omega_2 + \omega_1}{2}t = A_2(t)\sin\frac{\omega_2 + \omega_1}{2}t \qquad (8.95b)$$

The amplitudes

$$A_1(t) = \phi_o \cos\frac{\omega_2 - \omega_1}{2}t \qquad (8.96a)$$

$$A_2(t) = \phi_o \sin\frac{\omega_2 - \omega_1}{2}t \qquad (8.96b)$$

serve as the *envelopes* with a relatively slower frequency,

$$\frac{\omega_2 - \omega_1}{2} \ll \frac{\omega_2 + \omega_1}{2}$$

since $\omega_1 \approx \omega_2$.

Notice that the beat phenomenon also occurs in other engineering disciplines such as in electrical engineering communications. These envelopes, $A_1(t)$ and $A_2(t)$, are called the *modulating signals*, and the terms

$$\cos\frac{\omega_2 + \omega_1}{2}t \qquad \text{and} \qquad \sin\frac{\omega_2 + \omega_1}{2}t$$

are called the *carrier signals*.

Finally, the normalized (dimensionless) responses are

$$\frac{\phi_1(t)}{\phi_o} = \underbrace{\cos\frac{\omega_2 - \omega_1}{2}t}_{\text{amplitude}}\cos\frac{\omega_2 + \omega_1}{2}t \qquad (8.97a)$$

$$\frac{\phi_2(t)}{\phi_o} = \underbrace{\sin\frac{\omega_2 - \omega_1}{2}t}_{\text{amplitude}}\sin\frac{\omega_2 + \omega_1}{2}t \qquad (8.97b)$$

Under certain conditions, a simple approximation may be made for the second natural frequency. From the binomial expansion (Appendix C),

$$(1 + \varepsilon)^n \approx 1 + n\varepsilon \qquad \text{if } \varepsilon \ll 1$$

we have

$$\omega_2 = \sqrt{\frac{g}{L} + 2\frac{kl^2}{mL^2}} = \sqrt{\frac{g}{L}\left(1 + 2\frac{kl^2/(mL)}{g}\right)^{1/2}}$$

$$\approx \sqrt{\frac{g}{L}}\left(1 + \frac{kl^2}{mgL}\right) \quad \text{if } \frac{2kl^2}{mgL} \ll 1 \tag{8.98}$$

Thus,

$$\omega_2 - \omega_1 \approx \sqrt{\frac{g}{L}} \cdot \frac{kl^2}{mgL} = \frac{kl^2}{m\sqrt{gL^3}} \tag{8.99}$$

$$\frac{\omega_2 - \omega_1}{2} \approx \frac{1}{2}\sqrt{\frac{g}{L}}\frac{kl^2}{mgL} \tag{8.100}$$

$$\frac{\omega_2 + \omega_1}{2} \approx \frac{1}{2}\left[2\sqrt{\frac{g}{L}} + \sqrt{\frac{g}{L}}\frac{kl^2}{mgL}\right] = \sqrt{\frac{g}{L}} + \frac{1}{2}\sqrt{\frac{g}{L}}\frac{kl^2}{mgL} \tag{8.101}$$

For plotting purposes, the following numerical values are used:

$$\frac{g}{L} = 100 \text{ (rad/s)}^2 \qquad \sqrt{\frac{g}{L}\frac{ka^2}{mgL}} = 1 \text{ rad/s} \implies \frac{ka^2}{mgL} = 0.1$$

Thus,

$$\omega_1 = \sqrt{\frac{g}{L}} = 10 \text{ rad/s}$$

$$(\omega_2)_{\text{approx}} \approx \sqrt{\frac{g}{L}}\left(1 + \frac{kl^2}{mgL}\right) = 11 \text{ rad/s}$$

$$(\omega_2)_{\text{exact}} = \sqrt{\frac{g}{L}\left(1 + 2\frac{kl^2}{mgL}\right)^{1/2}} = 10.9545 \text{ rad/s}$$

Notice that the approximated value for the second natural frequency, from the binomial expansion, is higher than the exact one.

Finally, with the numerical values, the exact normalized responses become

$$\frac{\phi_1(t)}{\phi_o} = \underbrace{\cos 0.4773t}_{\text{amplitude}} \cos 10.4773t$$

$$\frac{\phi_2(t)}{\phi_o} = \underbrace{\sin 0.4773t}_{\text{amplitude}} \sin 10.4773t$$

and the approximated normalized responses are

$$\frac{\phi_1(t)}{\phi_o} \approx \underbrace{\cos 0.5t}_{\text{amplitude}} \cos 10.5t$$

$$\frac{\phi_2(t)}{\phi_o} \approx \underbrace{\sin 0.5t}_{\text{amplitude}} \sin 10.5t$$

444 Dynamic Systems: Modeling and Analysis

The approximated responses are shown in Fig. 8.25, though the exact responses can be plotted with the same ease.

It is interesting to see how the solution can be obtained using the Laplace transform method. The exact result will be shown (Fig. 8.26) using the exact solution.

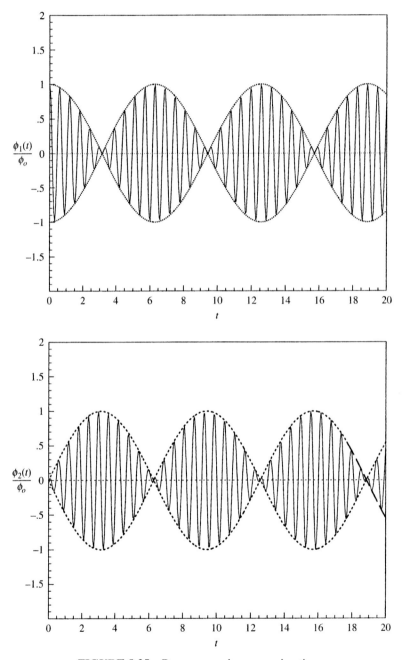

FIGURE 8.25 Responses using approximation.

CHAPTER 8: Introduction to Vibrations

Taking the Laplace transform

$$mL^2[s^2\Phi_1(s) - s\phi_1(0^-) - \dot{\phi}_1(0^-)] + (mgL + kl^2)\Phi_1(s) - kl^2\Phi_2(s) = 0 \quad (8.102a)$$

$$mL^2[s^2\Phi_2(s) - s\phi_2(0^-) - \dot{\phi}_2(0^-)] - kl^2\Phi_1(s) + (mgL + kl^2)\Phi_2(s) = 0 \quad (8.102b)$$

After applying the same initial conditions, the equations are expressed in the second-order matrix form as

$$\begin{bmatrix} mL^2s + mgL + kl^2 & -kl^2 \\ -kl^2 & mL^2s + mgL + kl^2 \end{bmatrix} \begin{Bmatrix} \Phi_1(s) \\ \Phi_2(s) \end{Bmatrix} = \begin{Bmatrix} mL^2 s\phi_o \\ 0 \end{Bmatrix} \quad (8.103)$$

Thus,

$$\Phi_1(s) = \frac{(mL^2s + mgL + kl^2)mL^2 s\phi_o}{\Delta(s)} \quad (8.104a)$$

$$\Phi_2(s) = \frac{kl^2 mL^2 s\phi_o}{\Delta(s)} \quad (8.104b)$$

where

$$\Delta(s) = (mL^2s + mgL + kl^2)^2 - (kl^2)^2$$

$$\Delta(s) = mL^2\left[mL^2 s^4 + 2(mgL + kl^2) + \frac{g}{L}(mgL + 2kl^2)\right] \quad (8.105)$$

Thus,

$$\Phi_1(s) = \frac{(mL^2s + mgL + kl^2)s\phi_o}{mL^2 s^4 + 2(mgL + kl^2) + (g/L)(mgL + 2kl^2)} \quad (8.106a)$$

$$\Phi_2(s) = \frac{kl^2 s\phi_o}{mL^2 s^4 + 2(mgL + kl^2) + (g/L)(mgL + 2kl^2)} \quad (8.106b)$$

Dividing the numerators and denominators by mgL, then repeating but multiplying by g/L yields

$$\Phi_1(s) = \frac{[s^2 + g/L(1 + kl^2/mgL)]s}{s^4 + 2g/L(1 + (kl^2/mgL))s^2 + (g/L)^2(1 + 2(kl^2/mgL))} \cdot \phi_o \quad (8.107a)$$

$$\Phi_2(s) = \frac{(g/L)(kl^2/mgL)s}{s^4 + 2(g/L)[1 + (kl^2/mgL)]s^2 + (g/L)^2[1 + 2(kl^2/mgL)]} \cdot \phi_o \quad (8.107b)$$

Thus,

$$\frac{\phi_1(t)}{\phi_o} = L^{-1}\left\{\frac{\Phi_1(s)}{\phi_o}\right\}$$

$$= L^{-1}\left\{\frac{s^3 + (g/L)[1 + (kl^2/mgL)]s}{s^4 + 2(g/L)[1 + (kl^2/mgL)]s^2 + (g/L)^2[1 + 2(kl^2/mgL)]} \cdot 1\right\} \quad (8.108a)$$

$$\frac{\phi_2(t)}{\phi_o} = L^{-1}\left\{\frac{(g/L)(kl^2/mgL)s}{s^4 + 2(g/L)[1 + (kl^2/mgL)]s^2 + (g/L)^2[1 + 2(kl^2/mgL)]} \cdot 1\right\} \quad (8.108b)$$

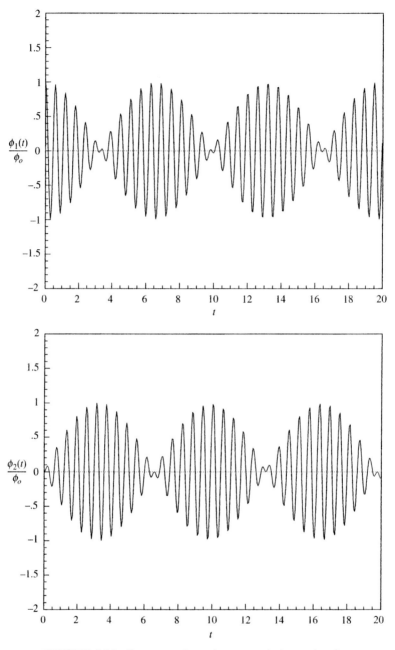

FIGURE 8.26 Responses from the exact solution, using the Laplace transform.

After substituting the numerical values, we obtain

$$\frac{\phi_1(t)}{\phi_o} = L^{-1}\left\{\frac{s^3 + 110s}{s^4 + 220s^2 + 12000} \cdot 1\right\}$$

$$\frac{\phi_2(t)}{\phi_o} = L^{-1}\left\{\frac{10s}{s^4 + 220s^2 + 12000} \cdot 1\right\}$$

It is interesting to look at the poles of this system whose characteristic equation is

$$s^4 + 220s^2 + 12000 = 0$$

The poles of this fourth-order system are solved as

$$s_{1,2} = \pm j10.0 \qquad s_{3,4} = \pm j10.9545$$

Thus the exact natural frequencies are

$$\omega_1 = 10.0 \qquad \omega_2 = 10.9545$$

which are the same as obtained previously. The exact responses are shown in Fig. 8.26.

PROBLEM SET 8.5

8.10. Consider the double pendulum system shown in Fig. P8.10. The differential equations are given as

$$\begin{bmatrix} mL^2 & 0 \\ 0 & mL^2 \end{bmatrix}\begin{Bmatrix} \ddot{\theta}_1 \\ \ddot{\theta}_2 \end{Bmatrix} + \begin{bmatrix} mgL + ka^2 & -ka^2 \\ -ka^2 & mgL + ka^2 \end{bmatrix}\begin{Bmatrix} \theta_1 \\ \theta_2 \end{Bmatrix} = \begin{Bmatrix} 0 \\ 0 \end{Bmatrix}$$

The natural frequencies are obtained as

$$\omega_1 = \sqrt{\frac{g}{L}} \qquad \omega_2 = \sqrt{\frac{g}{L} + \frac{2ka^2}{mL^2}}$$

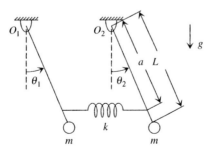

FIGURE P8.10 Pendulum system.

(a) Determine the exact system responses, $\theta_1(t)$ and $\theta_2(t)$, for the following initial conditions:

$$\theta_1(0^-) = 0 \text{ rad}, \qquad \theta_2(0^-) = \theta_o \text{ rad}, \qquad \dot{\theta}_1(0^-) = \dot{\theta}_2(0^-) = 0 \text{ rad/sec}$$

Use the Laplace transform method, and give the final answers in the *form* of the inverse Laplace transform. Provide the results in general terms (symbolic) first, then apply the following numerical values:

$$\frac{g}{L} = 100 \qquad \sqrt{\frac{g}{L}\frac{ka^2}{mgL}} = 1$$

(b) Determine the poles and the natural frequencies.
(c) Repeat part (a) for the following initial conditions:

$$\theta_1(0^-) = \theta_2(0^-) = 0 \text{ rad}, \qquad \dot{\theta}_1(0^-) = \omega_o = 1.0 \text{ rad/sec}, \qquad \dot{\theta}_2(0^-) = 0 \text{ rad/sec}$$

8.11. For each of the following cases, obtain the time-response plots for $\theta_1(t)$ and $\theta_2(t)$ by using any computer or software. The simulation time is suggested as from $t = 0$ to 20 s.

(a)
$$\frac{\theta_1(t)}{\theta_o} = L^{-1}\left\{\frac{10s}{s^4 + 220s^2 + 12000} \cdot 1\right\}$$

$$\frac{\theta_2(t)}{\theta_o} = L^{-1}\left\{\frac{s^3 + 110s}{s^4 + 220s^2 + 12000} \cdot 1\right\}$$

(b)
$$\frac{\theta_1(t)}{\theta_o} = \underbrace{\cos 0.4773t}_{\text{amplitude}} \cos 10.4773t$$

$$\frac{\theta_2(t)}{\theta_o} = \underbrace{\sin 0.4773t}_{\text{amplitude}} \sin 10.4773t$$

8.6 FREQUENCY RESPONSE OF VIBRATION SYSTEMS

The frequency response is an important subject in vibrations. When a sinusoidal forcing function is applied to a system that has very little damping, it may vibrate excessively if the forcing frequency is in the neighborhood of the natural frequency of the system. Sinusoidal input is also called *simple harmonic excitation*. This type of input is fundamental because it can be extended to periodic excitation using Fourier series analysis. Many important vibration systems are presented in this section, including simple mechanical systems, vibration isolation, systems with support motion, and rotating unbalanced mass.

Simple Mechanical System

Let us consider the simple system shown in Fig. 8.27, which is a single-degree-of-freedom system. The system consists of a block of mass m connected to a spring

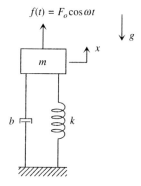

FIGURE 8.27 Simple system.

of stiffness k and a viscous damper of damping coefficient b. A sinusoidal forcing function $f(t)$ is applied to the block.

This system is a simplified model of a supporting structure on which an engine is mounted. Since the structure is relatively light compared to the engine, it is modeled as massless. The structure is also modeled as a combination of a translational spring and viscous damper in parallel. Although structural damping is the actual damping, the use of a viscous damping model is justified for a simple model. If the experiment is performed and the free response shows an exponential decayed envelope, then the model of viscous damper is valid.

The differential equation of motion is given readily as

$$m\ddot{x} + b\dot{x} + kx = f(t)$$

Note that the gravity, g, does not enter into the equation because the displacement x is measured from the static equilibrium position.

In terms of the natural frequency and the damping ratio, the equation becomes

$$\ddot{x} + 2\zeta\omega_n\dot{x} + \omega_n^2 x = \frac{f(t)}{m}$$

Note that the parameter m still appears in the equation, which is typical of the mass-spring-damper system. We have the usual definitions,

$$\omega_n = \sqrt{\frac{k}{m}}, \quad \zeta = \frac{b}{2\sqrt{km}}, \quad \frac{b}{m} = 2\zeta\omega_n$$

The TF (transfer function) is given as

$$G(s) = \frac{X(s)}{F(s)} = \frac{1/m}{s^2 + 2\zeta\omega_n s + \omega_n^2} \qquad (8.109)$$

Thus, the FRF (frequency response function) is

$$G(j\omega) = \frac{(1/m)(1/\omega_n^2)}{\omega_n^2/\omega_n^2 - \omega^2/\omega_n^2 + j2\zeta\omega_n\omega/\omega_n^2} = \frac{1/k}{1 - (\omega/\omega_n)^2 + j2\zeta\omega/\omega_n} \qquad (8.110)$$

Since the applied force, a sinusoidal function, is given as

$$f(t) = F_o \cos \omega t$$

the steady-state response is obtained as

$$x_{ss}(t; \omega) = \underbrace{F_o |G(j\omega)|}_{|x_{ss}|} \cos(\omega t + \phi) \tag{8.111}$$

where

$$|G(j\omega)| = \frac{1/k}{\sqrt{[1 - (\omega/\omega_n)^2]^2 + [2\zeta\omega/\omega_n]^2}} \tag{8.112}$$

$$\phi = -\tan^{-1} \frac{2\zeta\omega/\omega_n}{1 - (\omega/\omega_n)^2} \tag{8.113}$$

Thus the normalized magnitude of the response is

$$\frac{|x_{ss}|}{F_o/k} = \frac{1}{\sqrt{[1 - (\omega/\omega_n)^2]^2 + [2\zeta\omega/\omega_n]^2}} \tag{8.114}$$

Figure 8.28 shows plots of $|x_{ss}|/(F_o/k)$ versus ω/ω_n for various values of damping ratio ζ. If a curve of Fig. 8.28 has a peak value, this peak is called the *resonant peak* M_r and the corresponding frequency is called the *resonant frequency* ω_r, which is given as

$$\omega_r = \omega_n \sqrt{1 - 2\zeta^2} \qquad (0 \le \zeta \le 0.707) \tag{8.115}$$

or

$$\frac{\omega_r}{\omega_n} = \sqrt{1 - 2\zeta^2} \tag{8.116}$$

From this equation, we can see that when the damping ratio ζ is small, the resonant frequency ω_r is approximately the same as the natural frequency ω_n,

$$\omega_r \approx \omega_n \qquad (\zeta \approx 0)$$

However, for a larger value of ζ (but still smaller than 0.707), the resonant peak is shifted to the left of the natural frequency ω_n.

The value of the resonant peak M_r is

$$M_r = \frac{1}{2\zeta \sqrt{1 - \zeta^2}} \qquad (\zeta < 0.707) \tag{8.117}$$

$$M_r \approx \frac{1}{2\zeta} \qquad (\zeta \text{ is very small}) \tag{8.118}$$

$$M_r = 1 \qquad (\zeta = 0.707) \tag{8.119}$$

Note that the resonant peak does not occur when $\zeta > 0.707$.

In designing mechanical components or systems, resonant frequencies must be avoided. The system will vibrate excessively if the forcing frequency ω matches one of the natural frequencies of the system and if the inherent damping is small. Consequently, the system will fail because of fatigue.

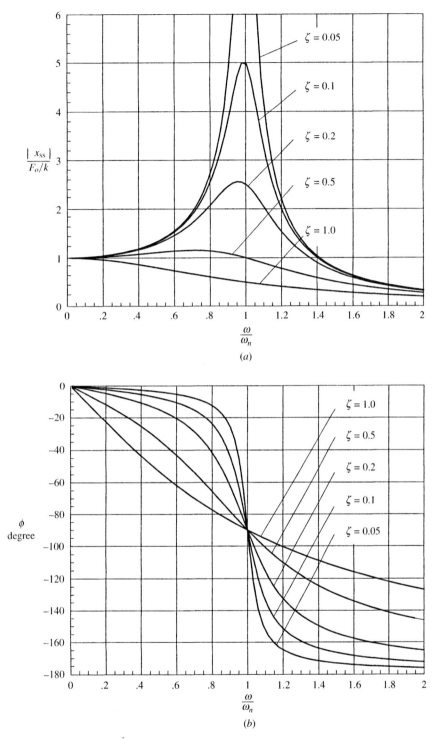

FIGURE 8.28 Normalized frequency response, linear scale: (*a*) magnitude, (*b*) phase.

Vibration Isolation

Consider the system shown in Fig. 8.29. The force applied on the mass m is $f(t) = F_o \cos \omega t$. As a result, the foundation experiences a transmitted force through the damper b and the spring k. The system should be designed such that this transmitted force is minimized.

The force transmitted to the foundation through the damper and the spring is

$$f_t = b\dot{x} + kx \tag{8.120}$$

After Laplace transformation,

$$F_t(s) = (bs + k)X(s) \tag{8.121}$$

Thus the transfer function is

$$G(s) = \frac{F_t(s)}{F(s)} = (bs + k)\frac{X(s)}{F(s)} = \frac{(bs + k)(1/m)}{s^2 + 2\zeta\omega_n s + \omega_n^2} = \frac{2\zeta\omega_n s + \omega_n^2}{s^2 + 2\zeta\omega_n s + \omega_n^2} \tag{8.122}$$

The FRF is

$$G(j\omega) = \frac{1 + j2\zeta\omega/\omega_n}{1 - (\omega/\omega_n)^2 + j2\zeta\omega/\omega_n} \tag{8.123}$$

then the steady-state response is

$$f_{t,ss}(t;\omega) = \underbrace{F_o|G(j\omega)|}_{|f_{t,ss}|} \cos(\omega t + \phi) \tag{8.124}$$

where

$$|G(j\omega)| = \frac{\sqrt{1 + (2\zeta\omega/\omega_n)^2}}{\sqrt{[1 - (\omega/\omega_n)^2]^2 + [2\zeta\omega/\omega_n]^2}} \tag{8.125}$$

$$\phi = \tan^{-1} 2\zeta\omega/\omega_n - \tan^{-1} \frac{2\zeta\omega/\omega_n}{1 - (\omega/\omega_n)^2} \tag{8.126}$$

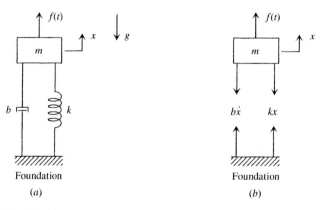

FIGURE 8.29 Vibration isolation: (a) system, (b) free-body diagram showing transmitted force.

Thus the normalized magnitude of the transmitted force is

$$\frac{|f_{t,ss}|}{F_o} = |G(j\omega)| = \frac{\sqrt{1 + (2\zeta\omega/\omega_n)^2}}{\sqrt{[1 - (\omega/\omega_n)^2]^2 + [2\zeta\omega/\omega_n]^2}} \tag{8.127}$$

The normalized frequency response is shown in Fig. 8.30 using a linear scale and in Fig. 8.31 as a Bode plot. For $\omega/\omega_n > \sqrt{2}$, the force transmitted decreases with increasing forcing frequency ω. Thus, the system should be designed such that its natural frequency ω_n is very small compared to the forcing frequency ω. For a fixed mass, a soft spring should be selected since $\omega_n = \sqrt{k/m}$. However, a softer spring will yield a higher static deflection. This is a trade-off in designing a vibration isolation system. Note that a relatively small value of damping is actually better as long as $\omega > \omega_n \sqrt{2}$ and the system is driven fast enough to avoid resonance. It is also interesting to see that the normalized magnitude is equal to 1 when $\omega = \omega_n \sqrt{2}$, for all values of damping.

System with Support Motion

Many applications in vibrations involve systems where the input function is displacement or acceleration. An example of such a system is an automobile suspension system. Figure 8.32 shows a simple model. The system consists of a mass m connected by a damper b and a spring k, in parallel. (The elasticity and damping of the tires are not modeled.) The disturbance input $y(t)$ represents the displacement of the tire due to the surface of the road, $y(t) = Y_o \cos \omega t$. The differential equation can be obtained readily as

$$m\ddot{x} + b\dot{x} + kx = c\dot{y} + ky \tag{8.128}$$

or in terms of the natural frequency and damping ratio,

$$\ddot{x} + 2\zeta\omega_n \dot{x} + \omega_n^2 x = 2\zeta\omega_n \dot{y} + \omega_n^2 y \tag{8.129}$$

Thus the TF is

$$G(s) = \frac{X(s)}{Y(s)} = \frac{2\zeta\omega_n s + \omega_n^2}{s^2 + 2\zeta\omega_n s + \omega_n^2} \tag{8.130}$$

and the FRF is

$$G(j\omega) = \frac{1 + j2\zeta\omega/\omega_n}{1 - (\omega/\omega_n)^2 + j2\zeta\omega/\omega_n} \tag{8.131}$$

which are identical with those given previously in this section (Vibration Isolation), thus the frequency response is the same (Figs. 8.30 and 8.31). The steady-state response is

$$x_{ss}(t;\omega) = \underbrace{Y_o|G(j\omega)|}_{|x_{ss}|} \cos(\omega t + \phi) \tag{8.132}$$

Thus the normalized magnitude of the steady-state response is

$$\frac{|x_{ss}|}{Y_o} = |G(j\omega)| = \frac{\sqrt{1 + (2\zeta\omega/\omega_n)^2}}{\sqrt{[1 - (\omega/\omega_n)^2]^2 + [2\zeta\omega/\omega_n]^2}} \tag{8.133}$$

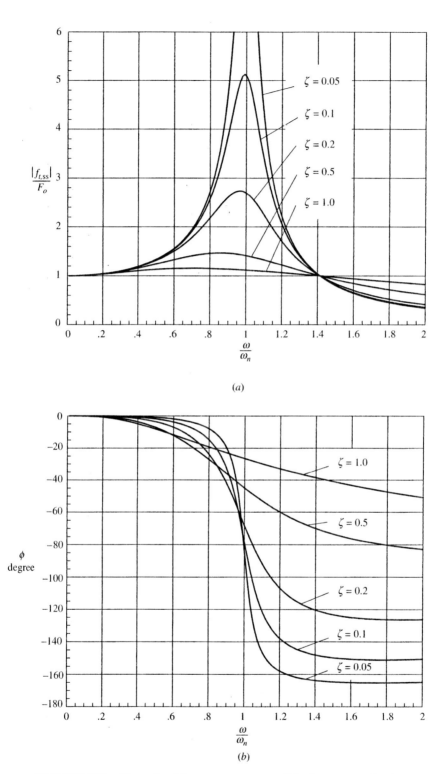

FIGURE 8.30 Normalized frequency response of the force-transmitted, linear scale: (*a*) magnitude, (*b*) phase.

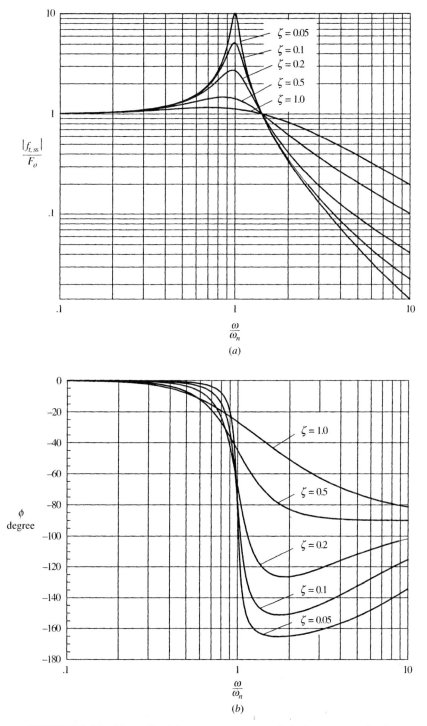

FIGURE 8.31 Normalized frequency response of the force-transmitted, Bode plot: (*a*) magnitude, (*b*) phase.

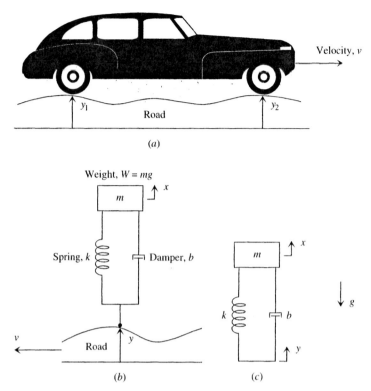

FIGURE 8.32 Automobile suspension system: (*a*) physical system, (*b*) simplified system, (*c*) simple model.

Rotating Unbalanced Mass

Consider the system of rotating unbalanced mass shown in Fig. 8.33. The system consists of a motor of mass M supported on a foundation by a damper b and a spring k in parallel. The rotor of the motor is not perfectly dynamically balanced (in rotation). The distance between the unbalanced mass m and the axis of rotation is e. The motion of this unbalanced mass creates an unbalanced force acting on the system. Applying Newton's second law for the system in the x direction,

$$+\uparrow \sum_{i=1}^{2} F_{ix} = \sum_{j=1}^{2} m_j a_{cjx}$$

$$-b\dot{x} - kx = M\ddot{x} + m(\ddot{x} + e\ddot{\theta}\cos\theta - e\dot{\theta}^2 \sin\theta) \qquad (8.134)$$

The equation is rearranged as

$$(M + m)\ddot{x} + me\ddot{\theta}\cos\theta - me\dot{\theta}^2 \sin\theta + b\dot{x} + kx = 0 \qquad (8.135)$$

If

$$\dot{\theta} = \omega = \text{constant}$$

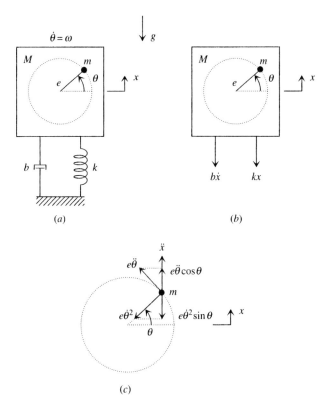

FIGURE 8.33 Vibration isolation: (a) system, (b) free-body diagram, (c) kinematics for m.

then
$$\ddot{\theta} = 0 \qquad \theta = \omega t$$
and the equation becomes
$$(M + m)\ddot{x} + b\dot{x} + kx = me\omega^2 \sin \omega t \qquad (8.136)$$
Note that the equation is equivalent to
$$(M + m)\ddot{x} + b\dot{x} + kx = me\omega^2 \sin \omega t = me(-\ddot{y}) \qquad (8.137)$$
where the applied forcing function is
$$y(t) = 1 \cdot \sin \omega t$$
Thus,
$$(-\ddot{y}) = \omega^2 \sin \omega t$$
The transfer function is
$$G(s) = \frac{X(s)}{Y(s)} = \frac{-me/(M + m)s^2}{s^2 + 2\zeta\omega_n s + \omega_n^2} \qquad (8.138)$$

and the FRF is

$$G(j\omega) = \frac{[me/(M+m)](\omega/\omega_n)^2}{1 - (\omega/\omega_n)^2 + j2\zeta\omega/\omega_n} \qquad (8.139)$$

where

$$\omega_n = \sqrt{\frac{k}{M+m}} \qquad (8.140)$$

$$\zeta = \frac{b}{2\sqrt{k(M+m)}} \qquad (8.141)$$

Thus, the steady-state response is

$$x_{ss}(t;\omega) = 1 \cdot \frac{[me/(M+m)](\omega/\omega_n)^2}{\sqrt{[1-(\omega/\omega_n)^2]^2 + [2\zeta\omega/\omega_n]^2}} \sin(\omega t + \phi) \qquad (8.142)$$

where

$$\phi = -\tan^{-1} \frac{2\zeta\omega/\omega_n}{1-(\omega/\omega_n)^2}$$

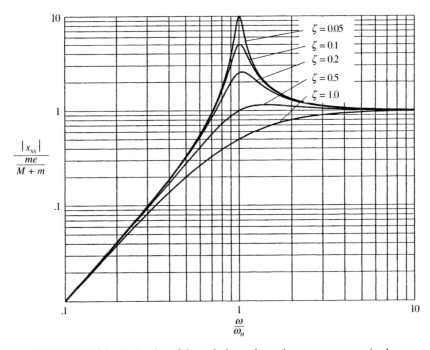

FIGURE 8.34 Bode plot of the unbalanced rotating system: magnitude.

Thus the normalized magnitude of the displacement is

$$\frac{|x_{ss}|}{me/(M+m)} = \frac{(\omega/\omega_n)^2}{\sqrt{[1-(\omega/\omega_n)^2]^2 + [2\zeta\omega/\omega_n]^2}} \quad (8.143)$$

The magnitude part of the Bode plot for the normalized frequency response is shown in Fig. 8.34. The phase part is the same as that of Fig. 8.28.

PROBLEM SET 8.6

8.12. Consider the simple system shown in Fig. P8.12, where the applied forcing function is $f(t) = F_0 \sin \omega t$. Obtain:
(a) the frequency-response function, and
(b) the steady-state response and the normalized magnitude of the displacement.

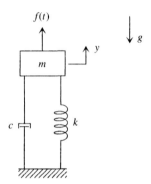

FIGURE P8.12 Simple system.

8.13. Consider the system shown in Fig. P8.13. The force applied on the mass M is $f(t) = F_0 \sin \omega t$. As a result, the foundation experiences a transmitted force through the damper

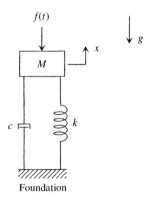

FIGURE P8.13 Vibration isolation system.

c and the spring k. The system should be designed such that this transmitted force is minimized. Obtain the steady-state response and the normalized magnitude of the transmitted force.

8.14. Many applications in vibrations involve systems where the input function is displacement (Fig. P8.14). An example is an automobile suspension system. The system consists of a mass m connected by a damper b and a spring k, in parallel. (The elasticity and damping of the tires are not modeled.) The disturbance input $y(t)$ represents the displacement of the tire due to the surface of the road, $y(t) = Y_0 \sin \omega t$. Determine the steady-state response and its normalized magnitude.

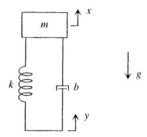

FIGURE P8.14 Simple model of an automobile suspension system.

8.15. Consider the system of rotating unbalanced mass shown in Fig. P8.15. The system consists of a motor of mass M supported on a foundation by a damper b and two springs k_1 and k_2, all in parallel. The rotor of the motor is not perfectly dynamically balanced (in rotation). The distance between the unbalanced mass m and the axis of rotation is e. The motion of this unbalanced mass creates an unbalanced force acting on the system.
(a) Derive the general nonlinear differential equation.
(b) Assuming $\dot{\theta} = \omega =$ constant, obtain the TF and the FRF.
(c) Obtain the steady-state response and its normalized magnitude.

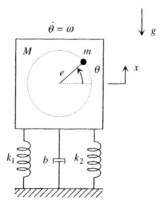

FIGURE P8.15 Vibration isolation system.

8.16. Consider the system of rotating unbalanced mass shown in Fig. P8.16. The system consists of a motor of mass M supported on a foundation by a damper b and a spring k in parallel. The rotor of the motor is not perfectly dynamically balanced (in rotation). The distance between the unbalanced mass m and the axis of rotation is e. The motion of this unbalanced mass creates an unbalanced force acting on the system. Derive the general nonlinear differential equation.

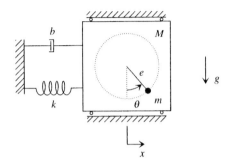

FIGURE P8.16 Vibration isolation system.

8.7 DAMPED VIBRATION ABSORBER

Figure 8.35 shows a **damped vibration absorber** (Den Hartog 1985) attached to a single-degree-of-freedom structure. It is also called a **dynamic vibration absorber**. The complete system is a fourth-order system, which may serve as an example of higher-order systems. The structure is of mass M and stiffness K, whereas the absorber is composed of a smaller mass m, a spring of stiffness k, and a dashpot (damper) of viscous damping coefficient c. The absorber is designed to provide a counter-balance force that acts against the disturbance force $f(t)$ so that the motion of the structure mass M can be minimized. This is the famous **passive vibration control** using a damped vibration absorber. The differential equations are given in the second-order matrix form as

$$\begin{bmatrix} M & 0 \\ 0 & m \end{bmatrix} \begin{Bmatrix} \ddot{x}_1 \\ \ddot{x}_2 \end{Bmatrix} + \begin{bmatrix} c & -c \\ -c & c \end{bmatrix} \begin{Bmatrix} \dot{x}_1 \\ \dot{x}_2 \end{Bmatrix} + \begin{bmatrix} K+k & -k \\ -k & k \end{bmatrix} \begin{Bmatrix} x_1 \\ x_2 \end{Bmatrix} = \begin{Bmatrix} f \\ 0 \end{Bmatrix} \quad (8.144)$$

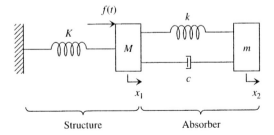

FIGURE 8.35 System with damped vibration absorber (dynamic vibration absorber).

Taking the Laplace transform, with zero initial conditions, the terms are collected as

$$\begin{bmatrix} Ms^2 + cs + K + k & -(cs + k) \\ -(cs + k) & ms^2 + cs + k \end{bmatrix} \begin{Bmatrix} X_1(s) \\ X_2(s) \end{Bmatrix} = \begin{Bmatrix} F(s) \\ 0 \end{Bmatrix} \quad (8.145)$$

Using Cramer's rule, we obtain

$$X_1(s) = \frac{ms^2 + cs + k}{\Delta(s)} F(s) \quad (8.146)$$

$$X_2(s) = \frac{cs + k}{\Delta(s)} F(s) \quad (8.147)$$

where

$$\Delta(s) = [Ms^2 + cs + K + k](ms^2 + cs + k) - (cs + k)^2$$

$$\Delta(s) = Mms^4 + (M + m)cs^3 + [Mk + m(K + k)]s^2 + cKs + Kk \quad (8.148)$$

Thus the transfer functions are

$$G_1(s) = \frac{X_1(s)}{F(s)}$$

$$= \frac{ms^2 + cs + k}{Mms^4 + (M + m)cs^3 + [Mk + m(K + k)]s^2 + cKs + Kk} \quad (8.149)$$

$$G_2(s) = \frac{X_2(s)}{F(s)}$$

$$= \frac{cs + k}{Mms^4 + (M + m)cs^3 + [Mk + m(K + k)]s^2 + cKs + Kk} \quad (8.150)$$

Frequency Response

The primary concern is to reduce the vibration of the structure mass M, but the system must be designed such that the absorber does not vibrate excessively. This is an important consideration in design since fatigue leads to failure. At this point, let us focus on the transfer function $G_1(s)$ whose FRF function is given as

$$G_1(j\omega) = \left. \frac{X_1(s)}{F(s)} \right|_{s=j\omega}$$

$$= \frac{k - m\omega^2 + jc\omega}{Kk + Mm\omega^4 - [Mk + m(K + k)]\omega^2 + jc\omega[K - (M + m)\omega^2]} \quad (8.151)$$

where the magnitude and phase are given, respectively, as

$$|G_1(j\omega)| = \frac{\sqrt{(k - m\omega^2)^2 + (c\omega)^2}}{\sqrt{\{Kk + Mm\omega^4 - [Mk + m(K + k)]\omega^2\}^2 + \{c\omega[K - (M + m)\omega^2]\}^2}} \quad (8.152)$$

$$\phi_1(\omega) = \angle G_1(j\omega)$$

$$= \tan^{-1}\frac{c\omega}{k - m\omega^2} - \tan^{-1}\frac{c\omega\left[K - (M+m)\omega^2\right]}{Kk + Mm\omega^4 - \left[Mk + m(K+k)\right]\omega^2} \quad (8.153)$$

Finally, the steady-state response for x_{1ss} is given as

$$x_{1ss}(t;\omega) = F_0|G_1(j\omega)|\cos(\omega t + \phi_1) \quad (8.154)$$

The magnitude of the response is

$$|x_{1ss}(t;\omega)| = F_0|G_1(j\omega)| \quad (8.155)$$

It is desired to express the result in terms of dimensionless groups. The normalized dimensionless displacement is

$$\frac{|x_{1ss}(t;\omega)|}{x_{st}} = \frac{|x_{1ss}(t;\omega)|}{F_0/K} = K|G_1(j\omega)| \quad (8.156)$$

Dropping the variable t, and changing ω to β, we have

$$\frac{|x_{1ss}(\beta)|}{x_{st}} = \frac{|x_{1ss}(\beta)|}{F_0/K} = K|G_1(j\beta)| \quad (8.157)$$

After performing a considerable amount of manipulation (given below), we obtain

$$KG_1(j\beta) = \frac{\alpha^2 - \beta^2 + j2\zeta\beta}{\alpha^2 + \beta^4 - \left[1 + (1+\mu)\alpha^2\right]\beta^2 + j2\zeta\beta\left[1 - (1+\mu)\beta^2\right]} \quad (8.158)$$

magnitude: $|KG_1(j\beta)| = K|G_1(j\beta)| =$

$$\frac{\sqrt{(\alpha^2 - \beta^2)^2 + (2\zeta\beta)^2}}{\sqrt{\left\{\alpha^2 + \beta^4 - \left[1 + (1+\mu)\alpha^2\right]\beta^2\right\}^2 + \left\{2\zeta\beta\left[1 - (1+\mu)\beta^2\right]\right\}^2}} \quad (8.159)$$

phase: $\phi_1(\beta) = \angle KG_1(j\beta) = \angle G_1(j\beta) =$

$$\tan^{-1}\frac{2\zeta\beta}{\alpha^2 - \beta^2} - \tan^{-1}\frac{2\zeta\beta\left[1 - (1+\mu)\beta^2\right]}{\alpha^2 + \beta^4 - \left[1 + (1+\mu)\alpha^2\right]\beta^2} \quad (8.160)$$

where the dimensionless groups are defined as follows:

mass ratio: $\quad\quad\quad\quad\quad\quad\quad\quad \mu = \dfrac{m}{M} \quad (8.161)$

frequency ratios: $\quad\quad\quad\quad \alpha = \dfrac{\omega_a}{\omega_n} = \dfrac{\sqrt{k/m}}{\sqrt{K/M}} \quad (8.162)$

$$\beta = \frac{\omega}{\omega_n} = \frac{\omega}{\sqrt{K/M}} \tag{8.163}$$

damping ratio:
$$\zeta = \frac{c}{2m\omega_n} = \frac{c}{2m\sqrt{K/M}} \tag{8.164}$$

displacement ratio:
$$\frac{|x_{1ss}(t;\omega)|}{x_{st}} = \frac{|x_{1ss}(t;\omega)|}{F_0/K} \tag{8.165}$$

Note that ω_n and ω_a denote the natural frequencies of the structure and the absorber, respectively.

We start out with

$$G_1(j\omega) = \frac{X_1(s)}{F(s)}\bigg|_{s=j\omega}$$

$$= \frac{k - m\omega^2 + jc\omega}{Kk + Mm\omega^4 - [Mk + m(K+k)]\omega^2 + jc\omega[K - (M+m)\omega^2]} = \frac{N}{D} \tag{8.166}$$

Dividing the numerator N by mK/M, we have

$$\frac{N}{mK/M} = \frac{k - m\omega^2 + jc\omega}{mK/M}$$

$$= \frac{k/m}{K/M} - \frac{\omega^2}{K/M} + j\frac{c}{m\sqrt{K/M}}\frac{\omega}{\sqrt{K/M}} = \frac{\omega_a^2}{\omega_n^2} - \frac{\omega^2}{\omega_n^2} + j2\zeta\frac{\omega}{\omega_n}$$

$$\frac{N}{mK/M} = \alpha^2 - \beta^2 + j2\zeta\beta$$

Similarly, dividing the denominator D by $m(K/M)K$, we have

$$\frac{D}{m(K/M)K} = \frac{Kk + Mm\omega^4 - [Mk + m(K+k)]\omega^2}{m(K/M)K}$$

$$+ j\frac{c\omega}{m(K/M)}\left[1 - \left(\frac{M}{K} + \frac{m}{K}\right)\omega^2\right] \tag{8.167}$$

Since these terms are a bit more complicated, let us consider each term separately:

$$\frac{Kk}{m(K/M)K} = \frac{k/m}{K/M} = \frac{\omega_a^2}{\omega_n^2} = \alpha^2$$

$$\frac{Mm\omega^4}{m(K/M)K} = \frac{\omega^4}{(K/M)(K/M)} = \frac{\omega^4}{\omega_n^4} = \beta^4$$

$$\frac{Mk\omega^2}{m(K/M)K} = \frac{(k/m)\omega^2}{(K/M)(K/M)} = \frac{\omega_a^2\omega^2}{\omega_n^2\omega_n^2} = \alpha^2\beta^2$$

$$\frac{mK\omega^2}{m(K/M)K} = \frac{\omega^2}{K/M} = \frac{\omega^2}{\omega_n^2} = \beta^2$$

$$\frac{mk\omega^2}{m(K/M)K} = \frac{(m/M)(k/m)\omega^2}{(K/M)(K/M)} = \frac{\mu\omega_a^2\omega^2}{\omega_n^2\omega_n^2} = \mu\alpha^2\beta^2$$

$$\frac{c\omega}{mK/M} = \frac{c}{m\sqrt{K/M}}\frac{\omega}{\sqrt{K/M}} = 2\zeta\beta$$

$$\frac{M}{K}\omega^2 = \frac{\omega^2}{K/M} = \frac{\omega^2}{\omega_n^2} = \beta^2$$

$$\frac{m}{K}\omega^2 = \frac{m}{M}\frac{\omega^2}{K/M} = \mu\beta^2$$

Finally, the normalized FRF is expressed in terms of the dimensionless groups.
For *optimal response*, the following formulas are given (Den Hartog 1985):

Optimal tuning ratio $\quad\quad \alpha_{opt} = \dfrac{1}{1+\mu}$ (8.168)

Optimal damping ratio $\quad\quad \zeta_{opt} = \sqrt{\dfrac{3\mu}{8(1+\mu)^3}}$ (8.169)

For a selected mass ratio, μ, the optimal parameters, α_{opt} and ζ_{opt}, can be computed.
Thus, the optimal normalized FRF is

$$KG_1(j\beta)|_{opt} = \frac{\alpha_{opt}^2 - \beta^2 + j2\zeta_{opt}\beta}{\alpha_{opt}^2 + \beta^4 - [1+(1+\mu)\alpha_{opt}^2]\beta^2 + j2\zeta_{opt}\beta[1-(1+\mu)\beta^2]}$$

(8.170)

Magnitude:
$$|KG_1(j\beta)_{opt}| = \frac{\sqrt{(\alpha_{opt}^2 - \beta^2)^2 + (2\zeta_{opt}\beta)^2}}{\sqrt{\{\alpha_{opt}^2 + \beta^4 - [1+(1+\mu)\alpha_{opt}^2]\beta^2\}^2 + \{2\zeta_{opt}\beta[1-(1+\mu)\beta^2]\}^2}}$$

(8.171)

A family of curves with different values of the mass ratio μ for the magnitude can be generated.

In the design of practical systems, the absorber mass m should be relatively small compared to the structure mass M. As a case study, the following numerical value for the mass ratio is used

$$\mu = 0.25$$

(This value is used for convenience, in case the reader wishes to check Den Hartog's results.) Using the formula for optimal tuning and damping, we have

$$\alpha_{opt} = 0.80, \quad \zeta_{opt} = 0.21909$$

$$|KG_1(j\beta)_{opt}| = \frac{\sqrt{(0.64-\beta^2)^2 + (0.43818\beta)^2}}{\sqrt{(0.64+\beta^4-1.8\beta^2)^2 + [0.43818\beta(1-1.25\beta^2)]^2}}$$ (8.172)

The results are shown in Figs. 8.36 and 8.37.

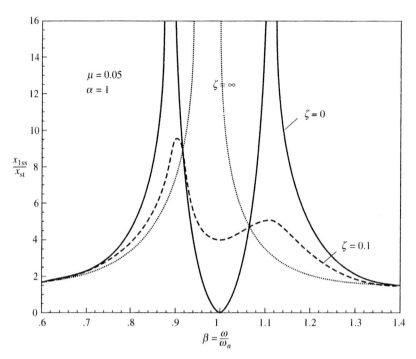

FIGURE 8.36 Frequency response of the main mass for various values of damping ratio.

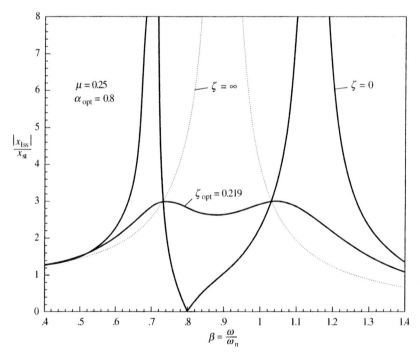

FIGURE 8.37 Frequency response of the main mass using optimal values for tuning and damping.

Transient Response

The optimal parameters given above are obtained strictly for frequency response. For transient response, these formulas may be used, but the parameters may not be absolutely optimal.

As a case study, the following numerical value for the mass ratio is used

$$\mu = 0.25$$

Using the formulas for optimal tuning and damping, we have

$$\alpha_{\text{opt}} = 0.80, \quad \zeta_{\text{opt}} = 0.21909$$

Thus if $M = 1$ kg and $K = 1$ N/m then $m = 0.25$ kg, $k = 0.080$ N/m, and $c = 0.10955$ N · s/m. The transfer functions are obtained as

$$\frac{X_1(s)}{F(s)} = \frac{0.25s^2 + 0.10955s + 0.16}{0.25s^4 + 0.13694s^3 + 0.45s^2 + 0.10955s + 0.16}$$

$$\frac{X_2(s)}{F(s)} = \frac{0.10955s + 0.16}{0.25s^4 + 0.13694s^3 + 0.45s^2 + 0.10955s + 0.16}$$

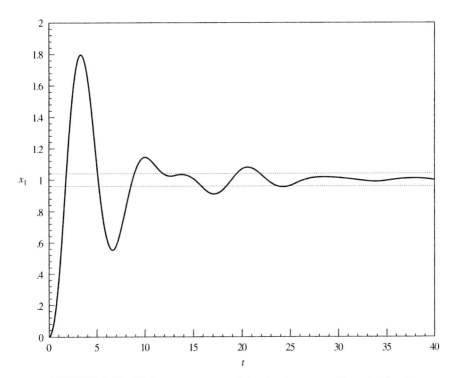

FIGURE 8.38 Unit-step response of the structure mass M, x_1 (optimal tuning and damping, $\mu = 0.25$, $K = M = 1$).

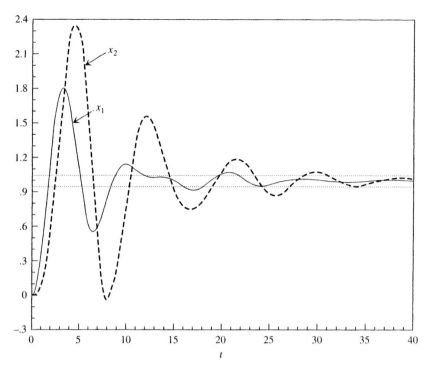

FIGURE 8.39 Unit-step responses of the system, x_1 and x_2 (optimal tuning and damping, $\mu = 0.25$, $K = M = 1$).

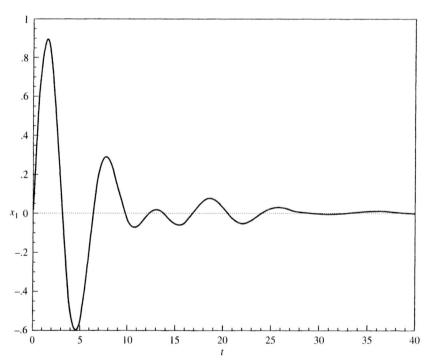

FIGURE 8.40 Unit-impulse response of the main mass M, x_1 (optimal tuning and damping, $\mu = 0.25$, $K = M = 1$).

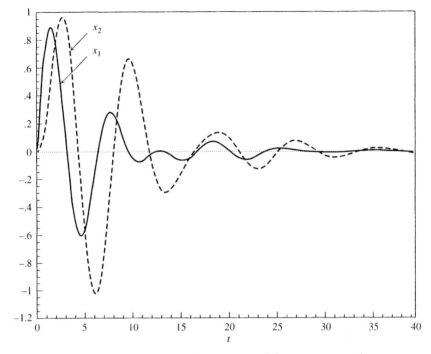

FIGURE 8.41 Unit-impulse responses of the system, x_1 and x_2 (optimal tuning and damping, $\mu = 0.25$, $K = M = 1$).

It is interesting to look at the poles of this system, whose characteristic equation is

$$0.25s^4 + 0.13694s^3 + 0.45s^2 + 0.10955s + 0.16 = 0$$

Using MATRIX$_x$, the poles of this fourth-order system are given as

$$s_{1,2} = -0.1101 \pm j0.7250$$
$$s_{3,4} = -0.1638 \pm j1.0786$$

Notice that the real parts of all the four poles are nearly the same, thus there are no dominant poles.

The *unit-step responses* are shown in Figs. 8.38–8.39, where the maximum percent overshoot $M_p^\%$ and the settling time $t_s|_{5\%}$ can be determined readily. Similarly, the *unit-impulse responses* are shown in Figs. 8.40–8.41.

PROBLEM SET 8.7

8.17. Figure P8.17 shows a single-degree-of-freedom structure with a damped vibration absorber attached. The structure is of mass m and stiffness k, whereas the absorber is composed of a smaller mass m_a, a spring of stiffness k_a, and a dashpot of viscous damping coefficient c_a. The absorber is designed to provide a counter-balance force that acts

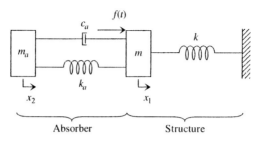

FIGURE P8.17

against the disturbance force $f(t) = F_o \cos \omega t$ so that the motion of the structure mass M can be minimized. This is the famous passive vibration control using a damped vibration absorber.
(a) Obtain the differential equations.
(b) Obtain the transfer functions for x_1 and x_2. What is the order of the system?
(c) Obtain the FRFs for x_1 and x_2. Determine the magnitudes and phases of the FRFs.
(d) Obtain the steady-state responses for x_1 and x_2.

8.18. Consider the damped vibration absorber shown in Fig. 8.35. The normalized FRF of the absorber is given as

$$KG_1(j\beta) = \frac{\alpha^2 - \beta^2 + j2\zeta\beta}{\alpha^2 + \beta^4 - [1 + (1+\mu)\alpha^2]\beta^2 + j2\zeta\beta[1 - (1+\mu)]\beta^2}$$

Determine the magnitude and phase of the FRF.

8.19. Using the dimensionless groups defined in the text, verify the following result

$$KG_1(j\beta) = \frac{\alpha^2 - \beta^2 + j2\zeta\beta}{\alpha^2 + \beta^4 - [1 + (1+\mu)\alpha^2]\beta^2 + j2\zeta\beta[1 - (1+\mu)]\beta^2}$$

8.20. Consider the damped vibration absorber system given in the text. Assume the following numerical values

$$\mu = \frac{m}{M} = 0.05, \quad \alpha = \frac{\omega_a}{\omega_n} = 1.0, \quad M = 1 \text{ kg}, \quad K = 1 \text{ N/m}$$

Determine the stiffness of the absorber.

8.21. Consider the damped vibration absorber system given in the text. Assume the following numerical values

$$\mu = \frac{m}{M} = 0.25, \quad \alpha = \frac{\omega_a}{\omega_n} = 0.8, \quad M = 1 \text{ kg}, \quad K = 1 \text{ N/m}$$

Determine the stiffness of the absorber.

8.22. Consider the damped vibration absorber system where the mass ratio is selected as $\mu = 15$ percent. Determine the optimal tuning and damping ratios.

8.23. Consider the damped vibration absorber system where the mass ratio is selected as $\mu = 25$ percent. Determine the optimal tuning and damping ratios.

8.24. Consider the damped vibration absorber system. Assume the following numerical values

$$\mu = \frac{m}{M} = 0.20, \quad M = 100 \text{ kg}, \quad K = 10 \text{ N/m}$$

(a) Determine the optimal tuning and damping ratios.
(b) Determine the stiffness of the absorber.

8.25. The normalized FRF is

$$KG_1(j\beta) = \frac{\alpha^2 - \beta^2 + j2\zeta\beta}{\alpha^2 + \beta^4 - [1 + (1 + \mu)\alpha^2]\beta^2 + j2\zeta\beta[1 - (1 + \mu)]\beta^2}$$

Assume the following numerical values

$$\mu = \frac{m}{M} = 0.10, \quad M = 100 \text{ kg}, \quad K = 10 \text{ N/m}$$

(a) Determine the optimal tuning and damping ratios.
(b) Determine the magnitude of the optimal normalized FRF as a function of β.

8.26. Consider the damped vibration absorber, where the mass ratio is 25 percent. Assume $M = 1$ kg and $K = 1$ N/m and the applied forcing function to the main mass M is a unit-impulse function. Compute the optimal tuning and damping ratios, then express the responses $x_1(t)$ and $x_2(t)$ in the form of the inverse Laplace transform.

8.27. Consider the damped vibration absorber, where the mass ratio is 25 percent. Assume $M = 1$ kg and $K = 1$ N/m and the applied forcing function to the main mass M is a step function of amplitude A. Compute the optimal tuning and damping ratios, then express the responses $x_1(t)$ and $x_2(t)$ in the form of the inverse Laplace transform.

8.28. The characteristic equation of a vibration system is

$$0.2s^4 + 0.14s^3 + 0.45s^2 + 0.11s + 0.16 = 0$$

Determine the poles by using any method.

8.29. The characteristic equation of a vibration system is

$$0.2s^4 + 0.1s^3 + 0.4s^2 + 0.1s + 0.2 = 0$$

Determine the poles by using any method.

8.8 MODAL ANALYSIS: SYSTEM RESPONSE

Although **modal analysis** may be beyond the scope of the reader for a course in modeling and analysis, it is included here for completeness. Modal analysis is a powerful method of obtaining closed-form solutions for vibration problems. After solving for the natural frequencies and mode shapes and establishing the orthonormality property, the response can be obtained by using **modal expansion.** Hence, the method is called *modal analysis*. Note that *eigenvectors* expansion is associated with discrete systems (lumped-parameter model) whereas **eigenfunctions** expansion is for continuous systems (distributed-parameter model).

Modal analysis is an extremely useful method because it provides closed-form solutions; however, it has its own disadvantages and limitations. The primary disadvantage is that the method is a bit complicated for most junior engineering students. The main limitation is that modal analysis cannot be used to solve for the response if the damping matrix is arbitrary. Under this condition, the damping matrix cannot be diagonalized because the modal matrix is not orthogonal with respect to the damping matrix. In other words, the orthogonality (or orthonormality) property for the damping matrix cannot be established. The method is only useful if the damping matrix is proportional to the mass matrix or the stiffness matrix (or proportional to both matrices so it can be diagonalized).

It should be mentioned that if the primary objective is to obtain response plots for MDOF systems, then the use of the Laplace transform method and suitable software is the most practical way.

It is desired to obtain the response of the vibration system whose differential equations are given in the general matrix form (second order) as

$$\mathbf{m\ddot{x} + c\dot{x} + kx = f}(t) \tag{8.173}$$

where $\mathbf{m}$ = mass matrix
$\mathbf{c}$ = damping matrix
$\mathbf{k}$ = stiffness matrix
$\mathbf{f}$ = applied forcing function matrix
$\mathbf{x}$ = displacement matrix

Eigenvalue Problem

The *eigenvalue problem* is defined as the *free, undamped vibration* problem. Thus, the matrix equation becomes

$$\mathbf{m\ddot{x} + kx = 0} \tag{8.174}$$

Previously, we showed how to obtain the eigensolutions, which are the solutions to eigenvalue problems. These **eigensolutions** are the natural frequencies (related to eigenvalues) and mode shapes (eigenvectors). The eigensolutions consist of eigenvalues and eigenvectors or eigenfunctions.

The natural frequencies are related to the eigenvalues, but they are not exactly the same. The word *eigen* is from the German root meaning *characteristic* in English. Thus the term *eigenvalues* means the same thing as *characteristic values*.

The term *mode shape* has exactly the same meaning as *eigenvector* or *eigenfunction*. For discrete systems, mode shapes are called eigenvectors. For continuous systems, mode shapes are called eigenfunctions or **mode functions.**

Eigensolutions: Natural Frequencies and Mode Shapes

The solution to Eq. (8.174) is in the form

$$\mathbf{x} = \mathbf{X}e^{j\omega t}$$

where ω and $\mathbf{X}$ are the natural frequency and the corresponding mode shape, respectively. Thus

$$\ddot{\mathbf{x}} = -\omega^2 \mathbf{X} e^{j\omega t} = -\omega^2 \mathbf{x}$$

Therefore,

$$\underbrace{[-\omega^2 \mathbf{m} + \mathbf{k}]}_{\det[\,\cdot\,\cdot\,]=0} \underbrace{\mathbf{X}}_{\neq 0} \underbrace{e^{j\omega t}}_{\neq 0} = 0 \qquad (8.175)$$

This is how we proceed to solve the eigenvalue problem for the eigensolutions.

Although continuous systems are beyond the scope of this text, a few things should be mentioned here. In this book we mainly deal with discrete systems, which are governed by ordinary differential equations (ODE). Continuous systems, which are governed by partial differential equations (PDE), are relatively more difficult to handle for most junior or senior students in engineering. An example of a continuous system, the transverse vibration of a flexible beam, is shown in Fig. 8.42. The governing differential equation of the transverse vibration of a flexible beam is given as

$$EI\frac{\partial^4 v(x,t)}{\partial x^4} + [\rho A + M\delta(x-l)]\frac{\partial^2 v(x,t)}{\partial t^2} = -f(x,t) \qquad 0 < x < L \quad (8.176)$$

The beam is of mass per unit length ρA and length L, with a concentrated mass M at some point along the length, say l. The displacement $v(x, t)$ is a function of both x and t. The applied force $f(t)$ is per unit length. An example of this type of load is the snow piled up on the beam in winter time. Other parameters are the usual definitions in the mechanics of materials: E is the modulus of elasticity (Young's modulus), I is the area moment of inertia, and δ is the Dirac delta function.

We now focus our attention on a fundamental class of discrete systems: the TDOF system. Reconsider the general free undamped vibration system of two

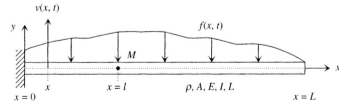

FIGURE 8.42 Transverse vibration of a flexible beam.

474   Dynamic Systems: Modeling and Analysis

degrees of freedom, whose differential equations are given as

$$\begin{bmatrix} m_{11} & m_{12} \\ m_{21} & m_{22} \end{bmatrix} \begin{Bmatrix} \ddot{x}_1 \\ \ddot{x}_2 \end{Bmatrix} + \begin{bmatrix} k_{11} & k_{12} \\ k_{21} & k_{22} \end{bmatrix} \begin{Bmatrix} x_1 \\ x_2 \end{Bmatrix} = \begin{Bmatrix} 0 \\ 0 \end{Bmatrix}$$

From previous work, the natural frequencies are given by

$$\begin{matrix} \omega_1^2 \\ \omega_2^2 \end{matrix} = \frac{k_{11}m_{22} + k_{22}m_{11} - 2k_{12}m_{12}}{2(m_{11}m_{12} - m_{12}^2)}$$

$$\mp \sqrt{\left[\frac{k_{11}m_{22} + k_{22}m_{11} - 2k_{12}m_{12}}{2(m_{11}m_{12} - m_{12}^2)}\right]^2 - \frac{k_{11}k_{22} - k_{12}^2}{m_{11}m_{22} - m_{12}^2}}$$

For mode shapes, we have

$$\begin{Bmatrix} X_1 \\ X_2 \end{Bmatrix}_1 = \begin{Bmatrix} 1 \\ X_2/X_1 \end{Bmatrix}_1$$

$$\begin{Bmatrix} X_1 \\ X_2 \end{Bmatrix}_2 = \begin{Bmatrix} 1 \\ X_2/X_1 \end{Bmatrix}_2$$

where

$$\left(\frac{X_2}{X_1}\right)_1 = \frac{k_{11} - m_{11}\omega_1^2}{k_{12} - m_{12}\omega_1^2}$$

$$\left(\frac{X_2}{X_1}\right)_2 = \frac{k_{11} - m_{11}\omega_2^2}{k_{12} - m_{12}\omega_2^2}$$

Orthonormality

Orthonormality has been treated in Section 8.3, and the essential results are reproduced here for convenience.

$$\Phi^T m \Phi = I = [1] = \text{diagonal matrix} \qquad (8.76a)$$
$$\Phi^T k \Phi = [\omega_r^2] = \text{diagonal matrix} \qquad (8.76b)$$

where **I** is the identity matrix.

System Response via Modal Analysis

We are now ready to obtain the exact solutions for the response using modal analysis. The method is broken down into three steps:

1. Obtain the eigensolutions (solutions of the eigenvalue problems): natural frequencies and mode shapes.
2. Obtain the orthonormality (normalized orthogonality).
3. Obtain the solution by modal expansion.

We will show how to obtain the response for two classes of vibration: free vibration and forced vibration. Free vibration is defined as the vibration caused only by the initial conditions, whereas forced vibration is caused by an applied forcing function.

Undamped free vibration

Consider the following system, whose differential equations are given in the second-order matrix form as

$$\begin{bmatrix} mL^2 & 0 \\ 0 & mL^2 \end{bmatrix} \begin{Bmatrix} \ddot{\theta}_1 \\ \ddot{\theta}_2 \end{Bmatrix} + \begin{bmatrix} mgL + ka^2 & -ka^2 \\ -ka^2 & mgL + ka^2 \end{bmatrix} \begin{Bmatrix} \theta_1 \\ \theta_2 \end{Bmatrix} = \begin{Bmatrix} 0 \\ 0 \end{Bmatrix} \quad (8.177)$$

and the initial conditions are assumed as

$$\theta_1(0^-) = \theta_o \qquad \dot{\theta}_1(0^-) = \theta_2(0^-) = \dot{\theta}_2(0^-) = 0 \quad (8.178)$$

In compact form, the matrix equation becomes

$$\mathbf{M}\ddot{\boldsymbol{\theta}} + \mathbf{K}\boldsymbol{\theta} = \mathbf{0} \quad (8.179)$$

Step 1. Eigensolutions. As shown previously, the natural frequencies are obtained as

$$\omega_1 = \sqrt{\frac{g}{L}}$$

$$\omega_2 = \sqrt{\frac{g}{L} + 2\frac{k}{m}\frac{a^2}{L^2}}$$

and the modal matrix is

$$\mathbf{U} = [\mathbf{u}_1 \quad \mathbf{u}_2] = \begin{bmatrix} \begin{Bmatrix} 1 \\ r \end{Bmatrix}_1 & \begin{Bmatrix} 1 \\ r \end{Bmatrix}_2 \end{bmatrix}$$

For this problem,

$$r_1 = 1, \qquad r_2 = -1$$

Step 2. Normalized Orthogonality (Orthonormality). After the normalization, the modal matrix $\mathbf{U}$ becomes $\boldsymbol{\Phi}$,

$$\mathbf{U} \to \boldsymbol{\Phi}$$

$$\boldsymbol{\Phi} = \begin{bmatrix} \alpha_1 \begin{Bmatrix} 1 \\ r \end{Bmatrix}_1 & \alpha_2 \begin{Bmatrix} 1 \\ r \end{Bmatrix}_2 \end{bmatrix} \quad (8.180)$$

where the two constants α_1 and α_2 are given by the orthonormality condition.

Step 3. Modal Analysis—Eigenfunction Expansion.

$$\boldsymbol{\theta} = \boldsymbol{\Phi}\boldsymbol{\eta} \quad (8.181a)$$

$$\ddot{\boldsymbol{\theta}} = \boldsymbol{\Phi}\ddot{\boldsymbol{\eta}} \quad (8.181b)$$

$$\begin{Bmatrix} \theta_1 \\ \theta_2 \end{Bmatrix} = \begin{bmatrix} \alpha_1 \begin{Bmatrix} 1 \\ r \end{Bmatrix}_1 & \alpha_2 \begin{Bmatrix} 1 \\ r \end{Bmatrix}_2 \end{bmatrix} \begin{Bmatrix} \eta_1 \\ \eta_2 \end{Bmatrix} \tag{8.182}$$

where η_1 and η_2 are called **natural coordinates.** Thus,

$$\theta_1 = \alpha_1 \eta_1 + \alpha_2 \eta_2 \tag{8.183a}$$

$$\theta_2 = \alpha_1 r_1 \eta_1 + \alpha_2 r_2 \eta_2 \tag{8.183b}$$

After substituting,

$$\mathbf{M\Phi\ddot{\eta}} + \mathbf{K\Phi\eta} = \mathbf{0} \tag{8.184}$$

Multiplying the above equation by the transpose of $\mathbf{\Phi}$

$$\underbrace{\mathbf{\Phi}^T \mathbf{M} \mathbf{\Phi}}_{\mathbf{I}} \ddot{\eta} + \underbrace{\mathbf{\Phi}^T \mathbf{K} \mathbf{\Phi}}_{[\omega_r^2]} \eta = \mathbf{\Phi}^T \mathbf{0} = \mathbf{0} \tag{8.185}$$

Thus,

$$\ddot{\eta} + [\omega_r^2]\eta = \mathbf{0} \tag{8.186}$$

or

$$\ddot{\eta}_1 + \omega_1^2 \eta_1 = 0 \tag{8.187a}$$

$$\ddot{\eta}_2 + \omega_2^2 \eta_2 = 0 \tag{8.187b}$$

Therefore, the natural coordinates, which are the homogeneous solutions, are

$$\eta_1(t) = A_1' \cos\omega_1 t + B_1' \sin\omega_1 t \tag{8.188a}$$

$$\eta_2(t) = A_2' \cos\omega_2 t + B_2' \sin\omega_2 t \tag{8.188b}$$

or

$$\eta_1(t) = C_1 \cos(\omega_1 t + \phi_1) \tag{8.189a}$$

$$\eta_2(t) = C_2 \cos(\omega_2 t + \phi_2) \tag{8.189b}$$

Note that Eq. (8.188) is more convenient for evaluating the coefficients, whereas Eq. (8.189) is better for plotting purposes.

Back to the physical coordinates

$$\begin{Bmatrix} \theta_1 \\ \theta_2 \end{Bmatrix} = \begin{bmatrix} \alpha_1 \begin{Bmatrix} 1 \\ r \end{Bmatrix}_1 & \alpha_2 \begin{Bmatrix} 1 \\ r \end{Bmatrix}_2 \end{bmatrix} \begin{Bmatrix} \eta_1 \\ \eta_2 \end{Bmatrix}$$

Thus,

$$\theta_1(t) = 1 \cdot (A_1 \cos\omega_1 t + B_1 \sin\omega_1 t) + 1 \cdot (A_2 \cos\omega_2 t + B_2 \sin\omega_2 t) \tag{8.190a}$$

$$\theta_2(t) = r_1 \cdot (A_1 \cos\omega_1 t + B_1 \sin\omega_1 t) + r_2 \cdot (A_2 \cos\omega_2 t + B_2 \sin\omega_2 t) \tag{8.190b}$$

Notice that the constants of normalization, α_1 and α_2, are absorbed in the unprimed coefficients A_i and B_i. For convenience, in evaluating these coefficients, the time derivatives are written as

$$\dot{\theta}_1(t) = \omega_1(-A_1 \sin \omega_1 t + B_1 \cos \omega_1 t)$$
$$+ \omega_2(-A_2 \sin \omega_2 t + B_2 \cos \omega_2 t) \quad (8.191a)$$

$$\dot{\theta}_2(t) = r_1\omega_1(-A_1 \sin \omega_1 t + B_1 \cos \omega_1 t)$$
$$+ r_2\omega_2(-A_2 \sin \omega_2 t + B_2 \cos \omega_2 t) \quad (8.191b)$$

Applying the initial conditions with $r_1 = 1$ and $r_2 = -1$,

$$\theta_1(0) = \theta_o = A_1 + A_2$$
$$\theta_2(0) = 0 = A_1 - A_2$$
$$\dot{\theta}_1(0) = 0 = \omega_1 B_1 + \omega_2 B_2$$
$$\dot{\theta}_2(0) = 0 = \omega_1 B_1 - \omega_2 B_2$$

Thus,

$$A_1 = A_2 = \frac{\theta_o}{2}$$

$$B_1 = B_2 = 0$$

Knowing that

$$\frac{\cos(b-a)t + \cos(a+b)t}{2} = \cos at \cos bt \quad (8.192a)$$

$$\frac{\cos(b-a)t - \cos(a+b)t}{2} = \sin at \sin bt \quad (8.192b)$$

the system responses are given, finally, as

$$\theta_1(t) = \frac{\theta_o}{2}(\cos \omega_1 t + \cos \omega_2 t) = \theta_o \cos \frac{\omega_2 - \omega_1}{2} t \cdot \cos \frac{\omega_2 + \omega_1}{2} t \quad (8.193a)$$

$$\theta_2(t) = \frac{\theta_o}{2}(\cos \omega_1 t - \cos \omega_2 t) = \theta_o \sin \frac{\omega_2 - \omega_1}{2} t \cdot \sin \frac{\omega_2 + \omega_1}{2} t \quad (8.193b)$$

Damped forced vibration

The general procedure for solving the forced vibration problem, using modal analysis, follows. Consider the general system whose differential equations are given, in the second-order matrix form, as

$$m\ddot{x} + c\dot{x} + kx = f(t)$$

with the general initial conditions

$$x(0^-) = x_o \quad \dot{x}(0^-) = v_o \quad (8.194)$$

Step 1. Solve the eigenvalue problem (or free, undamped vibration problem) to obtain the natural frequencies, the mode shapes, and the modal matrix. The natural frequencies, $\omega_1, \omega_2, \ldots, \omega_r, \ldots, \omega_n$ and the corresponding mode shapes are obtained. Thus, the modal matrix is

$$U = [\mathbf{u}_1 \quad \mathbf{u}_2 \quad \cdots \quad \mathbf{u}_r \quad \cdots \quad \mathbf{u}_n] \tag{8.195}$$

Step 2. Obtain the orthonormality condition (or property). After the normalization, the modal matrix $\mathbf{U}$ becomes $\mathbf{\Phi}$,

$$\mathbf{U} \to \mathbf{\Phi}$$

$$\mathbf{\Phi} = [\alpha_1 \mathbf{u}_1 \quad \alpha_2 \mathbf{u}_2 \quad \cdots \quad \alpha_r \mathbf{u}_r \quad \cdots \quad \alpha_n \mathbf{u}_n] \tag{8.196}$$

where the n constants $\alpha_1, \alpha_2, \ldots, \alpha_r, \ldots, \alpha_n$ are given by the orthonormality condition, Eq. (8.170). Let us assume that we have

$$\mathbf{\Phi}^T \mathbf{c} \mathbf{\Phi} = \lceil 2\zeta_r \omega_r \rfloor$$

Step 3. Solve for the response using modal (eigenvector) expansion.

$$\mathbf{x} = \mathbf{\Phi} \boldsymbol{\eta} \tag{8.197a}$$

$$\ddot{\mathbf{x}} = \mathbf{\Phi} \ddot{\boldsymbol{\eta}} \tag{8.197b}$$

After substituting,

$$\mathbf{m}\mathbf{\Phi}\ddot{\boldsymbol{\eta}} + \mathbf{c}\mathbf{\Phi}\dot{\boldsymbol{\eta}} + \mathbf{k}\mathbf{\Phi}\boldsymbol{\eta} = \mathbf{f}(t) \tag{8.198}$$

Multiplying the above equation by the transpose of $\mathbf{\Phi}$,

$$\underbrace{\mathbf{\Phi}^T \mathbf{m} \mathbf{\Phi}}_{\mathbf{I}} \ddot{\boldsymbol{\eta}} + \underbrace{\mathbf{\Phi}^T \mathbf{c} \mathbf{\Phi}}_{\lceil 2\zeta_r \omega_r \rfloor} \dot{\boldsymbol{\eta}} + \underbrace{\mathbf{\Phi}^T \mathbf{k} \mathbf{\Phi}}_{\lceil \omega_r^2 \rfloor} \boldsymbol{\eta} = \underbrace{\mathbf{\Phi}^T \mathbf{f}(t)}_{\mathbf{N}(t)} \tag{8.199}$$

where

$$\mathbf{N}(t) = \mathbf{\Phi}^T \mathbf{f}(t) \tag{8.200}$$

Thus,

$$\ddot{\boldsymbol{\eta}} + \lceil 2\zeta_r \omega_r \rfloor \dot{\boldsymbol{\eta}} + \lceil \omega_r^2 \rfloor \boldsymbol{\eta} = \mathbf{N}(t) \quad r = 1, 2, \ldots, n \tag{8.201}$$

or

$$\ddot{\eta}_1 + 2\zeta_1 \omega_1 \dot{\eta}_1 + \omega_1^2 \eta_1 = N_1(t)$$

$$\ddot{\eta}_2 + 2\zeta_2 \omega_2 \dot{\eta}_2 + \omega_2^2 \eta_2 = N_2(t)$$

$$\vdots$$

$$\ddot{\eta}_r + 2\zeta_r \omega_r \dot{\eta}_r + \omega_r^2 \eta_r = N_r(t)$$

$$\vdots$$

$$\ddot{\eta}_n + 2\zeta_n \omega_n \dot{\eta}_n + \omega_n^2 \eta_n = N_n(t)$$

The solution for $\eta_r(t)$ can be obtained by various methods learned previously. The convolution integral is useful if the functions in the forcing matrix $\mathbf{f}(t)$ are arbitrary (see Chapter 7). Notice that the homogeneous solution for the first mode, for

example, is
$$\eta_{1h}(t) = e^{-\sigma_1 t}\left(A_1 \cos \omega_{d1} t + B_1 \sin \omega_{d1} t\right) \tag{8.202}$$
where
$$\sigma_1 = \zeta_1 \omega_1 \qquad \omega_{d1} = \omega_1 \sqrt{1 - \zeta_1^2} \tag{8.203a}$$
The initial conditions for the natural coordinates are given by
$$\dot{\mathbf{x}}(0) = \boldsymbol{\Phi}\dot{\boldsymbol{\eta}}(0) \tag{8.203b}$$
or
$$\boldsymbol{\eta}(0) = \boldsymbol{\Phi}^{-1}\mathbf{x}(0) \tag{8.204a}$$
$$\dot{\boldsymbol{\eta}}(0) = \boldsymbol{\Phi}^{-1}\dot{\mathbf{x}}(0) \tag{8.204b}$$
Thus, the complete response for the natural coordinates can be obtained.

Finally, we get back to the physical coordinates for the system response,
$$\mathbf{x}(t) = \boldsymbol{\Phi}\boldsymbol{\eta}(t)$$

8.9 VIBRATION TESTING

The study of experimental vibration, or *vibration testing,* provides a significant contribution to our knowledge. Vibration is one of the major areas in engineering where analytical and experimental techniques complement each other immensely. With experimental vibration, we can effectively determine the nature of vibration response of certain machines or structures.

Considerable efforts have been devoted to **modal testing,** also called **experimental modal analysis.** This activity can help us to verify theoretical models and their corresponding predictions. There are two basic approaches in modal testing: *hammer testing* and *shaker testing.* Hammer testing is fast, while shaker testing is more controlled. As with other things, there are pros and cons between the two. With either approach, a dynamic signal analyzer plays a vital role. At a minimum, the dynamic signal analyzer must be a two-channel FFT (fast fourier transform) analyzer. One channel is for the input and the other is for the response (output). It takes in the data, then computes the FRF (frequency response function) and provides the Bode plots and other useful information.

From these results the theoretical models can be validated or correlated. In certain applications where the structure is complex, performing a modal test is more effective than building a computer model using the FEM (finite element method). The reader is urged to consult the many excellent publications in this area of activity.

SUMMARY

Eigensolutions are the solutions to eigenvalue problems, and a free, undamped vibration is an eigenvalue problem. Eigensolutions consist of *eigenvalues* and *eigenvectors* (for discrete or lumped-parameter systems) or eigenfunctions (for continuous or distributed-parameter systems). The natural frequencies of a vibration system are related to the *eigenvalues*, whereas the mode shapes are the same as *eigenvectors*/eigenfunctions. The natural frequencies are obtained by first solving for the eigenvalues. The zero frequencies indicate the rigid-body modes, whereas the nonzero ones are the natural frequencies of the vibrational modes.

The solution of a free, undamped vibration is in the form

$$\mathbf{x} = \mathbf{X} e^{j\omega t}$$

where ω and $\mathbf{X}$ are the natural frequency and the corresponding mode shape, respectively. This is a direct result from the method of separation of variables.

Modal matrix, with its orthogonality properties, can be used to check for the correctness of the eigensolutions. The modal matrix $\mathbf{U}$ is orthogonal with respect to the mass matrix $\mathbf{m}$ and the stiffness matrix $\mathbf{k}$,

$$\mathbf{U}^T \mathbf{m} \mathbf{U} = [M_r]$$

$$\mathbf{U}^T \mathbf{k} \mathbf{U} = [K_r]$$

where M_r, K_r, and $\omega_r = \sqrt{K_r/M_r}$ are the modal mass, modal stiffness, and natural frequency, respectively. The subscript r denotes the rth mode. The matrix $\mathbf{U}$ is generally *not* orthogonal with respect to itself if the mass and stiffness matrices are not proportional to the identity matrix,

$$\mathbf{U}^T \mathbf{U} \neq \text{diagonal matrix} \qquad \text{if } \mathbf{k} \neq \alpha \mathbf{I} \text{ or } \mathbf{m} \neq \beta \mathbf{I}$$

where α and β are proportional constants.

When the modal matrix is normalized for modal analysis, the orthogonality becomes orthonormality. The modal matrix is changed from $\mathbf{U}$ to $\mathbf{\Phi}$, thus

$$\mathbf{\Phi}^T \mathbf{m} \mathbf{\Phi} = \mathbf{I} = [1]$$

$$\mathbf{\Phi}^T \mathbf{k} \mathbf{\Phi} = [\omega_r^2]$$

Note that the normalization of the mode shapes for modal analysis is different than that for mode-shapes sketching.

If the damping matrix is proportional to either the mass matrix $\mathbf{m}$ or the stiffness matrix $\mathbf{k}$, or to both,

$$\mathbf{c} = \alpha \mathbf{m} \text{ or } \mathbf{c} = \beta \mathbf{m} \text{ or } \mathbf{c} = \alpha \mathbf{m} + \beta \mathbf{m}$$

where, again, α and β are proportional constants, then the modal matrix is orthogonal or orthonormal with respect to the damping matrix

$$\mathbf{U}^T \mathbf{c} \mathbf{U} = [C_r]$$

$$\mathbf{\Phi}^T \mathbf{c} \mathbf{\Phi} = [2\zeta_r \omega_r]$$

where C_r is the modal damping coefficient of the rth mode.

The log decrement method can be used to determine the amount of damping of an underdamped second-order system with viscous damping. The relation between the log decrement δ and damping ratio ζ is given as

$$\delta = \ln \frac{x_1}{x_2} = \frac{2\pi\zeta}{\sqrt{1-\zeta^2}}, \qquad \zeta = \frac{\delta}{\sqrt{4\pi^2 + \delta^2}}$$

where x_1 and x_2 are the two consecutive displacements separated by one period.

The beat phenomenon occurs when the two natural frequencies of a vibration system are nearly equal. This happens when the two modes are weakly coupled, as in the pendulum system discussed in Section 8.5. The two natural frequencies of this system (Fig. 8.24) are

$$\omega_1 = \frac{g}{L} \qquad \omega_2 = \sqrt{\frac{g}{L} + \frac{2kl^2}{mL^2}}$$

The second natural frequency, ω_2, is slightly larger than the first, ω_1, if $2kl^2/mL^2$ is very small compared to g/L.

The steady-state response of a fundamental mass-spring-damper system is given as follows.

DE: $\qquad m\ddot{x} + b\dot{x} + kx = f(t), \qquad f(t) = F_0 \cos\omega t$

$$\ddot{x} + 2\zeta\omega_n \dot{x} + \omega_n^2 x = \frac{f(t)}{m}$$

$$\omega_n = \sqrt{\frac{k}{m}}, \qquad \zeta = \frac{b}{2\sqrt{km}}, \qquad \frac{b}{m} = 2\zeta\omega_n$$

TF: $\qquad G(s) = \dfrac{X(s)}{F(s)} = \dfrac{1/m}{s^2 + 2\zeta\omega_n s + \omega_n^2}$

FRF: $\qquad G(j\omega) = \dfrac{(1/m)(1/\omega_n^2)}{(\omega_n^2/\omega_n^2) - (\omega^2/\omega_n^2) + j(2\zeta\omega_n\omega/\omega_n^2)}$

$$= \frac{1/k}{1 - (\omega/\omega_n)^2 + j2\zeta(\omega/\omega_n)}$$

Steady-state response: $\quad x_{ss}(t;\omega) = \underbrace{F_0|G(j\omega)|}_{|x_{ss}|} \cos(\omega t + \phi)$

where

$$|G(j\omega)| = \frac{1/k}{\sqrt{\left[1 - (\omega/\omega_n)^2\right]^2 + \left[2\zeta(\omega/\omega_n)\right]^2}}, \qquad \phi = -\tan^{-1}\frac{2\zeta(\omega/\omega_n)}{1 - (\omega/\omega_n)^2}$$

Normalization: $\quad \dfrac{|x_{ss}|}{F_0/k} = \dfrac{1}{\sqrt{\left[1 - (\omega/\omega_n)^2\right]^2 + \left[2\zeta(\omega/\omega_n)\right]^2}}$

Resonant frequency: $\omega_r = \omega_n \sqrt{1 - 2\zeta^2} \qquad (0 \leq \zeta < 0.707)$

Resonant peak: $\quad M_r = \dfrac{1}{2\zeta\sqrt{1-\zeta^2}} \qquad (0 \leq \zeta < 0.707)$

The damped vibration absorber (Fig. 8.35) is often used in design as a passive vibration control, as discussed in Section 8.7. The following formulas are used to design the optimal absorber:

optimal tuning ratio: $\quad\alpha_{opt} = \dfrac{1}{1+\mu}$

optimal damping ratio: $\quad\zeta_{opt} = \sqrt{\dfrac{3\mu}{8(1+\mu)^3}}$

where

$$\mu = \frac{m}{M}, \quad \alpha = \frac{\omega_a}{\omega_n} = \frac{\sqrt{k/m}}{\sqrt{K/M}}, \quad \zeta = \frac{c}{2m\omega_n} = \frac{c}{2m\sqrt{K/M}}$$

PROBLEMS

8.30. Consider the system shown in Fig. P8.30. Assuming small motions, determine the differential equations.

8.31. Consider the system shown in Fig. P8.31. Assuming small motions, determine the differential equations.

8.32. Obtain the differential equations of the system shown in Fig. P8.32.

8.33. Obtain the differential equations of the system shown in Fig. P8.33.

8.34. Obtain the differential equations for the vibration system shown in Fig. P8.34.

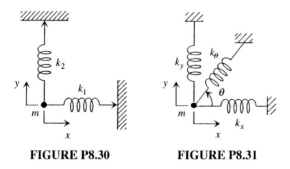

FIGURE P8.30   FIGURE P8.31

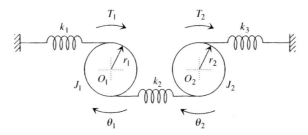

FIGURE P8.32

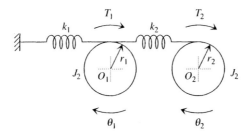

FIGURE P8.33

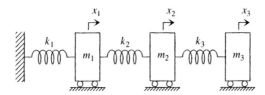

FIGURE P8.34

8.35. Consider the vibration system shown in Fig. P8.35.
  (a) Obtain the natural frequencies.
  (b) Obtain and sketch the mode shapes.

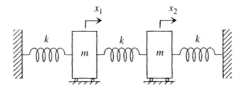

FIGURE P8.35

8.36. Consider the vibration system shown in Fig. P8.36.
  (a) Obtain the natural frequencies.
  (b) Obtain and sketch the mode shapes.

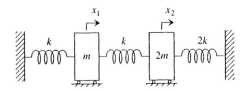

FIGURE P8.36

8.37. Consider the vibration system shown in Fig. P8.37. Obtain the following:
(a) the natural frequencies,
(b) the (normalized) mode shapes,
(c) the modal matrix **U**, which is defined as

$$\mathbf{U}^T \mathbf{m} \mathbf{U} = [M_r], \qquad \mathbf{U}^T \mathbf{k} \mathbf{U} = [K_r]$$

(d) the normalized modal matrix $\mathbf{\Phi}$, which is defined as

$$\mathbf{\Phi}^T \mathbf{m} \mathbf{\Phi} = \mathbf{I} = [1], \qquad \mathbf{\Phi}^T \mathbf{k} \mathbf{\Phi} = [\omega_r^2]$$

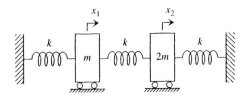

FIGURE P8.37

8.38. Consider the vibration system shown in Fig. P8.38. Similarly to Problem 8.37, obtain the following:
(a) the natural frequencies,
(b) the (normalized) mode shapes,
(c) the modal matrix **U**, and
(d) the normalized modal matrix $\mathbf{\Phi}$.

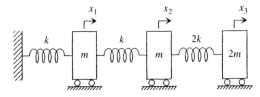

FIGURE P8.38

8.39. Consider the simple mechanical system shown in Fig. P8.39, which is subject to the following sinusoidal input $f(t) = F_0 \cos \omega t$. Obtain the steady-state response for x.

8.40. Repeat Problem 8.39, but use the following:

$$f(t) = F_0 \cos(\omega t + \theta)$$

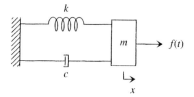

FIGURE P8.39

8.41. Consider the mechanical system shown in Fig. P8.41, which is subject to the following sinusoidal input $f(t) = F_0 \sin \omega t$. Obtain the steady-state response for x_1.

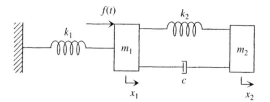

FIGURE P8.41

8.42. Reconsider the system in Problem 8.41, but obtain the steady-state response for x_2.

8.43. The transfer function and the sinusoidal input of a vibration system are given below. Determine the steady-state response.

$$G(s) = \frac{X(s)}{F(s)} = \frac{-5}{s^2 + 5s + 1}, \qquad f(t) = 2\cos\left(5t + \frac{\pi}{6}\right)$$

8.44. The transfer function and the sinusoidal input of a dynamic system are given as shown. Determine the steady-state response.

$$G(s) = \frac{X(s)}{F(s)} = \frac{1}{s(s^2 + 5s + 1)}, \qquad f(t) = 2\sin 3t$$

8.45. Obtain the roots of the following characteristic equation of a vibration system:

$$0.25s^4 + 0.13694s^3 + 0.45s^2 + 0.10955s + 0.16 = 0$$

8.46. Consider the vibration system shown in Fig. P8.46. The system is initially at rest, and suddenly the force $f(t) = F_0 u_s(t)$ is applied, where F_0 and $u_s(t)$ are a constant and a unit-step function, respectively. Obtain the time responses, $x_1(t)$ and $x_2(t)$.

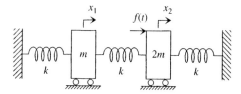

FIGURE P8.46

CHAPTER 9

Block Diagram Representation

The focus of this chapter is on the introduction and implementation of block diagrams representing dynamic systems. The more complicated a system is in nature, the more components become involved and naturally, signal flow has a tendency to become more complex. Up to this point, dynamic systems have been described by their governing equations in the form of ordinary differential equations, and several techniques were introduced in order to treat them. However, for systems with higher orders of complexity, block diagram representation will prove to be a useful and powerful tool to provide a better insight and understanding of the operations involved, as well as the interconnection between the components. It should be noted that many dynamic systems, although possibly quite different in nature, may be represented by block diagrams with the same basic structure. Furthermore, block diagram representation of a dynamic system is *not unique;* that is, different diagrams may be constructed corresponding to the same system, all containing the same information about the dynamic behavior of the system. Block diagrams are constructed based on the relationships between the Laplace transforms of the variables involved. In this chapter, block diagram *reduction techniques*, such as *Mason's rule*, will be introduced enabling one to obtain transfer functions relating different input/output pairs directly from the block diagram.

9.1 BLOCK DIAGRAMS

DEFINITION 9.1. A **block diagram** representation of a system is an interconnection of blocks, each of which represents some mathematical operation performed by a certain component, in such a way that the diagram, in its entirety, agrees with the system's mathematical model. Schematically, each block is identified with an associated transfer function reflecting a certain mathematical operation.

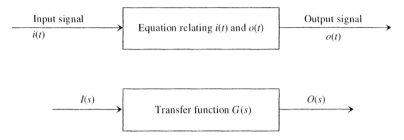

FIGURE 9.1   Schematics of a block.

In Fig. 9.1, a **member** (or **element**) of a block diagram is shown. Here, $i(t)$ and $o(t)$ represent the input and the output associated with the block, respectively, and may be related through an algebraic equation, a differential equation, or an integral equation. In any event, through Laplace transformation, the relation between $I(s)$ and $O(s)$ is desired. Following the definition of the transfer function in Section 3.4, we then have

$$G(s) = \frac{O(s)}{I(s)} \tag{9.1}$$

which is also known as the **gain** of the block.

Block Diagram Operations

The operations most frequently used in block diagrams include algebraic summation of signals, amplification of a signal, and multiplication of transfer functions.

Summing junction

The output is the algebraic sum of signals coming toward it. A plus or minus sign associated with the signal signifies whether it is to be added or subtracted. A typical situation involving four signals is shown in Fig. 9.2. The arrowhead pointing away from the junction is the outcome of the operations performed by the summing junction, i.e., $Y(s) = X_1(s) - X_2(s) + X_3(s)$. Moreover, all input and output signals must be of identical units. A summing junction can have as many inputs as possible, but only a single output signal.

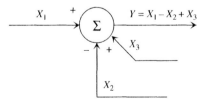

FIGURE 9.2   Summing junction.

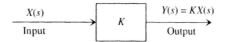

FIGURE 9.3 Constant gain block.

FIGURE 9.4 Transfer function.

Constant gain

Figure 9.3 shows the situation where the output, $Y(s)$, of a block is a constant multiple of its input, $X(s)$. In that case, $Y(s) = KX(s)$, where K is a constant.

Transfer function

Given a linear system subjected to zero initial conditions, the transfer function is the ratio of the Laplace transforms of the output and the input signals. In this case, as shown in Fig. 9.4, we have $O(s) = G(s)I(s)$.

EXAMPLE 9.1. Suppose the input-output equation of a certain dynamic system is given by

$$\ddot{y} + 3\dot{y} + y = Kx$$

where K is a constant. Assuming zero initial conditions, obtain the block relating the input and the output.

Solution. Since we are interested in the relationship between the input and the output, the block must be characterized by the transfer function. The system's transfer function is obtained via Laplace transformation, as

$$G(s) = \frac{Y(s)}{X(s)} = \frac{K}{s^2 + 3s + 1}$$

Subsequently, the block representation is obtained as shown in Fig. 9.5.

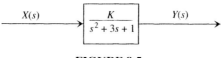

FIGURE 9.5

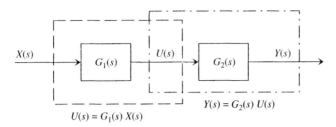

FIGURE 9.6 Transfer functions in series.

Transfer functions combined in series

Consider the series combination of two blocks represented by transfer functions, $G_1(s)$ and $G_2(s)$, as shown in Fig. 9.6. Note that the signal $U(s)$, the output of the first block, acts as the input to the second block. In order to obtain the overall transfer function $Y(s)/X(s)$, we make use of the definition of the transfer function at each block, i.e.,

$$\text{First block} \quad G_1(s) = \frac{U(s)}{X(s)} \quad \Longrightarrow \quad U(s) = G_1(s)X(s)$$

$$\text{Second block} \quad G_2(s) = \frac{Y(s)}{U(s)} \quad \Longrightarrow \quad Y(s) = G_2(s)U(s)$$

Substituting the expression of $U(s)$ given in the first relation into the second, one obtains

$$Y(s) = G_2(s)G_1(s)X(s) \quad \Longrightarrow \quad \frac{Y(s)}{X(s)} = G_2(s)G_1(s) \tag{9.2}$$

As a result, the series combination of the two blocks in Fig. 9.6 may now be replaced by the single-block representation of Fig. 9.7. Of course, this result may be extended to any finite number of blocks combined in series.

Transfer functions combined in parallel

Consider the two blocks represented by transfer functions $G_1(s)$ and $G_2(s)$ arranged in a parallel combination as shown in Fig. 9.8. The input to both blocks is $X(s)$, while the overall output is $Y(s)$. Note that $X(s)$, in its entirety, is an input to the block of $G_1(s)$, as well as the block of $G_2(s)$. Point P in Fig. 9.8 is referred to as a **branch point**.

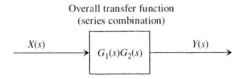

FIGURE 9.7 Equivalent block for a series combination.

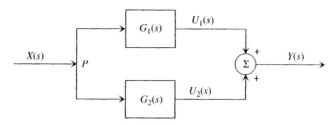

FIGURE 9.8 Transfer functions in parallel.

It is desired to replace this combination with a single-block representation relating the overall input, $X(s)$, to the overall output, $Y(s)$, as in Fig. 9.1. To this end, we need to apply the definition of the transfer function, as well as the property of a summing junction, as

Summing junction $Y(s) = U_1(s) + U_2(s)$

Transfer functions $U_1(s) = G_1(s)X(s),$ $U_2(s) = G_2(s)X(s)$

Substitution from the second relations into the first yields

$$Y(s) = G_1(s)X(s) + G_2(s)X(s) = [G_1(s) + G_2(s)]X(s)$$

as a result of which, we have

$$\frac{Y(s)}{X(s)} = G_1(s) + G_2(s) \tag{9.3}$$

indicating that the overall transfer function is simply the sum of the transfer functions of the individual blocks. As a result, the parallel combination in Fig. 9.8 can be replaced by its equivalent, as shown in Fig. 9.9.

Integration

If a signal, $y(t)$, is defined at any $t > 0$ as

$$y(t) = \int_0^t x(\tau)\,d\tau \tag{9.4}$$

then, by Theorem 1.6, its Laplace transform is determined as $Y(s) = (1/s)X(s)$. This implies that $Y(s)/X(s) = 1/s$, which leads to the block diagram in Fig. 9.10.

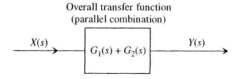

FIGURE 9.9 Equivalent block for a parallel combination.

CHAPTER 9: Block Diagram Representation   491

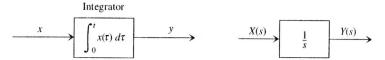

FIGURE 9.10 Integral operation and the corresponding block representation.

EXAMPLE 9.2. Represent the block diagram of Fig. 9.11 as a single block relating $X(s)$ and $Y(s)$, and then obtain the overall transfer function.

Solution. First, the parallel combination is simplified via Eq. (9.3) and the equivalent block is obtained as in Fig. 9.12. Next, replace the parallel combination in Fig. 9.11 by the configuration of Fig. 9.12 to obtain what appears as a series combination of two blocks, Fig. 9.13. Since this is a series combination, the equivalent transfer function is simply the product of the individual transfer functions, Eq. (9.2), i.e.,

$$\frac{Y(s)}{X(s)} = \left(\frac{1}{s+3} + \frac{s+2}{s+5}\right)\left(\frac{3}{s+1}\right) = \frac{3(s^2 + 6s + 11)}{(s+1)(s+3)(s+5)}$$

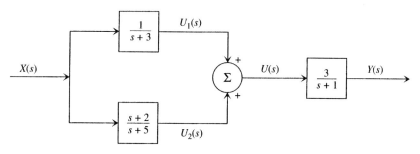

FIGURE 9.11 Block diagram considered in Example 9.2.

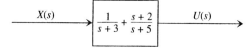

FIGURE 9.12

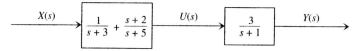

FIGURE 9.13

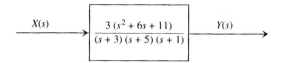

FIGURE 9.14 Block representation equivalent to Fig. 9.11.

Ultimately, the original block diagram of Fig. 9.11 is readily seen to be equivalent to the representation in Fig. 9.14.

Closed-Loop Systems

In the event that an output signal is *fed back* to be compared with an input signal, one speaks of a **closed-loop (feedback) system**. This comparison, performed by a summing junction and the configuration of a simple feedback system, is illustrated in Fig. 9.15, where $E(s)$ denotes the **error signal** and is defined as $E(s) = U(s) - Y(s)$. While feeding back an output, it is essential that it be converted to the same form as that of the input. To this end, the system in Fig. 9.15 is now modified so that it includes a **feedback element**, characterized by transfer function $H(s)$, as shown in Fig. 9.16. When the feedback element is unity, i.e., $H(s) = 1$, the closed loop is called a **negative unity feedback** (Fig. 9.15).

In Fig. 9.16, the product $G(s)H(s)$ is known as the **open-loop transfer function (OLTF)**. This quantity is alternatively defined as the ratio

Open-loop transfer function $\quad \dfrac{H(s)Y(s)}{E(s)} \quad$ (9.5)

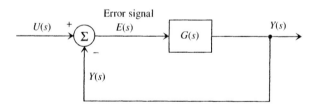

FIGURE 9.15 Unity negative feedback system.

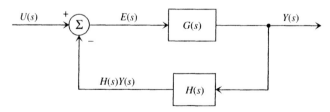

FIGURE 9.16 Closed-loop system including a feedback element.

From Fig. 9.16, we notice that $Y(s) = G(s)E(s)$. Insertion into Eq. (9.5), and simplification, yields

$$\frac{H(s)\overbrace{[G(s)E(s)]}^{Y(s)}}{E(s)} = H(s)G(s)$$

which is indeed what we earlier defined as the *open-loop transfer function*. On the other hand, the **feedforward transfer function** is defined as the ratio

Feedforward transfer function $\quad \dfrac{Y(s)}{E(s)} = G(s) \quad$ (9.6)

The overall transfer function, $Y(s)/U(s)$, for the closed-loop system is known as the **closed-loop transfer function (CLTF)** and is determined as follows. Directly from Fig. 9.16, we have

$$Y(s) = G(s)E(s) \quad (9.7a)$$

$$E(s) = U(s) - H(s)Y(s) \quad (9.7b)$$

Substitute $E(s)$, given by Eq. (9.7b), in Eq. (9.7a) to obtain

$$Y(s) = G(s)[U(s) - H(s)Y(s)]$$

Collecting terms,

$$[1 + G(s)H(s)]Y(s) = G(s)U(s)$$

Ultimately, the desired transfer function is determined, as

Closed-loop transfer function $\quad \dfrac{Y(s)}{U(s)} = \dfrac{G(s)}{1 + G(s)H(s)} \quad$ (9.8)

9.2 BLOCK DIAGRAM FROM GOVERNING EQUATIONS

Having provided the basic concepts and results regarding block diagrams, the natural next step is to learn how to construct a block diagram to represent the governing equation(s) of a dynamic system. The procedure is entirely based on the Laplace transformation of the governing equations, together with the assumption of zero initial conditions. Upon transformation of each of the resulting equations, a block diagram, known as a **partial block**, is constructed. Ultimately, these partial blocks are suitably assembled to represent the system model as a whole. The following example contains some basic steps involved in such construction.

EXAMPLE 9.3. Consider the simple mechanical system in Fig. 9.17 consisting of a spring and a damper. The motion of the spring is characterized by x (input) and the massless point A as y (output). From the equation of motion, construct a block diagram with overall input, $X(s)$, and overall output, $Y(s)$.

Solution. From the free-body diagram, the equation of motion is written as

$$b\dot{y} + ky = kx \quad (9.9)$$

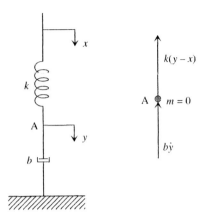

FIGURE 9.17 Mechanical system of Example 9.3.

The desired (overall) block diagram must have an input of $X(s)$ and an output of $Y(s)$. As a first step, take the Laplace transform of Eq. (9.9), assuming zero initial conditions, and rearrange to obtain

$$sY(s) = \frac{k}{b}[X(s) - Y(s)] \qquad (9.10)$$

The corresponding partial block diagram, in agreement with Eq. (9.10), is then constructed as shown in Fig. 9.18. First, notice that the operation $X(s) - Y(s)$ is conducted by a summing junction. Next, since the result is multiplied by k/b, the output signal of the junction must go through a block of constant gain. Subsequently, according to Eq. (9.10), the outcome is the signal $sY(s)$. However, by assumption, the system output is y, so $Y(s)$ must be extracted from the signal, $sY(s)$. This, of course, is possible through one integration of this signal, as shown in Fig. 9.19. Combine the latter with the partial block diagram of Fig. 9.18 to construct the overall block diagram, as shown in Fig. 9.20.

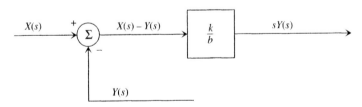

FIGURE 9.18 Partial block diagram corresponding to Eq. (9.10).

FIGURE 9.19

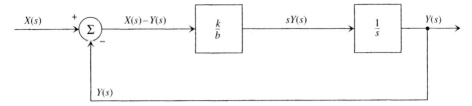

FIGURE 9.20 Block diagram of the mechanical system in Fig. 9.17.

Block Diagram from State-Space Model

Based on the detailed discussions of Chapter 3, state-space form is merely a convenient way of representing the governing equations (model) of a dynamic system. Therefore, the fundamental ideas of block diagram construction developed through Example 9.3 still hold. The following example contains some general steps that are typically undertaken while treating the state-space form.

EXAMPLE 9.4. Consider a dynamic system represented in state-space form as follows.

$$\begin{cases} \dot{\mathbf{x}} = \mathbf{A}\mathbf{x} + \mathbf{B}u \\ y = \mathbf{C}\mathbf{x} \end{cases}$$

where $\mathbf{x} = \begin{Bmatrix} x_1 \\ x_2 \end{Bmatrix}$, $\mathbf{A} = \begin{bmatrix} 0 & 1 \\ -2 & -3 \end{bmatrix}$, $\mathbf{B} = \begin{bmatrix} 0 \\ 1 \end{bmatrix}$, $\mathbf{C} = \begin{bmatrix} 1 & 1 \end{bmatrix}$, $u =$ scalar

Construct the corresponding block diagram representation.

Solution. Interpreting the state and output equations component-wise yields

$$\dot{x}_1 = x_2 \tag{9.11a}$$

$$\dot{x}_2 = -2x_1 - 3x_2 + u \tag{9.11b}$$

$$y = x_1 + x_2 \tag{9.11c}$$

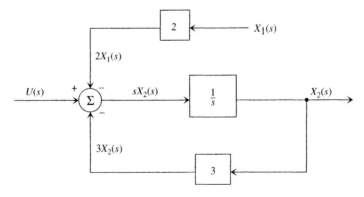

FIG. 9.21 Partial diagram representing Eq. (9.12).

FIGURE 9.22 Integration block representing Eq. (9.13).

To begin with, it is understood that u denotes the system input, while $y = x_1 + x_2$ is the scalar output. With this in mind, it is then clear that the overall block diagram must have $U(s)$ and $Y(s)$ as its input and output, respectively. Laplace transformation of Eq. (9.11b), using zero initial conditions, yields

$$sX_2(s) = -2X_1(s) - 3X_2(s) + U(s) \tag{9.12}$$

The right-hand side of this equation indicates that there must be a summing junction, the output signal of which is $sX_2(s)$. Then, the signal $X_2(s)$ may be extracted via one integration. The partial block diagram associated with Eq. (9.12) is then constructed accordingly, as illustrated in Fig. 9.21. Next, taking the Laplace transform of Eq. (9.11a) results in

$$sX_1(s) = X_2(s) \tag{9.13}$$

for which the partial block diagram is constructed as shown in Fig. 9.22. The Laplace transform of the output equation, Eq. (9.11c), results in

$$Y(s) = X_1(s) + X_2(s) \tag{9.14}$$

for which the corresponding partial block diagram is obtained as in Fig. 9.23. Combination of the three partial block diagrams illustrated in Figs. 9.21–9.23 in a suitable

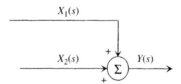

FIGURE 9.23. Block diagram for Eq. (9.14).

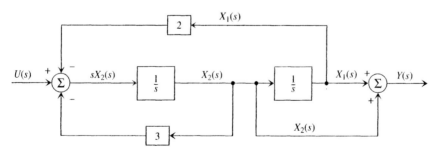

FIG. 9.24 Block diagram for the state-space form in Example 9.4.

manner yields the overall block diagram representation of the state-space formulation above; see Fig. 9.24.

Multiple outputs

Examples 9.3 and 9.4 treated systems with a single input and a single output. The approach implemented throughout this section, however, is capable of handling systems with multiple inputs and outputs. The following example involves a system with a single input and two outputs.

EXAMPLE 9.5. Consider a system for which the state equation is as in Example 9.4 and the output equation is defined as

$$\mathbf{y} = \mathbf{Cx}, \quad \mathbf{C} = \begin{bmatrix} 1 & 0 \\ 0 & 1 \end{bmatrix}$$

Construct the corresponding block diagram.

Solution. From the output equation it can be seen that the output *vector* consists of two components, $X_1(s)$ and $X_2(s)$. However, this is the only feature of the present system that differs from that in the previous example. The steps of the procedure employed in the previous example still hold true except for the partial block diagram associated with the output equation. The resulting overall block diagram for the present system is shown in Fig. 9.25.

Block Diagram from Input-Output Equation

We now turn our attention to block diagram construction from a system's input-output equation. The key is to first obtain the state-space model, via the techniques in Section 3.2, directly from the I/O equation. Subsequently, the information provided by Examples 9.4 and 9.5 can be utilized for the block diagram construction.

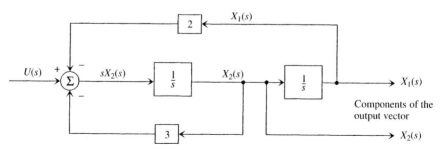

FIGURE 9.25 Block diagram representation of the system in Example 9.5.

EXAMPLE 9.6. The input-output equation for a certain dynamic system is given as

$$\ddot{y} + 2\dot{y} + 3y = 2\dot{u} + u \tag{9.15}$$

where u and y represent the system input and output, respectively. Construct the block diagram.

Solution. Using the methodology introduced in Section 3.5, the state variables are selected as $x_1 = y$ and $x_2 = \dot{y}$, which will result in the following state-space form

$$\dot{x}_1 = x_2 \tag{9.16a}$$
$$\dot{x}_2 = -2x_2 - 3x_1 + u \tag{9.16b}$$
$$\underbrace{y = 2x_2 + x_1}_{\text{output}} \tag{9.16c}$$

Note that this is a single-output system; the block diagram representation of the model is created as in Example 9.4. That is, take the Laplace transform of Eqs. (9.16a) and (9.16b), assuming zero initial conditions, and construct the partial block diagram, as shown in Fig. 9.26. The Laplace transform of the output equation, Eq. (9.16c), results in

$$Y(s) = X_1(s) + 2X_2(s)$$

for which the corresponding partial block diagram is obtained as in Fig. 9.27. Combination of Figs. 9.26 and 9.27 will result in the overall block diagram shown in Fig. 9.28.

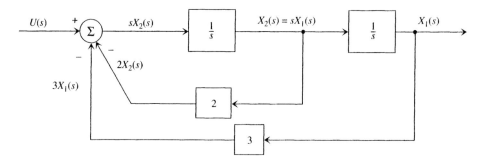

FIGURE 9.26 Partial diagram representing Eqs. (9.16a) and (9.16b).

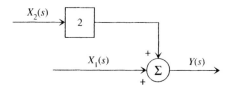

FIGURE 9.27 Block diagram for the output equation.

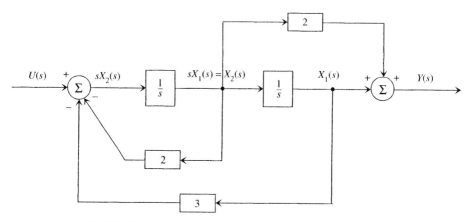

FIGURE 9.28 Block diagram representation of Eq. (9.15).

9.3 BLOCK DIAGRAM REDUCTION

In the process of analyzing the model of a dynamic system, knowledge of the system's transfer function is often required. As extensively discussed in previous chapters, the transfer function enables the analyst to study the system's response to various types of inputs. As systems get more complicated, so do their block diagrams, which makes the determination of the transfer function more difficult and challenging. To circumvent the problems associated with complex block diagrams, it is essential to introduce and employ certain rules to reduce the algebra involved. In doing so, any internal (feedback) loop must be replaced by an equivalent transfer function, and branch points and summing junctions must be moved appropriately. During this process, the product of transfer functions in a feedforward path and around a loop must remain unchanged. Some basic rules concerning block diagram algebra are outlined below.

Moving a Branch Point

Consider the block diagram in Fig. 9.29. The objective is to move the branch point, P, to the right side of the block, $G(s)$. In the meantime, we need to make certain that signals $Y_1(s)$ and $Y_2(s)$ remain unchanged before and after this operation. Note that

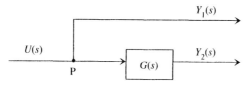

FIGURE 9.29 Branch point.

prior to the move, we have

$$Y_1(s) = U(s)$$

$$Y_2(s) = G(s)U(s)$$

Then, in order for the two signals to be preserved, the branch point is moved as shown in Fig. 9.30. It is then readily observed that the information on $Y_1(s)$ and $Y_2(s)$ has remained unchanged, since

$$Y_1(s) = \frac{1}{G(s)}\{U(s)G(s)\} = U(s)$$

$$Y_2(s) = G(s)U(s)$$

Moving a Summing Junction

Suppose that the summing junction in Fig. 9.31 is to be moved to the left side of the block $G_1(s)$. The signal $Y(s)$ must be preserved under the operation. Prior to the move, we have

$$Y(s) = G_1(s)U_1(s) + G_2(s)U_2(s)$$

In order for $Y(s)$ to maintain this value, we propose the strategy in Fig. 9.32. As a result,

$$Y(s) = G_1(s)\left[U_1(s) + U_2(s)\frac{G_2(s)}{G_1(s)}\right] = G_1(s)U_1(s) + G_2(s)U_2(s)$$

which agrees with what was known prior to the move.

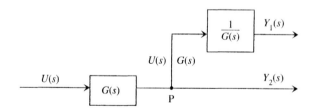

FIGURE 9.30 Branch point relocated.

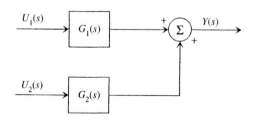

FIGURE 9.31

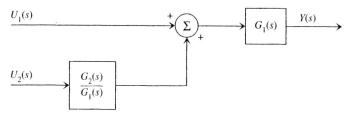

FIGURE 9.32

Negative and Positive Feedback

DEFINITION 9.2. If a certain output signal is fed back and subtracted from the input, the closed loop is referred to as **negative feedback** (as in Fig. 9.16). If subtraction is replaced with addition, it is called **positive feedback**.

In the event that a closed loop, such as that in Fig. 9.16, appears in the overall block diagram of a dynamic system, it can be replaced by a single equivalent block. This block is then characterized by the equivalent transfer function obtained in Eq. (9.8), i.e.,

Negative feedback $\quad \dfrac{Y(s)}{U(s)} = \dfrac{G(s)}{1 + G(s)H(s)}$

As a result, we have what is presented in Fig. 9.33. In the case of positive feedback, it is readily verified that the only difference in the transfer function is the replacement of the positive sign in the denominator by a negative sign, i.e.,

Positive feedback $\quad \dfrac{Y(s)}{U(s)} = \dfrac{G(s)}{1 - G(s)H(s)} \qquad (9.17)$

Thus, the equivalent single block for the closed loop with positive feedback is as shown in Fig. 9.34.

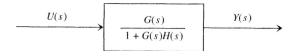

FIGURE 9.33 Equivalent block for the negative feedback of Fig. 9.16.

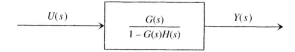

FIGURE 9.34 Equivalent block for the positive feedback.

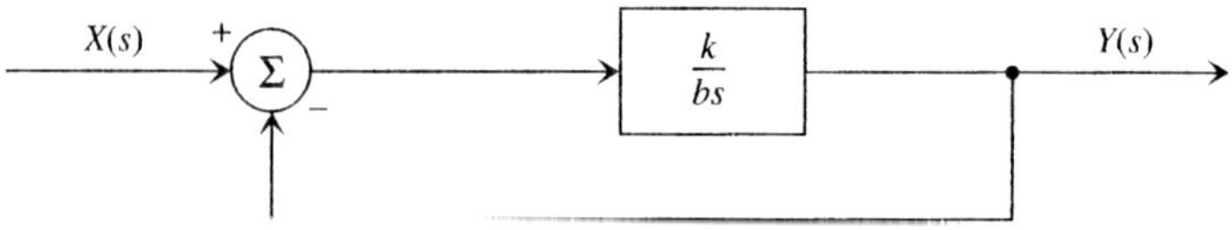

502 Dynamic Systems: Modeling and Analysis

FIGURE 9.35

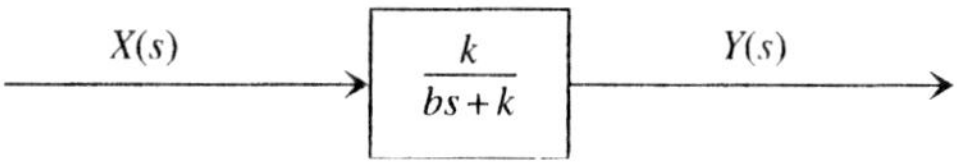

FIGURE 9.36

EXAMPLE 9.7. The block diagram representation of the mechanical system of Example 9.3 was found to be a unity negative feedback, and was shown in Fig. 9.20. Reduce the block diagram and then determine the closed-loop transfer function (CLTF).

Solution. First, observe that the two blocks in the feedforward path are in a series combination, so they can be replaced by a single block whose transfer function is the product of the two individual transfer functions. The resulting diagram is shown in Fig. 9.35. The reduced block diagram is clearly a unity negative feedback, so it can be replaced by its equivalent as shown in Fig. 9.36. The closed-loop transfer function is determined via Eq. (9.8) with $G(s) = k/(bs)$ and $H(s) = 1$; that is,

$$\frac{Y(s)}{X(s)} = \frac{k/(bs)}{1 + k/(bs)} = \frac{k}{bs + k}$$

This result is in complete agreement with what would be obtained via Laplace transformation of the equation of motion, Eq. (9.9).

EXAMPLE 9.8. Consider the block diagram for the state-space form of the system in Example 9.4, now shown in Fig. 9.37. Perform the block diagram reduction and obtain the CLTF subsequently.

Solution. In order to reduce the given diagram, we follow the steps listed below:

Step 1. Consider the negative feedback enclosed by dashed lines (Fig. 9.37) and replace it by its equivalent block diagram with transfer function

$$\frac{1/s}{1 + 3/s} = \frac{1}{s + 3}$$

which results in Fig. 9.38.

Step 2. Move the branch point, P, to the right side of the block $1/s$ to obtain the diagram in Fig. 9.39.

Step 3. The two blocks, enclosed by dashed lines, are in a series combination. Replace them by one block with their product as the transfer function, as shown in Fig. 9.40.

CHAPTER 9: Block Diagram Representation 503

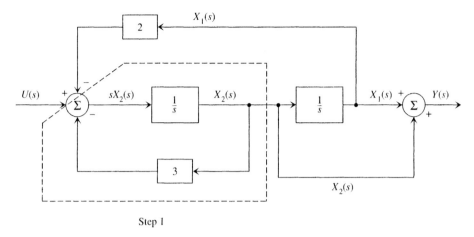

Step 1

FIGURE 9.37

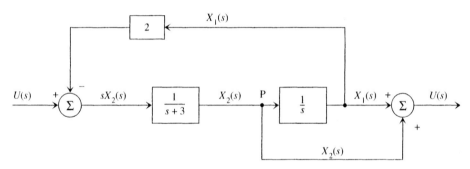

FIGURE 9.38

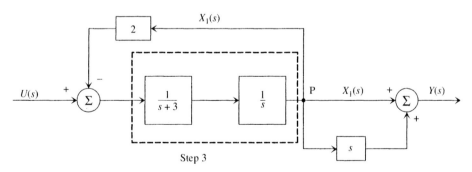

Step 3

FIGURE 9.39

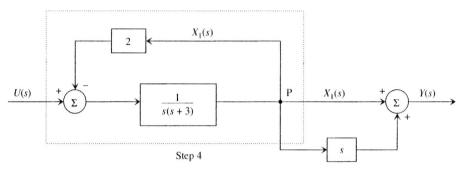

FIGURE 9.40

Step 4. The negative feedback loop shown in Fig. 9.40 is replaced by a single block with transfer function

$$\frac{1/[s(s+3)]}{1 + 2/[s(s+3)]} = \frac{1}{s(s+3) + 2}$$

so that the diagram is reduced to Fig. 9.41.

Step 5. The portion of the diagram in the dashed box consists of two blocks in parallel combination, one with a transfer function of s and the other, 1. Thus, the last block diagram reduces to Fig. 9.42.

Step 6. Finally, the two blocks in the forward path are combined in series; hence, the equivalent transfer function is the product of the individual functions. The closed-

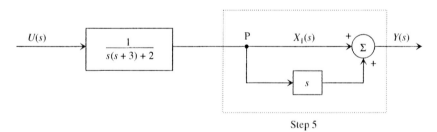

FIGURE 9.41

FIGURE 9.42

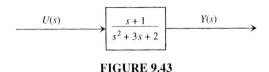

FIGURE 9.43

loop transfer function may now be directly obtained as

$$\frac{Y(s)}{U(s)} = \frac{s+1}{s^2+3s+2}$$

and the block diagram of Fig. 9.37 is replaced by a single block, as shown in Fig. 9.43.

Mason's Rule—Special Case

The overall gain of a block diagram can be calculated via what is known as **Mason's rule.** So far, we have encountered situations where a block diagram contained one or more loops. With the information available, each loop was replaced by its equivalent block using the loop's transfer function, given by either Eq. (9.8) or (9.17), depending on whether the feedback was positive or negative. A **forward path** is one that originates from the overall input and leads to the overall output, without ever moving in the opposite direction. This means that, for instance, the path cannot go through the same variable twice. A **loop path** (or a **loop**) is one that starts from a certain variable and returns to the same variable. The gain of a path, whether forward or loop, is defined as the product of the gains of the individual blocks that make up the path. We first introduce the *special case* of Mason's rule, applicable when *all loops and forward paths touch*. Later, we will discuss the general formulation that does not have this limitation. The special version of Mason's rule states that

$$\text{overall gain} = \frac{\sum \text{gains of forward paths}}{1 - \sum \text{gains of loop paths}} \qquad (9.18)$$

Note that the minus sign in Eq. (9.18) is based on the assumption that every single loop in the diagram is a positive feedback. In the event that one or more negative feedbacks are present, the sign attached to the gain of that particular loop changes to a positive.

EXAMPLE 9.9. Find the closed-loop transfer function for a system with the block diagram representation shown in Fig. 9.44.

Solution. A close look at the diagram reveals that there are two forward paths, starting from the input and leading to the output. Also, there are two loop paths, both returning to the same summing junction. Furthermore, notice that *all paths are closely*

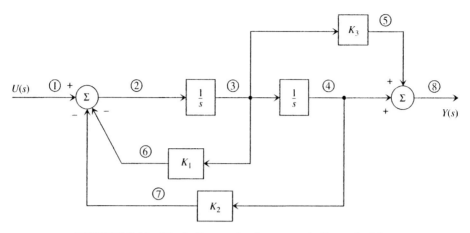

FIGURE 9.44 Block diagram for the system in Example 9.9.

coupled (touch), so the special case of Mason's rule, Eq. (9.18), applies. First, let us identify the gains of all forward and loop paths.

Forward paths		Loop paths	
Path	Gain	Path	Gain
12348	$1/s^2$	12361	K_1/s
12358	K_3/s	123471	K_2/s^2

Since each of the two loops is a negative feedback, a sign adjustment in Eq. (9.18) is required. Insert these quantities into Eq. (9.18) to obtain the closed-loop transfer function as

$$\frac{Y(s)}{U(s)} = \frac{K_3/s + 1/s^2}{1 + K_1/s + K_2/s^2} = \frac{K_3 s + 1}{s^2 + K_1 s + K_2}$$

EXAMPLE 9.10. Find the closed-loop transfer function corresponding to the block diagram of Example 9.8.

Solution. In Example 9.8, the method of block diagram reduction required a considerable amount of effort before the closed-loop transfer function could be determined. However, this diagram certainly meets the conditions required by the special case of Mason's rule, namely, all loops and forward paths touch. Once again, two loops and two forward paths can be identified as in the previous example. Substitution of the appropriate gains in Eq. (9.18) yields the closed-loop transfer function, as

$$\frac{Y(s)}{U(s)} = \frac{1/s + 1/s^2}{1 + 3/s + 2/s^2} = \frac{s + 1}{s^2 + 3s + 2}$$

which is in complete agreement with the result obtained earlier.

Mason's rule, when applicable, is clearly the best possible method for determining the closed-loop transfer function from a system's block diagram representation. Although the special form of this rule, defined by Eq. (9.18), can be implemented for block diagrams representing many physical systems, there are systems that do not meet the required conditions; for instance, a block diagram containing loops that do not touch, as illustrated by the following case study.

EXAMPLE 9.11. Obtain the closed-loop transfer function for the block diagram in Fig. 9.45.

Solution. Mason's rule, Eq. (9.18), does *not* apply since two loops are detected that are not coupled. However, both of these loops are negative feedback; hence, they may be replaced by their equivalent blocks with transfer functions

$$\frac{G_1}{1 + H_1 G_1} \quad \text{and} \quad \frac{G_2}{1 + H_2 G_2}$$

Consequently, Fig. 9.45 reduces to the diagram depicted by Fig. 9.46. Since three blocks are combined in series, the product of their individual gains gives the transfer function of the equivalent single block. Thus, the block diagram in Fig. 9.46 reduces to the diagram shown in Fig. 9.47. As the final step, notice that this diagram consists of two blocks in a parallel combination. Based on earlier results, the transfer function

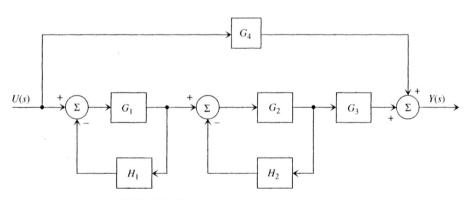

FIGURE 9.45 Block diagram in Example 9.11.

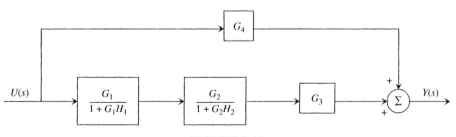

FIGURE 9.46

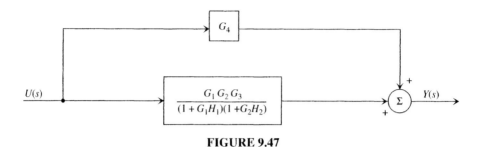

FIGURE 9.47

of the equivalent block is the sum of the individual gains; consequently,

$$\frac{Y(s)}{U(s)} = G_4 + \frac{G_1 G_2 G_3}{(1 + G_1 H_1)(1 + G_2 H_2)}$$

Mason's Rule—General Form

Assume that the input and output of a system are denoted by u and y, respectively. The transfer function is determined by

$$\frac{Y(s)}{U(s)} = \frac{1}{D} \sum_i G_i D_i \quad (9.19)$$

where $D = 1 - \sum$ gains of single loops $+ \sum$ gain product of all *nontouching* two-loops $- \sum$ gain product of all *nontouching* three-loops $- \cdots$
G_i = gain of the ith forward path
D_i = value of D when the block diagram is restricted to the portion not touching the ith forward path

Note: The plus and minus signs in the expression of D are based on the assumption that loops are positive feedback. In the event that a certain loop is a negative feedback, the sign attached to its gain must be appropriately adjusted.

EXAMPLE 9.12. Recall the block diagram in Example 9.11, now shown in Fig. 9.48, to which the special version of Mason's rule did not apply. Use the general form of Mason's rule, Eq. (9.19), to obtain the closed-loop transfer function.

Solution. In doing so, we need to make the distinction between the gains $G_1, G_2, \ldots$ appearing as block gains in Fig. 9.45, and the gain G_i appearing in the general formulation of Mason's rule.

In order to implement Eq. (9.19), we first need to determine the range of the summation index, i. Since there are two forward paths, 123456 10 and 17 10, we

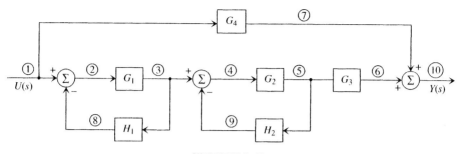

FIGURE 9.48

have $i = 1, 2$. Next, the quantities involved in Eq. (9.19) are identified. Since there exist a total of two single loops, 12381 and 34593, at most the first three terms in D need to be considered. Thus,

$$D = 1 - \sum \text{gains of single loops} + \sum \text{gain product of all nontouching two loops}$$

$$= 1 + \underbrace{G_1 H_1 + G_2 H_2}_{\text{loop gains}} + \underbrace{G_1 H_1 G_2 H_2}_{\text{nontouching 2 loops}}$$

In accordance with the foregoing *Note*, the signs attached to the single loop gains have been switched to positive. Next, the gains of the two forward paths just identified are calculated from Eq. (9.19) as

$$G_1 = G_1 G_2 G_3$$

$$G_2 = G_4$$

Finally, let us consider D_1 in Eq. (9.19). D_1 is that portion of D not touching the first forward path, 123456 10. But we notice that this forward path touches all paths of D; hence,

$$D_1 = 1 - 0 = 1$$

D_2 is that portion of D not touching the second forward path, 17 10. From Fig. 9.48, it is clear that this forward path does not touch any path of D, so

$$D_2 = D$$

Substitution of these quantities into Eq. (9.19) yields the closed-loop system transfer function as

$$\frac{Y(s)}{U(s)} = \frac{G_1 G_2 G_3 + G_4[1 + G_1 H_1 + G_2 H_2 + G_1 H_1 G_2 H_2]}{1 + G_1 H_1 + G_2 H_2 + G_1 H_1 G_2 H_2}$$

$$= \frac{G_1 G_2 G_3}{(1 + G_1 H_1)(1 + G_2 H_2)} + G_4$$

which is precisely what was determined earlier in Example 9.11.

SUMMARY

The *series* combination of blocks with gains $G_1, G_2, \ldots, G_n$ can be replaced with an equivalent single block whose gain is $G_1 G_2 \ldots G_n$. The equivalent single block for a *parallel* combination is $G_1 + G_2 + \cdots + G_n$. A *closed-loop* (or *feedback*) system refers to a situation where an output signal is fed back for comparison with an input signal. When the feedback element is unity, the system is called a unity feedback. The closed-loop transfer function (CLTF) of a negative feedback with a feedforward transfer function $G(s)$ and a feedback element $H(s)$ is given by

$$\frac{G(s)}{1 + G(s)H(s)}$$

For positive feedback, the plus sign changes to a minus sign. Block diagrams corresponding to each of the governing equations of a system are called *partial block diagrams*. Suitable assemblage of partial blocks forms the complete diagram.

A *branch point* P can be relocated according to the schematics in Figs. 9.29 and 9.30. A *summing junction* can be moved as shown in Figs. 9.31 and 9.32.

A *forward path* is originated from the system input and leads to the system output, without ever moving in the opposite direction. A *loop* starts from a certain variable and returns to the same variable. In the event that all forward paths and loops touch, a special case of *Mason's rule* states that the overall gain is given by

$$\frac{\sum \text{forward path gains}}{1 - \sum \text{loop gains}}$$

and is based on the assumption that every loop is a positive feedback. If a loop is negative, the corresponding term in the denominator will have a positive sign attached to it.

PROBLEMS

9.1. Consider the two block diagrams shown in Figs. P9.1(a) and P9.1(b). For each case, use Mason's rule to obtain the closed-loop transfer function. How do the results compare? In case they agree, which rule of block diagram reduction can be used to justify this?

9.2. Obtain a block diagram representation for the following state-space form with a single input u

$$\begin{cases} \dot{\mathbf{x}} = \mathbf{A}\mathbf{x} + \mathbf{B}u \\ y = \mathbf{C}\mathbf{x} \end{cases}, \quad \mathbf{A} = \begin{bmatrix} -2 & 4 \\ -1 & -3 \end{bmatrix}, \quad \mathbf{B} = \begin{bmatrix} 2 \\ 0 \end{bmatrix}, \quad \mathbf{C} = [0 \quad 1], \quad \mathbf{x} = \begin{Bmatrix} x_1 \\ x_2 \end{Bmatrix}$$

9.3. Repeat Problem 9.2 when the output matrix is $\mathbf{C} = [1 \quad 2]$.

9.4. Repeat Problem 9.2 when the output matrix is

$$\mathbf{C} = \begin{bmatrix} 1 & 0 \\ 0 & 1 \end{bmatrix}$$

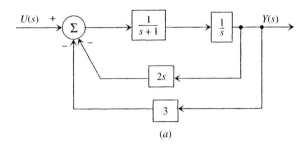

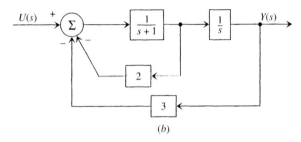

FIGURE P9.1

9.5. The equations of motion for a mechanical system are described by
$$\begin{cases} m\ddot{x}_1 + b_1\dot{x}_1 + k_1(x_1 - x_2) = f(t) \\ b_2\dot{x}_2 - k_1(x_1 - x_2) + k_2 x_2 = 0 \end{cases}$$
where x_1 and x_2 denote the displacements of a block of mass m and a massless point, respectively. Assume that the system input is the applied force $f(t)$, and the output is x_2.
(a) Determine the state-space representation for the governing equations.
(b) Construct a block diagram.
(c) Obtain the system transfer function via the block diagram.

9.6. Repeat parts (a) and (b) of Problem 9.5, assuming that the output vector is
$$\mathbf{y} = \begin{Bmatrix} x_1 \\ x_2 \end{Bmatrix}$$

9.7. Obtain the governing equations for the RC circuit shown in Fig. P9.7. Then, construct a block diagram. The input and output voltages, e_i and e_o, are assumed to be the system input and output, respectively.

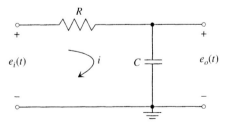

FIGURE P9.7

9.8. The input-output equation of a certain system is described by $\dddot{y} + 2\ddot{y} + \dot{y} + 3y = 2\ddot{u} + u$.
 (a) Express the input-output equation in state-space form.
 (b) Find a block diagram representation for the state-space model.
 (c) Using block diagram reduction and Mason's rule, obtain the closed-loop transfer function. Does your answer agree with the transfer function directly obtained from the I/O equation?

9.9. For the block diagram shown in Fig. P9.9, find the closed-loop transfer function via
 (a) Block diagram reduction
 (b) Mason's rule

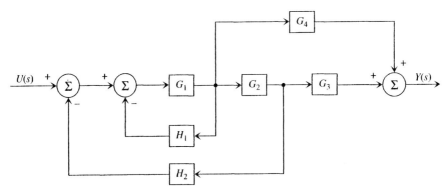

FIGURE P9.9

9.10. The equation of motion for a rotational (mechanical) system follows
$$J\ddot{\theta} + B\dot{\theta} + K\theta = \tau(t)$$
where θ and $\tau(t)$ denote the angular displacement and applied torque, respectively. Physical parameters J, B, and K are constants. System output is assumed to be the angular displacement.
 (a) By choosing suitable state variables, obtain the state-variable equations.
 (b) Obtain the block diagram representation for this system.
 (c) Directly from the block diagram, find the closed-loop transfer function.

9.11. Find the unit-step response of the system whose block diagram representation is shown in Fig. P9.11.

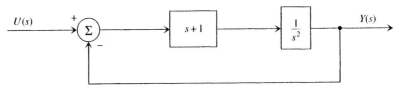

FIGURE P9.11

9.12. Replace the partial block diagram of Fig. P9.12 with one in which the summing junction is moved to the right of the block $G(s)$.

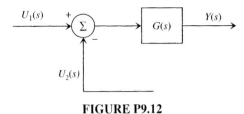

FIGURE P9.12

9.13. Consider the two block diagrams in Figs. P9.13(a) and (b). In Figure P9.13(b), what is the gain of the block that has been left blank so that the two diagrams are equivalent?

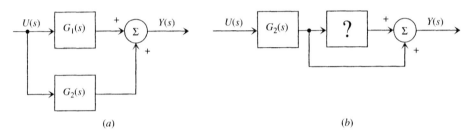

FIGURE P9.13

9.14. Find the system transfer function, $Y(s)/X(s)$, corresponding to the block diagram in Fig. P9.14.

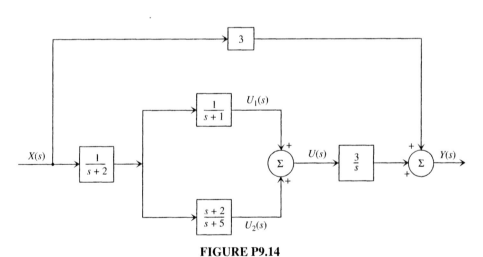

FIGURE P9.14

9.15. The transfer function for a certain dynamic system is given by

$$\frac{Y(s)}{U(s)} = \frac{3}{s^3 + 3s^2 + s + 1}$$

514   Dynamic Systems: Modeling and Analysis

(a) Write the input-output equation and determine the state-space form accordingly.
(b) Construct a block diagram.
(c) Using Mason's rule, find the overall transfer function. How does this compare with the transfer function just given?

9.16. In the block diagram of Figure P9.16, assume that $\zeta < 1$.
  (a) Find the closed-loop transfer function.
  (b) Determine the system's unit-step response.

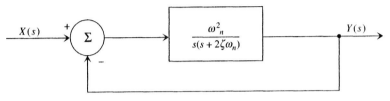

FIGURE P9.16

9.17. Reconsider the block diagram of Figure P9.16 and assume $\zeta < 1$. Determine the system's unit-impulse response.

9.18. Repeat Problem 9.17 for the case when $\zeta > 1$.

9.19. Consider the block diagram representation of the system shown in Fig. P9.19 in which $W(s)$ may be thought of as a disturbance input. Using Mason's rule, find
  (a) The transfer function $Y(s)/X(s)$ that relates the input $X(s)$ to the output $Y(s)$, assuming that $w = 0$
  (b) The transfer function $Y(s)/W(s)$ that relates the disturbance input $W(s)$ to the output $Y(s)$, with the assumption that $x = 0$

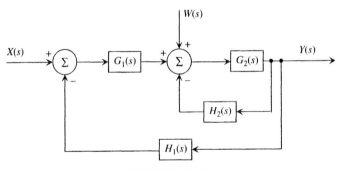

FIGURE P9.19

9.20. Consider the block diagram in Fig. P9.19. Suppose that

$$G_1(s) = \frac{2}{s}, \quad G_2(s) = \frac{1}{s}, \quad H_1(s) = 2, \quad H_2(s) = 3$$

(a) Assuming $x = 0$, determine $Y(s)/W(s)$.
(b) Given that w is a unit step, find the corresponding response, y.

9.21. Repeat Problem 9.20 for the case when w is a unit ramp.

9.22. Consider the block diagram in Fig. P9.22. Obtain the overall transfer function $Y(s)/U(s)$ via
   (a) Block diagram reduction
   (b) The general form of Mason's rule

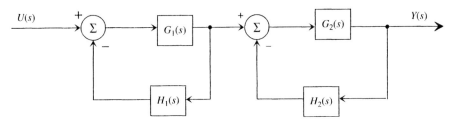

FIGURE P9.22

9.23. In Figure P9.22, let $G_1(s) = k = G_2(s)$ and $H_1(s) = s + 1$, $H_2(s) = 2s + 1$, where k is a positive constant.
   (a) Find the overall transfer function.
   (b) Determine the system's unit-ramp response.

9.24. Repeat Problem 9.23 for the case when the system's unit-step response is sought.

CHAPTER 10

Introduction to Control Systems

This chapter provides the reader with the elementary aspects of control systems. Included are the fundamentals of control: the Routh-Hurwitz stability criterion and the Ziegler-Nichols methods of tuning. Various types of controllers are discussed: proportional control, proportional-plus-derivative control, proportional-plus-integral control, and proportional-plus-integral-plus-derivative control. The chapter also contains important concepts of dynamic systems, which are essential for the study of dynamics and control.

10.1 INTRODUCTION

Control systems are ubiquitous and play a critical role in many aspects of our lives. They exist in many systems of engineering and sciences. Thanks to high technology, we are able to fly thousands of miles in a matter of hours in safe and comfortable airplanes. In these airplanes, many control systems are employed to maintain certain conditions. The goal of an automatic control is to use feedback to control the output of a dynamic process by comparing it to a desired reference input accurately, despite any external disturbance inputs. As another example, when making long journeys by car, some of us use cruise-control features to maintain constant speeds, which would be tiring if done manually.

This chapter begins with definitions of elementary terms. A historical perspective and examples that illustrate the usefulness of control systems follow. The examples may provide a strong motivation for studying control theory and its applications.

10.2 DEFINITIONS

Open-loop systems are simple and not very expensive, whereas closed-loop systems are more complex and usually cost more. However, closed-loop systems are far superior in performance to open-loop ones. In control systems, the following terms and concepts are useful.

Closed-loop systems. All closed-loop systems have feedback loops in which the signals from the sensors are fed back to influence the control input. An automobile cruise-control system serves as an excellent example. When the car goes uphill or downhill, the speed is adjusted and maintained at a constant value by feedback mechanisms. Another example of a closed-loop system is a heating-ventilating-air-conditioning (HVAC) system. The temperature inside a building is maintained at a constant level by feedback mechanisms regardless of whether the building heat is lost or gained.

Disturbance inputs. A disturbance input, as the words suggest, is an input signal that affects the system response, but that we have no control over. For example, while we are driving on a freeway, an earthquake suddenly occurs. We have no control over the earthquake signals, although we do have control of the resulting motions of the car by means of feedback mechanisms.

Negative feedbacks and positive feedbacks. Negative feedbacks are useful because they try to stabilize the system, whereas positive feedbacks would tend to make the system become unstable.

Open-loop systems. Unlike closed-loop systems, open-loop systems have no feedback. Open-loop systems are control systems where the control action is operated by a timer that is independent of output. The functions of such systems are not influenced by external conditions. An example is a sprinkler system that waters a lawn. The waterings are predetermined for certain times and time intervals. Whether the sun shines or a storm is present (the lawn is dry or wet), the sprinkler system operates exactly the same way.

Plants and processes. The term *plant* refers to a device, component, or system, such as a motor or a chemical plant, whereas *process* refers to an operation, such as a mechanical or biochemical process. They are, however, often used interchangeably in the modeling of dynamic systems and in control engineering.

Process control or regulator. In this type of control, the output follows the input, which is usually constant.

Servomechanisms. A servomechanism is a control system whose mechanical variables are controlled. These variables may be mechanical position, velocity, or acceleration. For example, an accelerometer with a built-in control system is a servomechanism. In this type of control, the output follows the input, which varies.

10.3 ESSENTIAL COMPONENTS OF CONTROL SYSTEMS

The essential components of a control system are

- Plant
- Sensor(s)
- Actuator(s)
- Controller

Any control system must have at least one sensor and one actuator. Sensors and actuators are generally expensive, so control-system designers often try to minimize their use.

10.4 HISTORICAL PERSPECTIVE

The following time line serves as a brief chronological history of development in automatic control, which gives a useful perspective:

1728	Watt	Flyball governor
1868	Maxwell	Stability analysis of flyball governor
1877	Routh-Hurwitz	Stability criterion
1890	Liapunov (Lyapunov)	Stability of nonlinear systems
1927	Black	Feedback electronic amplifier
1932	Nyquist	Nyquist stability criterion
1938	Bode	Frequency response methods and Bode plots
1942	Ziegler-Nichols	PID tuning
1947	Nichols	Nichols chart
1948	Evans	Root locus
1957	Bellman	Dynamic programming

The flyball governor is probably one of the most fundamental and important control systems in mechanical engineering (Fig. 10.1). In this speed control system, the engine is the **plant** and its speed is the **controlled output.** Suppose the engine is operating at an equilibrium point, a desired speed or **reference input.** Suddenly, a small variation in the load occurs due to a **disturbance input.** When the load is decreased, the engine speeds up. In this system, the ball linkage is used as a **sensor** to measure the engine speed (controlled output). As the balls of the governor fly out, the levers cause the butterfly valve to allow less steam to the engine. The lever mechanism and the butterfly valve can be considered as the **controller** and the **actuator,** respectively. The end result is to decrease the engine speed to the original value. If the load is increased, the reverse situation occurs. The engine speed decreases, and the ball makes a smaller cone. Thus, the levers make the butterfly valve open wider to allow more steam to enter the engine, in an effort to restore the speed to the original value.

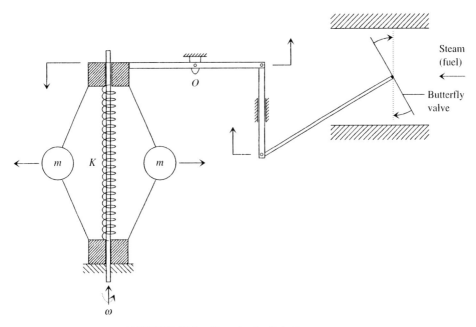

FIGURE 10.1 Sketch of a flyball governor.

10.5 GENERAL BLOCK DIAGRAM FOR CONTROL SYSTEMS

Transfer functions are used to represent a mathematical model of each block, especially in the block-diagram format. On the block diagram all the signals are transfer functions. For clarity, the time functions are represented by lowercase letters, whereas the transfer functions are represented by capital letters. For example, the time function reference input is $r(t)$, and its Laplace transform is $R(s)$. Notice that s is the *Laplace-transform variable,* and it is also called the *complex frequency.*

Basic principles of feedback must be well understood before studying various methods in the analysis and design of control systems. This can be best achieved by understanding the fundamentals of control first. A general block diagram of control systems is given in Fig. 10.2. In this system, the variables are defined as

$G_p(s)$ = transfer function of the plant (process)
$G_c(s)$ = transfer function of the controller
$G_a(s)$ = transfer function of the actuator
$H(s)$ = transfer function of the sensor (feedback)
$R(s)$ = reference input
$U(s)$ = control input
$Y(s)$ = controlled output
$D(s)$ = disturbance input
$E_a(s)$ = actuating error

$$E_a(s) = R(s) - H(s)Y(s)$$

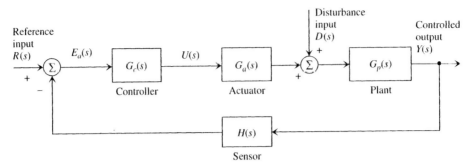

FIGURE 10.2 General block diagram of a control system.

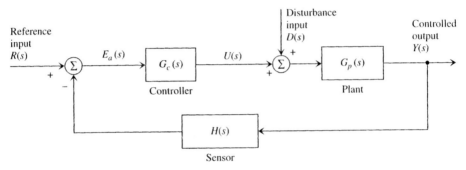

FIGURE 10.3 General block diagram of control systems (without actuator block).

$E(s) =$ (true) error

$$E(s) = R(s) - Y(s)$$

Notice that the actuating error and the (true) error become the same for unity feedback, $H(s) = 1$.

A simplified block diagram is often used where the actuator block is removed (Fig. 10.3). In this case, the transfer function of the actuator is assumed to be unity, $G_a(s) = 1$. If the dynamics of the actuator cannot be ignored, its transfer function may be embedded in the plant or in the controller, for block-diagram representation.

10.6 FURTHER EXAMPLES OF CONTROL SYSTEMS

Heating Control Systems: an Engineering Application

When the weather gets cold, many of us use a heating system with feedback control at home. Figure 10.4 shows a building, a physical model of an automatic control heating system, and the corresponding block diagram. A typical thermostat is constructed by using a bimetallic strip that expands or contracts when the temperature increases or decreases. The thermostat breaks or makes the electrical contact

CHAPTER 10: Introduction to Control Systems   521

(a)

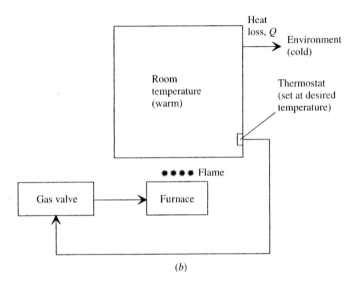

(b)

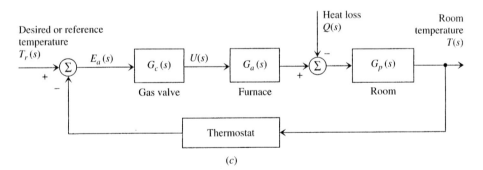

(c)

FIGURE 10.4 A heating system using negative feedback: (a) building, (b) physical model, (c) block diagram.

accordingly. In this system, the room temperature is maintained at or near the desired temperature by a negative feedback mechanism. Within some tolerable range, when the room temperature gets below the desired temperature, say 70°F, the thermostat (sensor) activates the gas valve (controller) and the furnace (actuator) to provide heat to the house. When the room temperature rises above the desired temperature, the thermostat breaks the contact, which in turn shuts off the valve and furnace to stop the furnace from adding heat to the house. Since the house (plant) loses heat to the cold environment, a negative sign is placed at the summing junction.

Heating Control Systems: Homeostasis, a Biological Condition

Homeostasis is defined as the capability of living organisms to maintain internal conditions, within some tolerable range, when external conditions change. For example, birds have negative feedback mechanisms to maintain body temperature when outside temperatures drop significantly. One of the mechanisms is to decrease heat loss and to increase metabolic heat. The bird's temperature sensors signal the brain, which in turn sends signals to integument (skin) cells. These cells control feather movements and create feather fluffing, which retains body heat. In addition, the body receives metabolic heat. When the body temperature rises above the desired temperature, the temperature sensors deactivate the feather fluffing, and the metabolic heat also decreases. Therefore, the desired body temperature is maintained.

10.7 POLES AND ZEROS

Knowledge of the location of poles and zeros is critical in the analysis and design of dynamic and control systems. Poles and zeros define transfer functions. The pole-and-zero location of a system of any order can be shown using the **complex s-plane**.

After the transfer function of a dynamic system is determined, the poles and zeros can be obtained readily. Recall from Chapter 1 that to determine the zero(s) of a dynamic system, we set the numerator of the transfer function to zero, $N(s) = 0$, and then solve for the roots. These roots are called the **zeros**. Similarly, if we equate the denominator of the transfer function with zero, $D(s) = 0$, and then solve for the roots; these roots are called the **poles**. In the complex s-plane, the poles are marked with an × and the zero is represented by a circle.

Again, note that the denominator of the transfer function is the characteristic polynomial (CP). When it is set to zero, the resulting equation is called the characteristic equation (CE), $D(s) = 0$.

As an example, consider the following transfer function of a dynamic system:

$$\frac{X(s)}{Y(s)} = \frac{N(s)}{D(s)} = \frac{s + 0.5}{s^2 + 3s + 2} = \frac{s + 0.5}{(s + 1)(s + 2)}$$

Setting $N(s) = 0$,

$$N(s) = s + 0.5 = 0$$

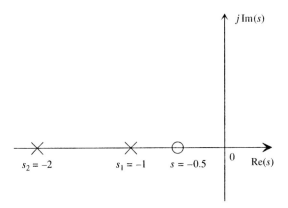

FIGURE 10.5 Location of poles and zero.

Thus, the zero is $s = -0.5$. The characteristic polynomial is $D(s)$, and setting $D(s) = 0$,

$$D(s) = (s + 1)(s + 2) = 0$$

Hence, the poles are $s_1 = -1$ and $s_2 = -2$. The poles and the zero are shown in Fig. 10.5.

First-Order Systems: Time Constant

If the differential equation of a first-order system is given as

$$\tau \dot{v} + v = f(t) \tag{10.1}$$

then the characteristic equation is

$$\tau s + 1 = 0 \tag{10.2}$$

Thus, the pole is

$$s = -\frac{1}{\tau} \tag{10.3}$$

Since the time constant τ must be positive, it is given as

$$\tau = \frac{1}{|s|} \tag{10.4}$$

The pole location of a first-order system is shown in Fig. 10.6, in the complex s-plane. As an example, let us consider a first-order mechanical system (Fig. 10.7). The differential equation of motion is given in the standard input-output form as

$$m\ddot{x} + c\dot{x} = f(t) \tag{10.5}$$

In order to transform Eq. (10.5) into the standard form, Eq. (10.1), let $\dot{x} = v$. As a result,

$$m\dot{v} + cv = f(t) \tag{10.6}$$

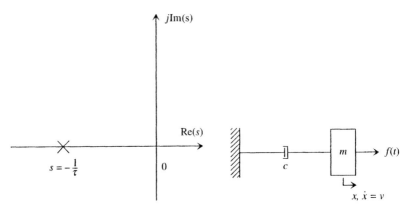

FIGURE 10.6 Pole location of a first-order system.

FIGURE 10.7 First-order system.

which is of the first order, in v. Dividing both sides of the equation by c, we have

$$\frac{m}{c}\dot{v} + v = \frac{f(t)}{c} \tag{10.7}$$

or

$$\tau \dot{v} + v = \frac{f(t)}{c} \tag{10.8}$$

where the *time constant* τ is defined as

$$\tau = \frac{m}{c} \tag{10.9}$$

EXAMPLE 10.1. Obtain the time constant of the following differential equation.

$$8\dot{v} + 2v = f(t)$$

Solution. Dividing both sides of the equation by 2,

$$4\dot{v} + v = \frac{f(t)}{2}$$

Thus, the time constant is

$$\tau = 4$$

Second-Order Systems: Natural Frequency and Damping Ratio

Consider the general differential equation of second-order systems given as

$$\ddot{x} + 2\zeta\omega_n\dot{x} + \omega_n^2 x = f(t) \tag{10.10}$$

The resulting characteristic equation is

$$s^2 + 2\zeta\omega_n s + \omega_n^2 = 0 \tag{10.11}$$

and the poles are the roots of this equation,

$$\begin{matrix} s_1 \\ s_2 \end{matrix} = -\zeta\omega_n \pm \omega_n \sqrt{\zeta^2 - 1} \tag{10.12}$$

For $\zeta > 1$ (an *overdamped* system), the roots (poles) are

$$\begin{matrix} s_1 \\ s_2 \end{matrix} = -\zeta\omega_n \pm \omega_n \sqrt{\zeta^2 - 1} \tag{10.13}$$

For $\zeta = 1$ (a *critically damped* system),

$$\begin{matrix} s_1 \\ s_2 \end{matrix} = -\omega_n \tag{10.14}$$

For $0 < \zeta < 1$ (an *underdamped* system),

$$\begin{matrix} s_1 \\ s_2 \end{matrix} = -\zeta\omega_n \pm \omega_n \sqrt{-(1-\zeta^2)} = -\zeta\omega_n \pm j\omega_n \sqrt{1-\zeta^2} \tag{10.15}$$

For convenience, the following are often used:

$$\sigma = \zeta\omega_n, \qquad \omega_d = \omega_n\sqrt{1-\zeta^2} \qquad (\zeta < 1) \tag{10.16}$$

Thus, the poles of the system are written as

$$\begin{matrix} s_1 \\ s_2 \end{matrix} = -\sigma \pm j\omega_d \tag{10.17}$$

For $\zeta = 0$ (an *undamped* system),

$$\begin{matrix} s_1 \\ s_2 \end{matrix} = \pm j\omega_n \tag{10.18}$$

Note that the overdamped and critically damped cases actually belong to one group, whereas the underdamped and undamped systems belong to another. The time responses of the overdamped and critically damped systems have no oscillation; on the other hand, those of the underdamped and undamped systems have numbers of oscillations.

Figure 10.8 shows the pole location of a second-order system with various cases of damping ratio. As an example, let us consider a simple mechanical system of spring, mass, and damper (Fig. 4.19), which is a classical second-order system. The differential equation of motion is given as

$$m\ddot{x} + c\dot{x} + kx = f(t) \tag{10.19}$$

Dividing both sides of the equation by m, we have

$$\ddot{x} + \frac{c}{m}\dot{x} + \frac{k}{m}x = \frac{f(t)}{m} \tag{10.20}$$

The *natural frequency* ω_n and *damping ratio* ζ are defined as

$$\omega_n = \sqrt{\frac{k}{m}} \qquad \zeta = \frac{c}{c_{cr}} = \frac{c}{2m\omega_n} = \frac{c}{2\sqrt{km}} \tag{10.21}$$

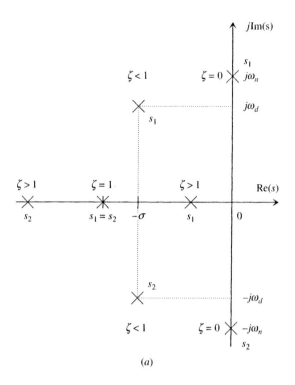

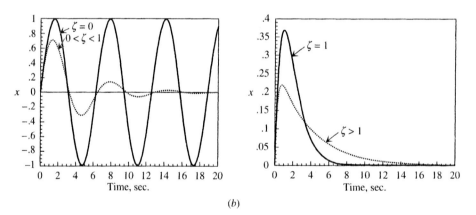

FIGURE 10.8 Second-order system subject to a unit-impulse function: (*a*) pole location, (*b*) time response. For these unit-impulse responses, the exact values for $\zeta < 1$ and $\zeta > 1$ are $\zeta = 0.25$ and 2, respectively.

and c_{cr} is the *critical* damping. It is called critical damping because the damping ratio, ζ, is equal to 1.0 when $c = c_{cr}$. Thus,

$$\frac{c}{m} = 2\zeta\omega_n, \qquad \frac{k}{m} = \omega_n^2 \qquad (10.22)$$

Therefore, the differential equation becomes

$$\ddot{x} + 2\zeta\omega_n \dot{x} + \omega_n^2 x = \frac{f(t)}{m} \qquad (10.23)$$

Higher-Order Systems

Complex dynamic systems are usually of higher order: third order, fourth order, and higher. No matter how complex the systems are, they are composed of two fundamental building blocks: first order and second order.

The system shown in Fig. 10.9 serves as an example of a complex dynamic system. It is a two-degree-of-freedom (TDOF) system, a subclass of multiple-degree-of-freedom (MDOF) systems. We will show that it is a fourth-order system. Note that the physical system shown in the preceding section is single-degree-of-freedom (SDOF) and second-order.

The differential equations are given in the second-order matrix form as

$$\begin{bmatrix} m_1 & 0 \\ 0 & m_2 \end{bmatrix} \begin{Bmatrix} \ddot{x}_1 \\ \ddot{x}_2 \end{Bmatrix} + \begin{bmatrix} c_1 + c_2 & -c_2 \\ -c_2 & c_2 \end{bmatrix} \begin{Bmatrix} \dot{x}_1 \\ \dot{x}_2 \end{Bmatrix} + \begin{bmatrix} k_1 + k_2 & -k_2 \\ -k_2 & k_2 \end{bmatrix} \begin{Bmatrix} x_1 \\ x_2 \end{Bmatrix} = \begin{Bmatrix} f_1 \\ f_2 \end{Bmatrix}$$

After taking the Laplace transform, with zero initial conditions, and then using Cramer's rule, we obtain

$$X_1(s) = \frac{m_2 s^2 + c_2 s + k_2}{\Delta(s)} F_1(s) + \frac{c_2 s + k_2}{\Delta(s)} F_2(s)$$

$$X_2(s) = \frac{c_2 s + k_2}{\Delta(s)} F_1(s) + \frac{m_1 s^2 + (c_1 + c_2)s + k_1 + k_2}{\Delta(s)} F_2(s)$$

where $\Delta(s)$ is the characteristic polynomial of this fourth-order system, which is given as

$$\Delta(s) = m_1 m_2 s^4 + [m_1 c_2 + m_2(c_1 + c_2)]s^3 + [m_1 k_2 + c_1 c_2 + m_2(k_1 + k_2)]s^2 + (c_1 k_2 + c_2 k_1)s + k_1 k_2$$

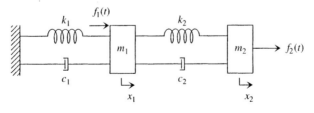

FIGURE 10.9 A TDOF and fourth-order system.

528 Dynamic Systems: Modeling and Analysis

The corresponding characteristic equation is

$$m_1 m_2 s^4 + [m_1 c_2 + m_2(c_1 + c_2)]s^3 + [m_1 k_2 + c_1 c_2 + m_2(k_1 + k_2)]s^2$$
$$+ (c_1 k_2 + c_2 k_1)s + k_1 k_2 = 0$$

(10.24)

The roots of higher-order equations can be obtained by using a *symbolic* software package such as Maple or Mathematica. If the numerical values of the parameters are given, then the roots are determined readily using a *numerical* software package such as MATRIX$_x$ or Xmath or MATLAB. Note that quadratic equations can be solved easily by using the quadratic formula, whereas for higher-order equations the formulas (see a mathematical handbook) are so complicated that it is impractical for hand calculations.

For higher-order systems, such as this one, it is not obvious what the time constants, natural frequencies, and damping ratios are. One would need to solve for the roots of the characteristic equation, and then determine them appropriately. The concept of poles and zeros, which is essential in understanding control systems, was discussed previously.

EXAMPLE 10.2. Find the roots of the following characteristic equation.

$$2s^4 + 11s^3 + 28s^2 + 42s + 24 = 0$$

Solution. Using a computer software package, such as MATRIX$_x$, the poles of this fourth-order system are obtained as

$$s_1 = -1.0966, \qquad s_{2,3} = -0.9593 \pm j1.8644, \qquad s_4 = -2.4848$$

Damping Ratio and ζ-Line

Figure 10.10 shows the essential information regarding the damping ratio ζ for underdamped second-order systems. For these underdamped systems, the ζ-line is defined as the straight line connecting the origin to the pole s_1, in the complex s-plane. From the geometry shown in Fig. 10.10, we can obtain a simple relationship for the damping ratio ζ and its associated angle θ,

$$\sin \theta = \frac{\zeta \omega_n}{\omega_n} = \zeta$$

or

$$\theta = \sin^{-1} \zeta$$

(10.25)

If the value of the damping ratio ζ is given, the angle θ may be obtained readily, and vice versa. For example, if the damping ratio is given as $\zeta = 0.707$, then the

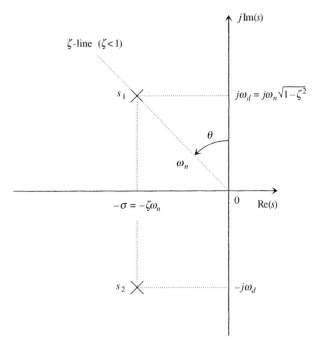

FIGURE 10.10 Damping ratio and ζ-line for underdamped second-order systems.

angle is computed as $\theta = 45°$. On the other hand, if the angle is given, say $\theta = 30°$, then the damping ratio is computed as $\zeta = 0.5$.

EXAMPLE 10.3. Consider the mechanical system shown in Fig. 10.11.

(a) Obtain the poles and show their locations in the complex s-plane.
(b) Obtain the differential equation in a_1 where $a_1 = \ddot{x}_1$. What is the order of the system?

Solution

(a) The differential equations are given in the second-order matrix form as

$$\begin{bmatrix} m & 0 \\ 0 & m \end{bmatrix} \begin{Bmatrix} \ddot{x}_1 \\ \ddot{x}_2 \end{Bmatrix} + \begin{bmatrix} k & -k \\ -k & k \end{bmatrix} \begin{Bmatrix} x_1 \\ x_2 \end{Bmatrix} = \begin{Bmatrix} f \\ 0 \end{Bmatrix}$$

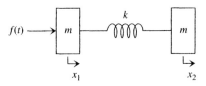

FIGURE 10.11 TDOF system.

Taking the Laplace transform, with zero initial conditions, we have

$$\begin{bmatrix} ms^2 + k & -k \\ -k & ms^2 + k \end{bmatrix} \begin{Bmatrix} X_1(s) \\ X_2(s) \end{Bmatrix} = \begin{Bmatrix} F(s) \\ 0 \end{Bmatrix}$$

The characteristic equation is

$$(ms^2 + k)^2 - k^2 = 0$$

Thus,

$$(ms^2 + k)^2 = k^2$$
$$ms^2 + k = \pm k$$

Case "+": $ms^2 + k = k$

$$ms^2 = 0$$
$$s_1 = s_2 = 0$$

Case "−": $ms^2 + k = -k$

$$ms^2 = -2k$$
$$s_{3,4} = \pm j\sqrt{\frac{2k}{m}}$$

The poles are shown in Fig. 10.12.

(b) Using Cramer's rule, $X_1(s)$ is obtained as

$$X_1(s) = \frac{(ms^2 + k)F(s)}{\Delta(s)}$$

where

$$\Delta(s) = (ms^2 + k)^2 - k^2 = (m^2s^2 + 2km)s^2$$

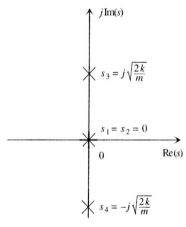

FIGURE 10.12 TDOF system.

Thus,

$$A_1(s) = \frac{(ms^2 + k)}{m^2s^2 + 2km}F(s)$$

where $A_1(s) = s^2 X_1(s)$ since $a_1(t) = \ddot{x}_1$. Interpreting the above in time domain, the differential equation is thus given as

$$m^2 \ddot{a}_1 + 2kma_1 = m\ddot{f} + kf$$

which is of second order, in a_1.

10.8 TRANSIENT RESPONSE SPECIFICATIONS

This section contains certain fundamentals of the **time response,** or the response in the time domain, of dynamic systems for control system design and analysis. If the input is a step function, the response is called the **step response.** Similarly, if the input is a unit-impulse function, the response is called the **unit-impulse response.** Important concepts include settling time and maximum percent overshoot.

The time response of first- and second-order systems has been treated extensively in Chapters 1 and 7 mainly for the purpose of modeling and analysis. We now look at it at a different angle for control system analysis.

First-Order Systems

Consider the mass-viscous damper system as shown in Fig. 10.7.

Step response

Consider the case when the applied force is a step function, $f(t) = Au_s(t)$. The differential equation of motion of the system is given as

$$m\dot{v} + cv = f(t)$$

or

$$\tau\dot{v} + v = \frac{f(t)}{c} = A\frac{u_s(t)}{c}$$

The step response with zero initial conditions is shown in Fig. 10.13. Whereas the concept of time constant was thoroughly discussed in Section 7.2 (Fig. 7.2), the idea is revisited here for the transient response specifications of first-order systems pertaining to control systems. In reference to Fig. 10.13, the role of time constant in the response curve is once again evident. For instance, after four time constants the response curve reaches 98 percent of its steady-state value.

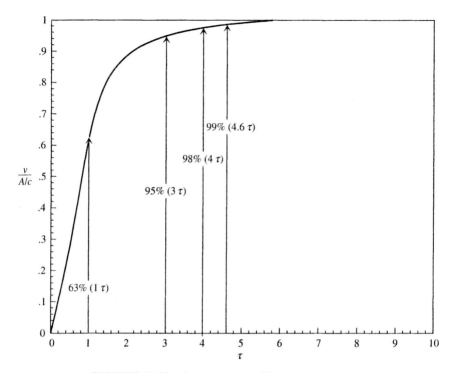

FIGURE 10.13 Step response of first-order systems.

Impulse response

If the applied force is an impulse function, we have

$$m\dot{v} + cv = f(t) = A\delta(t)$$

or

$$\tau\dot{v} + v = \frac{f(t)}{c} = \frac{A\delta(t)}{c}$$

The impulse response of the first-order system with zero initial conditions is shown in Fig. 10.14.

Second-Order Systems

Let us consider a simple mechanical system (Fig. 10.15) that is a classical second-order mechanical system. The differential equation of motion is given as

$$m\ddot{x} + c\dot{x} + kx = f(t)$$

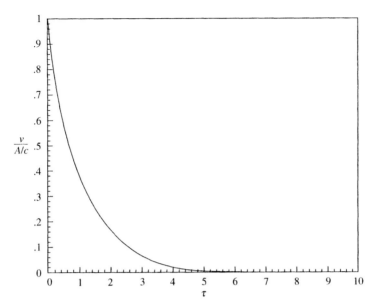

FIGURE 10.14 Impulse response of first-order systems.

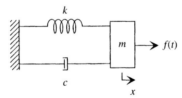

FIGURE 10.15 Mechanical system as a second-order system.

or

$$\ddot{x} + 2\zeta\omega_n \dot{x} + \omega_n^2 x = \frac{f(t)}{m}$$

Assuming the system is *underdamped* ($0 < \zeta < 1$), which is the most interesting, the poles are given as

$$\begin{matrix} s_1 \\ s_2 \end{matrix} = -\zeta\omega_n \pm j\omega_n \sqrt{1-\zeta^2} = -\sigma \pm j\omega_d$$

Figure 10.16 shows the location of the poles in the complex s-plane.

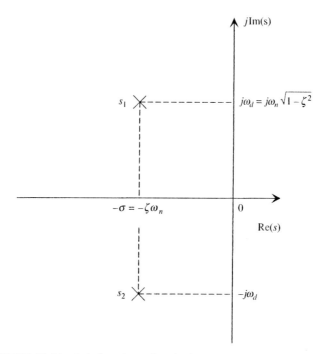

FIGURE 10.16 Pole locations of underdamped second-order systems.

Step response

Consider the case when the applied force is a step function, $f(t) = Au_s(t)$. The differential equation of motion is given as

$$m\ddot{x} + c\dot{x} + kx = f(t)$$

or

$$\ddot{x} + 2\zeta\omega_n\dot{x} + \omega_n^2 x = \frac{f(t)}{m} = A\frac{u_s(t)}{m}$$

Figure 10.17 shows the unit-step response associated with fixed parameter values, $\zeta = 0.25$ and $\omega_n = 1 = m = A$. All underdamped, second-order systems exhibit similar behavior to that in Fig. 10.17 when subject to a unit-step input. That is, they experience a maximum peak and eventually settle to within a small percentage of the final value.

A thorough knowledge of a linear system's unit-step response enables one to determine its response to other types of input. To learn more about the unit-step response, we next present what are known as transient-response specifications, which, in particular, play a significant role in the design of control systems. In the preceding equation, it is common practice to assume that $f(t) = ku_s(t)$ so that

$$\ddot{x} + 2\zeta\omega_n\dot{x} + \omega_n^2 x = \omega_n^2 u_s(t)$$

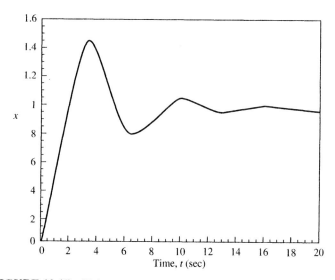

FIGURE 10.17 Unit-step response of the underdamped second-order system ($\zeta = 0.25, \omega_n = m = 1$).

with the corresponding second-order transfer function in the form

$$\frac{X(s)}{U_s(s)} = \frac{\omega_n^2}{s^2 + 2\zeta\omega_n s + \omega_n^2}$$

Because this transfer function is in the standard form, it can represent any second-order dynamic system. In what follows, the assumption is that the system is underdamped ($0 < \zeta < 1$). Therefore, the unit-step response is given by Eq. (7.35) with a slight modification; the initial conditions are zero and the factor $1/\omega_n^2$ is replaced with 1. As a result,

$$x(t) = 1 - e^{-\zeta\omega_n t}\left[\cos\omega_d t + \frac{\zeta}{\sqrt{1-\zeta^2}}\sin\omega_d t\right]$$

which is graphically represented in Fig. 10.18. As a consequence of the standard form of the transfer function, the final value is unity. In relation to Fig. 10.18 we define the following terms.

Maximum percent overshoot ($M_p^\%$). The vertical distance between the maximum peak of the response curve and the horizontal line from 1 (final value) is called the maximum overshoot (M_p) and is defined as

$$M_p = e^{-\pi\zeta/\sqrt{1-\zeta^2}}$$

The maximum percent overshoot is subsequently defined as $M_p^\% = 100 M_p$.

Peak time. The time required for the response curve to reach the maximum overshoot is known as the peak time (t_p). The analytical expression of the peak time

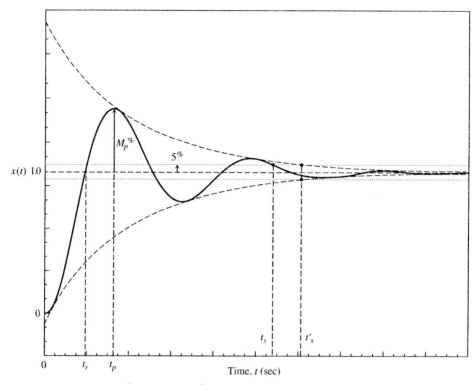

FIGURE 10.18 Transient response specifications of the underdamped second-order system.

can readily be obtained by seeking the value of t that maximizes the unit-step response defined by the function $x(t)$. Solving $\dot{x}(t) = 0$ leads to

$$t_p = \frac{\pi}{\omega_d}$$

With this information, the derivation of the expression of M_p is straightforward.

Settling time. The time required for the response curve to reach and stay within a certain strip (or band) about the final value is referred to as the settling time (t_s). In practice, different criteria may be used to define the settling time. These include 1 percent, 2 percent, and 5 percent criteria. Figure 10.18 shows a 5 percent criterion where the strip is of a width of 0.05 on each side of 1. The exponential-decay curves (dashed line) define the envelope corresponding to the response curve. Because the trigonometric functions in the expression of $x(t)$ are bounded in magnitude, the term $e^{-\zeta\omega_n t}$ characterizes these curves. As a result, comparing with the general form $e^{-t/\tau}$, we conclude that each of these curves has a time constant of $\tau = 1/(\zeta\omega_n)$. It can be shown that the settling time associated with the three criteria mentioned earlier are given as

$$t_s\big|_{1\%} \approx t'_s\big|_{1\%} = 4.6\tau = \frac{4.6}{\zeta\omega_n} = \frac{4.6}{\sigma} \qquad (\zeta < 1)$$

$$t_s\big|_{2\%} \approx t'_s\big|_{2\%} = 4\tau = \frac{4}{\zeta\omega_n} = \frac{4}{\sigma} \qquad (\zeta < 1)$$

$$t_s\big|_{5\%} \approx t'_s\big|_{5\%} = 3\tau = \frac{3}{\zeta\omega_n} = \frac{3}{\sigma} \qquad (\zeta < 1)$$

where t'_s denotes the time at which the envelope crosses the strip.

Rise time. For an underdamped, second-order system the time required for the response curve to rise from 0 to 100 percent of the final value is called the rise time (t_r). Analytically, the expression for the rise time can be derived as

$$t_r = -\frac{1}{\omega_d}\tan^{-1}\frac{\sqrt{1-\zeta^2}}{\zeta}$$

Impulse response

When the input is an impulse function, $f(t) = A\delta(t)$. Thus, the differential equation of motion with a unit-impulse input ($A = 1$) is given as

$$m\ddot{x} + c\dot{x} + kx = f(t) = \delta(t)$$

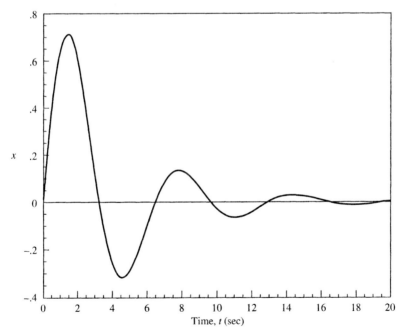

FIGURE 10.19 Unit-impulse response of the underdamped second-order system ($\zeta = 0.25$, $\omega_n = m = 1$).

or

$$\ddot{x} + 2\zeta\omega_n\dot{x} + \omega_n^2 x = \frac{f(t)}{m} = \frac{\delta(t)}{m}$$

The unit-impulse response of the underdamped system is shown in Fig. 10.19.

10.9 DOMINANT POLE CONCEPT

The concept of **dominant pole** is based on the dominant time constant. The **dominant time constant** is the time constant corresponding to the pole that is closest to the $j\omega$-axis. The reason is that this pole corresponds to the largest time constant, or slowest time response. All the real negative poles that are far away from the $j\omega$-axis do not contribute significantly to the system response. Similarly, the contribution of the complex conjugate pairs whose real parts are far away from the $j\omega$-axis is insignificant. This may be rationalized as the *out-of-sight, out-of-mind* concept.

The time constant τ is primarily defined for first-order systems; however, it can be used for second-order systems. Moreover, since first-order and second-order subsystems are the building blocks, the time constant τ can be applied naturally to complex higher-order systems. These higher-order systems can be approximated as simple first-order or second-order systems under certain conditions. This approximation gives insight about the behavior of such complex systems.

First-Order and Second-Order Systems

We can borrow the concept of a time constant τ from first-order systems and apply it to second-order systems. This idea is shown in Fig. 10.20, where (*a*) is for first-order systems and (*b*) is for underdamped second-order systems. The double-arrowed line is used to denote the length, a positive quantity. Note that the time constant τ, damping ratio ζ, and natural frequency ω_n are all positive, whereas the pole s in Fig. 10.20*a* is negative.

Higher-Order Systems

The time response of a complex linear system is contributed by its many parts with their corresponding time constants. The time response of a linear system is the sum of the time responses of the parts. If a complex system has one dominant pole, it behaves approximately as a first-order system. Similarly, if a higher-order system has a pair of complex-conjugate dominant poles, it behaves as a second-order system.

For example, Fig. 10.21 shows the pole configuration of a third-order system. Since the dominant pole is s_1, the response of this system can be approximated as that of a first-order system that has time constant τ.

As another example, Fig. 10.22 shows the pole configuration of a fifth-order system. Since the pair of complex conjugates s_1 and s_2 are closest to the $j\omega$-axis,

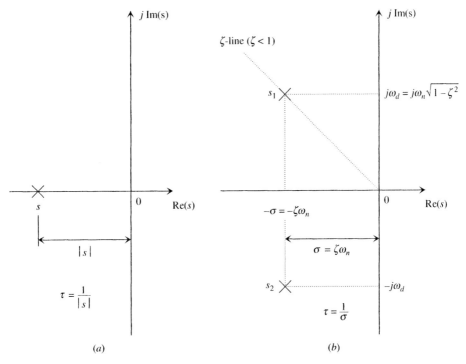

FIGURE 10.20 Time constant τ for: (a) first-order system and (b) underdamped second-order system.

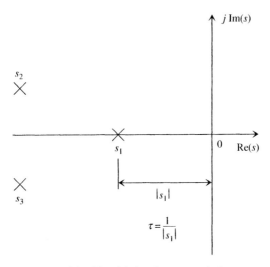

FIGURE 10.21 The third-order system behaves as a first-order system.

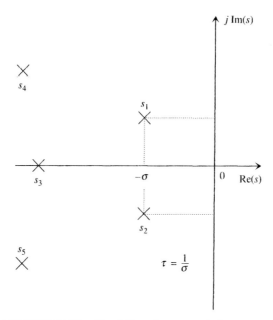

FIGURE 10.22 The fifth-order system behaves as a second-order system.

these poles are the most dominant ones. Thus, the response of this system behaves as if it were a second-order system with these two poles.

10.10 ROUTH-HURWITZ STABILITY CRITERION

The stability of a dynamic system can be determined by many means. The characteristic equation can be solved to obtain the roots first, and then the stability of the system is determined based on the location of the roots. For systems of third- or higher-order, there is no simple formula for solving the equation. One would have to resort to numerical methods or use a calculator or computer to find the roots. Fortunately, with the Routh-Hurwitz stability criterion, a simple yes-or-no answer regarding stability can be obtained easily.

General Systems

The characteristic equation of a general dynamic system or control system can be written as

$$s^n + a_1 s^{n-1} + a_2 s^{n-2} + \cdots + a_{n-1} s + a_n = 0$$

The stability of the system can be determined using the Routh-Hurwitz stability criterion.

CHAPTER 10: Introduction to Control Systems 541

ROUTH-HURWITZ STABILITY CRITERION. The system is stable iff both of the following conditions are met:

1. All the coefficients of the characteristic equation are positive.
2. All the elements of the first column of the Routh-Hurwitz table are positive.

Note that the word *iff* means *if and only if*. The Routh-Hurwitz table can be constructed as

$$
\begin{array}{c|cccc}
s^n & 1 & a_2 & a_4 & a_6 \\
s^{n-1} & a_1 & a_3 & a_5 & a_7 \\
s^{n-2} & b_1 & b_2 & b_3 & \ldots \\
s^{n-3} & c_1 & c_2 & c_3 & \ldots \\
\vdots & \vdots & \vdots & \vdots & \\
s^2 & & & & \\
s^1 & & & & \\
s^0 & a_n & & &
\end{array}
$$

where

$$b_1 = -\frac{1}{a_1}\begin{vmatrix} 1 & a_2 \\ a_1 & a_3 \end{vmatrix} = \frac{a_1 a_2 - a_3}{a_1}, \qquad b_2 = -\frac{1}{a_1}\begin{vmatrix} 1 & a_4 \\ a_1 & a_5 \end{vmatrix} = \frac{a_1 a_4 - a_5}{a_1},$$

$$b_3 = -\frac{1}{a_1}\begin{vmatrix} 1 & a_6 \\ a_1 & a_7 \end{vmatrix} = \frac{a_1 a_6 - a_7}{a_1}$$

$$c_1 = -\frac{1}{b_1}\begin{vmatrix} a_1 & a_3 \\ b_1 & b_2 \end{vmatrix} = \frac{a_3 b_1 - a_1 b_2}{b_1}, \qquad c_2 = -\frac{1}{b_1}\begin{vmatrix} a_1 & a_5 \\ b_1 & b_3 \end{vmatrix} = \frac{a_5 b_1 - a_1 b_3}{b_1},$$

$$c_3 = -\frac{1}{b_1}\begin{vmatrix} a_1 & a_7 \\ b_1 & b_4 \end{vmatrix} = \frac{a_7 b_1 - a_1 b_4}{b_1}$$

Fourth-Order Systems

The Routh-Hurwitz table for the fourth-order characteristic equation

$$s^4 + a_1 s^3 + a_2 s^2 + a_3 s + a_4 = 0$$

is constructed as

$$
\begin{array}{c|ccc}
s^4 & 1 & a_2 & a_4 \\
s^3 & a_1 & a_3 & \\
s^2 & b_1 & b_2 & \\
s^1 & c_1 & & \\
s^0 & a_4 & &
\end{array}
$$

where

$$b_1 = -\frac{1}{a_1}\begin{vmatrix} 1 & a_2 \\ a_1 & a_3 \end{vmatrix} = \frac{a_1 a_2 - a_3}{a_1}, \qquad b_2 = -\frac{1}{a_1}\begin{vmatrix} 1 & a_4 \\ a_1 & 0 \end{vmatrix} = \frac{a_1 a_4}{a_1} = a_4$$

$$c_1 = -\frac{1}{b_1}\begin{vmatrix} a_1 & a_3 \\ b_1 & b_2 \end{vmatrix} = \frac{b_1 a_3 - b_2 a_1}{b_1}, \qquad c_2 = 0$$

For stability, both of the following conditions must be satisfied:

1. $a_1, a_2, a_3, a_4 > 0$ (necessary condition)
2. $b_1 > 0$:

$$a_1 a_2 > a_3$$

and $c_1 > 0$:

$$b_1 a_3 - b_2 a_1 > 0$$

$$\left(\frac{a_1 a_2 - a_3}{a_1}\right) a_3 - a_4 a_1 > 0$$

$$(a_1 a_2 - a_3) a_3 - a_4 a_1^2 > 0$$

In summary, the fourth-order system is stable if all the following conditions are satisfied:

1. $a_1, a_2, a_3, a_4 > 0$ (necessary condition)
2. $a_1 a_2 > a_3$ and $(a_1 a_2 - a_3) a_3 - a_4 a_1^2 > 0$

Third-Order Systems

Consider the stability condition for a general third-order dynamic system whose characteristic equation is given as

$$s^3 + a_1 s^2 + a_2 s + a_3 = 0$$

For this problem, the Routh-Hurwitz table is constructed as

s^3	1	a_2	0
s^2	a_1	a_3	0
s^1	b_1	b_2	
s^0	a_3		

where

$$b_1 = -\frac{1}{a_1}\begin{vmatrix} 1 & a_2 \\ a_1 & a_3 \end{vmatrix} = \frac{a_1 a_2 - a_3}{a_1}, \qquad b_2 = -\frac{1}{a_1}\begin{vmatrix} 1 & 0 \\ a_1 & 0 \end{vmatrix} = 0$$

Note: The last element of the first column of the Routh-Hurwitz table is always the same as the last coefficient of the characteristic equation. For this problem,

$$c_1 = -\frac{1}{b_1}\begin{vmatrix} a_1 & a_3 \\ b_1 & 0 \end{vmatrix} = \frac{a_3 b_1 - 0}{b_1} = a_3$$

Therefore, the following stability conditions are given for third-order systems:

1. a_1, a_2, and a_3 are all positive
2. $a_1 a_2 - a_3 > 0$

Since the characteristic equation of a general third-order system is given as

$$1s^3 + \underbrace{a_1 s^2 + a_2 s}_{\text{inner}} + a_3 = 0$$

$$\underbrace{}_{\text{outer}}$$

a simple rule is established as follows.

THE INNER-OUTER STABILITY RULE. A third-order system is stable iff

1. All the coefficients are positive: $a_1 > 0$, $a_2 > 0$, $a_3 > 0$
2. Inner product > outer product: $a_1 a_2 > a_3$

This rule is simple to remember if one uses a mnemonic device such as *inner strength is greater than outer beauty.*

Second-Order Systems

The general characteristic equation of second-order systems is given as

$$s^2 + a_1 s + a_2 = 0$$

The Routh-Hurwitz table is constructed as

s^2	1	a_2
s^1	a_1	
s^0	a_2	

The second-order system is stable if a_1 and a_2 are positive (necessary condition). Notice that the second condition, i.e., all the elements of the first column of the Routh-Hurwitz table (shaded column) must be positive, is automatically satisfied by the first condition (necessary condition).

PROBLEM SET 10.1–10.10

10.1. State the essential components for a closed-loop control system.

10.2. Determine the poles and zeros of the following systems. Also, indicate the stability.

(a) $G(s) = \dfrac{s+4}{s^2+3s+2}$

(b) $G(s) = \dfrac{s-5}{s^2+3s+1}$

(c) $G(s) = \dfrac{1}{(s^2+3s+2)(3s+2)}$

10.3. The characteristic equations of the first- and second-order systems are given. For each of the systems, determine the pole(s) and indicate whether the system is stable or unstable.

(a) $2s + 1 = 0$
(b) $s + 6 = 0$
(c) $\tau s + 1 = 0 \quad (\tau > 0)$
(d) $\tau s + 1 = 0 \quad (\tau < 0)$
(e) $2s^2 + 3s + 5 = 0$

10.4. The characteristic equations of the general second-order systems are given. For each of the systems, determine the pole(s) and indicate whether the system is stable or unstable.

(a) $s^2 + 2\zeta\omega_n s + \omega_n^2 = 0 \quad (\zeta \geq 0, \omega_n > 0)$
(b) $s^2 + 2\zeta\omega_n s + \omega_n^2 = 0 \quad (\zeta < 0, \omega_n > 0)$

10.5. For each of the systems whose transfer functions are given, determine the pole(s) and zero, then indicate whether the system is stable or unstable.

(a) $\dfrac{s+5}{s^2+3s+4}$

(b) $\dfrac{s-10}{s^2+3s+1}$

(c) $\dfrac{1}{s^2+1.5s+2.5}$

10.6. For the following characteristic equation, determine whether the system is stable or unstable.

$$1 + \dfrac{s+5}{s^2+3s+4} = 0$$

10.7. The characteristic equations of third-order systems are given. For each of the systems, determine whether it is stable.

(a) $3s^3 + 4s^2 + 5s + 2 = 0$
(b) $s^3 + 4s^2 + 2s + 10 = 0$

10.8. Determine whether the system is stable or unstable whose characteristic equation is given as

(a) $1 + \dfrac{s+5}{3s^3+5s^2+3s+4} = 0$

(b) $1 + \dfrac{s+5}{s(s^2+s+5)} = 0$

10.9. State the stability condition for the system whose characteristic equation is given as

$$1 + \left(K_p + \frac{K_i}{s} + K_d s\right)\frac{1}{s(s^2 + 10s + 50)} = 0$$

where K_p, K_i, and K_d are arbitrary constants.

10.10. Consider a closed-loop system where the transfer function of the plant is given. In this system, a negative unity feedback is used. State the stability condition.

$$G_c(s) = K, \qquad G_p(s) = \frac{s+3}{s(s+1)(s+5)}$$

10.11. The characteristic equation of a second-order system is given in general form as

$$s^2 + 2\zeta\omega_n s + \omega_n^2 = 0$$

Determine the angle for the ζ-line if $\zeta = 0.5$.

10.12. Starting with

$$M_p^\% = 100e^{-\pi\zeta/\sqrt{1-\zeta^2}}$$

show that

$$\zeta = \frac{1}{\sqrt{1 + [\pi/\ln(M_p^\%/100)]^2}}$$

10.13. For each of the following systems, determine the dominant pole(s) and the corresponding dominant time constant τ.

(a) $G(s) = \dfrac{s+4}{s^2 + 3s + 2}$

(b) $G(s) = \dfrac{s-5}{(s+10)(s^2 + 3s + 1)}$

(c) $G(s) = \dfrac{s+5}{(s+0.1)(s^2 + 3s + 4)}$

(d) $G(s) = \dfrac{s+10}{(s+10)(s^2 + 3s + 4)}$

10.11 CONTROLLER TYPES AND ACTIONS

There are many types of control (or controller), but the classification of the primary ones is given as follows:

- Proportional control (P control)
- Derivative control (D control)
- Integral control (I control)
- Proportional-plus-derivative control (PD control)
- Proportional-plus-integral control (PI control)
- Proportional-plus-integral-plus-derivative control (PID control)

Proportional Control (P Control)

Proportional control, or **P control,** is a simple gain. The proportional control is the simplest type of control and should be tried first in control system design. If the system is still unstable or if the performance needs improvement, the designer may wish to consider other types. The transfer function of a P controller is defined as

$$G_c(s) = K_p = K$$

where $K_p = K$, a proportional constant. Thus

$$U(s) = G_c(s)E_a(s) = KE_a(s)$$

so the time function $u(t)$ is proportional to $e_a(t)$

$$u(t) = Ke_a(t)$$

Therefore, this type is called a proportional controller.

Derivative Control (D Control)

Derivative control, or **D control,** adds damping into the system and thus provides stability. The transfer function of a derivative controller is defined as

$$G_c(s) = K_d s$$

where K_d is constant. Thus,

$$U(s) = G_c(s)E_a(s) = K_d s E_a(s)$$

so $u(t)$ is proportional to the derivative of $e_a(t)$,

$$u(t) = K_d \dot{e}_a(t)$$

Integral Control (I Control)

Integral control, or **I control,** reduces the steady-state error of the system but tends to increase instability. The transfer function of an integral controller is defined as

$$G_c(s) = \frac{K_i}{s}$$

where K_i is constant. Thus,

$$U(s) = G_c(s)E_a(s) = \frac{K_i}{s}E_a(s)$$

so $u(t)$ is proportional to the integral of $e_a(t)$,

$$u(t) = K_i \int_0^t e_a(t)\,dt$$

Proportional-plus-Derivative Control (PD Control)

A PD control is a combination of P control and D control. The transfer function of this type of a controller is defined as

$$G_c(s) = K_p + K_d s \quad \text{or} \quad G_c(s) = K(1 + T_d s)$$

where K_p, K_d, and K are constant, and $K_p = K$, $K_d = KT_d$. Thus,

$$U(s) = G_c(s)E_a(s) = (K_p + K_d s)E_a(s)$$

or the actuating input $u(t)$ is linear in the actuating error $e_a(t)$ itself and its derivative,

$$u(t) = K_p e_a(t) + K_d \dot{e}_a(t)$$

Proportional-plus-Integral Control (PI Control)

A PI control is a combination of P control and I control. When the proportional and integral actions are combined we have a PI controller,

$$G_c(s) = K_p + \frac{K_i}{s} \quad \text{or} \quad G_c(s) = K\left(1 + \frac{1}{T_i s}\right)$$

where K_p, K_i, and K are constant, and $K_p = K$, $K_i = K/T_i$. Thus,

$$U(s) = G_c(s)E_a(s) = \left(K_p + \frac{K_i}{s}\right)E_a(s)$$

or the actuating input $u(t)$ is proportional to the actuating error $e_a(t)$ itself and its integral,

$$u(t) = K_p e_a(t) + K_i \int_0^t e_a(t)\,dt$$

Proportional-plus-Integral-plus-Derivative Control (PID Control)

A PID control is a combination of P control, I control, and D control. Similar to PD control and PI control, when the proportional (P), the derivative (D), and integral (I) actions are combined we have a PID controller

$$G_c(s) = K_p + \frac{K_i}{s} + K_d s \quad \text{or} \quad G_c(s) = K\left(1 + \frac{1}{T_i s} + T_d s\right)$$

where K_p, K_i, K_d, and K are constant, and $K_p = K$, $K_i = K/T_i$, $K_d = KT_d$. Thus,

$$U(s) = G_c(s)E_a(s) = \left(K_p + \frac{K_i}{s} + K_d s\right)E_a(s)$$

The actuating input $u(t)$ is the sum of three elements as given by

$$u(t) = K_p e_a(t) + K_i \int_0^t e_a(t)\,dt + K_d \dot{e}_a(t)$$

The PID controller is one of the most versatile controllers used in the controls industry.

10.12 STEADY-STATE ERROR

For *transient* or *nonsinusoidal* input and stable systems, the steady-state error is given by the final-value theorem as

$$e_{ss} = \lim_{t\to\infty} e(t) = \lim_{s\to 0} sE(s) \tag{10.26}$$

Impulse functions and step functions are examples of nonsinusoidal input. At sufficiently large time, the response to a nonsinusoidal input settles down to a steady-state value, or a final value; thus, the final-value theorem is valid. In other words, it is valid because the limit exists. Note that for *sinusoidal* input, the steady-state response is a function of time—not a value—thus the final-value theorem is invalid. The limit simply does not exist.

In this section, steady-state errors are discussed using two different approaches: the fundamentals and the concept of system types. For the latter, the results are also given in tabulated form. For complicated and large systems, the results can be convenient for table lookup; however, for simple systems it is better to start from scratch and *not* use the concept of system types and tabulated results.

A general block diagram of negative feedback systems can be shown as in Fig. 10.23. The **error** (*actual* error or *true* error) is defined as

$$E(s) = R(s) - Y(s) \tag{10.27}$$

For linear systems, when both reference and disturbance inputs are present, the error is given as

$$E(s) = R(s) - \left[\left.\frac{Y(s)}{R(s)}\right|_{D(s)=0} R(s) + \left.\frac{Y(s)}{D(s)}\right|_{R(s)=0} D(s)\right] \tag{10.28}$$

where

$$Y(s) = \left.\frac{Y(s)}{R(s)}\right|_{D(s)=0} R(s) + \left.\frac{Y(s)}{D(s)}\right|_{R(s)=0} D(s) \tag{10.29}$$

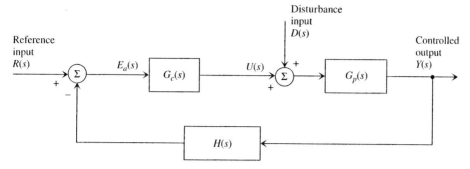

FIGURE 10.23 General block diagram including disturbance input.

The error can also be obtained by considering one input at a time. For nonzero reference input and zero disturbance input, the error is

$$E_1(s) = R(s) - Y(s) = \left[1 - \frac{Y(s)}{R(s)}\bigg|_{D(s)=0}\right] R(s) \qquad (10.30)$$

and for zero reference input, $R(s) = 0$, and nonzero disturbance input,

$$E_2(s) = R(s) - Y(s) = -\frac{Y(s)}{D(s)}\bigg|_{R(s)=0} D(s) \qquad (10.31)$$

Using **superposition** when both reference and disturbance inputs are present,

$$E(s) = E_1(s) + E_2(s) = \left[1 - \frac{Y(s)}{R(s)}\bigg|_{D(s)=0}\right] R(s) - \frac{Y(s)}{D(s)}\bigg|_{R(s)=0} D(s) \qquad (10.32)$$

which is the same result as given by Eq. (10.28).

To have a better focus, assume that the disturbance input is zero, $D(s) = 0$. The block diagram is reduced to Fig. 10.24. The (actual) error is

$$E(s) = R(s) - Y(s) = \left[1 - \frac{Y(s)}{R(s)}\right] R(s) \qquad (10.33)$$

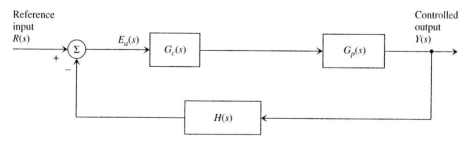

FIGURE 10.24 General block diagram without disturbance input.

and the closed-loop transfer function is

$$\frac{Y(s)}{R(s)} = \frac{G_c(s)G_p(s)}{1 + G_c(s)G_p(s)H(s)} \quad (10.34)$$

Introducing the transfer function into the error equation,

$$E(s) = \left[1 - \frac{G_c(s)G_p(s)}{1 + G_c(s)G_p(s)H(s)}\right]R(s) = \frac{1 + G_c(s)G_p(s)[H(s) - 1]}{1 + G_c(s)G_p(s)H(s)}R(s) \quad (10.35)$$

The **actuating error** is defined as the difference between the input signal and the feedback

$$E_a(s) = R(s) - H(s)Y(s) \quad (10.36)$$

Thus,

$$E_a(s) = \left[1 - H(s)\frac{G_c(s)G_p(s)}{1 + G_c(s)G_p(s)H(s)}\right]R(s) = \frac{1}{1 + G_c(s)G_p(s)H(s)}R(s) \quad (10.37)$$

Therefore, the actuating error $E_a(s)$ and the "actual" error $E(s)$ are identical only for the case of *negative unity feedback*, $H(s) = 1$. If the subscript a is dropped, it implies that the system has negative unity feedback (a hidden assumption).

For negative unity feedback, the steady-state error of the system subject to a nonsinusoidal input can also be shown as

$$e_{ss} = \lim_{t \to \infty} e(t) = \lim_{s \to 0} sE(s) = \lim_{s \to 0} s\frac{1}{1 + G_{ol}(s)}R(s) \quad (10.38)$$

where the open-loop transfer function $G_{ol}(s)$ is defined as

$$G_{ol}(s) = G_c(s)G_p(s) \quad (10.39)$$

EXAMPLE 10.4. Consider the mechanical system shown in Fig. 10.25, where $u(t)$ is the control input. It is desired to use a control system with a P controller and a negative unity feedback.

(a) Draw a block diagram for the closed-loop system.
(b) Assuming the reference input is a unit-step function, determine the steady-state error.

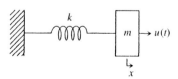

FIGURE 10.25 A system.

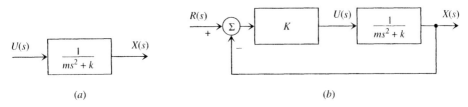

FIGURE 10.26 Block diagram: (a) before closing the loop, (b) after closing the loop.

Solution

(a) The differential equation of motion is given as

$$m\ddot{x} + kx = u(t)$$

Thus the transfer function is

$$\frac{X(s)}{U(s)} = \frac{1}{ms^2 + k}$$

The block diagram of the system is shown in Fig. 10.26.

(b) The steady-state error is

$$e_{ss} = \lim_{t \to \infty} e(t) = \lim_{s \to 0} sE(s)$$

where

$$E(s) = R(s) - X(s) = \left[1 - \frac{X(s)}{R(s)}\right]R(s)$$

$$\frac{X(s)}{R(s)} = \frac{K/(ms^2 + k)}{1 + K/(ms^2 + k)} = \frac{K}{ms^2 + k + K}$$

$$E(s) = \left[1 - \frac{K}{ms^2 + k + K}\right]R(s)$$

Since the reference input is a unit-step function, $r(t) = u_s(t) \Longrightarrow R(s) = 1/s$, the steady-state error is

$$e_{ss} = \lim_{s \to 0} s\left(1 - \frac{K}{ms^2 + k + K}\right)\frac{1}{s} = 1 - \frac{K}{k + K} = \frac{k}{k + K}$$

For this system, with this type of input, increasing the "gain" K will decrease the steady-state error e_{ss}.

EXAMPLE 10.5. Consider the mechanical system shown in Fig. 10.27, where $u(t)$ and $d(t)$ are the control force input and disturbance force input, respectively. A simple closed-loop system is tried with a negative unity feedback and a PD controller. (The D action adds damping into the system.)

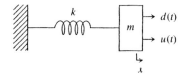

FIGURE 10.27 System with both control input and disturbance input.

(a) Draw the system block diagram both before and after closing the loop.
(b) Determine the steady-state error for the following cases of disturbance input: (i) unit-impulse function, (ii) unit-step function, (iii) unit-ramp function. The references input is zero.

Solution

(a) The differential equation of motion is given as

$$m\ddot{x} + kx = u(t) + d(t)$$

Thus, the transfer function is

$$X(s) = \frac{1}{ms^2 + k}[U(s) + D(s)]$$

The block diagram of the uncontrolled system (before closing the loop) is shown in Fig. 10.28. After closing the loop, the block diagram is shown in Fig. 10.29.

(b) The steady-state error is

$$e_{ss} = \lim_{t \to \infty} e(t) = \lim_{s \to 0} sE(s)$$

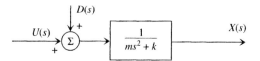

FIGURE 10.28 Block diagram of the uncontrolled system (before closing the loop).

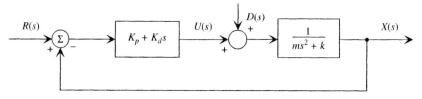

FIGURE 10.29 Block diagram of the controlled system (after closing the loop).

where

$$E(s) = R(s) - X(s) = \left[0 - \frac{X(s)}{D(s)}\right]D(s) = -\left[\frac{X(s)}{D(s)}\right]D(s)$$

$$\frac{X(s)}{D(s)} = \frac{1/(ms^2 + k)}{1 + (K_p + K_d s)/(ms^2 + k)} = \frac{1}{ms^2 + K_d s + k + K_p}$$

$$E(s) = -\left[\frac{1}{ms^2 + K_d s + k + K_p}\right]D(s)$$

Thus,

$$e_{ss} = \lim_{s \to 0} sE(s) = \lim_{s \to 0} s\left(-\frac{1}{ms^2 + K_d s + k + K_p}\right)D(s)$$

Case i: unit-impulse input, $d(t) = \delta(t) \implies D(s) = 1$:

$$e_{ss} = \lim_{s \to 0} s\left(-\frac{1}{ms^2 + K_d s + k + K_p}\right)1 = 0$$

Case ii: unit-step input, $d(t) = u_s(t) \implies D(s) = 1/s$:

$$e_{ss} = \lim_{s \to 0} s\left(-\frac{1}{ms^2 + K_d s + k + K_p}\right)\frac{1}{s} = -\frac{1}{k + K_p}$$

Case iii: unit-ramp input, $d(t) = u_r(t) \implies D(s) = 1/s^2$:

$$e_{ss} = \lim_{s \to 0} s\left(-\frac{1}{ms^2 + K_d s + k + K_p}\right)\frac{1}{s^2} = -\infty$$

PROBLEM SET 10.11–10.12

10.14. Consider the mechanical system shown in Fig. P10.14, where $f(t) = u(t)$ is the control force input. It is desired to use a control system with a P controller and a negative unity feedback.

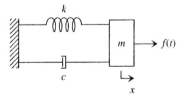

FIGURE P10.14

(a) Draw a block diagram for the closed-loop system.
(b) Assuming the reference input is a unit-step function, determine the steady-state error.

10.15. Consider the mechanical system shown in Fig. P10.15, where $u(t)$ and $d(t)$ are the control force input and disturbance force input, respectively. The control input $u(t)$ is given by a negative unity feedback and a PD controller.

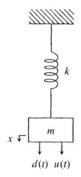

FIGURE P10.15

(a) Before and after closing the loop, draw the system block diagram for each case.
(b) Determine the steady-state error when the reference input is a unit-step function and the disturbance input is zero.
(c) Determine the steady-state error when the reference input is zero and the disturbance input is a unit-step function.
(d) Determine the steady-state error when the reference input is a unit-step function and the disturbance input is a unit-impulse function.

10.13 METHODS OF ZIEGLER-NICHOLS TUNING

Ziegler-Nichols tuning rules are extremely useful in designing PID controllers. These tuning rules assume that the control system is already installed on the plant. The transfer function of a PID controller is given previously as

$$G_c(s) = K\left(1 + \frac{1}{T_i s} + T_d s\right)$$

where the numerical values of the parameters K, T_i, and T_d are to be determined. It should be emphasized that the parameters of PID controllers can be determined even when the plant dynamics are not known. That is when the transfer function of the plant cannot be determined analytically.

Obtaining the exact optimal values is often difficult, if not impossible. However, the Ziegler-Nichols tuning rules provide a means of obtaining the **suboptimal** values. These values are not really the best; thus, they are called suboptimal as opposed to optimal. Starting with these values and by trial and error, the control engineer can fine tune the controller further, if necessary, to get closer to the optimal values.

There are two methods of implementing Ziegler-Nichols tuning rules. Since the first method is rather limited, it will only be briefly mentioned. The second method will be presented in detail.

The first method is based on the step response of the plant. However, this step response must be an S-shaped curve for the method to be applicable. If the plant has integrator(s), the method *cannot* be used because the step response will not be S-shaped. As an example, consider a plant that has an integrator, whose transfer function is given as

$$G(s) = \frac{X(s)}{F(s)} = \frac{1}{s(s+1)(s+2)}$$

The Laplace transform of the unit-step response is given as

$$X(s) = \frac{1}{s^2(s+1)(s+2)}$$

Next, from the partial-fraction-expansion method,

$$X(s) = \frac{1}{s^2(s+1)(s+2)} = \frac{A}{s^2} + \frac{B}{s} + \frac{C}{s+1} + \frac{D}{s+2}$$

where the constants A, B, C, and D can be obtained readily. Without evaluating these constants, the response is given as

$$x(t) = At + B + Ce^{-t} + De^{-2t} \quad (t > 0)$$

The first and second terms on the right-hand side represent a ramp function and a step function, respectively, whereas the last two terms are the exponential decayed functions. The ramp function is dominant and is responsible for the step response not being S-shaped, so the first method cannot be used.

The second method is based on the system response, which is due to the nonzero initial condition, where the controller is first set to proportional control. The only requirement for this method to be valid is that the proportional gain, K, must have a value that will make the system become unstable. Reconsider the system whose plant is given above. The closed-loop characteristic equation for the unity, negative feedback with proportional control is given as

$$1 + K \frac{1}{s(s+1)(s+2)} = 0$$

$$s^3 + 3s^2 + 2s + K = 0$$

Since this is a third-order system, the inner-outer stability rule yields the stability condition as

$$6 > K \quad (K > 0)$$

Thus $K_{max} = 6$ is the value beyond which the system will become unstable. Therefore, the second method is valid here.

The suboptimal values for the PID-controller parameters are given by the *second method of Ziegler-Nichols tuning* as

$$K = 0.6 K_{cr} \tag{10.40}$$

$$T_i = \frac{T_{cr}}{2} \tag{10.41}$$

$$T_d = \frac{T_{cr}}{8} \tag{10.42}$$

where the critical period, T_{cr}, is defined as

$$T_{cr} = \frac{2\pi}{\omega_{cr}} \tag{10.43}$$

In determining the critical gain K_{cr}, the controller must be set to proportional control only, $G_c(s) = K$. The critical gain is defined as the maximum value of K beyond which the system becomes unstable. When $K = K_{cr}$, the system response due to a nonzero initial condition (or an impulse input) is an undamped sinusoidal wave. The amplitude of the response is constant, neither growing nor diminishing. From this constant-amplitude oscillation, T_{cr} and ω_{cr} are defined as the critical period and critical (circular) frequency, respectively. In analytical work, ω_{cr} is obtained by replacing $s = j\omega$ in the closed-loop characteristic equation.

How to design a PID controller. The following steps show how to design a PID controller, or how to determine the numerical values for K, T_i, and T_d. Figure 10.30 shows the block diagram (without disturbance input) of the control system with a PID controller. In this system, the PID controller transfer function is given as

$$G_c(s) = K_p + \frac{K_i}{s} + K_d s = K\left(1 + \frac{1}{T_i s} + T_d s\right) \tag{10.44}$$

and, as an illustrative example, we use

$$G_p(s) = \frac{1}{s(s^2 + 8s + 20)}, \quad H(s) = 1$$

Step 1. Let T_i approach infinity and T_d equal zero. The controller transfer function becomes

$$G_c(s) = K$$

Notice that for analysis or design on paper, the integral and derivative actions are set to zero, whereas for a real physical system, the integral and

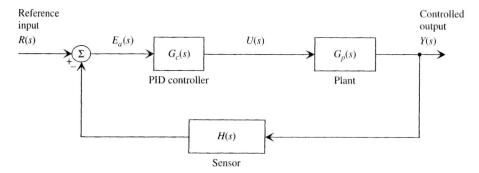

FIGURE 10.30 Block diagram with PID controller.

derivative parts of the controller are actually disconnected. Thus, the closed-loop characteristic equation becomes

$$1 + K\frac{1}{s(s^2 + 8s + 20)} = 0$$

$$s(s^2 + 8s + 20) + K = 0$$

$$s^3 + 8s^2 + 20s + K = 0$$

Step 2. Obtain the critical values ω_{cr}, K_{cr}, and T_{cr} by introducing $s = j\omega$ into the preceding equation. We have

$$s = j\omega, \qquad s^2 = -\omega^2, \qquad s^3 = j\omega(-\omega^2)$$

Collecting terms and equating the real and imaginary parts of both sides yields

$$K - 8\omega^2 + j\omega(20 - \omega^2) = 0$$

$$K - 8\omega^2 = 0, \qquad \omega(20 - \omega^2) = 0$$

Thus,

$$\omega = 0, \qquad K = 8\omega^2 = 0,$$

$$\omega = \sqrt{20} = 4.472, \qquad K = 8\omega^2 = 160$$

Therefore, the critical values when the system begins to become unstable are

$$\omega_{cr} = 4.472, \qquad K_{cr} = 160$$

$$T_{cr} = \frac{2\pi}{\omega_{cr}} = 1.405$$

Step 3. Obtain the suboptimal values for K, T_i, and T_d for the PID controller.
Using the formula, the suboptimal values for the PID-controller parameters are obtained as

$$K = 0.6K_{cr} = 96$$

$$T_i = 0.5T_{cr} = 0.703$$

$$T_d = 0.125T_{cr} = 0.176$$

Optimizing PID Parameters in Practice

After obtaining the suboptimal values of the PID parameters, the controls engineer can fine tune the PID controller to get closer to the optimal values. The following can be used as a guide for this purpose.

- On a real physical system, make small changes in parameter values and then study the results before making more changes. Make sure you know how the system will respond in limiting or emergency cases.

- Increasing the proportional gain K on the system increases the controller's sensitivity to change. However, if the gain K is increased too far, the system may become unstable.
- Decreasing the integral time constant T_i increases the integral control action, which reduces the steady-state error of the closed-loop system.
- Increasing the derivative time constant T_d increases the derivative control action, which increases the effective damping of the closed-loop system.

10.14 SPEED CONTROL SYSTEM

The electromechanical system (Fig. 10.31) can serve as an example, or a case study, of speed control systems. The system consists of an armature-controlled DC motor that is connected to a disk by a shaft. The motor has resistance R_a and inductance L_a. The mass moment of inertia of the disk is J, and that of the rotor is negligible. The disturbance load torque is $T_L(t)$, and the applied (input) voltage is v_a. Assume that the shaft connecting the rotor and the disk is *massless* and *rigid*. Assuming that the disturbance load torque is a unit-step function, discuss the control-system design using a PD control and a negative unity feedback.

From Chapter 5, the differential equations are given as

$$J\dot{\omega} + B\omega - K_t i_a = T_L$$
$$L_a \frac{di_a}{dt} + R_a i_a + K_e \omega = v_a$$
(10.45)

and the transfer functions are obtained as

$$G_{11}(s) = \frac{\Omega(s)}{T_L(s)} = \frac{L_a s + R_a}{\Delta(s)}, \qquad G_{12}(s) = \frac{\Omega(s)}{V_a(s)} = \frac{K_t}{\Delta(s)}$$
$$G_{21}(s) = \frac{I_a(s)}{T_L(s)} = -\frac{K_e}{\Delta(s)}, \qquad G_{22}(s) = \frac{I_a(s)}{V_a(s)} = \frac{Js + B}{\Delta(s)}$$
(10.46)

where

$$\Delta(s) = JL_a s^2 + (BL_a + JR_a)s + BR_a + K_e K_t$$
(10.47)

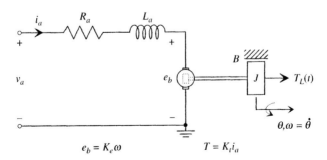

FIGURE 10.31 Electromechanical system.

Although both the speed and the current are the independent variables of the system, only the speed is important here. For simplicity, the following assumptions are made:

$$J = L_a = 1$$
$$B = R_a = 0$$
$$K_e = K_t = 1$$

Thus,

$$\Delta(s) = s^2 + 1$$

and the transfer functions of interest become

$$\frac{\Omega(s)}{T_L(s)} = \frac{s}{s^2 + 1}, \quad \frac{\Omega(s)}{V_a(s)} = \frac{1}{s^2 + 1} \quad (10.48)$$

The uncontrolled system (before closing the loop) is considered first; the block diagram is shown in Fig. 10.32. Assuming the disturbance load torque is a unit-step function, we have

$$T_L(s) = \frac{1}{s}$$

The Laplace transform of the speed is given as

$$\Omega(s) = \frac{s}{s^2 + 1} T_L(s) = \frac{s}{s^2 + 1} \frac{1}{s} = \frac{1}{s^2 + 1} \quad (10.49)$$

Therefore,

$$\omega(t) = L^{-1}\{\Omega(s)\} = \sin t \quad (10.50)$$

Since the speed is a sinusoid with constant amplitude, it will never reach a steady-state value.

Next, a closed-loop system is tried with a PD control and a unity negative feedback. The block diagram is modified as shown in Fig. 10.33. The characteristic equation of the closed-loop system is

$$1 + (K_p + K_d s)\frac{1}{s^2 + 1} = 0$$

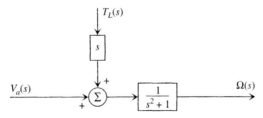

FIGURE 10.32 Block diagram of the uncontrolled system (before closing the loop).

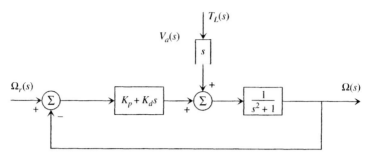

FIGURE 10.33 Block diagram of the controlled system (after closing the loop).

or
$$s^2 + K_d s + 1 + K_p = 0$$

Recall that the standard second-order characteristic equation is given as
$$s^2 + 2\zeta\omega_n s + \omega_n^2 = 0$$

Comparing these two equations term by term, we have
$$K_d = 2\zeta\omega_n, \qquad 1 + K_p = \omega_n^2$$

This is a second-order system, so select the damping ratio as $\zeta = 0.707$. Suppose the control-system design requirement is that the 5 percent-criterion settling time must be less than or equal to two seconds. Determine numerical values for K_d and K_p that will satisfy the design requirements.

In general, the settling times are given as

$$t_s\big|_{1\%} = 4.6\tau \tag{10.51}$$

$$t_s\big|_{2\%} = 4\tau \tag{10.52}$$

$$t_s\big|_{5\%} = 3\tau \tag{10.53}$$

Note that for first-order systems, τ is the ordinary time constant, i.e., the response reaches 98 percent of the steady-state value at 4τ, and so on. For underdamped second-order systems, τ is given by

$$\tau = \frac{1}{\sigma} = \frac{1}{\zeta\omega_n} \qquad (\zeta < 1) \tag{10.54}$$

For higher-order systems, τ is the dominant *generalized time constant*. The 5% criterion is often used for convenience,

$$t_s\big|_{5\%} = \frac{3}{\sigma} = \frac{3}{\zeta\omega_n} \qquad (\zeta < 1) \tag{10.55}$$

The requirement for this system is

$$t_s\big|_{5\%} = \frac{3}{\zeta\omega_n} \leq 2$$

Thus,
$$2\zeta\omega_n \geq 3$$

Therefore,

$$K_d \geq 3$$

The damping ratio is given, $\zeta = 0.707$, so

$$\omega_n \geq \left[\frac{3}{2\zeta} = \frac{3}{2(0.707)} = 2.122\right]$$

Since $K_p = \omega_n^2 - 1$ and $\omega_n \geq 2.122$, we have

$$K_p \geq 3.501$$

Now for a solution, pick $\omega_n = 3$. Thus, $K_p = 8$ and $K_d = 4.242$. After closing the loop, the transfer function becomes

$$\frac{\Omega(s)}{T_L(s)} = \frac{s/(s^2+1)}{1 + (K_p + K_d s)/(s^2+1)} = \frac{s}{s^2 + K_d s + 1 + K_p} = \frac{s}{s^2 + 4.242s + 9}$$

Thus, the speed is given as

$$\Omega(s) = \frac{1}{s^2 + 4.242s + 9}$$

Knowing that

$$L\left\{\frac{\omega_n^2}{s^2 + 2\zeta\omega_n s + \omega_n^2}\right\} = \frac{\omega_n}{\sqrt{1-\zeta^2}} e^{-\zeta\omega_n t} \sin \omega_n \sqrt{1-\zeta^2}\, t \quad (10.56)$$

we manipulate the variable as

$$\Omega(s) = \frac{1}{9} \frac{9}{s^2 + 4.242s + 9} \quad (10.57)$$

The speed as a function of time is, therefore, obtained as

$$\omega(t) = 0.471 e^{-2.122t} \sin 2.122 t \quad (10.58)$$

which is a decayed sinusoid. The closed-loop control system has improved the system performance by changing from a sinusoidal function with constant amplitude to a well-damped one. The steady-state value is zero for the closed-loop case.

PROBLEM SET 10.13–10.14

10.16. A PID controller and a negative unity feedback loop is used for a closed-loop system whose plant transfer function is given. Obtain the optimal (or suboptimal parameters) of the PID controller. (Hint: Use the second method of Ziegler-Nichols tuning.)

$$G_p(s) = \frac{1}{s(s^2 + 10s + 50)}$$

10.17. Consider the control system shown in Fig. P10.17 in which a PID controller is used. The transfer functions of the controller, the plant, and the sensor are given, respectively, as

$$G_c(s) = K\left(1 + \frac{1}{T_i s} + T_d s\right), \quad G_p(s) = \frac{1}{s[(s+3)^2 + 9]}, \quad H(s) = 1$$

Design or select the optimal (or suboptimal) values for K, T_i, and T_d.

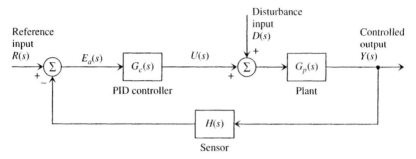

FIGURE P10.17 Block diagram of control system with PID controller.

10.18. The differential equation of a mechanical system is given as

$$\dot{x} + 50x = 500f - 100f_L$$

where x is the displacement, f is the applied force, and f_L is the disturbance load force.
(a) Obtain the transfer functions.
(b) Show a block diagram of the system (before closing the loop).

Now, a control-system design is considered: a PI control and a negative unity feedback loop.

(c) Show a block diagram of the closed-loop system (after closing the loop).
(d) Determine the numerical values of the PI controller assuming $\zeta = 0.707$ and $t_s|_{5\%} \leq 5$ sec.
(e) Obtain the steady-state error if the disturbance load force is a unit-step function and the reference input is zero.

SUMMARY

The essential components of a feedback control system are: plant, sensor(s), actuator(s), and controller. Any feedback control system must have at least one sensor and one actuator. Sensors and actuators are generally expensive, so they should be minimized in the design.

The locations of poles and zeros are important in the analysis and design of control systems. Poles are the roots of the characteristic equation, $D(s) = 0$, where $D(s)$ is the denominator of the transfer function. Zeros are the roots of $N(s) = 0$, where $N(s)$ is the numerator of the transfer function.

The Routh-Hurwitz stability criterion can be used to determine the stability of a dynamic system. The criterion states that the system is stable iff both of the following conditions are met:

1. All the coefficients of the characteristic equation are positive.
2. All the elements of the first column (shaded) of the Routh-Hurwitz table are positive.

Third-order system

The characteristic equation is given as

$$s^3 + \underbrace{a_1 s + a_2 s}_{\text{inner}} + a_3 = 0$$

(with outer brace spanning s^3 and a_3)

The inner-outer stability rule. The *third-order system* is stable iff

1. All the coefficients are positive: $a_1 > 0, a_2 > 0, a_3 > 0$
2. Inner product $>$ outer product: $a_1 a_2 > a_3$

Fourth-order system

The characteristic equation is given as

$$s^4 + a_1 s^3 + a_2 s^2 + a_3 s + a_4 = 0$$

The system is stable iff all the following conditions are satisfied:

1. $a_1, a_2, a_3, a_4 > 0$ (necessary condition)
2. $a_1 a_2 > a_3$ and $(a_1 a_2 - a_3) a_3 - a_4 a_1^2 > 0$

Steady-state error

For *nonsinusoidal* input and stable system, the steady-state error is given by the final-value theorem (FVT) as

$$e_{ss} = \lim_{t \to \infty} e(t) = \lim_{s \to 0} sE(s)$$

The general expression for $E(s)$ when both reference and disturbance inputs are present is

$$E(s) = R(s) - Y(s) = \left[1 - \frac{Y(s)}{R(s)}\bigg|_{D(s)=0}\right] R(s) - \frac{Y(s)}{D(s)}\bigg|_{R(s)=0} D(s)$$

Controller types

The controller types and the corresponding transfer functions are summarized below.

Controller type	Controller transfer function, $G_c(s)$		
P	K		
PD	$K_p + K_d s$	or	$K(1 + T_d s)$
PI	$K_p + \dfrac{K_i}{s}$	or	$K\left(1 + \dfrac{1}{T_i s}\right)$
PID	$K_p + \dfrac{K_i}{s} + K_d s$	or	$K\left(1 + \dfrac{1}{T_i s} + T_d s\right)$

Control actions

Proportional (P) control action. P control is the simplest of all types of control. For a certain input, the steady-state error of a closed-loop system can be reduced by increasing the proportional gain K (if the error is finite). However, the system may become unstable if K is increased too far.

Derivative (D) control action. D control adds damping into the system. The D action stabilizes the system and reduces the maximum percent overshoot (transient-time response) and resonant peak (frequency response).

Integral (I) control action. I control is used primarily to reduce the steady-state error. But this is at the expense of increasing instability of the system.

PID controller design and the second method of Ziegler-Nichols tuning

The transfer function of a PID controller is given as

$$G_c(s) = K\left(1 + \frac{1}{T_i s} + T_d s\right)$$

The suboptimal values of the parameters are given by

$$K = 0.6 K_{cr} \qquad T_i = \frac{T_{cr}}{2} \qquad T_d = \frac{T_{cr}}{8}$$

where $T_{cr} = 2\pi/\omega_{cr}$.

PROBLEMS

10.19. The characteristic equations of the first- and second-order systems are given. For each system, determine whether it is stable.
(a) $5s + 1 = 0$
(b) $s - 60 = 0$
(c) $2s^2 - 3s + 5 = 0$
(d) $2s^2 + 10s + 5 = 0$
(e) $1 + \dfrac{s+1}{s^2 + 3s + 4} = 0$
(f) $1 + \dfrac{s+5}{s^2 + 4} = 0$

10.20. The characteristic equations of third- and higher-order systems are given. For each system, determine whether is stable.
(a) $s^3 + 4s^2 + 5s + 20 = 0$
(b) $s^3 + 4s^2 + 20s + 10 = 0$
(c) $2s^4 + 6s^3 + 2s + 10 = 0$
(d) $1 + \dfrac{s+5}{3s^3 + 2s^2 + 3s + 4} = 0$

10.21. State the condition of stability for each of the systems whose characteristic equations are given.
(a) $\tau s + 1 = 0$
(b) $s^2 + 2\zeta\omega_n s + \omega_n^2 = 0$
(c) $s^3 + as^2 + bs + c = 0$
(d) $s^4 + as^3 + bs^2 + cs + d = 0$

10.22. The transfer functions of the plants are given. A closed-loop system is designed: a P control and a negative unity feedback loop. State the condition of stability for each system.

(a) $\dfrac{1}{s+1}$

(b) $\dfrac{s+1}{(s+5)(s^2+8s+100)}$

(c) $\dfrac{1}{s(s^2+s+5)}$

(d) $\dfrac{s+3}{s(s+1)(s+5)}$

10.23. A PID controller and a negative unity feedback loop is used for a closed-loop system whose plant transfer function is given. Obtain the suboptimal parameters of the controller. (*Hint:* Use the Ziegler-Nichols second method of tuning.)

$$G_p(s) = \dfrac{1}{s^3 + 5s + 10}$$

10.24. The differential equation of a mechanical system is given as

$$\ddot{x} + 0.2\dot{x} + x = 4f(t) + 10f_L(t)$$

where x is the displacement, f is the applied force, and f_L is the disturbance load force.
(a) Obtain the transfer functions.
(b) Show a block diagram of the system (before closing the loop).

Consider a control-system design with a PD control and a negative unity feedback loop.

(c) Show a block diagram of the closed-loop system (after closing the loop).
(d) Determine the numerical values of the PD controller assuming $\zeta = 0.707$ and $t_s|_{5\%} \leq 6$ sec.
(e) Obtain the steady-state error if the reference input is a unit-step function and the disturbance input is zero.

APPENDIX A

Tables

TABLE A.1
The International System (SI) of units

SI basic units		
Quantity	**Unit name**	**Symbol or Unit**
Length	meter	m
Mass	kilogram	kg
Time	second	s
Electric current	ampere	A
Temperature	kelvin	K
Luminous intensity	candela	cd
Amount of substance	mole	mol
SI auxiliary (supplementary) units		
Plane angle	radian	rad
Solid angle	steradian	sr
SI derived units		
(Circular) frequency (ω)		rad/s
Acceleration (a)		m/s^2
Activity (of a radioactive source)		s^{-1}
Angular acceleration (α)		rad/s^2
Angular velocity (ω)		rad/s
Area (A)		m^2
Area moment of inertia (I, J)		m^4
Density, mass density (ρ)		kg/m^3
Dynamic viscosity (μ)		N · s/m
Electric capacitance (C)	farad	F	A · s/V
Electric charge (q)	coulomb	C	A · s
Electric current (i)	ampere	A	C/s
Electric field strength		V/m

TABLE A.1 (Continued)
The International System (SI) of units

Quantity	Unit name	Symbol	or Unit
	SI derived units		
Electric resistance (R)	ohm	Ω	V/A
Energy (E)	joule	J	N · m
Entropy (s)			J/K
Force (f)	newton	N	kg · m/s^2
Frequency (f)	hertz	Hz	s^{-1}
Illuminance	lux	lx	lm/m^2
Inductance (L)	henry	H	V · s/A
Kinematic viscosity ($\nu = \mu/\rho$)			m^2/s
Luminance			cd/m^2
Luminous flux	lumen	lm	cd · sr
Magnetic field strength			A/m
Magnetic flux	weber	Wb	V · s
Magnetic flux density	tesla	T	Wb/m^2
Magnetomotive force	ampere	A	
Mass moment of inertia (I, J)			kg · m^2
Power	watt	W	J/s
Pressure, mechanical stress	pascal	Pa	N/m^2
Radiant intensity			W/sr
Specific heat (c_p, c_v, c)			J/(kg · K)
Specific volume (v)			m^3/kg
Surface coefficient of heat transfer(h)			W/(s · m^2 · K)
Temperature (T)			K
Thermal conductivity (k)			W/(s · m · K)
Velocity, speed (v)			m/s
Voltage (v, e)	volt	V	W/A
Volume (V)			m^3
Wave length (λ)			m^{-1}
Work (W), heat (Q)	joule	J	N · m

Prefix	Symbol	Factor
	SI prefixes	
tera	T	10^{12}
giga	G	10^9
mega	M	10^6
kilo	k	10^3
hecto	h	10^2
deka	da	10
deci	d	10^{-1}
centi	c	10^{-2}
milli	m	10^{-3}
micro	μ	10^{-6}
nano	n	10^{-9}
pico	p	10^{-12}
femto	f	10^{-15}
atto	a	10^{-18}

TABLE A.2
Conversion factors

Convection heat transfer coefficient	$1 \text{ W}/(\text{m}^2 \cdot °\text{C}) = 0.1761 \text{ Btu}/(\text{hr} \cdot \text{ft}^2 \cdot °\text{F})$
	$1 \text{ Btu}/(\text{hr} \cdot \text{ft}^2 \cdot °\text{F}) = 5.6782 \text{ W}/(\text{m}^2 \cdot °\text{C})$
Density	$1 \text{ g/cm}^3 = 62.43 \text{ lb}_m/\text{ft}^3$
Energy	$1 \text{ J} = 1 \text{ N} \cdot \text{m}$
	$1 \text{ cal} = 4.184 \text{ J}$
Force	$1 \text{ lb}_f = 4.45 \text{ N}$
Length	$1 \text{ in.} = 2.54 \text{ cm}$
	$1 \text{ ft} = 0.3048 \text{ m}$
	$1 \text{ mi.} = 5{,}280 \text{ ft} = 1{,}609 \text{ m}$
	$1 \text{ µm} = 10^{-6} \text{ m}$
Mass	$1 \text{ lb}_m = 0.45359237 \text{ kg} = 16 \text{ oz}$
	$1 \text{ slug} = 32.174 \text{ lb}_m$
	$1 \text{ ton} = 2{,}000 \text{ lb}_m$
Mole	$1 \text{ kg} \cdot \text{mol} = 10^3 \text{ mol} (\text{g} \cdot \text{mol})$
Power	$1 \text{ W} = 1 \text{ J/s} = 3.413 \text{ Btu/hr}$
Pressure	$1 \text{ atm} = 760 \text{ mm Hg} = 760 \text{ torr} = 1.0132 \times 10^{-5} \text{ Pa}$
	$1 \text{ torr} = 1.316 \times 10^{-3} \text{ atm}$
	$1 \text{ Pa} = 1 \text{ N/m}^2$
Temperature	$°\text{C} = (°\text{F} - 32)/1.8$
	$°\text{F} = °\text{C}(1.8) + 32$
	$\text{K} = °\text{C} + 273.16$
	$°\text{R} = °\text{F} + 459.69$
	$1 \text{ K} = 1.8 \, °\text{R}$
Thermal conductivity	$1 \text{ W}/(\text{m} \cdot °\text{C}) = 0.5778 \text{ Btu}/(\text{hr} \cdot \text{ft} \cdot °\text{F})$
	$1 \text{ Btu}/(\text{hr} \cdot \text{ft}^2 \cdot °\text{F}) = 1.7307 \text{ W}/(\text{m} \cdot °\text{C})$
Volume	$1 \text{ liter(L)} = 1 \text{ dm}^3 = 10^3 \text{ cm}^3 = 1.0564 \text{ quart} = 0.0353 \text{ ft}^3$
	$1 \text{ ft}^3 = 28.316 \text{ L}$
	$1 \text{ cm}^3 (\text{cc}) = 1 \text{ ml}$
	$1 \text{ gal} = 3.785 \text{ L} = 4 \text{quarts}$
	$1 \text{ quart} = 2 \text{ pints} = 67.2 \text{ in.}^3 = 0.9466 \text{ L}$
	$1 \text{ pint} = 16 \text{ ounces}$
	$1 \text{ ounce (oz)} = 1.734 \text{ in.}^3$

TABLE A.3
Physical constants

Universal gas constant	R_u	$= 8.3143 \text{ J}/(\text{mol} \cdot \text{K})$
		$= 8.3143 \text{ kJ}/(\text{kg} \cdot \text{mol} \cdot \text{K})$
		$= 1{,}545.3 \text{ ft} \cdot \text{lb}_f/(\text{lb}_m \cdot \text{mol} \cdot °\text{R})$
		$= 1.986 \text{ cal}/(\text{mol} \cdot \text{K})$
		$= 1.986 \text{ Btu}/(\text{lb}_m \cdot \text{mol} \cdot °\text{R})$
		$= 0.08206 \text{ L} \cdot \text{atm}/(\text{mol} \cdot \text{K})$
Specific gas constant of air	R_{air}	$= 53.35 \text{ ft} \cdot \text{lb}_f/(\text{lb}_m \cdot °\text{R})$
Molecular mass of air	$M_{m,air}$	$= 28.97 \text{ ft} \cdot \text{lb}_f/(\text{lb}_m \cdot °\text{R})$
Speed of light (in vacuum)	c	$= 3 \times 10^8 \text{ m/s}$

TABLE A.4
Physical properties of gases at 300°K (27°C) at 1 atm

Material	ρ, kg/m^3	c_p, kJ/(kg · °C)	k, W/(m · °C)
Air	1.177	1.1006	0.0262
Helium, He	0.1986	5.200	0.1492
Hydrogen, H$_2$	0.08185	14.314	0.182
Oxygen, O$_2$	1.3007	0.9203	0.02676
Nitrogen, N$_2$	1.1421	1.0408	0.02620
Carbon dioxide, CO$_2$	1.7973	0.871	0.016572

TABLE A.5
Physical properties of liquids at 20°C

Material	ρ, kg/m^3	c, kJ/(kg · °C)	k, W/(m · °C)
Water, H$_2$O	997.4	4.179	0.604
Ammonia, NH$_3$	611.75	4.798	0.521
Carbon dioxide, CO$_2$	772.57	5.0	0.0872
Sulfur dioxide, SO$_2$	1,386.40	1.3653	0.199
Freon-12, CCl$_2$F$_2$	1,330.18	0.9659	0.073
Glycerin, C$_3$H$_5$(OH)$_3$	1,264.02	2.386	0.286
Ethylene glycol, C$_2$H$_4$(OH)$_2$	1,116.65	2.382	0.249
Engine oil	88.23	1.880	0.145
Mercury, Hg	13,579.04	0.1394	8.69

TABLE A.6
Physical properties of pure (solid) metals at 20°C

Material	ρ, kg/m^3	c, kJ/(kg · °C)	k, W/(m · °C)
Aluminum	2707	0.896	204
Copper	8954	0.3831	385
Iron	7897	0.452	73
Magnesium	1746	1.013	171
Nickel	8906	0.4459	90
Silver	10,524	0.2340	419
Tin	7304	0.2265	64
Zinc	7144	0.3843	112

APPENDIX B

Computer Simulation

The following examples show certain applications in computer simulation. These applications include the response of a dynamic system when subject to an input, as well as the numerical solution of a nonlinear dynamic system. In the former, the input may be transient (e.g., a unit-step function) or sinusoidal (e.g., a sine or cosine function).

Many commercial software packages are available for computer simulation: $MATRIX_X$, SystemBuild, Xmath, MATLAB, PRO-MATLAB, SIMULINK, Math-CAD, Maple V, Mathematica, and so on. These packages are used extensively in many academic institutions as well as in industry. $MATRIX_X$, SystemBuild, Xmath, MATLAB, PRO-MATLAB, and SIMULINK are well known for their usage in teaching and research in the fields of dynamics and control; Maple V and Mathematica enjoy a great reputation for symbolic manipulation; MathCAD is popular among many engineering students.

For a given dynamic system, after a mathematical model is obtained, a computer simulation can be performed readily using any of the aforementioned software packages. The choice of a particular software primarily depends upon (a) which one is available to the reader, (b) whether it is numerical or symbolic, and (c) the purpose of simulation, e.g., modeling and analysis of dynamic systems or control-system design. For illustration purposes, only the code for $MATRIX_X$ is shown in this appendix.

USEFUL NOTES FOR $MATRIX_X$

Comment lines (//). It is often helpful to include comments in the code. In $MATRIX_X$, comments are preceded by a forward double slash (//). Note that this symbol also signifies the implementation of a user-defined function (see Example B.4).

EXAMPLE B.1. TRANSIENT RESPONSE OF A SINGLE-DEGREE-OF-FREEDOM (SDOF) SYSTEM. A simple mass-spring-damper system, as shown in Fig. 4.12, is reconsidered (see Fig. B.1). The applied force is $f(t) = Pu_s(t)$, where P and $u_s(t)$ are a constant and a unit-step function, respectively. The differential equation is given in standard input-output form as

$$m\ddot{x} + c\dot{x} + kx = f(t)$$

Method 1: Transfer functions
The transfer functions for displacement and velocity are given as

$$G_x(s) = \frac{X(s)}{F(s)} = \frac{1}{ms^2 + cs + k}$$

$$G_v(s) = \frac{V(s)}{F(s)} = \frac{sX(s)}{F(s)} = \frac{s}{ms^2 + cs + k}$$

Method 2: State-space representation

$$\begin{Bmatrix} \dot{x}_1 \\ \dot{x}_2 \end{Bmatrix} = \begin{bmatrix} 0 & 1 \\ -\frac{k}{m} & -\frac{c}{m} \end{bmatrix} \begin{Bmatrix} x_1 \\ x_2 \end{Bmatrix} + \begin{Bmatrix} 0 \\ \frac{1}{m} \end{Bmatrix} u \quad \text{(state equation)}$$

$$\begin{Bmatrix} y_1 \\ y_2 \end{Bmatrix} = \begin{bmatrix} 1 & 0 \\ 0 & 1 \end{bmatrix} \begin{Bmatrix} x_1 \\ x_2 \end{Bmatrix} + \begin{Bmatrix} 0 \\ 0 \end{Bmatrix} u \quad \text{(output equation)}$$

where

$$\underbrace{\begin{Bmatrix} x_1 \\ x_2 \end{Bmatrix}}_{\text{mathematical}} = \underbrace{\begin{Bmatrix} x \\ \dot{x} \end{Bmatrix}}_{\text{physical}}, \quad u = f(t), \quad \underbrace{\begin{Bmatrix} y_1 \\ y_2 \end{Bmatrix}}_{\text{mathematical}} = \underbrace{\begin{Bmatrix} x \\ \dot{x} \end{Bmatrix}}_{\text{physical}}$$

For *each* of the two methods, obtain a computer plot that shows the two superimposed curves of displacement and velocity, using the following numerical values:

$m = 1$ kg, $\omega_n = 1$ rad/s, $\zeta = 0.25$, $P = 1$ N, t from 0 to 20 s

Solution See Figs. B.2(a) and (b).

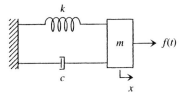

FIGURE B.1

APPENDIX B: Computer Simulation 573

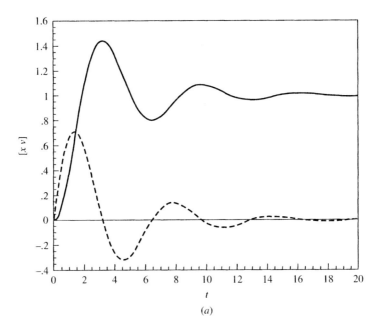

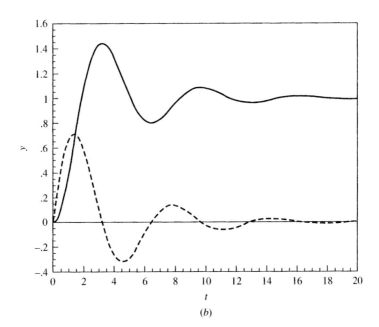

FIGURE B.2 Unit-step response: displacement (solid) and velocity (broken): (a) transfer-function approach, (b) state-space approach.

MATRIX$_X$ code: Transfer-function approach

```
// Specify the numerators, Nx and Nv, and the denominator, D
Nx=1;Nv=[1,0];D=[1 .5 1];
// Compute the 400 data points for the unit-step responses, x and v
[t,x]=step(Nx,D,20,400);
[t,v]=step(Nv,D,20,400);
// Plot the responses
pause
plot(t,[x v],'nogrid')
```

MATRIX$_X$ code: State-space approach

```
// Specify the matrices of the state-space representation
A=[0,1;-1,-.5];B=[0;1];C=[1,0;0,1];D=[0;0];
// Specify the system matrix S, and the number of states NS
S=[A,B;C,D];NS=2;
// Compute the 400 data points for the unit-step responses, x and v
[t,y]=step(S,NS,20,400);
// Plot the outputs
pause
plot(t,y,'nogrid')
```

Note: The two different approaches give identical results as expected. Only the labels on the ordinates are different.

EXAMPLE B.2. TRANSIENT RESPONSE OF A TWO-DEGREE-OF-FREEDOM (TDOF) SYSTEM. A simple mass-spring-damper system, as shown in Fig. 4.30, is reconsidered (see Fig. B.3). The applied forces are given as $f_1(t) = f(t) = P\delta(t)$, $f_2(t) = 0$, where P and $\delta(t)$ are a constant and unit-impulse function, respectively. The differential equation is given in standard input-output form as

$$\begin{bmatrix} m_1 & 0 \\ 0 & m_2 \end{bmatrix} \begin{Bmatrix} \ddot{x}_1 \\ \ddot{x}_2 \end{Bmatrix} + \begin{bmatrix} c & -c \\ -c & c \end{bmatrix} \begin{Bmatrix} \dot{x}_1 \\ \dot{x}_2 \end{Bmatrix} + \begin{bmatrix} k_1+k_2 & -k_2 \\ -k_2 & k_2 \end{bmatrix} \begin{Bmatrix} x_1 \\ x_2 \end{Bmatrix} = \begin{Bmatrix} f_1 = f(t) \\ f_2 = 0 \end{Bmatrix}$$

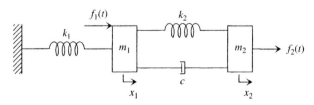

FIGURE B.3

Method 1: Transfer functions
The transfer functions are given as

$$G_1(s) = \frac{X_1(s)}{F(s)} = \frac{m_2 s^2 + cs + k_2}{\Delta(s)}$$

$$G_2(s) = \frac{X_2(s)}{F(s)} = \frac{cs + k_2}{\Delta(s)}$$

where

$$\Delta(s) = \left(m_1 s^2 + cs + k_1 + k_2\right)\left(m_2 s^2 + cs + k_2\right) - (cs + k_2)^2$$

$$= m_1 m_2 s^4 + (m_1 + m_2) c s^3$$

$$+ [(k_1 + k_2) m_2 + m_1 k_2] s^2 + c k_1 s + k_1 k_2$$

Method 2: State-space representation

$$\begin{Bmatrix} \dot{x}_1 \\ \dot{x}_2 \\ \dot{x}_3 \\ \dot{x}_4 \end{Bmatrix} = \begin{bmatrix} 0 & 0 & 1 & 0 \\ 0 & 0 & 0 & 1 \\ -\dfrac{k_1 + k_2}{m_1} & \dfrac{k_2}{m_1} & -\dfrac{c}{m_1} & \dfrac{c}{m_1} \\ \dfrac{k_2}{m_2} & -\dfrac{k_2}{m_2} & \dfrac{c}{m_2} & -\dfrac{c}{m_2} \end{bmatrix} \begin{Bmatrix} x_1 \\ x_2 \\ x_3 \\ x_4 \end{Bmatrix} + \begin{Bmatrix} 0 \\ 0 \\ \dfrac{1}{m_1} \\ 0 \end{Bmatrix} u \quad \text{(state equation)}$$

$$\begin{Bmatrix} y_1 \\ y_2 \end{Bmatrix} = \begin{bmatrix} 1 & 0 & 0 & 0 \\ 0 & 1 & 0 & 0 \end{bmatrix} \begin{Bmatrix} x_1 \\ x_2 \\ x_3 \\ x_4 \end{Bmatrix} + \begin{Bmatrix} 0 \\ 0 \end{Bmatrix} u \quad \text{(output equation)}$$

where

$$\underbrace{\begin{Bmatrix} x_1 \\ x_2 \\ x_3 \\ x_4 \end{Bmatrix}}_{\text{mathematical}} = \underbrace{\begin{Bmatrix} x_1 \\ x_2 \\ \dot{x}_1 \\ \dot{x}_2 \end{Bmatrix}}_{\text{physical}}, \quad u = f(t), \quad \underbrace{\begin{Bmatrix} y_1 \\ y_2 \end{Bmatrix}}_{\text{mathematical}} = \underbrace{\begin{Bmatrix} x_1 \\ x_2 \end{Bmatrix}}_{\text{physical}}$$

Obtain a computer plot that shows the two superimposed curves of displacements, using the following numerical values:

$$m_1 = 0.5 \text{ kg}, \quad m_2 = 1 \text{ kg}, \quad k_1 = k_2 = 1 \text{ N/m},$$

$$c = 0.25 \text{ N} \cdot \text{s/m}, \quad P = 10 \text{ N}, \quad t \text{ from 0 to 30 s}$$

Solution See Fig. B.4.

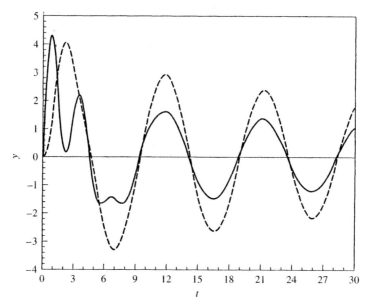

FIGURE B.4 Impulse response: x_1 (solid) and x_2 (broken) (from state-space approach).

MATRIX$_X$ code: Transfer-function approach

```
// Specify the numerical values
m1=.5;m2=1;c=.25;k1=1;k2=1;P=10;tf=30;n=400;
// Specify the numerators, N1 and N2, and the denominator, D
N1=[m2,c,k2];N2=[c,k2];D=[m1*m2,(m1+m2)*c,(k1+k2)*m2+m1*k2,c*k1,k1*k2];
// Compute the 400 data points for the impulse responses, x1 and x2
[t,x1]=impu(N1,D,tf,n);x1=P*x1;
[t,x2]=impu(N2,D,tf,n);x2=P*x2;
// Plot the responses
pause
plot(t,[x1 x2],'nogrid')
```

MATRIX$_X$ code: State-space approach

```
// Specify the numerical values
m1=.5;m2=1;c=.25;k1=1;k2=1;P=10;tf=30;n=400;
// Specify the matrices of the state-space representation
A=[0,0,1,0;0,0,0,1;-(k1+k2)/m1,k2/m1,-c/m1,c/m1;k2/m2,-k2/m2,c/m2,-c/m2];
B=[0;0;1;0];C=[1,0,0,0;0,1,0,0];D=[0;0];
// Specify the system matrix S, and the number of states NS
S=[A,B;C,D];NS=4;
// Compute the n data points for the impulse responses, x and v
[t,y]=impu(S,NS,tf,n);y=P*y;
// Plot the outputs
pause
plot(t,y,'nogrid')
```

EXAMPLE B.3. FREQUENCY RESPONSE OF A SYSTEM WITH A DAMPED VIBRATION ABSORBER. A system with a damped vibration absorber, shown in Fig. 8.35, is reconsidered here (see Fig. B.5). The applied force is a sinusoidal function, $f(t) = P\cos\omega t$. The normalized FRF, which is expressed in dimensionless groups, is

$$\frac{X_1(j\beta)}{x_{st}} = KG_1(j\beta) = \frac{\alpha^2 - \beta^2 + j2\zeta\beta}{\alpha^2 + \beta^4 - [1 + (1+\mu)\alpha^2]\beta^2 + j2\zeta\beta[1 - (1+\mu)\beta^2]}$$

where $x_{st} = P/K$. The magnitude is

$$\frac{|x_{1ss}(\beta)|}{x_{st}} = |KG_1(j\beta)|$$

$$= \frac{\sqrt{(\alpha^2 - \beta^2)^2 + (2\zeta\beta)^2}}{\sqrt{\{\alpha^2 + \beta^4 - [1 + (1+u)\alpha^2]\beta^2\}^2 + \{2\zeta\beta[1 - (1+\mu)\beta^2]\}^2}}$$

Obtain a plot of three superimposed magnitude curves, using

$$\mu = 1 \quad \text{(mass ratio)}$$

$$\alpha_{opt} = \frac{1}{1+\mu} \quad \text{(optimal tuning)}$$

and

(1) $\zeta = 0$
(2) $\zeta = \infty$
(3) $\zeta_{opt} = \sqrt{\frac{3\mu}{8(1+\mu)^3}}$ (optimal damping)

Solution See Fig. B.6.

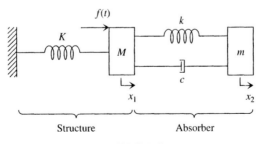

FIGURE B.5

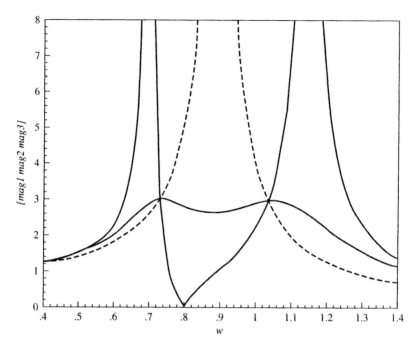

FIGURE B.6 Unedited version of Figure 8.37.

MATRIX$_X$ code

```
// Set the range of normalized frequency, beta (b) = omega (w).
w=[.4:.01:1.4]';b=w;
// Set the values of mass ratio mu (m), tuning ratio alpha (a).
m=.25;
a=1/(1+m);
pause
// Set the value of the first damping ratio to zero, zeta1
(z1) = 0.
z1=0;
// Specify the first numerator and denominator, n1 and d1, then
compute the first magnitude, mag1.
n1=a**2-b**2+jay*2*z1*b;
d1=a**2+b**4-(1+(1+m)*a**2)*b**2+jay*2*z1*b*.(1-(1+m)*b**2);
mag1=abs(n1)/.abs(d1);
// Set the value of the second damping ratio to a large number
(for infinity), zeta1 (z1) = 10.
z2=10;
// Specify the second numerator and denominator, n2 and d2, then
compute the second magnitude, mag2.
n2=a**2-b**2+jay*2*z2*b;
d2=a**2+b**4-(1+(1+m)*a**2)*b**2+jay*2*z2*b*.(1-(1+m)*b**2);
mag2=abs(n2)/.abs(d2);
```

```
// Compute the third numerator and denominator, n3 and d3, then
   compute the third magnitude, mag3.
z3=sqrt(3*m/(8*(1+m)**3));
n3=a**2-b**2+jay*2*z3*b;
d3=a**2+b**4-(1+(1+m)*a**2)*b**2+jay*2*z3*b*.(1-(1+m)*b**2);
mag3=abs(n3)/.abs(d3);
// Plot the three superimposed curves
pause
plot(w,[mag1 mag2 mag3],'ymax=8 nogrid')
```

EXAMPLE B.4. SOLUTION OF A FIRST-ORDER NONLINEAR DIFFERENTIAL EQUATION VIA RUNGE-KUTTA FOURTH-ORDER. Consider a first-order, nonlinear differential equation subject to zero initial condition, in the form

$$\dot{x} + x^3 = 1, \qquad x(0) = 0, \qquad 0 \le t \le 3.5$$

Using the fourth-order Runge-Kutta method, determine the numerical values of the variable x in the indicated time interval.

Solution See Fig. B.7.

MATRIX$_X$ code:

```
//This MATRIX_X code is based on the fourth-order Runge-Kutta method
//to solve the nonlinear system of Example 3.28 in the text
h=0.25;
t=[0:.25:3.5];
//Define initial conditions
x(1)=0;xnew(1)=x(1);
x=x(1);
//Perform function evaluations
for j=1:14,...
k1=h*func1(t(j),x);...
k2=h*func1(t(j)+h/2,x+0.5*k1);...
k3=h*func1(t(j)+h/2,x+0.5*k2);...
k4=h*func1(t(j)+h,x+k3);...
//Update the variable
x=x+(k1+2*k2+2*k3+k4)/6;...
xnew(j+1)=x;
//Rename xnew
x=xnew;
//Plot the numerical data
plot(t,x)
The  user-defined function, called func1, utilized in the main code
above
//f1=func1(t,x)
f1=1-x**3;
retf
```

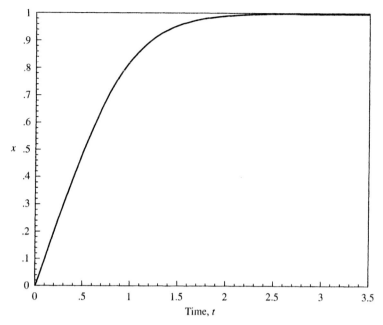

FIGURE B.7 Numerical solution of the nonlinear equation in Example B.4.

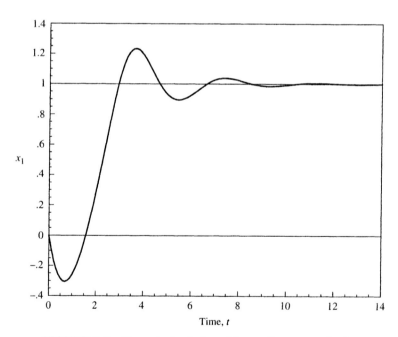

FIGURE B.8 Numerical solution of the nonlinear equation in Example B.5.

APPENDIX B: Computer Simulation

EXAMPLE B.5. SOLUTION OF A SECOND-ORDER NONLINEAR DIFFERENTIAL EQUATION VIA RUNGE-KUTTA. Consider a second-order, nonlinear differential equation subject to given initial condition, in the form

$$\ddot{x} + \dot{x} + x^3 = 1, \qquad x(0) = 0, \qquad \dot{x}(0) = -1, \qquad 0 \le t \le 14$$

Using the fourth-order Runge-Kutta method, determine the numerical values of the variable x in the indicated time interval.

Solution See Fig. B.8.

MATRIX$_X$ code:

```
//This MATRIX_x code is based on the fourth-order Runge-Kutta method
// to solve the nonlinear system of Example 3.29 in the text
h=0.1;
t=[0:.1:14];
//Define the initial conditions
x1(1)=0;x2(1)=-1;
x1=x1(1);x2=x2(1);x1new(1)=x1(1);x2new(1)=x2(1);
//Perform function evaluations via RK4
for j=1:140,...
k(1,1)=h*x2;...
k(1,2)=h*func2(x1,x2);...
k(2,1)=h*(x2+k(1,2)/2);...
k(2,2)=h*func2(x1+k(1,1)/2,x2+k(1,2)/2);...
k(3,1)=h*(x2+k(2,2)/2);...
k(3,2)=h*func2(x1+k(2,1)/2,x2+k(2,2)/2);...
k(4,1)=h*(x2+k(3,2));...
k(4,2)=h*func2(x1+k(3,1),x2+k(3,2));...
//Update the values of the variables
x1=x1+(k(1,1)+2*k(2,1)+2*k(3,1)+k(4,1))/6;...
x2=x2+(k(1,2)+2*k(2,2)+2*k(3,2)+k(4,2))/6;...
x1new(j+1)=x1;...
x2new(j+1)=x2;
//Rename x1new
x1=x1new;
//Plot x1 versus t
plot(t,x1)
```

The user-defined function, called func2, utilized in the main code above.

```
//f2=func2(x,y)
f2=1-x**3-y;
retf
```

APPENDIX C

Useful Formulas

EULER'S IDENTITIES

$$e^{jx} = \cos x + j \sin x \qquad e^{-jx} = \cos x - j \sin x$$

$$\cos x = \frac{e^{jx} + e^{-jx}}{2} \qquad \sin x = \frac{e^{jx} - e^{-jx}}{2j}$$

TAYLOR'S SERIES EXPANSION

$$\cos \theta = 1 - \frac{\theta^2}{2!} + \frac{\theta^4}{4!} - \cdots$$

$$\sin \theta = \theta - \frac{\theta^3}{3!} + \frac{\theta^5}{5!} - \cdots$$

$$\tan \theta = \theta + \frac{\theta^3}{3} + \frac{2\theta^5}{15} + \cdots$$

TRIGONOMETRY

Pythagorean Identities

$$\sin^2 a + \cos^2 a = 1$$
$$\tan^2 a + 1 = \sec^2 a$$
$$1 + \cot^2 a = \csc^2 a$$

Appendix C: Useful Formulas

Expansion Formulas

$$\cos(a \pm b) = \cos a \cos b \mp \sin a \sin b$$
$$\sin(a \pm b) = \sin a \cos b \pm \sin b \cos a$$
$$\tan(a \pm b) = \frac{\tan a \pm \tan b}{1 \mp \tan a \tan b}$$

Explicitly,

$$\cos(a + b) = \cos a \cos b - \sin a \sin b \qquad \sin(a + b) = \sin a \cos b + \sin b \cos a$$
$$\cos(a - b) = \cos a \cos b + \sin a \sin b \qquad \sin(a - b) = \sin a \cos b - \sin b \cos a$$

Product Formulas

$$\cos a \cos b = \frac{\cos(a + b) + \cos(a - b)}{2} \qquad \sin a \sin b = \frac{\cos(a - b) - \cos(a + b)}{2}$$

$$\sin a \cos b = \frac{\sin(a + b) + \sin(a - b)}{2} \qquad \sin b \cos a = \frac{\sin(a + b) - \sin(a - b)}{2}$$

Double-Angle/Half-Angle Formulas

$$\sin 2a = 2 \sin a \cos a$$
$$\cos 2a = \cos^2 a - \sin^2 a = 1 - 2 \sin^2 a = 2 \cos^2 a - 1$$
$$\cos^2 a = \frac{1 + \cos 2a}{2} \qquad \sin^2 a = \frac{1 - \cos 2a}{2}$$

Law of Cosines

$$c^2 = a^2 + b^2 - 2ab \cos C$$

where a, b, c are the sides of a plane triangle, and C is the angle opposite of the side c.

Miscellaneous

$$\sin 3a = 3 \sin a - 4 \sin^3 a$$
$$A \cos \omega t + B \sin \omega t = A_o \cos(\omega t + \phi) \qquad A_o = \sqrt{A^2 + B^2}$$
$$\phi = -\tan^{-1} \frac{B}{A} \qquad (-90° \leq \phi \leq 0°)$$

$$A\cos\omega t + B\sin\omega t = A_o \sin(\omega t + \theta) \qquad A_o = \sqrt{A^2 + B^2}$$

$$\theta = \tan^{-1} \frac{A}{B} \qquad (0° \leq \theta \leq 90°)$$

HYPERBOLIC FUNCTIONS AND RELATIONS

$$\cosh x = \frac{e^x + e^{-x}}{2} \qquad \sinh x = \frac{e^x - e^{-x}}{2}$$

$$e^x = \cosh x + \sinh x \qquad e^{-x} = \cosh x - \sinh x$$

$$\cosh^2 x - \sinh^2 x = 1$$

LOGARITHM

Common Logarithm

$$\log y \equiv \log_{10} y \qquad \log 10 = 1$$

$$y = 10^x \implies \log y = x$$

Natural Logarithm

$$\ln y \equiv \log_e y \qquad \ln e = 1 \qquad e = 2.71828$$

$$y = e^x \implies \ln y = x$$

Decibel (dB)

$$x(dB) = 20 \log x \implies x = 10^{x(dB)/20}$$

Change of Base

$$\log_b y = \log_a y \cdot \log_b a$$

Logarithm of a Complex Number

$$z = x + jy = r(\cos\theta + j\sin\theta) = re^{j\theta} \implies \ln z = \ln r + j\theta$$

APPENDIX C: Useful Formulas

BINOMIAL SERIES

The *binomial series* for $(1 + x)^m$ is given as

$$(1 + x)^m = 1 + mx + \frac{m(m-1)}{2!}x^2 + \frac{m(m-1)(m-2)}{3!}x^3 + \cdots$$
$$+ \frac{m(m-1)(m-2)\cdots(m-n+1)}{n!}x^n + \cdots$$

where m can be any number (it need not be a positive integer). This series is also called a *binomial-series expansion* or Maclaurin series. For example,

$$(1 + x)^{1/2} = 1 + \frac{1}{2}x + \frac{1}{2}\left(\frac{1}{2} - 1\right)\frac{x^2}{2!} + \frac{1}{2}\left(\frac{1}{2} - 1\right)\left(\frac{1}{2} - 2\right)\frac{x^3}{3!} + \cdots \quad \text{if } |x| < 1$$

$$(1 + x)^{1/2} \approx 1 + \frac{1}{2}x \quad \text{if } |x| < 1$$

Notice that this approximation is on the high side since the series is alternating and the second-order term is negative

$$\frac{1}{2}\left(\frac{1}{2} - 1\right)\frac{x^2}{2!} = -\frac{1}{8}x^2$$

APPENDIX D

Answers to Odd-Numbered Problems

CHAPTER 1

Problem Set 1.1

1.1. $-5 + j$
1.3. $\frac{-3}{5} + j\frac{1}{5}$

Problem Set 1.2

1.5. $z = \sqrt{2}e^{j(\pi/4)}$

1.7. $\frac{\sqrt{10}}{2}e^{j161.56}$

1.9. $5(\cos 2\theta + j\sin 2\theta)$, $\theta = \tan^{-1}(\frac{1}{2})$

Problem Set 1.3

1.11. $F(s) = \dfrac{3s + 1}{s^2 + 2s + 1} = 3 \cdot \dfrac{s + \frac{1}{3}}{(s + 1)^2}$; zero: $s = -\frac{1}{3}$; poles: $s_{1,2} = -1, -1$

1.13. $H(s) = \dfrac{1}{s(5s + 1)} = \dfrac{1}{5s(s + 0.2)} = \dfrac{1}{5} \cdot \dfrac{1}{s(s + 0.2)}$; zero: none; poles: $s_{1,2} = 0, -0.2$

1.15. $\dfrac{-1}{20} - j\dfrac{3}{20}$, magnitude $= \dfrac{1}{\sqrt{40}}$, phase $= 180° + \tan^{-1} 3$

Problem Set 1.4

1.17. Nonlinear
1.19. $\dot{\omega} + 4\omega = 4\sin 2t$, $\tau = \frac{1}{4}$
1.21. $x(t) = e^{-t}\sin 2t$
1.23. $x(t) = A + B\cos 2t + C\sin 2t$
1.25. $x(t) = C_1 + C_2 e^{-2t} - \frac{1}{2}te^{-2t}$

Problem Set 1.5

1.27. $\dfrac{s}{s^2 + \omega^2} + j\dfrac{\omega}{s^2 + \omega^2}$

1.29. $e^\beta \dfrac{1}{s + \alpha}$

1.31. $\dfrac{2}{s^3}$

1.33. $\dfrac{2}{(s+1)^2 + 4}$

1.35. $2\sin 3t$

Problem Set 1.6

1.37. $F(s) = \dfrac{2}{s(s^2+4)}$

1.39. $\dfrac{1}{(s+1)^2}$

1.41. $\dfrac{1}{\omega}(1 - \cos\omega t)$

1.43. $\dfrac{1}{\omega}\sin\omega t$

1.45. $t + e^{-t} - 1$

Problem Set 1.7

1.47. $\frac{1}{2}e^{-t} + \frac{3}{2}e^{-3t} = \frac{1}{2}(e^{-t} + 3e^{-3t})$
1.49. $e^{-t} - (\cos 2t + \frac{1}{2}\sin 2t)e^{-2t}$
1.51. $\frac{1}{2}t - \frac{1}{4} + \frac{5}{4}e^{-2t}$

Problem Set 1.8

1.55. $\dfrac{e^{-s}}{s} - 2\dfrac{e^{-2s}}{s} + \dfrac{e^{-2s}}{s^2} - \dfrac{e^{-3s}}{s^2}$

1.57. $\dfrac{1}{s} \cdot \dfrac{(1-e^{-\pi s})^2}{1-e^{-2\pi s}} = \dfrac{(1-e^{-\pi s})^2}{s(1-e^{-\pi s})(1+e^{-\pi s})} = \dfrac{1-e^{-\pi s}}{s(1+e^{-\pi s})}$

Problem Set 1.9

1.59. $x(t) = \begin{cases} 0 & \text{for } 0 \le t < 1 \\ e^{-2(t-1)} & \text{for } t > 1 \end{cases}$

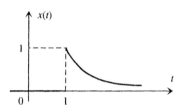

1.61. $x(t) = \begin{cases} 0 & \text{for } 0 \le t < 1 \\ 1 - e^{-(t-1)} & \text{for } t > 1 \end{cases}$

1.63. $x(t) = \begin{cases} 0 & \text{for } t = 0 \\ e^{-t}\left(\dfrac{-\sqrt{2}}{2}\sin\sqrt{2}t\right) + \dfrac{1}{2}\left[1 - \dfrac{\sqrt{3}}{2}e^{-t}\sin(\sqrt{2}t + \phi)\right] & \text{for } t > 0 \end{cases}$

$\phi = \tan^{-1}\sqrt{2}$

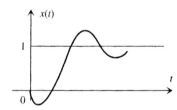

Problem Set 1.10

1.65. $f(0^+) = 0$. $f(t) = \tfrac{1}{4}(1 - \cos 2t)$

1.67. $1.1402, 74.7449°$

1.71. (a) $F(s) = \dfrac{-1}{s}(2e^{-2s} - e^{-s}) - \dfrac{1}{s^2}(e^{-2s} - e^{-s}) + \dfrac{1}{s}(e^{-3s} - e^{-s})$

(b) $f(t) = (t-1)u(t-1) - tu(t-2) + u(t-3)$

1.73. $\dfrac{2}{(s+1)^2 + 4}$

1.75. (a) $t + e^{-t} - 1$
1.77. $-2te^{-2t} - \frac{5}{2}e^{-2t} - \frac{1}{2}u_s(t) + 3e^{-t}$
1.79. FVT applies.

1.81. (a) $\dfrac{-a}{b}[(t-b)u(t) - (t-b)u(t-b)]$, $\dfrac{-a}{b}\left(\dfrac{1-e^{-sb}}{s^2} - \dfrac{b}{s}\right)$

1.83. (a) $x(t) = te^{-t} - e^{-t} + 1$, $t \geq 0$
(b) FVT applies. $x_{ss} = 1$
(c) $x(t) = -te^{-t} + 1 - e^{-t}$, $t \geq 0$

1.85. $R(s) = \dfrac{1}{(1 - e^{-\pi s})(1 + s^2)}$

1.87. $\omega(t) = \frac{1}{2}(1 - e^{-2t})$
1.89. $i(t) = u_s(t) - \frac{3}{2}e^{-t/3} + \frac{1}{2}e^{-t}$. FVT applies. $i_{ss} = 1$
1.91. $\angle F(j3) = 191.31°$, $|F(j3)| = 0.0654$

1.93. (a) $h(t) = \dfrac{R}{g}\left[u_s(t) - e^{-\frac{g}{RA}t}\right]$

(b) FVT applies. $h_{ss} = \dfrac{R}{g}$

1.95. $i(t) = \frac{1}{4}\{\sin 2t - \cos 2t + e^{-2t}\}$

1.97. $x(t) = 1 - e^{-0.5t}\left(\cos\dfrac{\sqrt{3}}{2}t + \dfrac{\sqrt{3}}{3}\sin\dfrac{\sqrt{3}}{2}t\right)$, $t \geq 0$

CHAPTER 2

Problem Set 2.1

2.1. Orthogonal, but not orthonormal
2.3. Neither orthogonal nor orthonormal

Problem Set 2.2

2.7. 0
2.9. 12

2.13. $A^{-1} = \begin{bmatrix} 1 & 2 & -1 \\ 0 & -1 & 2 \\ 0 & 0 & -1 \end{bmatrix}$

2.17. $x_1 = 0$, $x_2 = -1$, $x_3 = 2$

Problem Set 2.3

2.19. $\lambda_1 = \lambda_2 = 1, \lambda_3 = 0$

$$\mathbf{v}_1 = \begin{bmatrix} 0 \\ 1 \\ 4 \end{bmatrix}, \mathbf{v}_2 = \begin{bmatrix} 1 \\ 0 \\ -3 \end{bmatrix}, \mathbf{v}_3 = \begin{bmatrix} 0 \\ 0 \\ 1 \end{bmatrix}$$

2.21. $\lambda_1 = 1, \lambda_2 = -2$

$$\mathbf{v}_1 = \begin{bmatrix} -3 \\ 5 \end{bmatrix}, \mathbf{v}_2 = \begin{bmatrix} -1 \\ 2 \end{bmatrix}$$

2.23. **A** is Hermitian. Eigenvalues are real.

Problem Set 2.4

2.25. $\mathbf{P} = [\mathbf{v}_1 \quad \mathbf{v}_2 \quad \mathbf{v}_3] = \begin{bmatrix} 0 & 1 & 0 \\ 1 & 0 & 0 \\ 4 & -3 & 1 \end{bmatrix}$

2.29. $e^{\mathbf{A}t} = \begin{bmatrix} e^t & 0 & 0 \\ 0 & e^t & 0 \\ 3(1-e^t) & 4(e^t-1) & 1 \end{bmatrix}$

2.31. Linearly dependent

2.33. Linearly independent

2.35. $\mathbf{A}^{-1} = \begin{bmatrix} 0.2 & -0.3 & -0.7 \\ -0.2 & -0.7 & -1.3 \\ 0.2 & 0.2 & 0.8 \end{bmatrix}$

2.37. $\mathbf{C}^{-1}$ does not exist

2.41. $rk(\mathbf{A}) = 3$

2.43. $rk(\mathbf{C}) = 2$

2.47. Cramer's rule applies. $x_1 = 2, x_2 = -3$

2.49. $\mathbf{x} = \begin{Bmatrix} 0.75 \\ -0.25 \\ 1.50 \end{Bmatrix}$

2.51. $\mathbf{x} = \begin{Bmatrix} -2 \\ 0 \end{Bmatrix}$

2.53. $[\mathbf{C}^T \quad \mathbf{A}^T\mathbf{C}^T \quad (\mathbf{A}^T)^2\mathbf{C}^T] = \begin{bmatrix} 0 & -1 & 8 \\ 0 & -1 & 9 \\ 1 & -5 & 25 \end{bmatrix}$, rank = 3

2.55. $\lambda_{1,2,3} = 0, 2, -3$.

$$\mathbf{v}_1 = \begin{bmatrix} 0 \\ 3 \\ 2 \end{bmatrix}, \mathbf{v}_2 = \begin{bmatrix} 1 \\ 2 \\ 1 \end{bmatrix}, \mathbf{v}_3 = \begin{bmatrix} 0 \\ 0 \\ 1 \end{bmatrix}$$

2.57. $\mathbf{P} = [\mathbf{v}_1 \quad \mathbf{v}_2 \quad \mathbf{v}_3] = \begin{bmatrix} 0 & 1 & 0 \\ 3 & 2 & 0 \\ 2 & 1 & 1 \end{bmatrix}$

CHAPTER 3

Problem Set 3.1

3.1. $\begin{cases} \ddot{x}_1 = \overbrace{-2\dot{x}_1 - 2x_1 + x_2}^{f_1} \\ \ddot{x}_2 = \underbrace{x_1 - x_2}_{f_2} \end{cases}$

3.3. $\begin{bmatrix} J & 0 \\ 0 & m \end{bmatrix} \begin{Bmatrix} \ddot{\theta} \\ \ddot{x} \end{Bmatrix} + \begin{bmatrix} B & 0 \\ 0 & b \end{bmatrix} \begin{Bmatrix} \dot{\theta} \\ \dot{x} \end{Bmatrix} + \begin{bmatrix} K + kR^2 & -kR \\ -kR & k \end{bmatrix} \begin{Bmatrix} \theta \\ x \end{Bmatrix} = \begin{Bmatrix} Ru(t) \\ 0 \end{Bmatrix}$

Problem Set 3.2

3.5. $\dot{\mathbf{x}} = \mathbf{Ax} + \mathbf{B}u$, $\mathbf{x} = \begin{Bmatrix} x_1 \\ x_2 \\ x_3 \end{Bmatrix}$, $\mathbf{A} = \begin{bmatrix} 0 & 1 & 0 \\ 0 & 0 & 1 \\ -1 & -2 & -1 \end{bmatrix}$, $\mathbf{B} = \begin{Bmatrix} 0 \\ 0 \\ 2 \end{Bmatrix}$, $u = f(t)$

3.7. (a) $\dot{\mathbf{x}} = \mathbf{Ax} + \mathbf{B}u$, $\mathbf{x} = \begin{Bmatrix} x_1 \\ x_2 \\ x_3 \\ x_4 \end{Bmatrix}$, $\mathbf{A} = \begin{bmatrix} 0 & 0 & 1 & 0 \\ 0 & 0 & 0 & 1 \\ -\dfrac{K_1 + K_2}{J_1} & \dfrac{K_2}{J_1} & 0 & 0 \\ \dfrac{K_2}{J_2} & -\dfrac{K_3 + K_2}{J_2} & 0 & 0 \end{bmatrix}$, $\mathbf{B} = \begin{Bmatrix} 0 \\ 0 \\ \dfrac{1}{J_1} \\ 0 \end{Bmatrix}$,

$u = \tau$ $\mathbf{y} = \mathbf{Cx}$, $\mathbf{C} = [1 \ 0 \ 0 \ 0]$

(b) $\mathbf{y} = \mathbf{Cx}$, $\mathbf{y} = \begin{Bmatrix} \theta_1 \\ \theta_2 \end{Bmatrix}$, $\mathbf{C} = \begin{bmatrix} 1 & 0 & 0 & 0 \\ 0 & 1 & 0 & 0 \end{bmatrix}$

3.9. $J\left(\dfrac{K_2 + K_3}{K_2}\right)\ddot{\theta}_2 + \left[K_3 + \dfrac{K_1(K_2 + K_3)}{K_2}\right]\theta_2 = \tau$

$\dot{\mathbf{x}} = \mathbf{Ax} + \mathbf{B}u$, $\mathbf{x} = \begin{Bmatrix} x_1 \\ x_2 \end{Bmatrix}$, $\mathbf{A} = \begin{bmatrix} 0 & 1 \\ -\dfrac{K_2}{J(K_2 + K_3)}\left[K_3 + \dfrac{K_1(K_2 + K_3)}{K_2}\right] & 0 \end{bmatrix}$,

$\mathbf{B} = \begin{Bmatrix} 0 \\ \dfrac{K_2}{J(K_2 + K_3)} \end{Bmatrix}$, $u = \tau$

$\mathbf{y} = \mathbf{Cx}$, $\mathbf{C} = [1 \ 0]$

Problem Set 3.3

3.11. $\ddot{x}_1 + \frac{3}{2}\dot{x}_1 + x_1 = \frac{1}{2}f$

3.13. $m_1 m_2 x_2^{(4)} + [m_1 b_2 + m_2(b_1 + b_2)]\ddot{x}_2 + [m_1 k_2 + b_1 b_2 + (k_1 + k_2)m_2]\ddot{x}_2$
$\qquad + (b_2 k_1 + b_1 k_2)\dot{x}_2 + k_1 k_2 x_2 = m_1 \ddot{f} + (b_1 + b_2)\dot{f} + (k_1 + k_2)f$

Problem Set 3.4

3.15. $\dfrac{\Theta_o(s)}{\Theta_i(s)} = \dfrac{K}{Js^2 + Bs + K}$

3.17. $\dot{\mathbf{x}} = \mathbf{A}\mathbf{x} + \mathbf{B}u$, $\mathbf{x} = \begin{Bmatrix} i \\ \omega \end{Bmatrix}$, $\mathbf{A} = \begin{bmatrix} -\dfrac{R}{L} & -\dfrac{K_1}{L} \\ \dfrac{K_2}{J} & -\dfrac{B}{J} \end{bmatrix}$, $\mathbf{B} = \begin{Bmatrix} \dfrac{1}{L} \\ 0 \end{Bmatrix}$, $u = v$

(a) $G_1(s) = \dfrac{Js + B}{LJs^2 + (RJ + LB)s + RB + K_1 K_2}$

(b) $G_2(s) = \dfrac{K_2}{LJs^2 + (RJ + LB)s + RB + K_1 K_2}$

3.19. $G(s) = \dfrac{K}{Js^2 + Bs + K}$

Problem Set 3.5

3.21. $\ddot{y} + 3\dot{y} = \dot{u} + u$

(a) $\begin{cases} \dot{\mathbf{x}} = \mathbf{A}\mathbf{x} + \mathbf{B}u \\ y = \mathbf{C}\mathbf{x} \end{cases}$, $\mathbf{A} = \begin{bmatrix} 0 & 1 \\ 0 & -3 \end{bmatrix}$, $\mathbf{B} = \begin{Bmatrix} 0 \\ 1 \end{Bmatrix}$, $\mathbf{C} = \begin{bmatrix} 1 & 1 \end{bmatrix}$

(b) $G(s) = \dfrac{s+1}{s(s+3)}$

3.23. $\begin{cases} \dot{\mathbf{x}} = \mathbf{A}\mathbf{x} + \mathbf{B}u \\ y = \mathbf{C}\mathbf{x} \end{cases}$, $\mathbf{A} = \begin{bmatrix} 0 & 1 \\ -0.5 & -2.5 \end{bmatrix}$, $\mathbf{B} = \begin{Bmatrix} 0 \\ 1 \end{Bmatrix}$, $\mathbf{C} = \begin{bmatrix} 1 & 1 \end{bmatrix}$

3.25. (a) $2\dddot{x}_1 + 8\ddot{x}_1 + 7\dot{x}_1 + 2x_1 = \dot{f} + 3f$

(b) $\begin{cases} \dot{\mathbf{x}} = \mathbf{A}\mathbf{x} + \mathbf{B}u \\ y = \mathbf{C}\mathbf{x} \end{cases}$, $\mathbf{A} = \begin{bmatrix} 0 & 1 & 0 \\ 0 & 0 & 1 \\ -1 & -3.5 & -4 \end{bmatrix}$, $\mathbf{B} = \begin{Bmatrix} 0 \\ 0 \\ 1 \end{Bmatrix}$, $\mathbf{C} = \begin{bmatrix} 1.5 & 0.5 & 0 \end{bmatrix}$

Problem Set 3.6

3.27. $x^3 \approx \overline{x^3} + \overline{3x^2} \cdot \Delta x = 1 + 3\Delta x$

(a) $\begin{cases} y_{\text{nonlinear}} = (0.75)^3 = 0.4219 \\ y_{\text{linear}} = 1 + 3(x - \overline{x}) = 1 + 3(0.75 - 1) = 0.25 \end{cases}$, error = 41%

(b) $\begin{cases} y_{\text{nonlinear}} = (0.85)^3 = 0.6141 \\ y_{\text{linear}} = 1 + 3(0.85 - 1) = 0.55 \end{cases}$, error = 10%

(c) $\begin{cases} y_{\text{nonlinear}} = (0.95)^3 = 0.8574 \\ y_{\text{linear}} = 1 + 3(0.95 - 1) = 0.85 \end{cases}$, error = 0.86%

3.29. $\underbrace{f(x, y) = x^3 y}_{f_{\text{nonlinear}}} \approx \overline{f} + \left[\dfrac{\partial f}{\partial x}\right]_{(1,2)} \cdot \Delta x + \left[\dfrac{\partial f}{\partial y}\right]_{(1,2)} \cdot \Delta y = \overline{f}$

$+ [3x^2 y]_{(1,2)} \cdot \Delta x + [x^3]_{(1,2)} \cdot \Delta y = \underbrace{2 + 6\Delta x + \Delta y}_{f_{\text{linear}}}$

APPENDIX D: Answers to Odd-Numbered Problems   593

(a) $\begin{cases} f_{\text{nonlinear}} = (0.9)^3(2.1) = 1.5309 \\ f_{\text{linear}} = 2 + 6(0.9 - 1) + (2.1 - 2) = 1.5 \end{cases}$, error $= \dfrac{2.02\%}{}$

(b) $\begin{cases} f_{\text{nonlinear}} = (0.8)^3(2.1) = 1.0752 \\ f_{\text{linear}} = 2 + 6(0.8 - 1) + (2.1 - 2) = 0.9 \end{cases}$, error $= \dfrac{16.29\%}{}$

3.31. $\bar{x} = 1$, $\Delta \ddot{x} + \Delta \dot{x} + 2\Delta x = M \sin \omega t$, $\Delta x(0) = x(0) - 1 = -1$, $\Delta \dot{x}(0) = \dot{x}(0) = 1$

3.33. $\bar{x}_1 = -1$

$\begin{cases} \Delta \dot{x}_1 = -2\Delta x_1 - \Delta x_2 + \sin t \\ \Delta \dot{x}_2 = \Delta x_1 - \Delta x_2 \end{cases}$, $\Delta x_1(0) = 3$, $\Delta x_2(0) = 2$

Problems

3.35. (a) $\begin{cases} \dot{\mathbf{x}} = \mathbf{A}\mathbf{x} + \mathbf{B}u \\ y = \mathbf{C}\mathbf{x} + \mathbf{D}u \end{cases}$, $\mathbf{A} = \begin{bmatrix} 0 & 1 \\ -\frac{2}{3} & -2 \end{bmatrix}$, $\mathbf{B} = \begin{bmatrix} 0 \\ 1 \end{bmatrix}$, $\mathbf{C} = [1 \ \ 0]$, $D = 0$, $u = f$

(b) $\ddot{x}_1 + 2\dot{x}_1 + \frac{2}{3}x_1 = f(t)$

3.37. $\mathbf{A} = \begin{bmatrix} 0 & 1 & 0 \\ 0 & 0 & 1 \\ -5 & -4 & -3 \end{bmatrix}$, $\mathbf{B} = \begin{bmatrix} 0 \\ 0 \\ 1 \end{bmatrix}$, $\mathbf{C} = [1 \ \ 3 \ \ 1]$, $D = 0$

3.39. $\mathbf{A} = \begin{bmatrix} 0 & 1 & 0 \\ 0 & 0 & 1 \\ -3 & -2 & -1 \end{bmatrix}$, $\mathbf{B} = \begin{bmatrix} 0 \\ 0 \\ 2 \end{bmatrix}$, $\mathbf{C} = [1 \ \ 0 \ \ 0]$, $D = 0$

3.41. (a) $\bar{x} = 0$

(b) $m\Delta \ddot{x} + b\Delta \dot{x} = \Delta f$

3.43. $\mathbf{G}(s) = \begin{Bmatrix} G_1(s) \\ G_2(s) \end{Bmatrix}$, $G_1(s) = \dfrac{X_1(s)}{F(s)} = \dfrac{m_2 s^2 + cs + k}{\Delta(s)}$, $G_2(s) = \dfrac{X_2(s)}{F(s)} = \dfrac{cs + k}{\Delta(s)}$

$\Delta(s) = (m_1 s^2 + cs + k)(m_2 s^2 + cs + k) - (cs + k)^2$

3.45. (a) $\dfrac{I(s)}{E(s)} = \dfrac{s}{Ls^2 + Rs + \frac{1}{C}} = \dfrac{Cs}{LCs^2 + RCs + 1}$

(b) $\begin{cases} \dot{\mathbf{x}} = \mathbf{A}\mathbf{x} + \mathbf{B}u \\ y = \mathbf{C}\mathbf{x} + \mathbf{D}u \end{cases}$, $\mathbf{A} = \begin{bmatrix} 0 & 1 \\ -\dfrac{1}{LC} & -\dfrac{R}{L} \end{bmatrix}$, $\mathbf{B} = \begin{bmatrix} 0 \\ 1 \end{bmatrix}$, $\mathbf{C} = \begin{bmatrix} 1 & \dfrac{1}{L} \end{bmatrix}$, $D = 0$, $u = e(t)$

3.47. (a) $R_1 C_1 R_2 C_2 \ddot{q}_o + (R_2 C_1 + R_2 C_2 + R_1 C_1)\dot{q}_o + q_o = q_i$

(b) $\begin{cases} \dot{\mathbf{x}} = \mathbf{A}\mathbf{x} + \mathbf{B}u \\ y = \mathbf{C}\mathbf{x} \end{cases}$, $\mathbf{x} = \begin{Bmatrix} x_1 \\ x_2 \end{Bmatrix}$, $\mathbf{A} = \begin{bmatrix} 0 & 1 \\ -\dfrac{1}{R_1 C_1 R_2 C_2} & -\dfrac{R_2 C_1 + R_2 C_2 + R_1 C_1}{R_1 C_1 R_2 C_2} \end{bmatrix}$,

$\mathbf{B} = \begin{Bmatrix} 0 \\ \dfrac{1}{R_1 C_1 R_2 C_2} \end{Bmatrix}$, $\mathbf{C} = [1 \ \ 0]$

CHAPTER 4

Problem Set 4.1–4.4

4.1. $k_{eq} = k_1 + k_2$

4.3. $k_{eq} = \dfrac{k_1 k_2}{k_1 + k_2}$

4.5. $k_{eq} = k_3 + \dfrac{k_1 k_2}{k_1 + k_2}$

4.7. $b_{eq} = b_1 + b_2 + b_3$

4.9. $b_{eq} = \dfrac{b_1 b_2 b_3}{b_1 b_2 + b_2 b_3 + b_3 b_1}$

4.13. Holonomic

4.15. 2 DsOF; 2 indep. GCs: x_1 and x_2; 2 indep. DEs in x_1 and x_2

4.17. (a) $I_c = \tfrac{1}{2} m R^2$
  (b) 1 DOF; constraint: $x = R\theta$ (no slip)
  (c) 1 indep. GC: either x or θ; 1 indep. DE either in x or in θ.

4.19. 2 DsOF; 2 indep. GCs: x and θ; 2 indep. DEs in x and θ.

Problem Set 4.5

4.21. $m\ddot{y} + b\dot{y} + ky = f(t)$

4.23. $\begin{Bmatrix} \dot{x}_1 \\ \dot{x}_2 \end{Bmatrix} = \begin{bmatrix} 0 & 1 \\ -\dfrac{k}{m} & -\dfrac{b}{m} \end{bmatrix} \begin{Bmatrix} x_1 \\ x_2 \end{Bmatrix} + \begin{Bmatrix} 0 \\ \dfrac{1}{m} \end{Bmatrix} u$

(a) $y = \begin{bmatrix} 1 & 0 \end{bmatrix} \begin{Bmatrix} x_1 \\ x_2 \end{Bmatrix}$

(b) $y = \begin{bmatrix} 0 & 1 \end{bmatrix} \begin{Bmatrix} x_1 \\ x_2 \end{Bmatrix}$

4.25. $\dfrac{X(s)}{F(s)} = \dfrac{1}{ms^2 + k_3 + (k_1 k_2)/(k_1 + k_2)}$

4.27. (b) $\begin{bmatrix} M & 0 \\ 0 & m \end{bmatrix} \begin{Bmatrix} \ddot{x}_1 \\ \ddot{x}_2 \end{Bmatrix} + \begin{bmatrix} b & -b \\ -b & b \end{bmatrix} \begin{Bmatrix} \dot{x}_1 \\ \dot{x}_2 \end{Bmatrix} + \begin{bmatrix} k & -k \\ -k & k \end{bmatrix} \begin{Bmatrix} x_1 \\ x_2 \end{Bmatrix} = \begin{Bmatrix} 0 \\ f \end{Bmatrix}$

(c) $\begin{Bmatrix} \dot{x}_1 \\ \dot{x}_2 \\ \dot{x}_3 \\ \dot{x}_4 \end{Bmatrix} = \begin{bmatrix} 0 & 0 & 1 & 0 \\ 0 & 0 & 0 & 1 \\ -\dfrac{k}{M} & \dfrac{k}{M} & -\dfrac{b}{M} & \dfrac{b}{M} \\ \dfrac{k}{m} & -\dfrac{k}{m} & \dfrac{b}{m} & -\dfrac{b}{m} \end{bmatrix} \begin{Bmatrix} x_1 \\ x_2 \\ x_3 \\ x_4 \end{Bmatrix} + \begin{Bmatrix} 0 \\ 0 \\ 0 \\ \dfrac{1}{m} \end{Bmatrix} u$

$y = \begin{bmatrix} 1 & 0 & 0 & 0 \end{bmatrix} \begin{Bmatrix} x_1 \\ x_2 \\ x_3 \\ x_4 \end{Bmatrix} + 0 \cdot u$

APPENDIX D: Answers to Odd-Numbered Problems   595

4.29. $G_1(s) = \dfrac{X_1(s)}{F(s)} = \dfrac{bs + k}{\Delta(s)}$

$G_2(s) = \dfrac{X_2(s)}{F(s)} = \dfrac{Ms^2 + bs + k}{\Delta(s)}$

$\Delta(s) = (Ms^2 + bs + k)(ms^2 + bs + k) - (bs + k)^2$

$\Delta(s) = [Mms^2 + (M + m)bs + (M + m)k]s^2$

4.31. (b) $\begin{bmatrix} m_1 & 0 \\ 0 & m_2 \end{bmatrix} \begin{Bmatrix} \ddot{x}_1 \\ \ddot{x}_2 \end{Bmatrix} + \begin{bmatrix} c & -c \\ -c & c \end{bmatrix} \begin{Bmatrix} \dot{x}_1 \\ \dot{x}_2 \end{Bmatrix} + \begin{bmatrix} k_1 & -k_1 \\ -k_1 & k_1 + k_2 \end{bmatrix} \begin{Bmatrix} x_1 \\ x_2 \end{Bmatrix} = \begin{Bmatrix} f_1 \\ f_2 \end{Bmatrix}$

4.33. $G_{11}(s) = \left.\dfrac{X_1(s)}{F_1(s)}\right|_{F_2(s)=0} = \dfrac{m_2 s^2 + cs + k_1 + k_2}{\Delta(s)}$

$G_{12}(s) = \left.\dfrac{X_1(s)}{F_2(s)}\right|_{F_1(s)=0} = \dfrac{cs + k_1}{\Delta(s)}$

$G_{21}(s) = \left.\dfrac{X_2(s)}{F_1(s)}\right|_{F_2(s)=0} = \dfrac{cs + k_1}{\Delta(s)} \qquad G_{22}(s) = \left.\dfrac{X_2(s)}{F_2(s)}\right|_{F_1(s)=0} = \dfrac{m_1 s^2 + cs + k_1}{\Delta(s)}$

$\Delta(s) = (m_1 s^2 + cs + k_1)(m_2 s^2 + cs + k_1 + k_2) - (cs + k_1)^2$

$\Delta(s) = m_1 m_2 s^4 + (m_1 + m_2)cs^3 + [(m_1 + m_2)k_1 + m_1 k_2]s^2 + k_2 cs + k_1 k_2$

4.35. (b) $\begin{bmatrix} m_1 & 0 \\ 0 & m_2 \end{bmatrix} \begin{Bmatrix} \ddot{x}_1 \\ \ddot{x}_2 \end{Bmatrix} + \begin{bmatrix} c_1 + c_2 & -c_2 \\ -c_2 & c_2 \end{bmatrix} \begin{Bmatrix} \dot{x}_1 \\ \dot{x}_2 \end{Bmatrix} + \begin{bmatrix} k_1 + k_2 & -k_2 \\ -k_2 & k_2 \end{bmatrix} \begin{Bmatrix} x_1 \\ x_2 \end{Bmatrix} = \begin{Bmatrix} f_1 \\ f_2 \end{Bmatrix}$

4.37. $G_{11}(s) = \left.\dfrac{X_1(s)}{F_1(s)}\right|_{F_2(s)=0} = \dfrac{m_2 s^2 + c_2 s + k_2}{\Delta(s)} \qquad G_{12}(s) = \left.\dfrac{X_1(s)}{F_2(s)}\right|_{F_1(s)=0} = \dfrac{c_2 s + k_2}{\Delta(s)}$

$G_{21}(s) = \left.\dfrac{X_2(s)}{F_1(s)}\right|_{F_2(s)=0} = \dfrac{c_2 s + k_2}{\Delta(s)}$

$G_{22}(s) = \left.\dfrac{X_2(s)}{F_2(s)}\right|_{F_1(s)=0} = \dfrac{m_1 s^2 + (c_1 + c_2)s + k_1 + k_2}{\Delta(s)}$

$\Delta(s) = [m_1 s^2 + (c_1 + c_2)s + k_1 + k_2](m_2 s^2 + c_2 s + k_2) - (c_2 s + k_2)^2$

$\Delta(s) = m_1 m_2 s^4 + [(m_1 + m_2)c_2 + m_2 c_1]s^3 + [(m_1 + m_2)k_2 + m_2 k_1 + c_1 c_2]s^2$
$+ (c_1 k_2 + c_2 k_1)s + k_1 k_2$

4.39. (b) $\begin{bmatrix} m_1 & 0 & 0 \\ 0 & m_2 & 0 \\ 0 & 0 & m_3 \end{bmatrix} \begin{Bmatrix} \ddot{x}_1 \\ \ddot{x}_2 \\ \ddot{x}_3 \end{Bmatrix} + \begin{bmatrix} 0 & 0 & 0 \\ 0 & c & -c \\ 0 & -c & c \end{bmatrix} \begin{Bmatrix} \dot{x}_1 \\ \dot{x}_2 \\ \dot{x}_3 \end{Bmatrix}$

$+ \begin{bmatrix} k_1 + k_2 & -k_2 & 0 \\ -k_2 & k_2 + k_3 & -k_3 \\ 0 & -k_3 & k_3 \end{bmatrix} \begin{Bmatrix} x_1 \\ x_2 \\ x_3 \end{Bmatrix} = \begin{Bmatrix} 0 \\ 0 \\ f \end{Bmatrix}$

Problem Set 4.6–4.7

4.41. (a) $I_c = \frac{1}{2} mR^2$
(b) $I_{IC} \ddot{\theta} = mgR \sin \phi$
$I_{IC} = \frac{3}{2} mR^2$

4.43. $\begin{Bmatrix} \dot{x}_1 \\ \dot{x}_2 \end{Bmatrix} = \begin{bmatrix} 0 & 1 \\ 0 & 0 \end{bmatrix} \begin{Bmatrix} x_1 \\ x_2 \end{Bmatrix} + \begin{Bmatrix} 0 \\ \dfrac{mgR}{I_{IC}} \end{Bmatrix} u$

(a) $y = \begin{bmatrix} 1 & 0 \end{bmatrix} \begin{Bmatrix} x_1 \\ x_2 \end{Bmatrix}$

(b) $y = \begin{bmatrix} 0 & 1 \end{bmatrix} \begin{Bmatrix} x_1 \\ x_2 \end{Bmatrix}$

4.47. $\begin{Bmatrix} \dot{x}_1 \\ \dot{x}_2 \\ \dot{x}_3 \\ \dot{x}_4 \end{Bmatrix} = \begin{bmatrix} 0 & 0 & 1 & 0 \\ 0 & 0 & 0 & 1 \\ -\dfrac{K_1+K_2}{J_1} & \dfrac{K_2}{J_1} & -\dfrac{C}{J_1} & \dfrac{C}{J_1} \\ \dfrac{K_2}{J_2} & -\dfrac{K_2}{J_2} & \dfrac{C}{J_2} & -\dfrac{C}{J_2} \end{bmatrix} \begin{Bmatrix} x_1 \\ x_2 \\ x_3 \\ x_4 \end{Bmatrix} + \begin{Bmatrix} 0 \\ 0 \\ \dfrac{1}{J_1} \\ 0 \end{Bmatrix} u_1 + \begin{Bmatrix} 0 \\ 0 \\ 0 \\ \dfrac{1}{J_2} \end{Bmatrix} u_2$

$\begin{Bmatrix} y_1 \\ y_2 \end{Bmatrix} = \begin{bmatrix} 1 & 0 & 0 & 0 \\ 0 & 1 & 0 & 0 \end{bmatrix} \begin{Bmatrix} x_1 \\ x_2 \\ x_3 \\ x_4 \end{Bmatrix}$

4.49. $G_1(s) = \dfrac{\Theta(s)}{U(s)} = -\dfrac{m_o L s^2}{\Delta(s)}$

$G_2(s) = \dfrac{X(s)}{U(s)} = \dfrac{I_p s^2 - m_o g L}{\Delta(s)}$

$\Delta(s) = (I_p s^2 - m_o g L)(m_t s + b)s - (m_o L s^2)^2$

$\mathbf{G}(s) = \begin{Bmatrix} G_1(s) \\ G_2(s) \end{Bmatrix}$

4.51. $G_1(s) = \dfrac{\Theta(s)}{F(s)} = -\dfrac{mLs^2}{\Delta(s)}$

$G_2(s) = \dfrac{X(s)}{F(s)} = \dfrac{mL^2 s^2 + mgL}{\Delta(s)}$

$\Delta(s) = (mL^2 s^2 + mgL)(m_t s + c)s - (mLs^2)^2$

$\mathbf{G}(s) = \begin{Bmatrix} G_1(s) \\ G_2(s) \end{Bmatrix}$

Problem Set 4.8

4.53. $\begin{bmatrix} J_1 + J_2 + J_3\left(\dfrac{r_2}{r_3}\right)^2 & 0 \\ 0 & J_4 \end{bmatrix} \begin{Bmatrix} \ddot{\theta}_2 \\ \ddot{\theta}_4 \end{Bmatrix} + \begin{bmatrix} K_2\left(\dfrac{r_2}{r_3}\right)^2 & -K_2\dfrac{r_2}{r_3} \\ -K_2\dfrac{r_2}{r_3} & K_2 \end{bmatrix} \begin{Bmatrix} \theta_2 \\ \theta_4 \end{Bmatrix} = \begin{Bmatrix} T_a(t) \\ T_L(t) \end{Bmatrix}$

Problem Set 4.9

4.55. $T = \tfrac{1}{2} m \dot{x}^2$
$V = \tfrac{1}{2}(k_1 + k_2) x^2$
$m\ddot{x} + b\dot{x} + (k_1 + k_2)x = f(t)$

4.57. $T = \frac{1}{2}M\dot{x}_1^2 + \frac{1}{2}m\dot{x}_2^2$

$V = \frac{1}{2}Kx_1^2 + \frac{1}{2}k(x_1 - x_2)^2$

$\begin{bmatrix} M & 0 \\ 0 & m \end{bmatrix}\begin{Bmatrix} \ddot{x}_1 \\ \ddot{x}_2 \end{Bmatrix} + \begin{bmatrix} c & -c \\ -c & c \end{bmatrix}\begin{Bmatrix} \dot{x}_1 \\ \dot{x}_2 \end{Bmatrix} + \begin{bmatrix} K+k & -k \\ -k & k \end{bmatrix}\begin{Bmatrix} x_1 \\ x_2 \end{Bmatrix} = \begin{Bmatrix} f \\ 0 \end{Bmatrix}$

4.59. $T = \frac{1}{2}m_1\dot{x}_1^2 + \frac{1}{2}m_2\dot{x}_2^2 + \frac{1}{2}m_3\dot{x}_3^2$

$V = \frac{1}{2}k_1x_1^2 + \frac{1}{2}k_2(x_1-x_2)^2 + \frac{1}{2}k_3(x_2-x_3)^2$

$\begin{bmatrix} m_1 & 0 & 0 \\ 0 & m_2 & 0 \\ 0 & 0 & m_3 \end{bmatrix}\begin{Bmatrix} \ddot{x}_1 \\ \ddot{x}_2 \\ \ddot{x}_3 \end{Bmatrix} + \begin{bmatrix} 0 & 0 & 0 \\ 0 & c & -c \\ 0 & -c & c \end{bmatrix}\begin{Bmatrix} \dot{x}_1 \\ \dot{x}_2 \\ \dot{x}_3 \end{Bmatrix}$

$+ \begin{bmatrix} k_1+k_2 & -k_2 & 0 \\ -k_2 & k_1+k_2 & -k_3 \\ 0 & -k_3 & k_3 \end{bmatrix}\begin{Bmatrix} x_1 \\ x_2 \\ x_3 \end{Bmatrix} = \begin{Bmatrix} 0 \\ 0 \\ f(t) \end{Bmatrix}$

4.61. (a) $T = mL^2\dot{\theta}^2 + \frac{1}{2}mL^2\dot{\phi}^2 + mL^2\dot{\theta}\dot{\phi}\cos(\phi-\theta)$

$V = -2mgL\cos\theta - mgL\cos\phi$

$2mL^2\ddot{\theta} + mL^2\ddot{\phi}\cos(\phi-\theta) - mL^2\dot{\phi}^2\sin(\phi-\theta) + 2mgL\sin\theta = 0$

$mL^2\ddot{\phi} + mL^2\ddot{\theta}\cos(\phi-\theta) + mL^2\dot{\theta}^2\sin(\phi-\theta) + mgL\sin\theta = 0$

(b) $\begin{bmatrix} 2mL^2 & mL^2 \\ mL^2 & mL^2 \end{bmatrix}\begin{Bmatrix} \ddot{\theta} \\ \ddot{\phi} \end{Bmatrix} + \begin{bmatrix} 2mgL & 0 \\ 0 & mgL \end{bmatrix}\begin{Bmatrix} \theta \\ \phi \end{Bmatrix} = \begin{Bmatrix} 0 \\ 0 \end{Bmatrix}$

4.63. (a) $T = \frac{1}{2}\left(\frac{m_1}{3} + m_2\right)L_1^2\dot{\theta}_1^2 + \frac{1}{2}\frac{1}{3}m_2L_2^2\dot{\theta}_2^2 + \frac{m_2}{2}L_1L_2\dot{\theta}_1\dot{\theta}_2\cos(\theta_2-\theta_1)$

$V = -\left(\frac{m_1}{2} + m_2\right)gL_1\cos\theta_1 - m_2g\frac{L_2}{2}\cos\theta_2$

$\left(\frac{m_1}{3} + m_2\right)L_1^2\ddot{\theta}_1 + \frac{m_2}{2}L_1L_2\ddot{\theta}_2\cos(\theta_2-\theta_1)$

$- \frac{m_2}{2}L_1L_2\dot{\theta}_2^2\sin(\theta_2-\theta_1) + \left(\frac{m_1}{2} + m_2\right)gL_1\sin\theta_1 = 0$

$\frac{m_2}{3}L_2^2\ddot{\theta}_2 + \frac{m_2}{2}L_1L_2\ddot{\theta}_1\cos(\theta_2-\theta_1) + \frac{m_2}{2}L_1L_2\dot{\theta}_1^2\sin(\theta_2-\theta_1) + \frac{m_2}{2}gL_2\sin\theta_2 = 0$

(b) $\begin{bmatrix} \left(\frac{m_1}{3}+m_2\right)L_1^2 & m_2L_1\frac{L_2}{2} \\ m_2L_1\frac{L_2}{2} & \frac{m_2}{3}L_2^2 \end{bmatrix}\begin{Bmatrix} \ddot{\theta}_1 \\ \ddot{\theta}_2 \end{Bmatrix} + \begin{bmatrix} \left(\frac{m_1}{2}+m_2\right)gL_1 & 0 \\ 0 & m_2g\frac{L_2}{2} \end{bmatrix}\begin{Bmatrix} \theta_1 \\ \theta_1 \end{Bmatrix} = \begin{Bmatrix} 0 \\ 0 \end{Bmatrix}$

4.65. (a) $T = \frac{1}{2}(M+m+m_p)\dot{x}^2 + \frac{1}{2}\left(\frac{m}{3}+m_p\right)L^2\dot{\theta}^2 + \left(\frac{m}{2}+m_p\right)L\dot{x}\dot{\theta}\cos\theta$

$V = \left(m\frac{L}{2} + m_pL\right)g\cos\theta = \left(\frac{m}{2}+m_p\right)gL\cos\theta$

$(M+m+m_p)\ddot{x} + \left(\frac{m}{2}+m_p\right)L\ddot{\theta}\cos\theta - \left(\frac{m}{2}+m_p\right)L\dot{\theta}^2\sin\theta + b\dot{x} = f(t)$

$\left(\frac{m}{3}+m_p\right)L^2\ddot{\theta} + \left(\frac{m}{2}+m_p\right)L\ddot{x}\cos\theta - \left(\frac{m}{2}+m_p\right)gL\sin\theta = 0$

(b) $\begin{bmatrix} M + m + m_p & \frac{m}{2} + m_p \\ \frac{m}{2} + m_p & \frac{m}{3} + m_p \end{bmatrix} \begin{Bmatrix} \ddot{x} \\ L\ddot{\theta} \end{Bmatrix} + \begin{bmatrix} b & 0 \\ 0 & 0 \end{bmatrix} \begin{Bmatrix} \dot{x} \\ L\dot{\theta} \end{Bmatrix}$

$+ \begin{bmatrix} 0 & 0 \\ 0 & -\left(\frac{m}{2} + m_p\right)\frac{g}{L} \end{bmatrix} \begin{Bmatrix} x \\ L\theta \end{Bmatrix} = \begin{Bmatrix} f(t) \\ 0 \end{Bmatrix}$

Problems

4.67. $\begin{bmatrix} m & 0 \\ 0 & 0 \end{bmatrix} \begin{Bmatrix} \ddot{x}_1 \\ \ddot{x}_2 \end{Bmatrix} + \begin{bmatrix} b_1 & 0 \\ 0 & b_2 \end{bmatrix} \begin{Bmatrix} \dot{x}_1 \\ \dot{x}_2 \end{Bmatrix} + \begin{bmatrix} k_1 & -k_1 \\ -k_1 & k_1 + k_2 \end{bmatrix} \begin{Bmatrix} x_1 \\ x_2 \end{Bmatrix} = \begin{Bmatrix} f(t) \\ 0 \end{Bmatrix}$

4.69. $\begin{bmatrix} 0 & 0 \\ 0 & m \end{bmatrix} \begin{Bmatrix} \ddot{x}_1 \\ \ddot{x}_2 \end{Bmatrix} + \begin{bmatrix} b & 0 \\ 0 & 0 \end{bmatrix} \begin{Bmatrix} \dot{x}_1 \\ \dot{x}_2 \end{Bmatrix} + \begin{bmatrix} k & -k \\ -k & k \end{bmatrix} \begin{Bmatrix} x_1 \\ x_2 \end{Bmatrix} = \begin{Bmatrix} 0 \\ f(t) \end{Bmatrix}$

4.71. $\begin{bmatrix} m_1 & 0 & 0 \\ 0 & m_2 & 0 \\ 0 & 0 & m_3 \end{bmatrix} \begin{Bmatrix} \ddot{x}_1 \\ \ddot{x}_2 \\ \ddot{x}_3 \end{Bmatrix} + \begin{bmatrix} c & -c & 0 \\ -c & c & 0 \\ 0 & 0 & 0 \end{bmatrix} \begin{Bmatrix} \dot{x}_1 \\ \dot{x}_2 \\ \dot{x}_3 \end{Bmatrix}$

$+ \begin{bmatrix} k_1 + k_2 & -k_2 & 0 \\ -k_2 & k_2 + k_3 & -k_3 \\ 0 & -k_3 & k_3 \end{bmatrix} \begin{Bmatrix} x_1 \\ x_2 \\ x_3 \end{Bmatrix} = \begin{Bmatrix} 0 \\ 0 \\ f(t) \end{Bmatrix}$

CHAPTER 5

Problem Set 5.1–5.3

5.1. $L\frac{d^2i}{dt^2} + R\frac{di}{dt} + \frac{1}{C}i = \frac{dv_a}{dt}, \quad L = L_1 + L_2$

5.3. $R = \frac{R_1 R_2}{R_1 + R_2}$

(a) $L\frac{d^2i}{dt^2} + R\frac{di}{dt} + \frac{1}{C}i = \frac{dv_a}{dt}$

(b) $\frac{I_L(s)}{V_a(s)} = \frac{Cs}{LCs^2 + RCs + 1}$

(c) $\frac{V_L(s)}{V_a(s)} = \frac{LCs^2}{LCs^2 + RCs + 1}$

(d) $\frac{V_C(s)}{V_a(s)} = \frac{1}{LCs^2 + RCs + 1}$

5.5. (a) $\begin{bmatrix} C_1 & 0 \\ 0 & C_2 \end{bmatrix} \begin{Bmatrix} \dot{v}_1 \\ \dot{v}_o \end{Bmatrix} + \begin{bmatrix} \frac{1}{R_1} + \frac{1}{R_2} & -\frac{1}{R_2} \\ -\frac{1}{R_2} & \frac{1}{R_2} \end{bmatrix} \begin{Bmatrix} v_1 \\ v_o \end{Bmatrix} = \begin{Bmatrix} \frac{v_i}{R_1} \\ C_2 \dot{v}_i \end{Bmatrix}$

(b) $\dfrac{V_1(s)}{V_i(s)} = \dfrac{(R_1 + R_2)C_2 + 1}{R_1R_2C_1C_2s^2 + [(R_1 + R_2)C_2 + R_1C_1]s + 1}$

$\dfrac{V_o(s)}{V_i(s)} = \dfrac{R_1R_2C_1C_2s^2 + (R_1 + R_2)C_2s + 1}{R_1R_2C_1C_2s^2 + [(R_1 + R_2)C_2 + R_1C_1]s + 1}$

$\Delta(s) = (R_1R_2C_1s + R_2 + R_1)(R_2C_2s + 1) - R_1$

$\Delta(s) = \{R_1R_2C_1C_2s^2 + [(R_1 + R_2)C_2 + R_1C_1]s + 1\}R_2$

Problem Set 5.4

5.7. (b) $\begin{bmatrix} J & 0 \\ 0 & L_a \end{bmatrix} \begin{Bmatrix} \dot{\omega} \\ \dot{i}_a \end{Bmatrix} + \begin{bmatrix} B & -K_t \\ K_e & R_a \end{bmatrix} \begin{Bmatrix} \omega \\ i_a \end{Bmatrix} = \begin{Bmatrix} 0 \\ v_a \end{Bmatrix}$

5.9. (a) Armature-controlled DC motor.

(b) $\begin{bmatrix} J & 0 \\ 0 & L_a \end{bmatrix} \begin{Bmatrix} \dot{\omega} \\ \dot{i}_a \end{Bmatrix} + \begin{bmatrix} B & -K_t \\ K_e & R_a \end{bmatrix} \begin{Bmatrix} \omega \\ i_a \end{Bmatrix} = \begin{Bmatrix} 0 \\ v_a \end{Bmatrix}$

(c) $G_1(s) = \dfrac{\Omega(s)}{V_a(s)} = \dfrac{K_t}{\Delta(s)}$

$G_2(s) = \dfrac{I_a(s)}{V_a(s)} = \dfrac{Js + B}{\Delta(s)}$

$\Delta(s) = (Js + B)(L_as + R_a) + K_eK_t$

$\Delta(s) = JL_as^2 + (BL_a + JR_a)s + BR_a + K_eK_t$

5.11. $\begin{bmatrix} J_r & 0 & 0 & 0 \\ 0 & J & 0 & 0 \\ 0 & 0 & J & 0 \\ 0 & 0 & 0 & 0 \end{bmatrix} \begin{Bmatrix} \ddot{\theta}_1 \\ \ddot{\theta}_2 \\ \ddot{\theta}_3 \\ \ddot{i}_a \end{Bmatrix} + \begin{bmatrix} 0 & 0 & 0 & 0 \\ 0 & 0 & 0 & 0 \\ 0 & 0 & B & 0 \\ K_e & 0 & 0 & L_a \end{bmatrix} \begin{Bmatrix} \dot{\theta}_1 \\ \dot{\theta}_2 \\ \dot{\theta}_3 \\ \dot{i}_a \end{Bmatrix}$

$+ \begin{bmatrix} K & -K & 0 & -K_t \\ -K & 2K & -K & 0 \\ 0 & -K & K & 0 \\ 0 & 0 & 0 & R_a \end{bmatrix} \begin{Bmatrix} \theta_1 \\ \theta_2 \\ \theta_3 \\ i_a \end{Bmatrix} = \begin{Bmatrix} 0 \\ 0 \\ 0 \\ v_a \end{Bmatrix}$

Problem Set 5.5

5.13. $\dfrac{V_{L1}(s)}{V_a(s)} = \dfrac{L_1(L_2Cs^2 + 1)s}{L_1L_2Cs^3 + (L_1 + L_2)RCs^2 + L_1s + R}$

Problems

5.15. (a) $RCL_1\dfrac{d^2i}{dt^2} + (L_1 + L_2 + R^2C)\dfrac{di}{dt} + Ri = RC\dot{v}_a(t) + v_a(t)$

(b) $\dfrac{V_C(s)}{V_a(s)} = \dfrac{L_2s}{RCL_1s^2 + (L_1 + L_2 + R^2C)s + R}$

5.17. (a) $RCL\dfrac{d^2i}{dt^2} + L\dfrac{di}{dt} + Ri = LC\ddot{v}_a(t) + v_a(t)$

(b) $\dfrac{V_L(s)}{V_a(s)} = \dfrac{Ls}{RCLs^2 + Ls + R}$

5.19. (a) Two

(c) $\begin{bmatrix} C & -C \\ -C & C \end{bmatrix}\begin{Bmatrix} \dot{v}_1 \\ \dot{v}_2 \end{Bmatrix} + \begin{bmatrix} \dfrac{1}{R_1} & 0 \\ 0 & \dfrac{1}{R_2} \end{bmatrix}\begin{Bmatrix} \dot{v}_1 \\ \dot{v}_2 \end{Bmatrix} + \begin{bmatrix} \dfrac{1}{L_1} & 0 \\ 0 & \dfrac{1}{L_2} \end{bmatrix}\begin{Bmatrix} v_1 \\ v_2 \end{Bmatrix} = \begin{Bmatrix} \dfrac{\dot{v}_a}{R_1} \\ 0 \end{Bmatrix}$

(d) $\dfrac{V_{L2}(s)}{V_a(s)} = \dfrac{V_2(s)}{V_a(s)} = \dfrac{R_2 L_1 L_2 C s^3}{\Delta(s)}$

$\Delta(s) = (R_1 L_1 C s^2 + L_1 s + R_1)(R_2 L_2 C s^2 + L_2 s + R_2) - R_1 R_2 L_1 L_2 C^2 s^4$

$\Delta(s) = (R_1 + R_2)L_1 L_2 C s^3 + [L_1 L_2 + (L_1 + L_2)R_1 R_2 C]s^2 + (R_1 L_2 + R_2 L_1)s + R_1 R_2$

5.21. $i_{R1}(\infty) = \dfrac{v_a}{R_1}$, $i_{R2}(\infty) = \dfrac{v_a}{R_2}$

5.23. $RC\dot{v}_C + v_C = 0$
Initial condition: $v_C(0^-) = 12$

5.25. $RC\dot{v}_C + v_C = Ri$

5.27. $C\dot{v}_1 + \left(\dfrac{1}{R} + \dfrac{1}{R_i + R}\right)v_1 = \dfrac{1}{R_i}v_a$, $v_1 = v_R$

$\dot{v}_1 + v_1 = 12 \times 10^{-4}$

5.29. $\dfrac{V_o(s)}{V_a(s)} = \dfrac{R_2 + R_3}{R_2} \dfrac{1/(Cs)}{R_1 + (1/Cs)} = \dfrac{R_2 + R_3}{R_2} \dfrac{1}{R_1 Cs + 1}$

$R_1 C \dot{v}_o + v_o = \dfrac{R_2 + R_3}{R_2} v_a$

5.31. $G_1(s) = \dfrac{I_a(s)}{V_a(s)} = \dfrac{Js + B}{\Delta(s)}$

$G_2(s) = \dfrac{\Omega(s)}{V_a(s)} = \dfrac{K_t}{\Delta(s)}$

$\Delta(s) = (Js + B)(L_a s + R_a) + K_e K_t$

$\Delta(s) = JL_a s^2 + (BL_a + JR_a)s + BR_a + K_e K_t$

5.33. $\dfrac{I_a(s)}{V_a(s)} = \dfrac{I_a(s)}{V_a(s)}\bigg|_{T_L(s)=0} = \dfrac{Js + B}{JL_a s^2 + (JR_a + BL_a)s + BR_a + K_e K_t}$

$\Delta(s) = s(Js + B)(L_a s + R_a) + K_e K_t s$

$\Delta(s) = [JL_a s^2 + (JR_a + BL_a)s + BR_a + K_e K_t]s$

5.35. $\mathbf{x} = \begin{Bmatrix} x_1 \\ x_2 \\ x_3 \end{Bmatrix} = \begin{Bmatrix} \theta \\ \dot{\theta} \\ i_a \end{Bmatrix}$, $u = v_a(t)$, $y = \theta$

$$\begin{Bmatrix} \dot{x}_1 \\ \dot{x}_2 \\ \dot{x}_3 \end{Bmatrix} = \begin{bmatrix} 0 & 1 & 0 \\ 0 & -\dfrac{B}{J} & \dfrac{K_t}{J} \\ 0 & -\dfrac{K_e}{L_a} & -\dfrac{R_a}{L_a} \end{bmatrix} \begin{Bmatrix} x_1 \\ x_2 \\ x_3 \end{Bmatrix} + \begin{Bmatrix} 0 \\ 0 \\ \dfrac{1}{L_a} \end{Bmatrix} u$$

$$y = \begin{bmatrix} 1 & 0 & 0 \end{bmatrix} \begin{Bmatrix} x_1 \\ x_2 \\ x_3 \end{Bmatrix} + 0 \cdot u$$

5.37. (a) $\begin{bmatrix} J_r & 0 & 0 \\ 0 & J & 0 \\ 0 & 0 & 0 \end{bmatrix} \begin{Bmatrix} \ddot{\theta}_1 \\ \ddot{\theta}_2 \\ \ddot{i}_a \end{Bmatrix} + \begin{bmatrix} B & -B & 0 \\ -B & B & 0 \\ K_e & 0 & L_a \end{bmatrix} \begin{Bmatrix} \dot{\theta}_1 \\ \dot{\theta}_2 \\ \dot{i}_a \end{Bmatrix} + \begin{bmatrix} K & -K & -K_t \\ -K & K & 0 \\ 0 & 0 & R_a \end{bmatrix} \begin{Bmatrix} \theta_1 \\ \theta_2 \\ i_a \end{Bmatrix} = \begin{Bmatrix} 0 \\ 0 \\ v_a \end{Bmatrix}$

(b) $\dfrac{\Theta_2(s)}{V_a(s)} = \dfrac{K_t(Bs + K)}{\Delta(s)}$

$\Delta(s) = (J_r s^2 + Bs + K)(Js^2 + Bs + K)(L_a s + R_a)$
$\qquad + K_t K_e s(Js^2 + Bs + K) - (Bs + K)^2(L_a s + R_a)$

$\Delta(s) = \{L_a JJ_r s^4 + [(J + J_r)BL_a + JJ_r R_a]s^3 + [(J + J_r)(KL_a + BR_a)$
$\qquad + JK_t K_e]s^2 + [(J + J_r)KR_a + BK_t K_e]s + KK_t K_e\}s$

(c) $\mathbf{x} = \begin{Bmatrix} x_1 \\ x_2 \\ x_3 \\ x_4 \\ x_5 \end{Bmatrix} = \begin{Bmatrix} \theta_1 \\ \theta_2 \\ \dot{\theta}_1 \\ \dot{\theta}_2 \\ i_a \end{Bmatrix}, \qquad u = v_a(t), \qquad \mathbf{y} = \begin{Bmatrix} y_1 \\ y_2 \end{Bmatrix} = \begin{Bmatrix} \theta_1 \\ \theta_2 \end{Bmatrix}$

$$\begin{Bmatrix} \dot{x}_1 \\ \dot{x}_2 \\ \dot{x}_3 \\ \dot{x}_4 \\ \dot{x}_5 \end{Bmatrix} = \begin{bmatrix} 0 & 0 & 1 & 0 & 0 \\ 0 & 0 & 0 & 1 & 0 \\ -\dfrac{K}{J_r} & \dfrac{K}{J_r} & -\dfrac{B}{J_r} & \dfrac{B}{J_r} & -\dfrac{K_t}{J_r} \\ \dfrac{K}{J} & -\dfrac{K}{J} & \dfrac{B}{J} & -\dfrac{B}{J} & 0 \\ 0 & 0 & -\dfrac{K_e}{L_a} & 0 & -\dfrac{R_a}{L_a} \end{bmatrix} \begin{Bmatrix} x_1 \\ x_2 \\ x_3 \\ x_4 \\ x_5 \end{Bmatrix} + \begin{Bmatrix} 0 \\ 0 \\ 0 \\ 0 \\ \dfrac{1}{L_a} \end{Bmatrix} u$$

$$\begin{Bmatrix} y_1 \\ y_2 \end{Bmatrix} = \begin{bmatrix} 1 & 0 & 0 & 0 & 0 \\ 0 & 0 & 1 & 0 & 0 \end{bmatrix} \begin{Bmatrix} x_1 \\ x_2 \\ x_3 \\ x_4 \\ x_5 \end{Bmatrix} + \begin{Bmatrix} 0 \\ 0 \end{Bmatrix} u$$

CHAPTER 6

6.1. $\rho = 0.0740 \dfrac{\text{lb}_m}{\text{ft}^3}$

6.3. (b) $M_m = 29.02 \text{ g/mol}$

6.5. $A\dot{h} + \left(\dfrac{g}{R_1} + \dfrac{g}{R_2}\right)h = q_i$

$\tau = \dfrac{A}{[(1/R_1) + (1/R_2)]g}$

6.7. (b) $\begin{bmatrix} A_1 & 0 \\ 0 & A_2 \end{bmatrix}\begin{Bmatrix} \dot{h}_1 \\ \dot{h}_2 \end{Bmatrix} + \begin{bmatrix} \dfrac{g}{R_1} + \dfrac{g}{R_3} & -\dfrac{g}{R_3} \\ -\dfrac{g}{R_3} & \dfrac{g}{R_2} + \dfrac{g}{R_3} \end{bmatrix}\begin{Bmatrix} h_1 \\ h_2 \end{Bmatrix} = \begin{Bmatrix} q_i \\ 0 \end{Bmatrix}$

(c) $\begin{Bmatrix} x_1 \\ x_2 \end{Bmatrix} = \begin{Bmatrix} h_1 \\ h_2 \end{Bmatrix}, \quad u = q_i, \quad y = h_2$

$\begin{Bmatrix} \dot{x}_1 \\ \dot{x}_2 \end{Bmatrix} = \begin{bmatrix} -\left(\dfrac{1}{R_1} + \dfrac{1}{R_3}\right)\dfrac{g}{A_1} & \dfrac{g}{R_3 A_1} \\ \dfrac{g}{R_3 A_2} & -\left(\dfrac{1}{R_2} + \dfrac{1}{R_3}\right)\dfrac{g}{A_2} \end{bmatrix}\begin{Bmatrix} x_1 \\ x_2 \end{Bmatrix} + \begin{Bmatrix} \dfrac{1}{A_1} \\ 0 \end{Bmatrix} u$

$y = \begin{bmatrix} 0 & 1 \end{bmatrix}\begin{Bmatrix} x_1 \\ x_2 \end{Bmatrix} + 0 \cdot u$

6.9. (a) Yes

(b) $\dfrac{\rho V c}{h A_s}\dot{T} + T = T_\infty$

(c) $\tau = \dfrac{\rho V c}{h A_s} = 9{,}217$ s or 2.56 hr

6.11. (a) Yes

(b) $\dfrac{\rho V c}{h A_s}\dot{T} + T = T_\infty$

$\tau = \dfrac{\rho V c}{h A_s} = 4.18$ hr

CHAPTER 7

Problem Set 7.1–7.3

7.3. $h(t) = -2 + t + 3e^{-(t/2)}$

7.5. $\theta(t) = \dfrac{2}{\sqrt{3}}e^{-t}\sin\sqrt{3}\,t$

7.7. (a) $\dfrac{X_1(s)}{F(s)} = \dfrac{1}{s^2 + 2s + (2/3)}$

(b) $x_1(t) = \dfrac{3}{2}\left[1 + \dfrac{1}{1+\sqrt{3}}e^{-1-(1/\sqrt{3})t} + \dfrac{1}{1-\sqrt{3}}e^{-1+(1/\sqrt{3})t}\right]$. As $t \to \infty$, $x_1(t) \to \dfrac{3}{2}$.

7.9. $x(t) = \dfrac{1}{\omega_n^2}e^{-\sigma t}\left[\cos\omega_d t + \dfrac{\zeta}{\sqrt{1-\zeta^2}}\sin\omega_d t\right]$

APPENDIX D: Answers to Odd-Numbered Problems   603

Problem Set 7.4

7.11. (c) $x_1(t) = 0.7236e^{-0.3820t} - e^{-2t} + 0.2764e^{-2.6180t}$, $x_{1,ss} = 0$

$x_2(t) = 0.2764e^{-0.3820t} - e^{-2t} + 0.7236e^{-2.6180t}$, $x_{2,ss} = 0$

Problem Set 7.5–7.6

7.13. (a) $\dfrac{X(s)}{F(s)} = \dfrac{1}{2s+1}$

(b) $G(j\omega) = \dfrac{1}{1+j2\omega}$

(c) $x_{ss}(t) = \dfrac{F_0}{\sqrt{4\omega^2+1}}\sin[\omega t - \tan^{-1}2\omega]$

7.15. (1) $0 \le t \le 1$

$$x(t) = \int_0^t \frac{1}{\omega_n}\sin\omega_n(t-\tilde{t})\,d\tilde{t} = \frac{1}{\omega_n^2}(1-\cos\omega_n t) \quad \text{for} \quad 0 \le t \le 1$$

(2) $t > 1$

$$x(t) = \frac{1}{\omega_n^2}[(1-\cos\omega_n)\cos\omega_n(t-1) + \sin\omega_n\sin\omega_n(t-1)]$$

7.17. $x(t) = e^{-\sigma t}x_0\left(\cos\omega_d t + \dfrac{\sigma}{\omega_d}\sin\omega_d t\right) + \dfrac{1}{\omega_d\omega_n^2}[\omega_d - e^{-\sigma t}(\omega_d\cos\omega_d t + \sigma\sin\omega_d t)]$

Problem Set 7.7

7.19. (a) $\mathbf{x}(t) = e^{At}\mathbf{x}_0 = \begin{Bmatrix} 2e^{-t}-e^{-2t} \\ -2e^{-t}+2e^{-2t} \end{Bmatrix}$

(b) $L^{-1}\{(s\mathbf{I}-\mathbf{A})^{-1}\} = \begin{bmatrix} 2e^{-t}-e^{-2t} & e^{-t}-e^{-2t} \\ -2e^{-t}+2e^{-2t} & -e^{-t}+2e^{-2t} \end{bmatrix}$

Problem Set 7.8–7.9

7.21. $\mathbf{x} = \begin{Bmatrix} \dfrac{1}{5\sqrt{3}}\left(e^{\sqrt{3}t}-e^{-\sqrt{3}t}\right) - \dfrac{\sqrt{2}}{5}\sin\sqrt{2}t \\ \dfrac{2}{5\sqrt{3}}\left(e^{\sqrt{3}t}-e^{-\sqrt{3}t}\right) + \dfrac{1}{5\sqrt{2}}\sin\sqrt{2}t \end{Bmatrix}$

7.23. $f(t) = \dfrac{1}{2} - \dfrac{4}{\pi^2}\left[\cos\pi t + \dfrac{1}{9}\cos 3\pi t + \cdots\right]$

7.25. (a) $m\ddot{x} + b\dot{x} + kx = 0$, $x(0^-) = 0$, $\dot{x}(0^-) = 1$

(b) $\omega_n = \sqrt{\dfrac{k}{m}} = \sqrt{2} = 1.4$ rad/sec, $\zeta = \dfrac{b}{2\sqrt{mk}} = \dfrac{3}{2\sqrt{2}} > 1$

(c) Overdamped.

7.27. (a) $\dot{\omega} + 4\omega = u_s(t)$
(b) $\tau = 0.25$
(c) $\omega_{ss} = \frac{1}{4}$
(d) $4\tau = 1$ sec

7.29. $x(t) = \dfrac{1}{50}\left(t - \dfrac{1}{\sqrt{50}}\sin\sqrt{50}t\right)$, $t \geq 0$

7.31. $f(x) = \dfrac{1}{4} - \dfrac{2}{\pi^2}\left(\cos\pi x + \dfrac{1}{9}\cos 3\pi x + \cdots\right)$
$+ \dfrac{1}{\pi}\left(\sin\pi x - \dfrac{1}{2}\sin 2\pi x + \dfrac{1}{3}\sin 3\pi x - \cdots\right)$

7.35. $x = \dfrac{1}{5}\begin{Bmatrix} -2\cos 2t + e^t + e^{-t} \\ \cos 2t + 2e^t + 2e^{-t} \end{Bmatrix}$

CHAPTER 8

Problem Set 8.1–8.2

8.1. (a) $\omega_1 = 0.7654\sqrt{\dfrac{k}{m}}$

$\omega_2 = 1.8478\sqrt{\dfrac{k}{m}}$

(b) $\begin{Bmatrix} X_1 \\ X_2 \end{Bmatrix}_1 = \begin{Bmatrix} 1 \\ 2.4142 \end{Bmatrix}$

$\begin{Bmatrix} X_1 \\ X_2 \end{Bmatrix}_2 = \begin{Bmatrix} 1 \\ -0.4142 \end{Bmatrix}$

8.3. (a) $\omega_1 = \sqrt{\dfrac{g}{L}}$

$\omega_2 = \sqrt{\dfrac{g}{L} + \dfrac{18k}{16m}}$

(b) $\begin{Bmatrix} \Theta_1 \\ \Theta_2 \end{Bmatrix}_1 = \begin{Bmatrix} 1 \\ 1 \end{Bmatrix}$

$\begin{Bmatrix} \Theta_1 \\ \Theta_2 \end{Bmatrix}_2 = \begin{Bmatrix} 1 \\ -1 \end{Bmatrix}$

Problem Set 8.3

8.5. $U^T U = \begin{bmatrix} 3.6180 & 0 \\ 0 & 1.3820 \end{bmatrix}$

$$\mathbf{U}^T\mathbf{cU} = \begin{bmatrix} 1.38197c & 1.00001c \\ 1.00001c & 3.61803c \end{bmatrix}$$

Problem Set 8.4

8.9. (a) $\delta = 0.7985$
(b) $\zeta_{\text{exact}} = 0.1261$
(c) $\zeta_{\text{approx}} = 0.1271$, error $= +0.8\%$.

Problem Set 8.6

8.13. $f_{t,\text{ss}}(t;\omega) = \underbrace{F_o|G(j\omega)|}_{|f_{t,\text{ss}}|}\sin(\omega t + \phi)$

$$|G(j\omega)| = \frac{\sqrt{1 + [2\zeta(\omega/\omega_n)]^2}}{\sqrt{[1 - (\omega/\omega_n)^2]^2 + [2\zeta(\omega/\omega_n)]^2}}, \quad \phi = \tan^{-1} 2\zeta\frac{\omega}{\omega_n} - \tan^{-1}\frac{2\zeta(\omega/\omega_n)}{1 - (\omega/\omega_n)^2}$$

$$\frac{|f_{t,\text{ss}}|}{F_o} = |G(j\omega)|$$

8.15. (a) $(M + m)\ddot{x} + me\ddot{\theta}\cos\theta - me\dot{\theta}^2\sin\theta + b\dot{x} + (k_1 + k_2)x = 0$

(b) $G(s) = \dfrac{X(s)}{Y(s)} = \dfrac{-[me/(M+m)]s^2}{s^2 + 2\zeta\omega_n s + \omega_n^2}$

$$G(j\omega) = \frac{[me/(M+m)](\omega/\omega_n)^2}{1 - (\omega/\omega_n)^2 + j2\zeta(\omega/\omega_n)}$$

$$\omega_n = \sqrt{\frac{k_1 + k_2}{M + m}} \qquad \zeta = \frac{b}{2\sqrt{(k_1+k_2)(M+m)}}$$

(c) $x_{\text{ss}}(t;\omega) = 1 \cdot \dfrac{[me/(M+m)](\omega/\omega_n)^2}{\sqrt{[1 - (\omega/\omega_n)^2]^2 + [2\zeta(\omega/\omega_n)]^2}}\sin(\omega t + \phi)$

$$\phi = -\tan^{-1}\frac{2\zeta(\omega/\omega_n)}{1 - (\omega/\omega_n)^2}$$

$$\frac{|x_{\text{ss}}|}{me/(M+m)} = \frac{(\omega/\omega_n)^2}{\sqrt{[1 - (\omega/\omega_n)^2]^2 + [2\zeta(\omega/\omega_n)]^2}}$$

Problem Set 8.7

8.17. (a) $\begin{bmatrix} m & 0 \\ 0 & m_a \end{bmatrix}\begin{Bmatrix} \ddot{x}_1 \\ \ddot{x}_2 \end{Bmatrix} + \begin{bmatrix} c_a & -c_a \\ -c_a & c_a \end{bmatrix}\begin{Bmatrix} \dot{x}_1 \\ \dot{x}_2 \end{Bmatrix} + \begin{bmatrix} k + k_a & -k_a \\ -k_a & k_a \end{bmatrix}\begin{Bmatrix} x_1 \\ x_2 \end{Bmatrix} = \begin{Bmatrix} f \\ 0 \end{Bmatrix}$

(b) $G_1(s) = \dfrac{X_1(s)}{F(s)}$

$$= \frac{m_a s^2 + c_a s + k_a}{mm_a s^4 + (m + m_a)c_a s^3 + [mk_a + m_a(k + k_a)]s^2 + c_a k_a s + kk_a}$$

$G_2(s) = \dfrac{X_2(s)}{F(s)}$

$$= \frac{c_a s + k_a}{mm_a s^4 + (m + m_a)c_a s^3 + [mk_a + m_a(k + k_a)]s^2 + c_a k_a s + kk_a}$$

The order of the system is fourth.

(c) $G_1(j\omega) = \dfrac{X_1(s)}{F(s)}\bigg|_{s=j\omega}$

$= \dfrac{k_a - m_a\omega^2 + jc_a\omega}{kk_a + mm_a\omega^4 - [mk_a + m_a(k + k_a)]\omega^2 + jc_a\omega[k - (m + m_a)\omega^2]}$

$G_2(j\omega) = \dfrac{X_2(s)}{F(s)}\bigg|_{s=j\omega}$

$= \dfrac{k_a + jc_a\omega}{kk_a + mm_a\omega^4 - [mk_a + m_a(k + k_a)]\omega^2 + jc_a\omega[k - (m + m_a)\omega^2]}$

$|G_1(j\omega)| = \dfrac{\sqrt{(k_a - m_a\omega^2)^2 + (c_a\omega)^2}}{\sqrt{\{kk_a + mm_a\omega^4 - [mk_a + m_a(k + k_a)]\omega^2\}^2 + \{c_a\omega[k - (m + m_a)\omega^2]\}^2}}$

$|G_2(j\omega)| = \dfrac{\sqrt{k_a^2 + (c_a\omega)^2}}{\sqrt{\{kk_a + mm_a\omega^4 - [mk_a + m_a(k + k_a)]\omega^2\}^2 + \{c_a\omega[k - (m + m_a)\omega^2]\}^2}}$

$\phi_1(\omega) = \angle G_1(j\omega)$

$= \tan^{-1}\dfrac{c_a\omega}{k_a - m_a\omega^2} - \tan^{-1}\dfrac{c_a\omega[k - (m + m_a)\omega^2]}{kk_a + mm_a\omega^4 - [mk_a + m_a(k + k_a)]\omega^2}$

$\phi_2(\omega) = \angle G_2(j\omega) = \tan^{-1}\dfrac{c_a\omega}{k_a} - \tan^{-1}\dfrac{c_a\omega[k - (m + m_a)\omega^2]}{kk_a + mm_a\omega^4 - [mk_a + m_a(k + k_a)]\omega^2}$

(d) $x_{1ss}(t;\omega) = F_o|G_1(j\omega)|\cos(\omega t + \phi_1)$
$x_{2ss}(t;\omega) = F_o|G_2(j\omega)|\cos(\omega t + \phi_2)$

$|G_1(j\omega)| = \dfrac{\sqrt{(k_a - m_a\omega^2)^2 + (c_a\omega)^2}}{\sqrt{\{kk_a + mm_a\omega^4 - [mk_a + m_a(k + k_a)]\omega^2\}^2 + \{c_a\omega[k - (m + m_a)\omega^2]\}^2}}$

$|G_2(j\omega)| = \dfrac{\sqrt{k_a^2 + (c_a\omega)^2}}{\sqrt{\{kk_a + mm_a\omega^4 - [mk_a + m_a(k + k_a)]\omega^2\}^2 + \{c_a\omega[k - (m + m_a)\omega^2]\}^2}}$

$\phi_1(\omega) = \angle G_1(j\omega)$

$= \tan^{-1}\dfrac{c_a\omega}{k_a - m_a\omega^2} - \tan^{-1}\dfrac{c_a\omega[k - (m + m_a)\omega^2]}{kk_a + mm_a\omega^4 - [mk_a + m_a(k + k_a)]\omega^2}$

$\phi_2(\omega) = \angle G_2(j\omega) = \tan^{-1}\dfrac{c_a\omega}{k_a} - \tan^{-1}\dfrac{c_a\omega[k - (m + m_a)\omega^2]}{kk_a + mm_a\omega^4 - [mk_a + m_a(k + k_a)]\omega^2}$

8.21. $k = 0.16$ N/m
8.23. $\alpha_{opt} = 0.8$
$\zeta_{opt} = 0.219$
8.25. (a) $\alpha_{opt} = 0.9091$
$\zeta_{opt} = 0.1679$

(b) $|KG_1(j\beta)_{opt}| = \dfrac{\sqrt{(0.826 - \beta^2)^2 + (0.336\beta)^2}}{\sqrt{(0.826 + \beta^4 - 1.909\beta^2)^2 + [0.336\beta(1 - 1.1\beta^2)]^2}}$

8.27. $x_1(t) = L^{-1}\{X_1(s)\} = L^{-1}\left\{\dfrac{0.25s^2 + 0.10955s + 0.16}{0.25s^4 + 0.13694s^3 + 0.45s^2 + 0.10955s + 0.16} \cdot \dfrac{A}{s}\right\}$

$x_2(t) = L^{-1}\{X_2(s)\} = L^{-1}\left\{\dfrac{0.10955s + 0.16}{0.25s^4 + 0.13694s^3 + 0.45s^2 + 0.10955s + 0.16} \cdot \dfrac{A}{s}\right\}$

8.29. $s_{1,2} = \pm j, \quad s_{3,4} = -0.2500 \pm j0.9682$

Problems

8.31. $m\ddot{x} + (k_x + k_\theta \cos^2\theta)x = 0$
$m\ddot{x} + (k_y + k_\theta \sin^2\theta)y = 0$

8.33. $\begin{bmatrix} J_1 & 0 \\ 0 & J_2 \end{bmatrix} \begin{Bmatrix} \ddot{\theta}_1 \\ \ddot{\theta}_2 \end{Bmatrix} + \begin{bmatrix} (k_1 + k_2)r_1^2 & -k_2 r_1 r_2 \\ -k_2 r_1 r_2 & k_2 r_2^2 \end{bmatrix} \begin{Bmatrix} \theta_1 \\ \theta_2 \end{Bmatrix} = \begin{Bmatrix} T_1(t) \\ T_2(t) \end{Bmatrix}$

8.35. (a) $\omega_1 = \sqrt{\dfrac{k}{m}}$

$\omega_2 = \sqrt{\dfrac{3k}{m}} = 1.7321\sqrt{\dfrac{k}{m}}$

(b) $\begin{Bmatrix} X_1 \\ X_2 \end{Bmatrix}_1 = \begin{Bmatrix} 1 \\ 1 \end{Bmatrix}$

$\begin{Bmatrix} X_1 \\ X_2 \end{Bmatrix}_2 = \begin{Bmatrix} 1 \\ -1 \end{Bmatrix}$

8.37. (a) $\omega_1 = 0.7962\sqrt{\dfrac{k}{m}}$

$\omega_2 = 1.5382\sqrt{\dfrac{k}{m}}$

(b) $\begin{Bmatrix} X_1 \\ X_2 \end{Bmatrix}_1 = \begin{Bmatrix} 1 \\ 1.3660 \end{Bmatrix}$

$\begin{Bmatrix} X_1 \\ X_2 \end{Bmatrix}_2 = \begin{Bmatrix} 1 \\ -0.3660 \end{Bmatrix}$

(c) $\mathbf{U} = \begin{bmatrix} 1 & 1 \\ 1.3660 & -0.3660 \end{bmatrix}$

(d) $\mathbf{\Phi} = \dfrac{1}{\sqrt{m}} \begin{bmatrix} 0.4597 & 0.8881 \\ 0.6280 & -0.3250 \end{bmatrix}$

8.39. $x_{ss}(t) = F_o |G(j\omega)| \cos(\omega t + \phi)$

$|G(j\omega)| = \dfrac{1/k}{\sqrt{[1 - (\omega/\omega_n)^2]^2 + [2\zeta(\omega/\omega_n)]^2}}$

$\phi(j\omega) = -\tan^{-1}\dfrac{2\zeta(\omega/\omega_n)}{1 - (\omega/\omega_n)^2}$

8.41. $x_{1,ss}(t) = F_o|G(j\omega)|\sin(\omega t + \phi)$

$$|G(j\omega)| = \frac{\sqrt{A^2 + B^2}}{\sqrt{C^2 + D^2}}$$

$$\phi = \angle G(j\omega) = \tan^{-1}\left(\frac{B}{A}\right) - \tan^{-1}\left(\frac{D}{C}\right)$$

$A = k_2 - m_2\omega^2$ $\qquad\qquad\qquad\qquad\qquad B = c\omega$
$C = m_1 m_2 \omega^4 - [m_1 k_2 + m_2(k_1 + k_2)]\omega^2 + k_1 k_2 \qquad D = \omega[ck_1 - (m_1 + m_2)c\omega^2]$

8.43. $x_{ss}(t) = 0.288\cos(5t + 76.17°)$

8.45. $s_{1,2} = -0.1101 \pm j0.7250 \qquad s_{3,4} = -0.1638 \pm j1.0786$

CHAPTER 9

9.1. (a) $\dfrac{Y(s)}{U(s)} = \dfrac{1}{s^2 + 3s + 3}$

(b) $\dfrac{Y(s)}{U(s)} = \dfrac{1}{s^2 + 3s + 3}$

9.3.

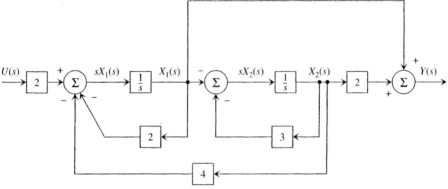

9.5. (a) $\begin{cases} \dot{\mathbf{x}} = \mathbf{A}\mathbf{x} + \mathbf{B}u \\ y = \mathbf{C}\mathbf{x} \end{cases}$, $\mathbf{A} = \begin{bmatrix} 0 & 0 & 1 \\ \dfrac{k_1}{b_2} & -\dfrac{k_1 + k_2}{b_2} & 0 \\ -\dfrac{k_1}{m} & \dfrac{k_1}{m} & -\dfrac{b_1}{m} \end{bmatrix}$, $\mathbf{B} = \begin{Bmatrix} 0 \\ 0 \\ \dfrac{1}{m} \end{Bmatrix}$,

$\mathbf{C} = [0 \ 1 \ 0]$, $u = f(t)$

(b)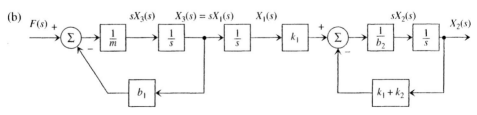

APPENDIX D: Answers to Odd-Numbered Problems   609

(c) $\dfrac{X_2(s)}{F(s)} = \dfrac{k_1}{s(ms + b_1)(b_2s + k_1 + k_2)}$

9.7. $E_i(s) \to \Sigma \to \boxed{\dfrac{1}{R}} \to I(s) \to \boxed{\dfrac{1}{Cs}} \to E_o(s)$ (with feedback)

9.9. $G(s) = \dfrac{G_1 G_2 G_3 + G_1 G_4}{1 + G_1 H_1 + G_1 G_2 H_2}$

9.11. $y(t) = 2(1 - e^{-0.5t} \cos \sqrt{0.75}\, t)$

9.13. $U(s) \to \boxed{G_2(s)} \to \boxed{\dfrac{G_1(s)}{G_2(s)}} \to \Sigma \to Y(s)$

9.15. (a) $\dddot{y} + 3\ddot{y} + \dot{y} + y = 3u$

$\begin{cases} \dot{\mathbf{x}} = \mathbf{A}\mathbf{x} + \mathbf{B}u \\ y = \mathbf{C}\mathbf{x} \end{cases}$, $\mathbf{x} = \begin{Bmatrix} x_1 \\ x_2 \\ x_3 \end{Bmatrix}$, $\mathbf{A} = \begin{bmatrix} 0 & 1 & 0 \\ 0 & 0 & 1 \\ -1 & -1 & -3 \end{bmatrix}$, $\mathbf{B} = \begin{Bmatrix} 0 \\ 0 \\ 3 \end{Bmatrix}$,

$\mathbf{C} = [1\ 0\ 0]$, $u = u$

(b) [block diagram]

(c) $\dfrac{Y(s)}{U(s)} = \dfrac{3}{s^3 + 3s^2 + s + 1}$

9.17. $y(t) = \dfrac{\omega_n}{\sqrt{1 - \zeta^2}} e^{-\zeta \omega_n t} \sin \omega_d t$, $\omega_d = \omega_n \sqrt{1 - \zeta^2}$

9.19. (a) $\dfrac{Y(s)}{X(s)} = \dfrac{G_1(s) G_2(s)}{1 + G_2(s) H_2(s) + G_1(s) G_2(s) H_1(s)}$

(b) $\dfrac{Y(s)}{W(s)} = \dfrac{G_2(s)}{1 + G_2(s) H_2(s) + G_1(s) G_2(s) H_1(s)}$

9.21. $y(t) = \dfrac{1}{4}\left[1 - e^{-(3/2)t}\left(\cos \dfrac{\sqrt{7}}{2} t + \dfrac{3}{\sqrt{7}} \sin \dfrac{\sqrt{7}}{2} t\right)\right]$

9.23. (a) $\dfrac{Y(s)}{U(s)} = \dfrac{k^2}{2k^2 s^2 + 3k(1 + k)s + (k + 1)^2}$

(b) $y(t) = A + Bt + Ce^{-\alpha t} + De^{-\beta t}$

$\alpha = \dfrac{1 + k}{k}$ and $\beta = \dfrac{1 + k}{2k}$, $A = -\dfrac{3}{16}\dfrac{\alpha}{\beta^4}$, $B = \dfrac{1}{4\beta^2}$,

$C = \dfrac{1/2}{\alpha^2(-\alpha + \beta)}$, $D = \dfrac{1/2}{\beta^2(\alpha - \beta)}$

CHAPTER 10

Problem Set 10.1–10.10

10.3. (a) $s = -0.5$, stable
(b) $s = -6$, stable
(c) $s = -1/\tau$ (negative), stable
(d) $s = -1/\tau$ (positive), unstable
(e) $s_{1,2} = \dfrac{-3 \pm j\sqrt{31}}{4}$, stable

10.5. (a) zero: $s = -5$; poles: $s_{1,2} = -1.5 \pm j1.323$; stable
(b) zero: $s = 10$; poles: $s_{1,2} = \dfrac{-3 \pm \sqrt{5}}{2}$; stable
(c) zero: none; poles: $s_{1,2} = \dfrac{-3 \pm j\sqrt{31}}{4}$; stable

10.7. (a) Stable
(b) Unstable

10.9. $K_d > -50$, $K_p > 0$, $K_i > 0$
$[10(50 + K_d) - K_p]K_p - 100K_i > 0$

10.11. $\theta = 30°$

10.13. (a) $\tau_{\text{dom}} = 1$
(b) $\tau_{\text{dom}} = 2.618$
(c) $\tau_{\text{dom}} = 10$
(d) $\tau_{\text{dom}} = 0.667$

Problem Set 10.11–10.12

10.15. (b) $e_{ss} = \dfrac{k}{k + K_p}$
(c) $e_{ss} = -\dfrac{1}{k + K_p}$
(d) $e_{ss} = \dfrac{k}{k + K_p}$

Problem Set 10.13–10.14

10.17. $K_{\text{opt}} = 64.8$, $(T_i)_{\text{opt}} = 0.7405$, $(T_d)_{\text{opt}} = 0.1851$

Problems

10.19. (a) Stable
(b) Unstable
(c) Unstable

(d) Stable
  (e) Stable
  (f) Stable
10.21. (a) $\tau > 0$
  (b) $\zeta, \omega_n > 0$
  (c) $a, b, c > 0; ab > c$
  (d) $a, b, c, d > 0; ab > c: (ab - c)c - da^2 > 0$
10.23. $K_{opt} = 27.6, (T_i)_{opt} = 0.993, (T_d)_{opt} = 0.248$

Bibliography

Belanger, P. R. (1995). *Control Engineering: A Modern Approach,* Saunders, Fort Worth, Tex.
Brewer, J. W. (1974). *Control Systems: Analysis, Design, and Simulation,* Prentice Hall, Englewood Cliffs, N.J.
Cannon, R. H., Jr. (1967). *Dynamics of Physical Systems,* McGraw-Hill, New York.
Close, C. M., and D. K. Frederick (1978). *Modeling and Analysis of Dynamic Systems,* Houghton Mifflin, Boston, Mass.
Esfandiari, R. S. (1995). *Applied Mathematics for Engineers,* McGraw-Hill, New York.
Fox, R. W., and A. T. McDonald (1978). *Introduction to Fluid Mechanics,* 2d ed., John Wiley & Sons, New York.
Franklin, G. F., J. D. Powell, and A. Emami-Naeini (1994). *Feedback Control of Dynamic Systems,* 3d ed., Addison-Wesley, Reading, Mass.
Friedland, B. (1986). *Control System Design: An Introduction to State-Space Methods,* McGraw-Hill, New York.
Greenwood, D. T. (1988). *Principles of Dynamics,* 2d ed., Prentice Hall, Englewood Cliffs, N.J.
Hale, F. J. (1988). *Introduction to Control System Analysis and Design,* 2d ed., Prentice Hall, Englewood Cliffs, N.J.
Holman, J. P. (1974). *Thermodynamics,* 2d ed., McGraw-Hill, New York.
Holman, J. P. (1976). *Heat Transfer,* 4th ed., McGraw-Hill, New York.
Huang, F. F. (1976). *Engineering Thermodynamics: Fundamentals and Applications,* Macmillan, New York.
Karnopp, B. H. (1974). *Introduction to Dynamics,* Addison-Wesley, Reading, Mass.
Karnopp, D. C., D. L. Margolis, and R. C. Rosenberg (1990). *System Dynamics: a Unified Approach,* 2d ed., John Wiley & Sons, New York.
Kuo, B. C. (1987). *Automatic Control Systems,* 7th ed., Prentice Hall, Englewood Cliffs, N.J.
Lindeburge, M. R. (1990). *Mechanical Engineering Reference Manual,* 8th ed., Professional Publications, Belmont Calif.
Meirovitch, L. (1986). *Elements of Vibration Analysis,* 2d ed., McGraw-Hill, New York.
Melsa, J. L., and D. G. Schultz (1969). *Linear Control Systems,* McGraw-Hill, New York.

Obert, E. F. (1973). *Internal Combustion Engines and Air Pollution,* Intext, New York.
Ogata, K. (1990). *Modern Control Engineering,* 2d ed., Prentice Hall, Englewood Cliffs, N.J.
Ogata, K. (1992). *System Dynamics,* 2d ed., Prentice Hall, Englewood Cliffs, N.J.
Palm, W. J. (1986). *Control Systems Engineering,* John Wiley & Sons, New York.
Potter, M. C., and J. F. Foss (1975). *Fluid Mechanics,* John Wiley & Sons, New York.
Raven, F. H. (1987). *Automatic Control Engineering,* 4th ed., McGraw-Hill, New York.
Rohsenow, W. M., and H. Choi (1961). *Heat, Mass, and Momentum Transfer,* Prentice Hall, Englewood Cliffs, N.J.
Rosenburg, R. C., and D. C. Karnopp (1983). *Introduction to Physical System Dynamics,* McGraw-Hill, New York.
Selkurt, E. E. (Ed.) (1971). *Physiology,* Little, Brown, Boston, Mass.
Shearer, J. L., A. T. Murphy, and H. H. Richardson (1971). *Introduction to System Dynamics,* Addison-Wesley, Reading, Mass.
Van de Vegte, J. (1990). *Feedback Control Systems,* 2d ed., Prentice Hall, Englewood Cliffs, N.J.
Vu, H. V. (1995). *Control of Dynamic Systems,* McGraw-Hill, New York.

INDEX

Acceleration, 173
Actuating error, 519
Actuator, 518
Adjoint matrix, 84–85
Algorithm, 155, 157
Analogous systems, 327
Angular acceleration, 175
Angular momentum, 213, 215, 217
Apparent force, 196
Applied forcing function, 334
Area moment of inertia, 175, 219
Argument, 4
Armature
  current, 281
  inductance, 281
  resistance, 281
  voltage, 281, 286
Armature-controlled DC motors, 279–280
Atom, 305
Auscultatory, 401
Avogadro's number, 305
Axisymmetric, 313

Beat phenomenon, 441, 481
Binomial coefficient, 98
Binomial expansion, 442
Binomial series, 98, 585
  expansion, 442, 585
Biot number, 304, 325–327, 330
Block diagram, 486
  reduction, 486, 499
  representation, 486
Bode plot, 365
Branch point, 489, 499, 510

Capacitance, 266, 290
Capacitor, 266
Carrier signals, 442
Cartesian form, 1
Cartesian coordinate system, 311
CE, *see* Characteristic equation
Center of gravity, 191
Center of mass, 191, 213
Centroidal mass moment of inertia, 220
Change of base, 584
Characteristic, 472
Characteristic length, 325, 330
Characteristic values, 14, 65, 89, 394, 472
Characteristic equation (CE), 14, 89, 134, 283, 340, 394, 522
Characteristic polynomial (CP), 89, 283, 522
Chemistry, 305
Closed-loop systems, 492, 517
Closed-loop transfer function (CLTF), 493
CLTF, *See* Closed-loop transfer function
Coefficient of (translational) viscous damping, 178
Coefficient of torsional viscous damping, 179, 281
Cofactor, 82
Column matrix (column vector), 77
Common logarithm, 365, 584
Companion matrix
  lower, 139
  upper, 140
Complete solution, 364
Complex analysis, 1
Complex conjugate, 2
Complex conjugation, 2

615

Complex frequency, 519
Complex function
    magnitude, 10
    phase, 10
Complex number
    addition, 2
    imaginary part, 1
    magnitude, 4
    multiplication, 2
    phase (argument), 4
    real part, 1
Complex powers, 6
Complex roots, 6
Complex s-plane, 522
Complex standard, 3
Complex variable, 9
Compound, 305
Compressibility, 304
Computer simulation, 571
Conduction, 323, 329
Configuration form, 108
Conservation of energy, 308, 329, 331
Conservation of mass, 321, 330
Constant gain, 488
Constant-pressure specific heat, 309
Constant-volume specific heat, 309
Constraints
    geometric (rolling), 183
    holonomic, 184
    nonholonomic, 184
    no-slip condition, 183
Control systems, 516
Controlled output, 518
Controller, 518, 545
    actions, 545, 564
    derivative, 545
    feedforward, 546
    integral, 545
    lag, 545
    lead, 545
    proportional, 545, 546
    types, 545, 563
Convection, 324, 329
Convergence of Fourier series, 389
Convolution, 34, 42
    integral, 43, 334, 370–371, 395
    method, 24, 42, 68
    symmetry, 43
Coulomb damping, 176–177, 193–194
Coupled, 122
CP, *see* Characteristic polynomial
Cramer's rule, 86, 527
Critical damping ratio, 369
Critical zone, 313
Critically damped, 341
Cylindrical coordinate system, 311

D'Alembert's principle, 196–198
Damped forced vibration, 477
Damped natural frequency, 395
Damped shaft, 284
Damped vibration absorber, 461
Damper, element, 176
Damping, 402, 431
    critical, 341
    ratio, 65, 339, 463, 481, 524
    matrix, 208
Dashpot, 176
DC, *see* Direct current
DC motor, 279, 280
    armature-controlled, 279, 280
    field-controlled, 279, 285
DE, *see* Differential equation
Decibel (dB), 365, 584
Decoupling, 122
Degree of freedom (DOF), 109, 181
    number of, 181–182
    single, 201
    two, 202
Den Hartog, 465
Density, 311
Derivative (D) control, 546
Derivatives, 65
Determinant, 83
Diagonal elements, 207
Diagonal exponential, 99
Diagonal matrix, 99, 430
Diagonalization, 96
Diatomic molecule, 305
Differential equation (DE), 11, 193
Dimension, 269
Dirac delta function, 26
Direct current (DC), 279
Direct transmission matrix, 117, 166
Dirichlet's theorem, 389
Displacement
    input, 200
    ratio, 464
Disturbance input, 517–518
Divergence, 311
DOF, *see* Degree of freedom
Dominant pole concept, 538
Dot notation, 174
Double-angle formula, 583
Dry air, 307
Dry friction, 176
Dynamic pressure, 311
Dynamic viscosity, 329
Dynamic vibration absorber, 461

Easy 5, 194
Eigen, 472
Eigenfunction, 472
  expansion, 475
Eigensolutions, 402, 473, 475
Eigenvalue problem, 89, 402, 472
Eigenvalues, 89, 402, 472, 479
Eigenvectors, 402, 472
Electrical circuits, 265
Electrical elements, 266
Electrical impedance, 290, 294,
Electrical systems, 265
Electromagnetic wave theory, 324
Electromechanical systems, 279, 296
Electromotive force (emf) voltage, 280
Electronic systems, 275
Electronic amplifier, 275
Element, 305, 487
Elemental equations, 173
Elemental relations
  electrical systems, 266
  electromechanical systems, 280
Emissivity, 325
Endergonic, 308
Endothermic, 308
Enthalpy, 310
Entrance region, 314
Envelopes, 442
Equal, 1
Equation of motion, 61
Equilibrium point, 144
Equivalence, 179
Error, 492, 519–520, 548–550
Euler constants, 385, 396
Euler-Fourier formulas, 385
Euler's formula, 4, 64
Euler's identities, 582
Even periodic functions, 388
Exergonic, 308
Exothermic, 308
Expansion formulas, 583
Exponential
  of a diagonal matrix, 99
  of a diagonalizable matrix, 99
  of a matrix, 97
Externally applied forces, 190
Extensive properties, 308

FBD, *see* Free-body diagram
Feedback element, 492
Feedforward transfer function, 493
FEM, *see* Finite-element method
Fictitious force, 196
Field
  current, 281
  inductance, 286
  resistance, 286
Field-controlled DC motors, 279, 285
Final value, 62
Final-value theorem (FVT), 62, 69
Finite-element method (FEM), 479
First law of thermodynamics, 308
First-order differential equation time constant,
  12, 13
First-order systems, 51
  unit-impulse response, 51
  unit-step response, 52
Fixed point, 213
Flexible shaft, 284
Fluid
  mechanics, 311
  resistance, 313, 315, 329
  systems, 304
Force matrix, 208
Forced response, 69, 334, 336
Forcing frequency, 10, 65, 362
Forward path, 505, 510
Fourier
  analysis, 385
  cosine series, 387, 396
  law of conduction, 323
  series, 385, 396
  sine series, 387–388
Fourier's equation, 323
Fourth-order Runge-Kutta method, 155, 167
Free energy, 309
Free response, 16, 335, 340, 462
Free variable, 90
Free-body diagram (FBD), 192
Frequency
  equation, 404
  ratios, 463
  response, 334, 362, 395
Frequency response function (FRF), 10, 64,
  395
FRF, *see* Frequency response function
FRF method, 362
Friction factor, 315
Fully developed laminar flow, 313
Fully developed region, 313
FVT, *see* Final-value theorem

Gaseous mixtures, 306
Gear-teeth ratio, 240
Gear-train systems, 238–244
Generalized coordinates, 108, 181–182
Generalized eigenvectors, 94
Generalized forces, 109
Generalized time constant, 560
Generalized velocities, 109

Global truncation error, 155
Gradient operator, 311
Graphical interpretation, 144
Gravity, 198–199

Half-angle formula, 583
Half-wave rectifier, 72
Hammer testing, 479
Harmonics, 385
Heat transfer, 323
Heating control systems
   engineering application, 520
   homeostasis, 522
Height, 311
Hermitian, 80, 94, 104
Higher-order systems, 527
Holonomic constraint, 184
Homogeneous solution, 18, 20
Homogeneous state equation, 375
Hyperbolic functions and relations, 584
Hysteresis damping, 176

IC, see Instantaneous center of rotation
I.C.s, see Initial conditions
Ideal gases, 306, 310
Identity matrix, 77
Imaginary parts, 1
Impedance, 290–291
   methods, 290
Impulse response, 51, 532, 537
Incompressible, 304
Increment variables, 144
Independent generalized coordinates, 244
Inductor, 266
Inductance, 266
Inertial force, 196
Inertial reference frame, 173
Initial conditions (I.C.s), 19, 20, 30
Initial value, 64
Initial-value problem, 32
Initial-value theorem (IVT), 61, 63, 69
Inner-outer stability rule, 543, 563
Input matrix, 116–117, 166
Input vector, 116, 165
Input-output (I/O) equation, 108, 125, 138
Instantaneous center of rotation (IC), 216
Integral (I) control, 546
Integration, 490
Integrator, 491
Integro-differential equation, 269
Intensive properties, 308
Internal energy, 309, 329
International system (SI) of units, 567
Interval extension, 389
Inverse Laplace transform, 23, 34, 68
Irreducible, 37–38

Irreducible polynomials, 38
Isothermal processes, 315
Isotropic, 311
IVT, see Initial-value theorem

Jump, 389

KCL, see Kirchhoff's current law
Kinematic viscosity, 311
Kinetic energy, 244–245
Kinetic friction force, 177, 194–195
Kinetic pressure, 311
Kirchhoff's current law (KCL), 265, 267, 296
Kirchhoff's laws, 267
Kirchhoff's voltage law (KVL), 265, 267, 296
Kronecker delta, 430
KVL, see Kirchhoff's voltage law

Lagrange's equations, 244
Lagrangian, 245
Laminar flows, 311
Laplace transform, 1, 22, 65
   approach, 376, 379
   convolution, 24
   derivatives, 30
   existence, 23
   integrals, 30
   inverse, 23, 28, 68
   linearity, 28
   necessary conditions for existence, 22
   operator, 22
   partial fractions, 24
   table, 66–67
   variable, 22, 519
Laplacian operator, 311
Law of cosines, 583
Left-hand limit, 389
L'Hôpital's rule, 26
Linear approximation, 151
Linear model, 150
Linear momentum, 173
Linear system, 363
Linearity, 22, 28, 65
Linearization, 144
Liquid-level systems, 320
Liquids, 310
Log decrement method, 437
Logarithm of a complex number, 584
Logarithmic decrement, 437, 481
Loop, 505, 510
   method, 267
   path, 505
Lower-companion matrix, 139, 166

Magnitude, 7
Maple V, 528, 571

Mason's rule, 505, 510
  general form, 508
  special case, 505
Mass
  center, 213
  element, 173
  matrix, 207
  ratio, 463
Mass moment of the disk, 281
Mass moment of the rotor, 281
Mass moments of inertia, 175, 219
Massless junctions, 205
Massless rod, 181
MathCAD, 528, 571
Mathematica, 528, 571
Mathematical modeling, 172, 317, 320, 325
MATLAB, 528, 571
Matrix, 77
  adjoint, 84
  cofactor, 82
  columns, 77
  components, 76
  convergent, 98
  diagonal, 81
  determinant, 83
  diagonalization, 95–96
  dot product, 77
  element, 77
  entry, 77
  equal, 77
  exponential, 97–98
  full, 95
  identity, 77
  inverse, 82
  length, 76
  linearly independent, 77
  lower triangular, 81
  minor, 82
  nonsingular, 82
  norm, 76
  nth-order vector, 76
  orthogonal, 77
  orthonormal, 77
  rank, 86
  rows, 77
  singular, 82
  size, 77
  square, 77
  transformation, 95
  transpose, 80
  upper triangular, 81
Matrix operations
  associate law, 78
  distributive law, 78
  elementary row operations, 91

  multiplication, 78
  sum, 78
$MATRIX_X$, 528, 571
Maximum peak, 369
Maximum percent overshoot, 535
Mechanical elements, 173
Mechanical impedance, 294
Member, 487
MIMO, see Multiple-input-multiple-output system
Minor, 144
Mixed-mean temperature, 324
Mixed systems, 226
Mixture, 305
Modal analysis, 471, 474
Modal damping coefficient, 431
Modal damping ratio, 431
Modal decomposition, 381, 396
Modal mass, 430
Modal matrix, 96, 382, 402, 427
Modal natural frequency, 430
Modal stiffness, 430
Modal testing, 479
Mode
  functions, 472
  shapes, 403, 472
Model, 172
Modeling, 172
  Coulomb damping, 193
Modulating signals, 442
Mol, 305
Mole, 305
Mole fractions, 306–307
Molecular mass, 306
Molecule, 305
Moment equation, 213
Moments of inertia, 217
Multiple outputs, 497
Multiple-input-multiple-output (MIMO) systems, 115

Natural coordinates, 476
Natural frequency, 403, 524, 525
  damped, 16
  undamped, 13
Natural logarithm, 365
Natural mode, 403
Navier-Stokes equations, 311
Negative damping, 177
Negative feedback, 501, 517
Neighborhood, 144
Newtonian mechanics, 173
Newtonian isotropic fluid, 311
Newton's equation, 324
Newton's laws, 172
Newton's second law, 190, 197

Node method, 267
Nonconservative forces, 247
Nonconservative generalized forces, 244
Nonholonomic constraint, 184–185
Nonhomogeneous general solution, 18
Nonhomogeneous linear ODEs, 18
Nonhomogeneous state equation, 378
Nonlinear model, 154
Nonuniqueness, 121
Normal force, 177
Normalized frequency, 369, 395
Normalized magnitude, 453
Normalized orthogonality, 475
Normalized responses, 442
Notation, 179
Numerical solution, 154

Odd periodic extensions, 390
Odd periodic functions, 388
ODE, see Ordinary differential equation
Off-diagonal elements, 208
OLTF, see Open-loop transfer function
Op amp, see Operational amplifier
Open-loop systems, 517
Open-loop transfer function (OLTF), 492
Operating point, 144, 149, 167
Operational amplifier (op amp), 275, 296
Optimal damping ratio, 465, 482
Optimal parameters, 467
Optimal response, 465
Optimal tuning ratio, 465, 482
Optimizing PID parameters, 557
Ordinary differential equation (ODE), 1, 11, 473
  classical (standard) form, 11
  constant coefficients, 11, 14
  first-order, 11
  general solution, 14
  linear, 11
  linearly independent, 14
  nonlinear, 11
  second-order, 12
Orthogonality, 427
Orthonormality, 430, 474
Output
  matrix, 117, 166
  vector, 116, 165
Overdamped, 341

Parallel, 179
Parallel-axis theorem, 215, 222, 262
Partial differential equation (PDE), 473
Partial fraction, 24
Partial fraction method, 34, 68
  complex conjugate poles, 37

distinct real poles, 35
repeated (multiple) real poles, 36
Partial pressure, 307
Partial volume, 307
Particular solution, 18, 20, 65
  forcing function, 19, 65
  method of undetermined coefficients, 18
Passive vibration control, 461
PDE, see Partial differential equation
Peak time, 535
Pendulum systems, 197, 220–223, 229–236
Periodic function, 48
PID, see Proportional-plus-integral-plus-derivative control
Piecewise continuous, 30
Planar motion, 215, 220
Plant, 517–518
Pneumatic capacitance, 316
Pneumatic systems, 315, 319
Pneumatic tank, 318
Polar form, 4
  magnitude, 4
  argument, 4
  phase, 4
Poles, 9
Poles and zeros,
  simple, 9
  multiplicity, 9
Pole-zero form, 9
Polytropic processes, 316
Positive damping, 177
Positive feedback, 501, 517
Potential energy, 244, 246
Pressure, 311
Principle of superposition, 359, 392
Process control, 517
Processes, 517
Proportional (P) control, 546
Proportional-plus-derivative (PD) control, 547
Proportional-plus-integral (PI) control, 547
Proportional-plus-integral-plus-derivative (PID) control, 547
Pulley, 228
Pure substance, 305
Pythagorean identities, 582

Quantum theory, 324

Radiation, 330
Ramp response, 338
Rank, 86
Rectangular form, 1, 2
Reference input, 518
Regulator, 517
Relative roughness, 315
Relativity, 173

Residues, 68, 363
Resistance, 266
Resistor, 266
Resonant frequency, 369, 395, 450, 481
Resonant peak, 369, 450, 481
Response
   to an arbitrary forcing function, 395
   to an arbitrary input, 370
   to a unit impulse, 345
   to a unit step, 350
Rest mass, 173
Reversed effective force, 196
Reynolds number, 312
RHP, see Right-half plane
Right-half plane (RHP), 69
Right-hand rule (RHR), 174
Rigid body, 214
Rise time, 537
RHR, see Right-hand rule
RLC circuit, 293
Rolling coin, 186
Rolling constraints, 183
Rolling disk, 183, 216
Rotating unbalanced mass, 456
Rotational damper, 179
Rotational mass, 175
Rotational spring, 176
Rotational systems, 213
Rotational vector, 174
Routh-Hurwitz stability criterion, 516, 540–541, 562
Row matrix (row vector), 77

SDOF, see Single-degree-of-freedom systems
Second-order differential equation
   damping ratio, 13
   natural frequency, 13
   standard form, 13
Second-order matrix form, 111, 165
Second-order systems, 54
   unit-impulse response, 55
   unit-step response, 58
Sensor, 518
Series, 180
Servomechanisms, 517
Settling time, 536
Sgn, see Signum function
Shaft, 378
Shaker testing, 479
Shear strain, 311
Shear stress, 310
Shift on s-axis, 28
Shift on t-axis, 44
Shifting, 65
SI, see International system of units
Sign convention, 192

Signum function, 194
Simple discontinuity, 389
Simple harmonic excitation, 448
Single-degree-of-freedom (SDOF) systems, 220, 401
Single-input-single-output (SISO) systems, 133, 166, 201
SISO, see Single-input-single-output systems
Skeleton approach, 206
Skew-Hermitian, 80, 94, 104
Sliding, 177
Solids, 310
Solving the state equation, 375
Sparse matrices, 140
Special functions, 24
   Dirac delta, 26
   exponential, 27
   periodic, 48
   unit impulse, 26
   unit pulse, 25
   unit ramp, 24
   unit sinusoidal, 26
   unit step, 24
Special matrices, 94
   Hermitian, 94
   skew-Hermitian, 94
   orthogonal, 94
   unitary, 94
Specific gas constant, 306, 329
Specific heats, 309
Speed control systems, 558
Speed of light, 173
Sphygmomanometer, 401
Spring element, 175
Springs
   in compression, 203
   in parallel, 179
   rotational, 176
   in series, 180
   in tension, 204
   translational, 175
Square matrix, 77
Stable, 209
Standard forms, 12, 65
State-space form, 117, 133, 165
State-space representation, 117
   output equation, 117
   state equation, 117
State-variable equations, 113, 119
Static friction force, 177
Steady-state error, 548, 563
Steady-state response, 335, 394
Steady-state value, 62, 327, 394
Stefan-Boltzmann constant, 325
Stefan-Boltzmann law, 324

Step function, 24
Step response, 51, 336, 531, 534
Stethoscope, 401
Structural damping, 176
Submatrix, 77
Summing junction, 487, 500, 510
Superposition, 549
Support motion, 453
Surface coefficient of heat transfer, 324
Symmetrical systems, 411
System response, 50, 474
    via Fourier series, 392
SystemBuild, 194, 571
Systems
    electrical, 265
    electromechanical, 279
    fluid, 304
    gear-train, 238
    liquid-level, 320, 321
    mixed mechanical, 226
    pneumatic, 315
    rotational mechanical, 213
    thermal, 304, 323
    translational mechanical, 190

Taylor series expansion, 144, 146, 582
TDOF, *see* Two-degree-of-freedom systems
Temperature, 323
TF, *see* Transfer function
Thermal conductivity, 323
Thermal radiation, 324
Thermal systems, 323
Thermodynamics, 305
    first law of, 308
Three-degree-of-freedom systems, 209, 241
Three-dimensional (3D) space, 214
Time constant, 12, 318, 321, 335, 523
Time invariance, 1, 22
TM, *see* Transfer matrix
Torsional stiffness, 176
Total response, 371
Total solution, 364
Trace, 78
Transfer function (TF), 9, 108, 128, 133, 166, 201
    matrix, 129
Transfer matrix (TM), 129, 136, 166
Transient response, 335, 356, 394, 467
    specifications, 531

Translational acceleration, 173
Translational mass, 173
Translational motion, 174
Translational spring, 175
Translational stiffness, 176
Translational viscous damper, 178
Trigonometric series, 385
Turbulent flows, 312
Two dimensions (2D), 215
Two-degree-of-freedom (TDOF) systems, 110, 202–205, 225
Two-tank liquid-level system, 321

Undamped, 341, 343
    free vibration, 475
    natural frequency, 65, 339
Underdamped, 341–342
Uniform circular disk, 220
Unit of, 269
Unitary matrix, 94
Unit-impulse (Dirac delta) function, 26
Unit-impulse response, 51, 69, 348, 395, 469
Unit-pulse function, 25
Unit-ramp function, 24
Unit-sinusoidal function, 26
Unit-step function, 24, 44
Unit-step response, 52, 69, 353, 469
Unit vector, 76
Universal gas constant, 306, 329
Unsymmetrical systems, 405
Upper-companion matrix, 140

Vacuum, 173
Vectors, 76
Vibration
    isolation, 452
    testing, 479
Viscous damping, 176
Voltage divider, 292
Voltages, 266

Xmath, 528, 571

Zero matrix, 77
ζ-line, 528
Ziegler-Nichols tuning rules, 516, 554
    first method, 555
    second method, 555, 564